The Focal Illustrated Dictionary of Telecommunications

The Focal Illustrated Dictionary of Telecommunications

Xerxes Mazda

Fraidoon Mazda

FOCAL PRESS

OXFORD JOHANNESBURG BOSTON MELBOURNE NEW DELHI SINGAPORE

Focal Press
An imprint of Butterworth-Heineman
Linacre House, Jordan Hill, Oxford OX2 8DP
225 Wildwood Avenue, Woburn, MA 01801-2041
A division of Reed Educational and Professional Publishing Ltd

A member of the Reed Elsevier plc group

First published 1999

British Library Cataloguing in Publication Data
A catalogue record for this book is available from the British Library

ISBN 0 240 51544 7

Printed in Great Britain by Biddles Limited, Guildford and King's Lynn

Preface

There are three aspects of this dictionary which we would like to cover in this preface: the scope of the dictionary, and its target readers; the layout of the entries; authors' acknowledgement for the source of some of the material.

We anticipate that the readers of this dictionary will either be relative newcomers to telecommunications, who wish to have a reference at hand to look up terms which they meet during their study or work, or telecommunications engineers who are specialists in a given area and wish to have information on a wider range of topics. With this twofold readership in mind the content of the dictionary has been selected to cover all the relevant telecommunications topics met with in industry today. Extensive use is made of tables and figures to convey information which could not otherwise have been included in a book of this size. Very little reference is made to generic electrical or electronic material, since to do so would fill several volumes. We have also avoided reference to products or services provided by organisations, unless this has gained widespread acceptance within the industry or has been adopted as a standard.

The entries within the dictionary have been grouped strictly in alphabetic order, with any space or special letters within words omitted. For example 'SDH frame' comes before 'S interface'. Numbers and Greek letters are entered according to their English spelling, so that, for example 'μ-law' comes before 'multicast'. The entries within the dictionary are in bold and where reference is made to a word, within an entry, which occurs in another part of the dictionary, this is given in italics to provide a trail for further study. However, if several words in sequence are shown in italics it may mean that they occur as separate entries within the dictionary.

Finally, we are very grateful to the publishers of the *Telecommunications Engineer's Reference Book*, second edition, published by Focal Press, edited by Fraidoon Mazda, for permission to use material from this book. The Reference Book was published in 1998 and contains a large amount of very up-to-date material covering the whole field of telecommunications, and our task was made much easier in having this excellent volume to consult. Our thanks also to the many authors who contributed to the Reference Book, for their excellent work.

Xerxes C. Mazda
Fraidoon F. Mazda

A

AAL: *ATM Adaptation Layer.*
AAN: *All Area Networking.*
AAR: *Automatic Alternative Routeing.*
abandoned call: A *call* which is terminated after a connection has been made but before the receiver goes *off-hook*, i.e. conversation occurs.
abbreviated address call: A *call* which requires a subset of the full *address* to initiate the call, the complete address being available from the *network* switch, such as a *packet assembler/disassembler.*
abbreviated answer: The reply to a *call*, made after communications has been established, in which the call sign of the calling station is not included.
abbreviated dialling: Feature of switches, such as a *PABX*, which allows users to use shortened numbers to place a *call*, the switch making the translation to the full number. See also *speed dialling.*
abort: An action, initiated automatically or by an operator, which causes all activity to be immediately stopped, e.g. the ending of *transmission* due to an unacceptably noisy *line.*
ABR: *Available Bit Rate.*
ABSBH: *Average Busy Season Busy Hour.*
absolute delay: It is the time between the generation and reception of a *signal waveform*, and is measured by reference to a particular point on the wave. Also referred to as *propagation delay*, it occurs in all *transmission* systems due to the delay caused by the *transmission medium* (copper, fibre and radio waves) and by intermediate systems such as switches. Delay can impede *voice* communications by causing *echo* and conversational delay. Overall delay can be calculated for a system by summing the delay caused by the separate parts. Guides on typical delays for planning are given in *ITU-T Recommendation* G.114.
absolute error: Errors in numbers can be represented as absolute or *relative errors*. Therefore if the actual number is 36845 but it is read as 36800 then the absolute error is $36845 - 36800 = 45$. If the number was originally represented as 36800 ± 50 then the absolute error is 50.
absolute gain: Absolute gain of a system is measured as the ratio of the *signal* level at the output to the signal level at the input. For an *antenna* it is often specified as the ratio of the power provided to an ideal radiator to that provided to the antenna under test for the same far field radiation intensity.
absorption: All *signals* travelling through a medium lose some of their energy, which is usually converted into heat. This phenomenon is known

as absorption. It occurs, for example, when light travels through an *optical fibre* and when radio waves travel through the *ionosphere*. All mediums show absorption, some absorbing all *wavelengths* equally whilst others absorb a greater amount of energy at certain wavelengths.

absorption coefficient: It is a measure of the amount of *absorption* of the transmitted energy, such as light or sound, within the *transmission medium*. For light of intensity P travelling in an *optical fibre*, if the intensity is reduced by dP over a distance of x then Lambert's law states that $dP/P = -Cx$ where C is the absorption coefficient. If P_0 is the intensity of the light at the starting point and P_x is the intensity after a propagation distance of x in the optical medium, then $P_x = P_0 e^{-Cx}$ where C is again the absorption coefficient.

absorption index: It is a measure of the loss caused by *absorption* when an *electromagnetic wave* propagates within a medium. It is given by $A_i = C\lambda/4\pi r$, where A_i is the absorption index, C is the *absorption coefficient*, λ is the wavelength in vacuum, and r is the refractive index of the medium.

absorption spectra: Impurities can be added to *transmission medium*, such as *optical fibre*, to vary its characteristics in relation to the *wavelenghts* which it absorbs, i.e. its absorption spectra.

Abstract Syntax Notation 1 (ASN.1): It is an *Open System Interconnect (OSI)* sponsored notation system for coding and transmitting *data*, and is independent of the architecture of the machine on which it runs. When used for designing *managed objects* an ASN.1 compiler would take the ASN.1 object definitions and provide the data structures and the source code routines needed to encode and decode management information relating to the *managed objects*.

ABT: *ATM Block Transfer.*

AC: *Alternating Current.*

ACA: *Australian Communications Authority.*

ACC: *Australian CCITT (ITU-T) Committee.*

ACCC: *Australian Competition and Consumer Commission.*

accentuated contrast: An image representation technique, used in systems such as *facsimile*, in which images above a certain threshold value are transmitted as white and those below this threshold are transmitted as black.

acceptance cone: It is the cone within which light falling on an optical *transmission medium*, such as *optical fibre*, will be communicated along its core due to *total internal reflection*. (Figure A.1) For incident rays of light having angles greater than that of the acceptance cone the light energy will be dissipated by *absorption* or *scattering* within the fibre cladding.

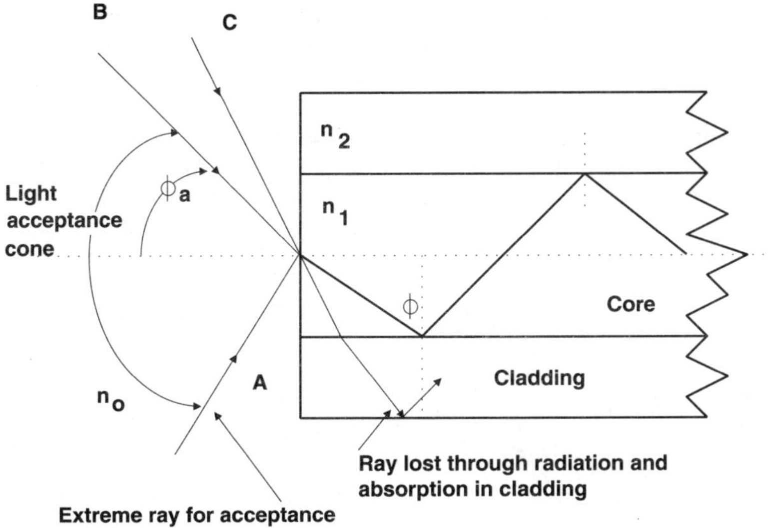

Figure A.1 Illustration of acceptance cone

accepted interference: Radio interference which is above the value defined by the standard bodies as *permissible interference* but which has been agreed as being acceptable by the parties affected.

access: Usually refers to *local access*. See also *access line*, *access loop* and *access network*.

access barred: Denial of access to a facility. For example access may be barred to a *closed user group* on failure of identity check. This is also referred to as *access denial*.

access barred ratio: The ratio of the number of times access to a facility has been denied or barred to the number of attempts made to access that facility. This is also referred to as *access denial ratio*.

access charge: Charges levied by operators for use of their *local network*. It may be a charge on the *users* themselves or on other operators, such as long distance (interexchange) carriers, for access to local users. See also *interconnect charge*.

access code: Additional numbers which need to be entered in order to access a special service or line, e.g. to access a long distance line belonging to another carrier, or to access a specified service on a *PABX*. The code may be a single digit, such as a 9 used to access an outside (*PSTN*) line from a PABX.

access control: The process for limiting access to certain features or parts of the network. This may be done manually or automatically by system software.

access denial: See *access barred.*

access denial ratio: See *access barred ratio.*

access group: A group of users who have equal access rights to a facility.

access line: The wire connecting the customer's terminal facility into the public network, e.g. to the first public switch. This is also referred to as the *access loop*, the *local exchange loop*, the *local loop* or the *access network.*

access loop: See *access line*. The original *local loop* was one wire and used the earth for the return path. This was too noisy and so two copper wires were introduced, to allow *speech circuits* to flow around; hence the use of the term 'loop'. A variety of mediums can be used for the local loop, such as copper, *optical fibre* (*Fibre To The Home*) and radio waves (*Wireless Local Loop*).

access network: See *access line* and *local loop.*

access point: The physical point at which the user connects into the *access line.*

access request: The signal sent by a user *terminal* to indicate a wish to start a communications session. An example is the *off-hook* access request sent when a *telephone handset* is lifted.

access time: The time from an *access request* being made to when it occurs.

ACCH: *Associated Control Channel.*

accommodation: Refers to the ability of the *human eye* to ensure that an object remains focused on the retina as it moves closer and farther away.

accommodation limit: The closest and farthest distance from the *eye* that an object can move and still remain focused.

account coding: Facility, mainly used with a *PABX*, where codes can be entered for *calls* in order to be able to bill a client for the call (incoming and outgoing). The amount can be 'forced', i.e. entered by the user, or can be 'at will', i.e. the user enters a client code only.

accounting management: It is one of five functions which *ISO* have defined as being required from a *network management* system. Accounting management aids in the preparation of bills for network users and for tracking their payment. It also helps in the sale of network resources. It is a set of facilities which enable charges to be determined for the use of the network resources and for costs to be identified and allocated to each resource. This management function depends on statistics provided by the objects on the network. Accounting management is frequently considered to include *inventory management.*

accounting rate: The cost of interconnecting traffic between operators, usually in different countries, is known as the accounting rate. It is based on measuring *traffic* which is generated in both directions between the two operators. If this traffic is approximately equal then the system works well. Simple accounting arrangements can now be used, such as 'bill and keep' in which each carrier bills and collects charges for all the calls which originate on their network, irrespective of where the calls terminate. However, if one operator generates much more traffic than it receives then it pays 50% of the accounting rate, known as the *settlement rate*, on the difference in traffic to the other operator.

Accredited Standards Committee (ASC): A group which has been accredited by the *American National Standards Institute* for standards production. Examples are the *IEEE*, the *EIA*, the *T1 committee* of the *Exchange Carriers Standards Association (ECSA)*, and the *X3 committee* of the *Computer Business Equipment Manufacturers Association (CBEMS)*.

accunet: It is the name of a digital service provided by AT&T for high speed *data* and high volume *voice transmission*. The Accunet T1.5 facility operates at 1.544Mbit/s.

ACD: *Automatic Call Distribution.*

AC-DC ringing: A telephony ringing system in which the alternating current is used to operate the ringer within the *telephone* and the direct current operates a relay which disconnects ringing once the telephone goes *off-hook*.

ACEC: *Advisory Committee on Electromagnetic Compatibility.*

ACET: *Advisory Committee on Electronics and Telecommunications.*

ACOS: *Advisory Committee On Safety.*

ACK: *Acknowledgement.*

Acknowledgement (ACK): An internationally recognised control character which is sent by the receiving station to the transmitting station to indicate that the previous block of *data* has been correctly received and the next block can be sent. See also *negative acknowledgement*.

acoustic coupler: A device for converting electrical signals to sound and sound to electrical signals. It is commonly used to transmit electrical data from a transmitting *DTE* by converting it into sound at a *telephone* handset and then sending the *telephony signals* over the *PSTN* before converting the signals from the receiving handset back into electrical signals at the receiving *DTE*.

acoustic noise: Unwanted *signals* which are in the *audio frequency* range.

acoustic resonance: Sound wave resonance which can occur between two parallel surfaces having a spacing equal to an odd number of half-wavelengths. If d is the spacing between the two surfaces and c is the velocity of light, then the frequency of resonance f is given by $f = nc/2d$ where n is equal to an integer, i.e. 1, 2, 3, etc. See also *standing wave*.

acoustic shock: A transient or power surge in a *telephone* system causing a surge of sound pressure in the listener's *ear*. Acoustic shock is more critical when an *headset* is being worn.

acoustooptics: The effect of sound waves on light, e.g. the deflection of light by using acoustic waves to vary the refractive index of the material in which the light is travelling.

acquisition radar: Radar which provides a scanning and surveillance function and passes information on objects detected on to a *tracking radar*.

acquisition time: In telecommunications *transmission* systems usually refers to the time needed for the system to lock on to a synchronising *signal*.

ACSE: *Association Control Service Element.*

AC signal: A *signal* with the characteristics of an *Alternating Current.*

AC signalling: A *signalling* system which uses in-band tones (2280Hz or 2600Hz) to convey the signalling information, sent during *call setup* or *call clear down* times.

ACTE: *Approvals Committee for Terminal Equipment.*

active filter: A *filter* which normally contains semiconductor devices and needs to be powered in order to operate effectively.

active line: In a television scanning system it is the *electron beam* trace on the screen which contains luminance information. For a telecommunications system an active line is a *transmission line* which is in use.

Active Optical Network (AON): An optical *access network* which uses active splitters and optical amplifiers to achieve higher split ratios, *bandwidth* and *range*. See also *Passive Optical Network (PON).*

ACTS: *Advanced Communications Technologies and Services.*

ACU: *Automatic Calling Unit.*

ada: A high level programming language, used by the US *Department of Defence* and specified in MIL-STD-1815.

adapter: A device for connecting cables of different size or having *connectors* with different number or arrangement of pins. Adapters are also used to connect two systems using different *data* rates or *transmission* codes.

adaptive antenna: An *antenna* which provides a focused beam directed towards the object, such as a mobile telephone. The beam can track the mobile as it moves. This provides a greater range than that available from conventional antennas and also minimises *interference* effects at the mobile. Also known as *smart antenna.*

adaptive channel allocation: *Transmission* technique in which the *capacity* of the *channels* are assigned to users on a demand basis, rather than each user being assigned a fixed amount.

Adaptive Delta Modulation (ADM): Digital *modulation* technique which overcomes the disadvantages associated with *delta modulation*, such as the reduction of the *SQNR* with decreasing *signal* level, by adjusting the step size as the signal level changes.

Adaptive Differential Pulse Code Modulation (ADPCM): An *encoding* technique which reduces the number of *bits* used in an analogue sample from eight to three or four, so resulting in compression of the transmitted *signal. ITU-T Recommendation* G.721 specifies an algorithm for a 32kbit/s ADPCM (8000 samples per second, each sample represented by 4 bits) which provides double the *capacity* compared to *PCM* transmission (8000 samples per second, each sample represented by 8 bits).

adaptive equaliser: An *equaliser* which automatically adapts to changing *line* conditions so as to maintain the integrity of the *transmission.* See also *Transversal Equaliser (TVE).*

adaptive high frequency radio: Radio system, operating in the *High Frequency (HF)* band, which automatically adjusts its characteristics (such as power level, frequency, etc.) to compensate for changing *transmission* conditions.

adaptive routing: *Data* routing method in which the routing conditions are automatically changed to take into account *network* conditions, such as *traffic* intensity, *line* characteristics, component failures, etc.

adaptive tree-walk protocol: In an ideal *multiple access* system users are divided in such a way that in any one group there is only one active user, so that the chance of a successful *transmission* is assured. The adaptive tree-walk protocol is one method for allocating users to the available slots within a transmission system, and it is illustrated in Figure A.2. Initially all users who wish to transmit data contend for a slot and the

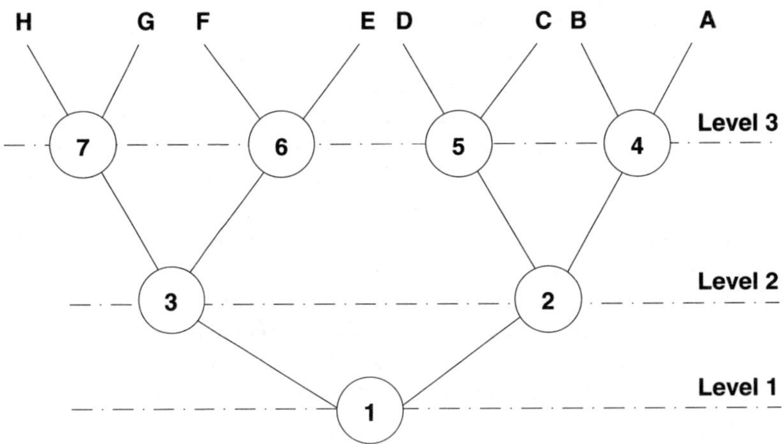

Figure A.2 Operation of an adaptive tree-walk protocol

protocol is at Level 1 and node 1. If a *collision* occurs then the protocol moves up the tree to Level 2 and node 2. Now only the users below this node (A and D) are permitted to contend for the slot. If a transmission is successful then the slot in the next packet is reserved for users under node 3. If, however, there was a contention at node 2 then the protocol moves further up the tree to Level 3 and node 4, limiting the users to A and B. By moving up the tree each time a collision occurs, the protocol eventually finds users who have information to transmit, and limits their number so that transmission is successful. To ensure fairness users on either side of the tree are searched in turn following a successful transmission or if there is no data available for transmission on one side.

ADC: *Analogue to Digital Converter.*

ADCCP: *Advanced Data Communications Control Protocol.*

Add-Drop Multiplexer (ADM): A *multiplexer* which has the facility for inserting or removing individual *channels* of *traffic* along the way.

ADDF: *Automatic Digital Distribution Frame.*

additive white gaussian noise: See *white noise.*

address: In a communication system it is that part of the *message* which specifies the source or destination of the body of the message. Also referred to as address field. All systems on a common *network* must have a unique address so that they can be identified.

address field extension: Additional information added to the *address* field so that it can transmit more information.

addressing: The technique for selecting the element on the *network* which is to be communicated with. Each *station* on a network has a unique *address* and this is referred to within a communication *message.*

Address Resolution Protocol (APR): It is an *Internet Protocol (IP)* used in *TCP/IP networks*, which dynamically binds high level *IP addresses* to low level physical *hardware* addresses.

adjacent channel: A *channel* which is next to another channel, either in terms of time (e.g. adjacent channels in a *time division multiplex system*) or *frequency* (e.g. adjacent frequencies in a *frequency division multiplex* or *wavelength division multiplex* system.)

adjacent channel interference: Unwanted *signals* from one *channel* affecting an *adjacent channel.*

adjacent channel selectivity: The ability of a receiver to differentiate between *signals* in a *channel* and an *adjacent channel.*

adjacent nodes: Two *nodes*, in a distributed *network*, which are connected directly together by one or more data *links*, without any intermediate nodes.

administrative management centre: A *network management centre* which controls all the functions on the network and is usually owned by the *network operator.*

administrative network: The part of the *network operator's* network which is used to convey control and management information to maintain the *network* (e.g. *fault* data, *billing* information, etc.), rather than carrying actual operational *traffic*.

ADM: *Add-Drop Multiplexer or Adaptive Delta Modulation.*

Administrative Unit (AU): Term used in the *Synchronous Digital Hierarchy (SDH)* as defined in *ITU-T Recommendations*. It is the level of the hierarchy at which circuit administration is carried out by the operator.

ADP: *Answering Detection Pattern.*

ADPCM: *Adaptive Differential Pulse Code Modulation*

ADPE: *Automatic Data Processing Equipment.*

ADSI: *Analogue Display Service Interface.*

ADSL: *Asymmetrical Digital Subscriber Line.*

Advanced Communications Technologies and Services (ACTS): A European research programme established under the Fourth Framework Programme on EU Research (1994–1998) which replaced the *RACE* programme. This is taking the *broadband* technologies of RACE and extending them into mobile and *Intelligent Networks (IN)*. The main areas of research, along with the estimated spend, are given in Table A.1. This also shows the horizontal actions and accompanying measures which serve to reinforce the emergence of consolidated programme results. ACTS projects are conducted by consortia and organisations drawn mainly from European Member States, but participation is also

Table A.1 Main areas of research within the ACTS programme

Technologies	*Funding (ECUs millions)*
Interactive Multimedia Services	150
Photonic Technologies	112
High Speed Networking	75
Mobility and Personal Communications	119
Intelligence in Networks and Service Engineering	100
Quality, Security and Safety of Communications Systems	43
Horizontal Actions and Accompanying Measures	31

open to organisations from other countries. Research is in the form of usage trials which ensure that the results are relevant to actual applications.

advanced communications technology satellite: A *satellite*, owned by *NASA*, which is used as a test bed for experimentation on satellite communications systems.

Advanced Data Communications Control Protocol (ADCCP): A Layer 2 or *data link level* protocol, specified by *ANSI*, which is used for *transmission* with *error control*. It grew out of IBM's *SNR SDLC* protocol and is specified by *ISO* as *HDLC*, and by the *ITU-T* as *LAP*.

Advanced Intelligent Network (AIN): It is an *Intelligent Network (IN)* standard proposed by *Bellcore* and mainly implemented in the USA. Its aim is to aid in modularisation and standardisation of interfaces so that intelligent network modules can be bought from different vendors, to meet time and price requirements, and still be made to work with each other.

Advanced Mobile Phone System (AMPS): It is a mobile *cellular radio system* developed in the USA, primarily by Bell Laboratories, as a successor to *IMTS*. It has been used in the USA, Canada, South America, Australia and some Asian countries. It uses the 800 MHz *frequency band* with 30 kHz *channel spacing*. Analogue *frequency modulation* is used for *speech transmission* with a frequency deviation of 12kHz. Use of this wider frequency deviation enables a greater *dynamic range* and provides greater protection against *co-channel interference*. *Manchester encoding* is used in the *signalling* between the mobile and the *base station* giving an effective *bit rate* of 20 kbit/s.

Advanced Project for information Exchange (APEX): A system for transferring documents electronically, including graphics and images, which is operated by the European Research Cooperative Action project.

Advanced Research Projects Agency (ARPA): An agency, part of the US *Department of Defence*, which developed the *ARPANET*.

Advanced Research Projects Agency Network (ARPANET): A *wideband data communications Packet Switched Network (PSN)* developed by the *ARPA* for connecting the USA *Department of Defence* sites. This later developed into the *National Science Foundation Network*, for linking academic sites, and the widely used *Internet* network.

Advanced Tactical Optical Fibre (ATOF): *Optical fibre* which is used in an aeroplane for communications between processors and sensors.

Advanced Television (ATV): Television which has improved capabilities, such as higher quality image, removal of cross-colour defects, elimination of flicker, and wide screens with high *aspect ratios*. See also *High Definition Television*.

Advisory Committee on Electromagnetic Compatibility (ACEC): One of the three committees of the Committee of Action which is responsible for regulating all the technical work within the *International Electrotechnical Commission*. The other two committees of the Committee of Action are the *Advisory Committee on Electronics and Telecommunications* and the *Advisory Committee On Safety*.

Advisory Committee on Electronics and Telecommunications: See Advisory Committee of Electromagnetic Compatibility.

Advisory Committee On Safety: See *Advisory Committee of Electromagnetic Compatibility*.

AEIA: Australian Electronics Industry Association.

aerial: That part of a radio system which receives radio *signals* (*radio waves*) or transmits radio signals. See also *antenna*.

aerial insert: That part of an underground communications cable which needs to go overground, suspended between towers, when it meets unsuitable terrain for its underground passage. (Figure A.3)

Aeronautical Mobile Satellite Service (AMSS): Radio communications system between air and ground for use by passengers and crew of commercial airlines. The radio *capacity* is in the L-band, provided by the *International Maritime Satellite Organisation (INMARSAT)*. There are two services: off-route, for aircraft outside national and international civil air routes, and route, for aircraft following these routes.

Aeronautical Multicom Service (AMS): A radio service for use by private aircraft only.

aeronautical utility land station: Radio station which is situated at an airport and is used to control aircraft on the ground and other ground vehicles.

AES: *Aircraft Earth Station*.

AF: *Audio Frequency*.

AFNOR: *Association Francaise de Normalisation*.

AGC: *Automatic Gain Control*.

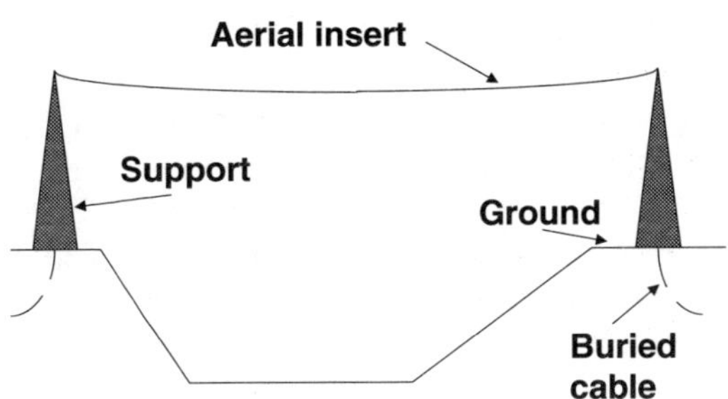

Figure A.3 An aerial insert

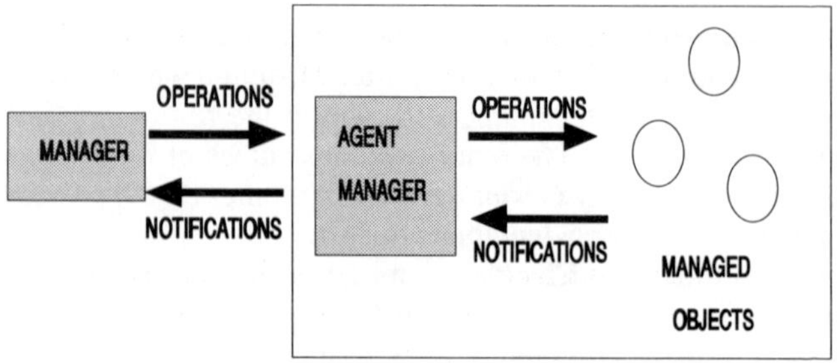

Figure A.4 A manager and agent process

agent process: A concept used within *OSI network management*. Since a manager cannot effectively control a large number of objects directly, it employs agents (also known as agent managers or agent processes) who control subsets. (Figure A.4.) Manager and agent applications in the same or different *open system architecture* exchange management information using *CMIS/CMIP*. The agent performs the following functions on behalf of the manager: handles the representation of the resources being controlled in the form of *managed objects*; gives the manager a management view of the objects being managed; handles management requests from the manager to the managed objects and responses from the objects to the manager. (These may be categorised as operations and notifications. The agent performs an amplifying and filtering role.)

Aggregate bit rate: Usually refers to the total *bit rate* of a transmitted *signal* in a multiplexed system. For example, the aggregate bit rate of the European *E1* transmitted signal is 2 Mbit/s and for the USA *T1* system it is 1.544 Mbit/s.

AI: *Artificial Intelligence.*

AIIA: Australian Information Industry Association.

AIIM: *Association for Information and Image Management.*

AIN: *Advanced Intelligent Network.*

AIOD: *Automatic Identified Outward Dialling.*

airborne radio relay: Equipment used on board an aircraft for the purpose of relaying radio *signals* and so boosting their strength.

airborne sound: Sound which is transmitted through the air rather than along some other medium such as a solid structure.

Aircraft Earth Station (AES): An Earth station, located on an aircraft, forming part of the *Aeronautical Mobile Satellite Service (AMSS).*

aircraft emergency frequency: An internationally agreed *frequency* used for safety related communications for aircraft.

air to air communications: Radio communications which occur between two airborne vehicles.

air interface: In a *mobile communication system*, such as *GSM*, it usually refers to the radio *link* between the mobile *terminal* and the *network*.

airtime: Usually refers to the amount of time for which the *subscriber* is to be billed in a *mobile communications system*.

air to ground communications: Radio communications which occur between an aircraft and a land based system.

air to ground radiotelephone service: A public-use telephone service between an aircraft and a land based station.

AIS: *Alarm Indication Signal* or *Automatic Intercept System*.

alarm centre: The centre which receives all the alarms on the network and therefore monitors it for abnormal operation.

Alarm Indication Signal (AIS): A signal which is sent, in a *transmission* system, to indicate that an error has occurred. It consists of an all ones pattern.

alarm signal: A *signal* which indicates an abnormal operation and is intended to draw the attention of an operator.

a-law: In *digital telephony* the analogue *voice signal* is converted to a *digital signal* within an *Analogue to Digital Converter (ADC)*. The digital signal is subject to *compression* prior to *transmission* and then *decompression* at the other end to reproduce the voice signal, a process know as *companding*. The signal compression *algorithm* used in Europe for this is known as a-law. It is non-linear, so giving more *quantisation* steps at lower signal levels (Figure A.5). This reduces the *dynamic range*

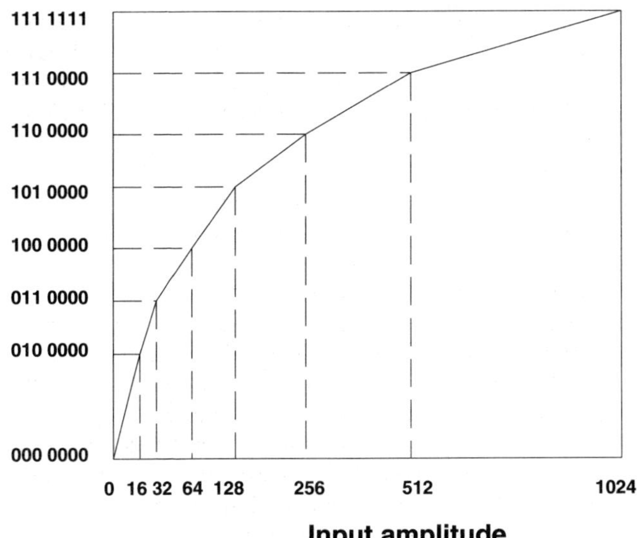

Figure A.5 ITU-T a-law characteristic

of the signal whilst at the same time providing a higher *signal to quantisation noise ratio* than that obtained from linear coding systems. USA and Japan use a different compression algorithm know as *μ-law*.

ALE: *Automatic Link Establishment.*

ALGOL: A high level computer programming language used primarily for mathematical related programmes.

algorithm: A set of rules which define the steps to be used for carrying out an action, e.g. a computer program algorithm for solving a problem.

aliasing: To convert an *analogue signal* to a *digital signal sampling* is done at twice the frequency of the analogue signal (the *Nyquist rate*). If the sampling rate is below this value then the lower *sideband* of the signal overlaps the *baseband* and the two cannot be separated. This phenomena is called aliasing.

All Area Networking (AAN): *Networks* which cover local and wide areas. Also used to imply combined use of a *Local Area Network (LAN)* and *Wide Area Network (WAN).*

Alliance for Telecommunications Industry Solutions (ATIS): Organisation of USA telecommunications manufacturers and service providers. It is involved in standards making through its T1 Committee. Previously known as the *Exchange Carriers Standards Association (ECSA).*

All Optical Network (AON): A network in which all the major functions are carried out optically, without any conversion to electrical signals. This includes an optical *access network*, *transport network*, and optical routing using an *optical cross connect (OXC).*

All Trunks Busy (ATB): *Transmission* in which all the *trunks* in a *trunk group* are busy and therefore not available for carrying further *traffic.*

ALOHA: A *multiple access* technique developed in the University of Hawaii for use with *satellite* based *transmission systems*. Its operation can be explained by the simplified flow diagram of Figure A.6. When-

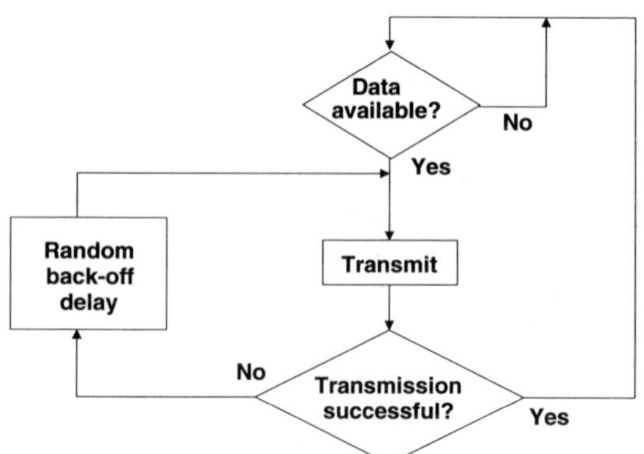

Figure A.6 Flow diagram of a pure ALOHA system

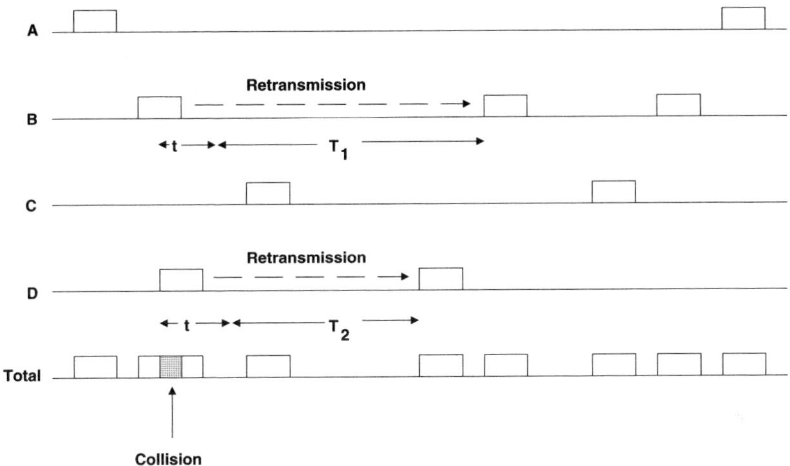

Figure A.7 Packet collision and retransmission in ALOHA

ever a source has *data* to send it does so without regard for any other data source which may also be using the *transmission medium*. It then waits the appropriate amount of time for an *acknowledgement*. If this is not obtained the source assumes that a *collision* has occurred. It then waits for a random amount of time before retransmitting the same information. A random wait or *back-off* time is required to ensure that two sources, which have just collided, do not again collide when they retransmit their data. Many *algorithms* are in use for calculating this random time. Figure A.6 illustrates the process of *packet collision* within ALOHA. If any part of two packets overlap (e.g. the first bit of one and the last bit of another) then they are both destroyed. This means that there is a vulnerable period equal to the sum of the two *packet* lengths (or twice a packet length, if all packets have the same length) when collision occurs. In Figure A.7 packets from users B and D are seen to overlap and collide. After this has occurred it is assumed that there is a finite time t which both senders take to detect that a collision has occurred. They then wait random periods of time (T_1 and T_2) before retransmitting their data, which in this case does not collide with any other data on the *line*.

ALOHA with capture: It is a modification to the basic *slotted ALOHA* technique for *multiple access*, and works on the principle that if two signals have different strengths then, on *collision*, the stronger one will be 'captured' by the receiver and will get through, so that only the weaker signal needs to be retransmitted. This, in effect, reduces the retransmissions by half and gives almost a 50% increase in *throughput* over slotted

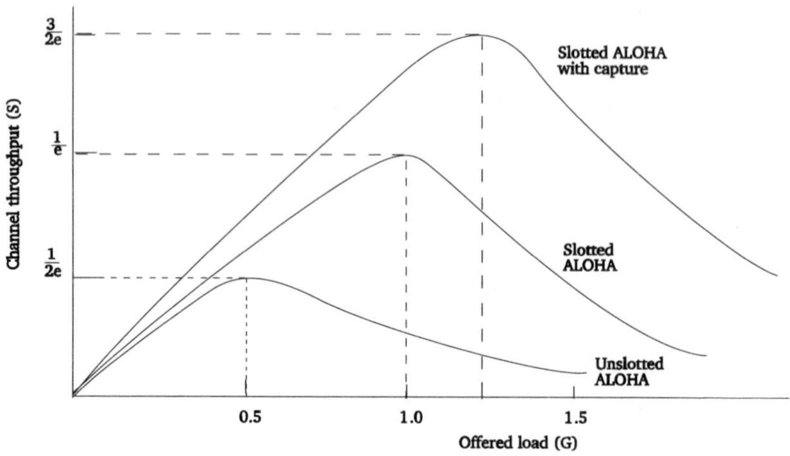

Figure A.8 Throughput for pure ALOHA, slotted ALOHA and ALOHA with capture

ALOHA. Figure A.8 compares the performance of pure *ALOHA* with *slotted ALOHA* and *ALOHA with capture*.

alphabetic code: A code which uses alphabets and control characters to represent data, but no numerals. Example is the *International Alphabet No. 5*. See also *alphanumeric code*.

alphabetic telegraphy: A common form of *telegraphy* which uses groups of pulses to represent letters, punctuation marks, etc. An alphabetic telegraphy code, such as the *International Telegraph Alphabet*, is used.

alphageometric coding: A coding system used to display *alphanumeric* characters and *graphics* in display systems, such as *videotext.*

alphamosaic coding: A coding system used to display *alphanumeric* characters and *graphics* in display systems, such as *videotext*, where the picture is built up by assembling small coloured pieces of equal size.

alphanumeric code: A coding system which uses alphabets and control characters, as well as numbers, to represent data.

alphanumeric pager: A *pager* which has the ability to display messages in *alphanumeric code*. Commercially available pagers can display 16, 32 or 80 *characters* and can store a much larger number of characters which are then scrolled through on the display.

ALT: *Alternate Local Transport company.*

alternate digit inversion: A technique used to prevent loss of *synchronisation* in equipment which use the incoming *bit stream* for maintaining synchronism. During periods of inactivity, when the bit stream consists

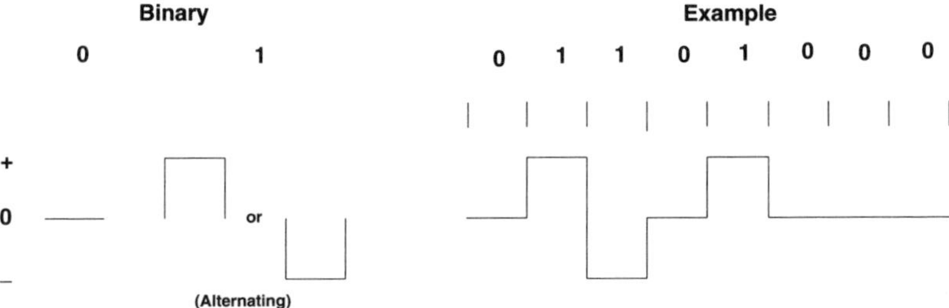

Figure A.9 Alternate Mark Inversion (AMI)

of zeros, bits in alternate *time slots* are inverted to provide a synchronisation source.

Alternate Local Transport company (ALT): A company that provides telecommunications services, primarily to businesses. These could include local access services, as provided by a *Competitive Access Provider (CAP).*

Alternate Mark Inversion (AMI): A *bipolar transmission* technique in which the logical zero is represented by a *space* and a logical one is represented alternatively by negative and positive polarity pulses (*marks*). (Figure A.9.)

Alternate Mark Inversion Violation: An error condition in the *Alternate Mark Inversion* coding system where two or more consecutive logical ones are represented by pulses of the same polarity. (Figure A.10)

alternate path routeing: *Routeing* of a signal along another *transmission path*, usually because its original path is congested or unavailable.

alternate routeing indicator: A *message*, sent to the originator of the signal being transmitted, to indicate that an *alternate path routeing* has been necessary.

alternating code: A code in which each *bit* can be represented by a positive or negative pulse, for example as in the *Alternate Mark Inversion* system.

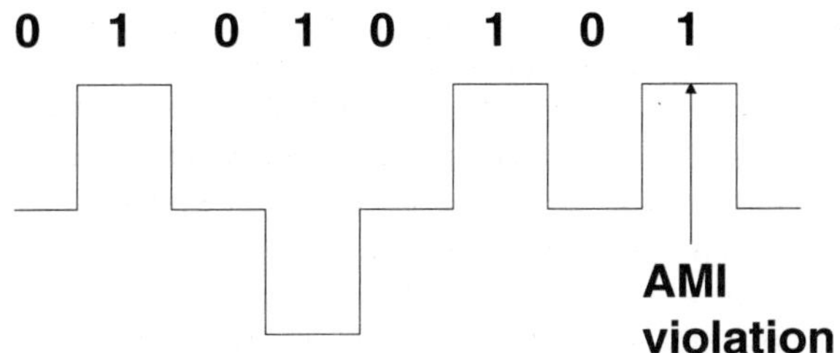

Figure A.10 Alternate Mark Inversion Violation

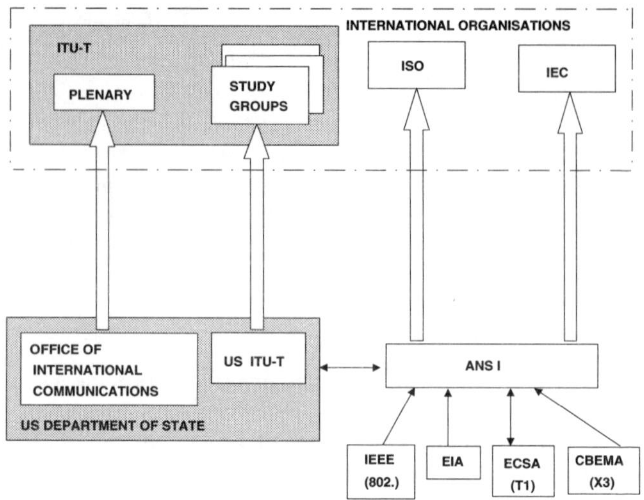

Figure A.11 Primary USA telecommunication standards bodies

Alternating Current (AC): Electrical current in which the polarity and magnitude are periodically changing. Usually it will be a *sine wave* although this may be distorted due to *harmonics*. In the UK the mains alternating current has a frequency of 50Hz (*hertz*) whereas in the USA it is 60Hz.

AM: *Amplitude Modulation*.

AMA: *Automatic Message Accounting*.

ambient noise: For acoustic systems it is the sound which is always present in the environment being considered. For communications systems it is the electrical *signal* which is present in the *transmission channel*, even when no data is being transmitted. These are also referred to as *background noise*, *white noise* and *Gaussian noise*.

American National Standards Institute (ANSI): ANSI is the principal standards making body in the USA. Figure A.11 shows its relationship to other USA and international bodies. ANSI was founded in 1918 by five engineering societies and three government agencies and is a non-profit making non-government organisation supported by diverse private and public sector organisations. It does not directly develop American National Standards (ANS) but accredits other groups to do so, taking their outputs, approving them, and assigning numbers to them before publication. These standards, like those from *ITU-T* and *ITU-R*, are voluntary, although they are frequently used for equipment procurement. ANSI has accredited several groups as *Accredited Standards Committees (ACS)*. It promotes US standards internationally and is the sole US representative on *ISO*, via the US National Committee (USNC), and on the *IEC*. It participates in most of the technical committees of both organisations.

American Standard Code for Information Interchange (ASCII): A variant of the *International Alphabet No. 5* developed by *ANSI* for information interchange between processors and communication systems. It uses seven bits to represent letters, numbers, and control characters. An eight bit is normally added as a *parity check*. The ASCII code is given in Table A.2.

Table A.2 American Standard Code for Information Interchange

$A_8A_{77}A_6A_5$	*0000*	*0001*	0010	0011	0100	0101	0110	0111
$A_4A_3A_2A_1$							,	
0000	NUL	DLE	SP	0	@	p	\	p
0001	SOH	DC1	!	1	A	Q	a	q
0010	STX	DC2	"	2	B	R	b	r
0011	ETX	DC3	#	3	C	S	c	s
0100	EOT	DC4	$	4	D	T	d	t
0101	ENQ	NAK	%	5	E	U	e	u
0110	ACK	SYN	&	6	F	V	f	v
0111	BEL	ETB	’	7	G	W	g	w
1000	BS	CAN	(	8	H	X	h	x
1001	HT	EM	)	9	I	Y	i	y
1010	LF	SUB	*	:	J	Z	j	z
1011	VT	ESC	+	;	K	[	k	{
1100	FF	FS	,	<	L	\	l	|
1101	CR	GS	-	=	M	]	m	}
1110	SO	RS	.	>	N		n	~
1111	SI	US	/	?	O	_–	o	DEL

American Wire Gauge (AWG): A standard for measuring the thickness of non-ferrous wires, such as aluminium and copper.

AMI: *Alternate Mark Inversion.*

AMIS: *Audio Messaging Interchange Specification.*

amplification by stimulated emission: The process used in a *LASER* for increasing the strength of an optical wave.

Amplified Spontaneous Emission (ASE): Unwanted emissions which occur, across a wide *wavelength* range, during the process of *Light Amplification by Stimulated Emission of Radiation (LASER).* These emissions are amplified and add to the *noise* in the output *signal* from the *amplifier*, as well as limiting amplifier *gain* by contributing to its *saturation* power.

amplifier: A device which provides an output *signal* which has a strength greater than that of the input signal, but is proportional to it. See also *amplifier gain.*

amplifier bandwidth: It is the *frequency band* over which the *amplifier* has been designed to operate. When handling *signals* outside of this band the characteristics of the amplifier will be worse than its specification.

amplifier gain: The *gain* of an *amplifier* is a measure of its capability to enhance the strength of a *signal.* It is calculated as the ratio of the output signal strength to the input signal strength. See also *attenuator.*

amplifying message: A *message* which contains auxiliary information in addition to that contained in an earlier transmission, usually explaining or adding to the original message.

amplitude: A measure of magnitude or strength of a *signal.* It is given as the peak deviation of a *waveform* from its mean value and is equal to half the *peak-to-peak* value. (Figure A.12.) See also *pulse amplitude.*

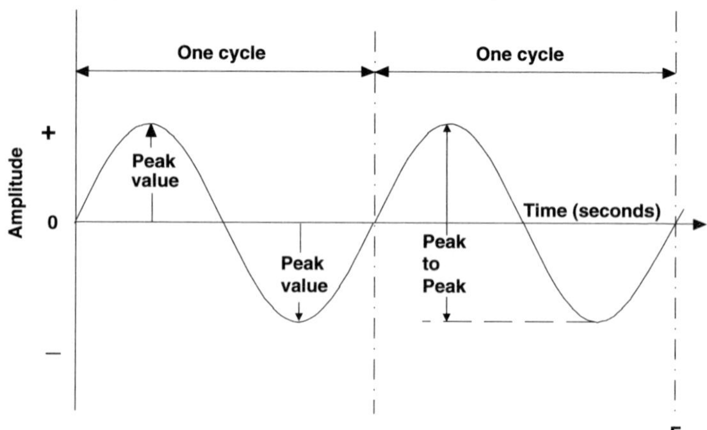

Figure A.12 Illustration of amplitude of a waveform

Amplitude and Phase Shift Keying (APSK): A signal *modulation* technique which encodes a *signal* by a combination of *amplitude modulation* and *phase modulation.*

amplitude distortion: *Distortion* of a transmitted *signal* such that its *amplitude* is different from that expected, the difference usually being caused by an a non-linearity in the *transmission path* or the presence of another interfering signal.

amplitude equaliser: A component used to modify the *amplitude* handling characteristics of a *transmission* system, often to compensate for any *amplitude distortion* caused to the *signal* it carries.

amplitude fading: Reduction in the *amplitude* of a *signal* due to losses in the *transmission medium.*

amplitude keying: A *keying* technique in which only the *amplitude* of the *carrier signal* is changed.

Amplitude Modulation (AM): One of three methods commonly used for *modulation* of a *signal*, the other two being *frequency modulation* and *phase modulation*. In amplitude modulation the *amplitude* of the *carrier signal*, which is usually a *sine wave*, is varied according to the characteristics of the *modulating signal*. Figure A.13 shows a *sine wave carrier signal* which is modulated by a square *modulating signal*, the *modulation depth* (i.e. the difference between the amplitudes of the modulated and unmodulated waves) being 100%.

Amplitude Shift Keying (ASK): A *keying* technique in which the *carrier signal* is switched on and off by a binary *modulating signal*, i.e. a 100% *modulation depth*. It is also known as *on-off keying*.

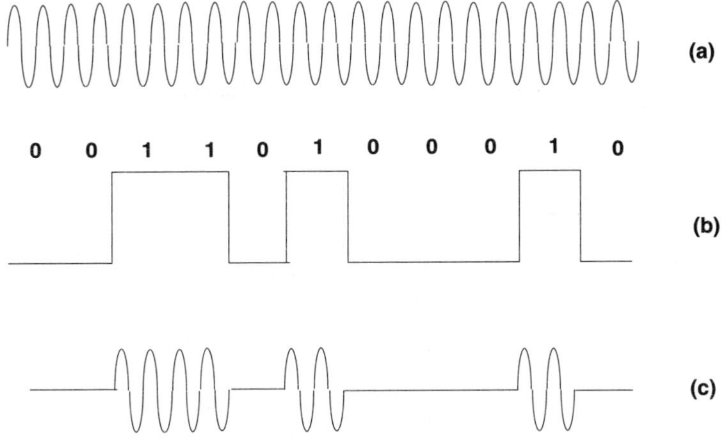

Figure A.13 Amplitude Modulation: (a) carrier; (b) modulating waveform; (c) modulated waveform

AMPS: *Advanced Mobile Phone System.*
AMS: *Aeronautical Multicom Service.*
AMSS: *Aeronautical Mobile Satellite Service.*
AMTS: *Automated Maritime Telecommunications System.*
Analogue Display Service Interface (ADSI): A Bellcore *protocol* for sending *data* over telephone lines for display on a *subscriber* set.
analogue exchange: Refers to the building which houses analogue switches and other telephony equipment used in the *transmission* and *switching* of *analogue signals.*
analogue multiplier: A device which produces an *analogue signal* output which is equal to the product of the analogue signals at its input terminals.
analogue network: In communications usually refers to older national *transmission networks* which carried *analogue signals* and used analogue components such as *analogue switches*. Figure A.14 shows the former North American analogue network which had four levels of *trunk exchanges*. The former UK analogue network is shown in Figure A.15 which had three levels of trunk exchanges.

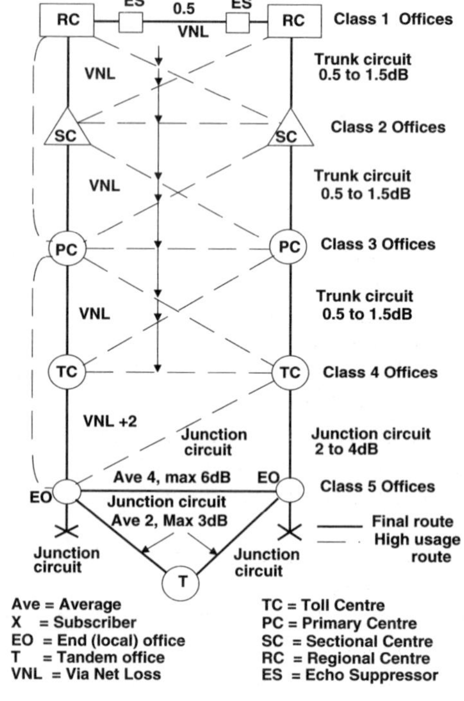

Figure A.14 North American former analogue network

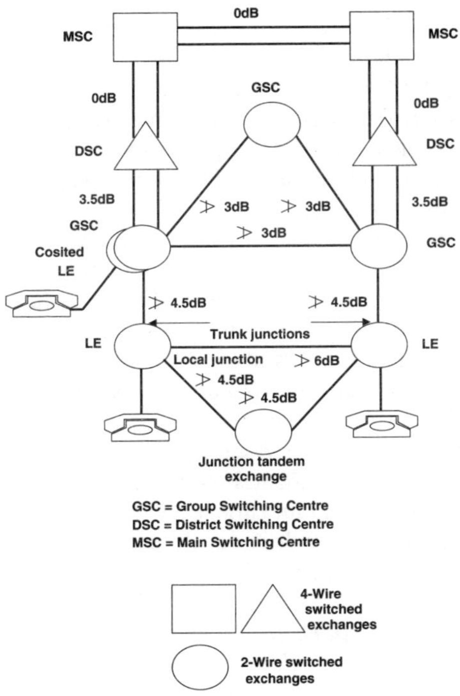

Figure A.15 UK former analogue network

analogue radio: A radio whose transmissions consist of *analogue signals* rather than *digital signals.*

analogue signal: A signal which varies continuously rather than in a discrete or discontinuous manner. An example of an analogue signal is the voice signal produced by a telephone before it is converted into digital form using *Pulse Code Modulation.*

Analogue Simultaneous Voice Data (ASVD): An *ITU* proposal for *transmission* of *voice* and *data* over conventional telephone lines.

analogue speech interpolation: *Speech interpolation* applied to an *analogue signal.*

analogue switch: A switching device which directly switches an *analogue signal* rather than first converting it into a *digital signal* and then back to an analogue signal after the switching process. Switches are used to interconnect two *subscribers* and are housed in buildings called *exchanges* or *telephone exchanges.* There were several types of *analogue exchanges*, such as the *step-by-step exchange*, the *crossbar exchange*, and the *reed electronic exchange.*

analogue telephone: A *telephone* which produces an *analogue signal* in response to the sound input from the speaker.

Analogue to Digital Converter (ADC): A device which converts an *analogue signal* into a *digital signal.* The same device also usually provides conversion in the reverse direction. Several techniques are used for this conversion, such as *successive approximation*, dual slope conversion and *flash conversion.*

analogue transmission: The *transmission* of an *analogue signal.*

Analysis and Forecasting Group: A subcommittee of *SOGT* which studies industrial developments in selected areas, such as *ISDN*, *broadband*, *cellular radio systems*, etc. It organises meetings at which representatives from members states of the *European Community* attend and provide opinions. The output are *Recommendations* which may be made mandatory within the Community.

ancillary equipment: Usually refers to equipment which is being controlled by another item of equipment. For example a printer which is under the control of a central computer connected on the same *network.* It can also refer to equipment which is not directly involved in providing a basic telephone service, such as an *answering machine* connected to a home telephone line.

ANDF: *Architecture Neutral Distribution Format.*

angled facet SOA: A *Semiconductor Optical Amplifier (SOA)* in which reflectivity is controlled by growing the ridge of the device at an angle (of about 7°) to the crystal plane of the material. This results in the bulk of the reflected power from the facet being radiated away and not being coupled back into the cavity, as in Figure A.16. See also *buried facet SOA.*

angle modulation: A form of *modulation* where the *phase* or *frequency* of the *carrier signal* is varied by the *modulating signal.* Both *phase modulation* and *frequency modulation* are examples of angle modulation. Frequency modulation can be produced by a phase modulator having a *modulating signal* which is an integral of the broadband signal, and

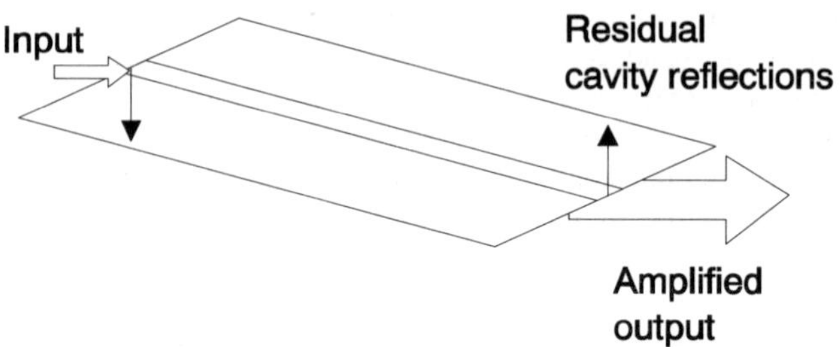

Figure A.16 Angle facet SOA

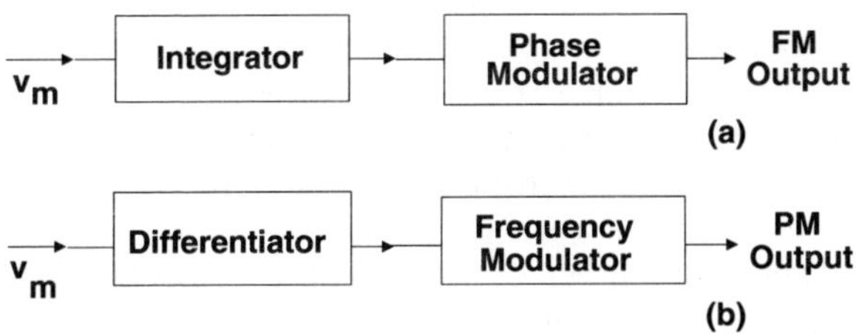

Figure A.17 Methods for generating angle modulation: (a) use of phase modulator to generate FM; (b) use of frequency modulator to generate PM

phase modulation can be produced using a frequency modulator and differentiating its *baseband* signal. (Figure A.17.)

angstrom: A unit of length used for optical measurements. One angstrom equals 10^{-10} metres.

angular misalignment loss: Optical power loss in *fibre optic* systems due to misalignment between two fibres at a *splice*, or between a fibre and its connector.

ANI: *Automatic Number Identification.*

anisochronous transmission: *Transmission* in which there is no fixed relationship between the time interval separating any two significant events and the time separating two other events. Unlike *synchronous transmission* there is no time relationship between the sending and receiving stations, i.e. no *master clock*. *Timing* or *synchronisation* between the two are established by the sender sending *start bits* and *stop bits* before and after a transmission. This is also known as *stop-start transmission* or *asynchronous transmission*. See also *isochronous transmission* and *synchronous transmission.*

anisotropic propagation medium: A propagation or *transmission medium* which has different characteristics for different signal conditions, e.g. differences in refractive index depending on the *polarisation* of a propagating wave.

anomalous transmission: Unexpected or abnormal propagation caused by discontinuities in the *transmission medium.*

ANSI: *American National Standards Institute.*

answerback: A signal sent by the receiving terminal to the sending terminal to indicate that it is ready to receive a *message*. Answerback is commonly used in *telex* and *ITU-T Recommendation* F.60 specifies the unique answerback *code* which each telex machine will have, made up of a *country code*, telex line number and *subscriber* name. In response to a *'who are you'* message from the transmitting telex terminal the receiving terminal will answerback with its unique code. This ensures that *transmission* is being made to the correct telex machine, so maintaining confidentiality of the transmitted message.

answerback unit: The unit within a receiving *telex* terminal which automatically generates the unique *answerback* code in response to a *'are you there'* signal from a sending terminal.

Answering Detection Pattern (ADP): A coded pattern sent by a transmitting *modem* operating with the V.42 *error correction* protocol to a receiving modem to check if the receiver is also using the same *protocol*. If this is the case the receiver will reply with a *Originator Detection Pattern* signal.

answering machine: A machine, connected to a telephone *network* and associated with a *telephone number*, which automatically answers the *call*, after a preset number of rings. The answering machine plays a recorded *message* before allowing the caller to leave a message which can be subsequently retrieved by the telephone *subscriber*.

answering sign: The position of the flag, in *semaphore* communication, in response to a *call*.

answering tone: A *signal* sent by a receiving *modem* to a sending modem to indicate that it is ready to accept a *transmission*.

answer signal: A *signal* sent by the receiving *terminal* to the sending terminal when the receiving terminal goes *off-hook*. The answer signal stops the *ringing* in the sending terminal and also starts the *billing* meter so that the cost of the call can be calculated, if applicable. (Some calls, such as those to the emergency services, are free.)

antenna: A device which transmits or receives *electromagnetic waves*, usually radio *signals*. The antenna either converts the electromagnetic wave into electrical energy (known as a *receiving antenna*) or converts electrical energy into electromagnetic waves (known as a *transmitting antenna*). The shape and size of an antenna determines its characteristics, such as the direction of radiation and *frequency band* handled. See also *aerial*.

antenna aperture: The effective area of the *antenna* which radiates *electromagnetic waves*.

antenna array: See *array antenna*.

antenna blind area: The area of space which cannot be scanned by an *antenna*, usually due to limitations in its physical design or *radiation pattern*.

antenna dissipative loss: Losses within the *antenna*, measured as the output from the antenna against that obtained by a perfect antenna, which result in *signal* power being adsorbed as heat rather than being radiated.

antenna effective area: That area of the *antenna* which collects or radiates electromagnetic energy. See also *antenna blind area*.

antenna efficiency: A performance measure of an *antenna*, given by the ratio of the total output radiated power to the total input electrical power.

antenna feed: The system used to conduct energy from the transmitter to the antenna, so that it can be radiated as *electromagnetic waves*. These can be of many different types, depending on the type of antenna. See also *horn feed* and *array feed*.

antenna gain: It is a measure of the ability of an *antenna* to concentrate energy or receive energy in a given direction. It is measured as the ratio of power input to the antenna under test to that required for a *reference antenna*, to produce the same *field strength* at a fixed distance and direction from the antenna. Gain is measured in *decibels*.

antenna matching: The process of matching the input impedance of the *antenna* to that of its *transmission line* (*antenna feed*) by adjusting the impedance of the antenna or that of the line. Matching usually occurs over a *frequency band*.

antenna mount: The physical structure on which the *antenna* is fixed.

antenna multiplier: A device which allows several different pieces of equipment to share a single *antenna*.

antenna rotation period: The period of time needed for the *antenna* to go through one complete cycle of rotation. This rotation can be achieved in many ways, such as physical movement of the antenna or by phasing of its arrays.

antenna sweep: The area of space, measured as an angle, which an *antenna* covers during the *antenna rotation period*. Antennas can sweep in a vertical or horizontal direction.

anti-aliasing filter: A *low pass filter* which prevents *aliasing* by limiting the bandwidth of the *signal* before it is sampled, to convert it from an *analogue signal* to a *digital signal*.

antinode: The point in a *standing wave* which has a peak or maximum pressure. (Figure A.18.)

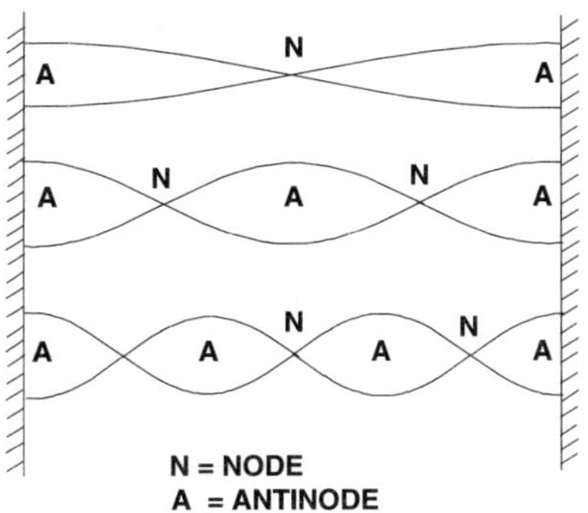

Figure A.18 Standing waves

antireflection coating: A thin film of dielectric or metal applied to an optical surface in order to reduce the *reflection* from the surface and to increase its *transmittance.*

anti-sidetone induction coil: In a *telephone* it is the *hybrid transformer* which controls *sidetone* and matches the impedances of the transmitter and receiver.

anti-singing filter: A device which prevents *singing* (unwanted oscillations) in the *circuit.*

anti-streaming device: A device, in equipment such as a modem, which enables it to ignore *'request to send'* messages from a terminal if it has been in communication with it for greater than a specified period of time.

AON: *Active Optical Network.*

AOWS: *Asia and Oceania Workshop for OSI Standardisation*

AP: *Application Process.*

APC: *Adaptive Predictive Coding.*

APD: *Avalanche photodiode.*

aperiodic antenna: An *antenna* which can operate over a wide *range* of *frequencies* whilst still presenting a relatively constant impedance to the input *signal.*

aperiodic signal: A *signal* which is not a *periodic signal*, i.e. repeats itself at fixed periods of time.

aperture: See *antenna aperture* and *numerical aperture.*

aperture distortion: The defect in a *facsimile* image caused by the shape and size of the transmitter *scanning spot* and the receiver recording spot.

APEX: *Advanced Project for Information Exchange (APEX).*

API: *Application Programme interface.*

apogee: The highest point in a *satellite orbit*, when it is farthest away from the Earth.

apogee motor: The final stage rocket which is fired when a *satellite* is being launched into *orbit.* It accelerates the satellite to a velocity of about 3 km/s and causes it to go into an approximately circular and *geostationary orbit.*

Application Layer: This is the top or seventh layer in the *OSI Basic Reference Model* and is directly involved in the user's applications. It contains the *Association Control Service Elements (ACSE)* which establishes the communications between *Application Processes (AP)*, and the *Application Service Elements (ASE).*

Application Processes (AP): Processes which run within an item of equipment and carry out the functions (applications) of that equipment. Application Processes can be self-contained within that equipment or may need to communicate with other APs within other equipment to complete their functions. *OSI* standards define procedures for effective communication between APs.

application programme: The programme in a processor which carries out the user's task. See *operating system.*

Application Programme Interface (API): A set of *software* calls and routines which may be accessed by the *application programme* in order to use other *network* provided services.

Application Service Elements: These are part of the seventh layer of the *OSI Basic Reference Model* and are used to carry out specific user operations. Examples include *Virtual Terminal (VT)*; *Commitment, Concurrency and Recovery (CC&R)*; File Transfer, Access and Management (FTAM); etc.

Application Specific Integrated Circuit (ASIC): Integrated circuits which have been designed for a specific application. Usually these take the form of a combination of standard functions which can be interconnected on the silicon die, at relative low cost and short time, to give the desired functionality. Most telecommunication system *hardware* is based on the use of ASICs.

application technology: The use of *core technologies* to achieve specific applications. Examples are *ATM*, *ISDN*, etc. (Figure A.19.)

Approvals Committee for Terminal Equipment (ACTE): European body responsible for approval of a *CTR*. ACTE represents the interests of all parties: operators, manufacturers and users. It is advised by *TRAC*.

APS: *Automatic Protection Switching.*

APSK: *Amplitude and Phase Shift Keying.*

architecture: The logical overall structure of the system, e.g. the architecture of a *telephone network.*

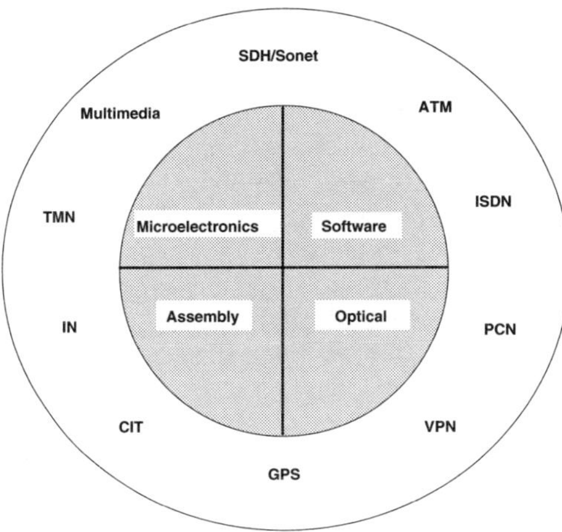

Figure A.19 Some core and application technologies

Architecture Neutral Distribution Format (ANDF): Scheme from the *Open Software Foundation (OSF)* to enable *software* to be produced in a single code to run on any *hardware*.

area code: The preliminary digits of a *telephone number* which indicates the geographical area of a country in which the *subscriber* is located.

area loss: The loss of transmitted optical *signal* which occurs when two *optical fibres* are connected together (for example at a *splice*) due to differences in their cross-sectional areas.

arithmetic mean: The average of a set of numbers found by adding them together and dividing by the number of items involved. Commonly called the mean. For example the mean of four numbers, 6, 8, 12 and 14 is equal to $(6+8+12+14)/4 = 10$.

Armstrong's phase modulator: A system for producing a *signal* with *phase modulation*, as shown in Figure A.20. The output from the *product modulator* is added to the *carrier signal* after it has been phase shifted by 90° which results in the phase modulated signal. This system is only suitable for small values of *modulation index*.

ARP: *Address Resolution Protocol.*

ARPA: *Advanced Research Projects Agency.*

ARPANET: *Advanced Research Projects Agency Network.*

ARQ: *Automatic Repeat Request.*

array antenna: An *antenna* made from several small discrete elements each of which radiates individually. The characteristics of the array are determined by the spacing of the elements and the *amplitude* and *phase* of the excitation currents in the elements.

arrester: A device which protects a system from damage caused by excessive energy surges, as can occur, for example, in lightning strikes, or due to an *electromagnetic pulse*.

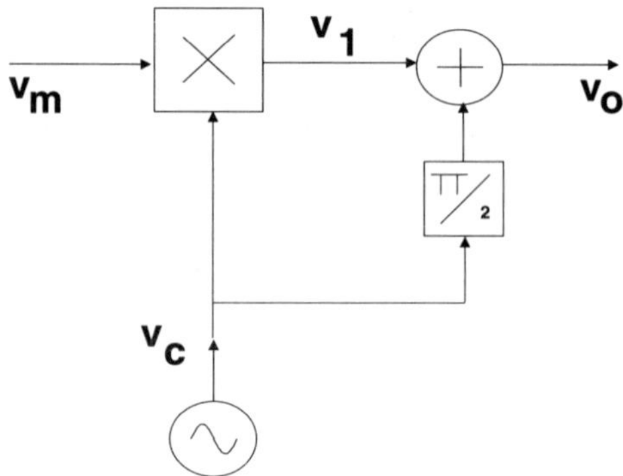

Figure A.20 Armstrong's phase modulator

ARS: *Automatic Route Selection.*

articulation test: A test made on a voice system to determine the ability of the listener to distinguish speech. It is measured by comparing the system under test against a standard reference system.

artificial ear: A system, with an acoustic impedance equal to that of an average human *ear*, which is used to calibrate *telephony* devices such as *headsets.*

Artificial Intelligence (AI): The characteristics of systems which allow them to mimic human intelligence, e.g. reasoning, decision making, etc. See also *expert systems.*

ARU: *Audio Response Unit.*

AS: *Autonomous System.*

ASC: *Accredited Standards Committee.*

ASCII: *American Standard Code for Information Interchange.*

ASE: Amplified Spontaneous Emission.

ASIC: *Application Specific Integrated Circuit.*

Asia and Oceania Workshop for OSI Standardisation (AOWS): One of the regional workshops, operating in Asia and Oceania, which proposes *profiles* to the *Joint Technical Committee on Information Technology* for ratification and adoption as *International Standard Profiles* on *OSI.*

ASK: *Amplitude Shift Keying.*

ASN.1: *Abstract Syntax Notation 1.*

aspect ratio: The ratio of width to height. Therefore the aspect ratio of a television screen is the ratio of the width to its height and that of a *facsimile* transmitted page is the ratio of its width to height.

ASR: *Automatic Send and Receive.*

assembler: A computer program which translates *code* written in *assembly language* into the basic *binary code* or *machine code* which a computer can understand.

assembly language: A low level computer programming language which uses mnemonics which can be translated on a one-to-one basis into *machine code* by an *assembler.*

assigned frequency: The frequency or *frequency band* which has been allocated to a *station* for use in its radio *broadcasts.*

associated control channel: One of the four basic categories of control *channel* specified in the *GSM* Recommendation issued by *ETSI.*

Association Control Service Element (ACSE): The ACSE is defined within the *Application Layer* of the *OSI Basic Reference Model.* It has functions to enable identification and approval of communicating end users; negotiation of the *protocol* functional units which will be used by the applications concerned; the establishment and release of application associations between processes. ACSE can be used directly by an application or it may be used by other Layer 7 *Application Service Elements (ASE).*

Association for Information and Image Management (AIIM): An international body, accredited by *ISO*, working in the field of electronic documentation.

Association Francaise de Normalisation: The standardisation authority in France. It was founded in 1926 and was made responsible for standardisation in France by the decree of 1984. It is responsible for preparing standards, and has its own permanent staff plus a large number who are sponsored by other organisations.

associative storage: Storage or memory used by a processor in which the memory is located or accessed by reference to what it contains rather than by its position or *address*. Also called *content addressable storage*.

ASVD: *Analogue Simultaneous Voice Data.*

asymmetrical access protocols: A *multiple access protocol* in which all users do not have equal probability for *transmission*. Example is the *adaptive tree-walk protocol*.

Asymmetrical Digital Subscriber Line (ADSL): A *Digital Subscriber Line* technique which provides a *broadband* downstream *channel*, (from the *central office* to the user) of between 1.5 Mbit/s and 6 Mbit/s over the copper *local loop*, with a *narrowband* upstream channel (user to *central office*) of between 16 kbit/s and 640 kbit/s. ADSL was launched in 1993 by AT&T to provide *video on demand* over telephone lines. It has been standardised by *ANSI* T1E1.4 (T1.413) with *ETSI* TM6 contributing to an Annex to cover European requirements. Four transport classes have been defined in the *ANSI* standard, as illustrated in Table A.3. Four

Table A.3 ADSL transport classes defined by ANSI

Parameter	*Transport class*						
	E1			*T1*			
	2M1	*2M2*	*2M3*	*1*	*2*	*3*	*4*
Max. capacity downstream (Mbit/s)	6.144	4.096	2.048	6.144	4.608	3.072	1.536
Max. capacity upstream (kbit/s)	640	608	176	640	608	608	176
Control (kbit/s)	64	64	16	64	64	64	16
POTS (kbit/s)	64	64	64	64	64	64	64

classes are based on multiples of 1.5 Mbit/s (*T1*) and three classes on multiples on 2.0 Mbit/s (*E1*). Each class specifies a maximum possible downstream and upstream bandwidth, taking factors such as line conditions, wire gauge, loop length, etc. into account. Classes 1 and 2M1 are for operating under best conditions and classes 4 and 2M3 for worst conditions. Progress in ADSL chipset development has allowed even faster rates than those in Table A.3 to be achieved in practice.

asymmetrical duplex transmission: The process of using two separate *transmission rates* to transmit *data* simultaneously over the same *transmission line*.

asynchronous data channel: A *data communications channel* in which no separate *timing* information is transferred between the sender and receiver. *Asynchronous transmission* occurs.

asynchronous multiplexer: An older type of *multiplexer* which handled *asynchronous data channels*.

asynchronous network: A *transmission network* which does not operate using a *synchronous* or *mesochronous clock*.

asynchronous satellite: A *satellite* whose rotation in its *orbit* is not affected by the rotation of the object around which it is moving.

asynchronous terminal: A *terminal* which operates using *asynchronous transmission*. It is also often referred to as an *ASCII* terminal or a *dumb terminal*.

Asynchronous Time Division Multiplexing (ATDM): A *Time Division Multiplexing (TDM)* technique which uses *asynchronous transmission*.

Asynchronous Transfer Mode (ATM): A *packet switching* communications standard which uses *packets* of constant length, called *ATM cells*. These cells are routed through the *network* by reference to *address* information rather than by their position in a *frame*. Operation is *connection mode* by setting up *virtual channels*. ATM is able to carry a mix of *traffic* types: *voice*, *data*, and *video*.

asynchronous transmission: A communication system in which there is no *timing* relationship between different elements. *Transmission* in an asynchronous system occurs with use of *start bits* and *stop bits*. See also *anisochronous system* and *synchronous transmission*.

ATB: *All Trunks Busy*.

ATC: *ATM Transfer Capabilities*.

ATDM: *Asynchronous Time Division Multiplexing*.

ATIS: *Alliance for Telecommunications Industry Solutions*.

ATM: *Asynchronous Transfer Mode*.

ATM Adaptation Layer (AAL): In the *B-ISDN* model this layer adapts the functions or services provided by the higher layers into the ATM bearer service. It comes between the ATM layer and the next higher layers in the user plane, the control plane and the management plane. The

ATM layer is subdivided into the Segmentation and Reassembly (SAR) sub-layer, which breaks the information in the higher layers into a size suitable for the ATM cells, and the Convergence Sub-layer (CS) which provides the AAL service. Further subdivisions of these are also specified in ITU standards. Five types of AAL are described in I.363. Type 1 is primarily intended for *Constant Bit Rate (CBR)* traffic, such as video, high quality audio, or voice. Type 2 is for the transmission of *Variable Bit Rate (VBR)* traffic, such as *compressed video* or compressed voice. Type 3/4 is intended for use with *bursty traffic* but Type 5 is used more often for this. Type 5 is for the general transmission of all data traffic, including bursty traffic, using the *Available Bit Rate (ABR)* or *Unspecified Bit Rate (UBR)* class of service.

ATM Block Transfer (ABT): One of four *ATM Transfer Capabilities* defined by the ITU-T. In it the user block structures the *data* stream and negotiates a *Current Bit Rate (CBR)* for each block.

ATM cell: *ATM* uses *cells* having a constant length and with a structure shown in Figure A.21. Each cell is 53 *bytes* in length with the *payload* occupying 48 bytes. The *header* consists of 5 bytes and is made up of the: *address* field, with the *virtual path identifier* and the *virtual channel identifier*; the *Payload Type Identifier (PTI)*; and an 8 *bit Cyclic Redun-*

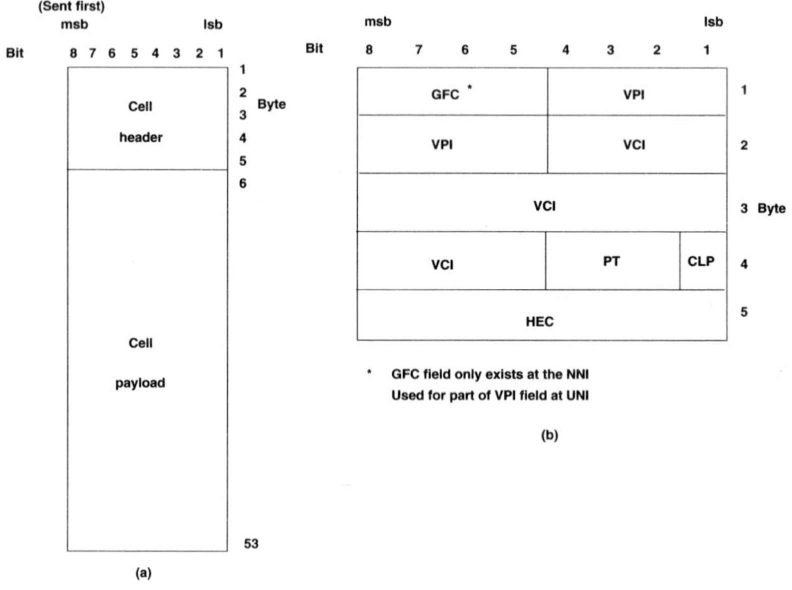

Figure A.21 ATM cell: (a) cell structure; (b) cell header

dancy Check field, which carries out *Header Error Control* and defines the start point for each cell.

ATM Forum: Formed in October 1991 by four telecommunications and computer vendors, membership quickly exceeded 750. The Forum's main aim is to accelerate the development and deployment of *ATM* products and services. It is recognised by *ITU-T* as a working group and it writes specifications which are passed on to the ITU-T. It is a non-profit making organisation with world headquarters in Mountain View, California and European Headquarters in Brussels. The Forum consists of three main sections: a world-wide Technical Committee with several working groups; three Marketing Committees (MACs) covering North America, Europe and Asia-Pacific, who aim to improve the acceptance and understanding of ATM by providing educational and marketing services; and the Enterprise Network Roundtable (ERN), set up in 1993, which works within the Forum's user group and aims to ensure that the technical specifications from the ATM Forum meet real end user needs.

atmospheric duct: A layer of air, usually close to the Earth's surface, which has a higher *refractive index* than the surrounding air. *Radio waves* propagate in the layer with less *attenuation* and are bent back into the duct by the lower refractive index atmosphere which surrounds it.

atmospheric noise: *Noise* caused in a radio based system by atmospheric effects, such as electrical discharges and lightning.

ATM Transfer Capabilities (ATC): *ITU-T* has defined four ATM Transfer Capabilities in order to standardise and simplify *connection management* in *ATM* systems. (*ITU-T Recommendation* I.371.) These are: *Statistical Bit Rate (SBR)*, *Available Bit Rate (ABR)*, *Deterministic Bit Rate (DBR)*, and *ATM Block Transfer (ABT)*. Each of these relates to a *Quality of Service (QoS)* level and a specified type of *traffic*.

ATOF: *Advanced Tactical Optical Fibre*.

attack dialling: The process of continuously *dialling* to establish a *call* if the *line* is busy, so that the call goes through the instant the line becomes free.

attendant: In telecommunications it refers to a switchboard operator, such as one associated with a *PABX*.

attendant access loop: A system in which *calls* are controlled manually by an *attendant*.

attendant conference: A feature of a switch which allows an *attendant* to set up a *teleconference* between several users.

attention code: A *signal* which is sent to a processing unit to request service.

attention key: A key or switch which enables an interrupt *signal* to be sent to the processing unit in order to terminate its current action and request service.

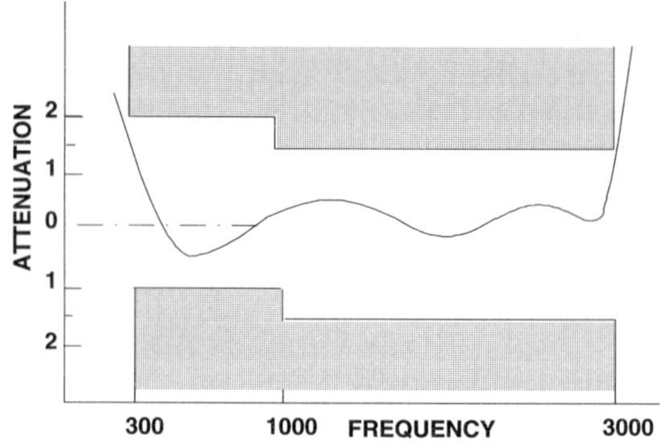

Figure A.22 Attenuation distortion voice channel

attention sign: The first flag position in a *semaphore* communications system, which indicates the start of transmission of the *message*.

attenuation: Loss of *signal power* during *transmission* through a medium. The amount of loss is usually dependent on the *wavelength* of the *signal*, the characteristics of the *transmission medium*, and the length of propagation. It is measured in *decibels*. Attenuation can occur during propagation of different signals, such as radio and light waves.

attenuation distortion: *Signals* are subject to different levels of *attenuation* depending on *frequency*. Therefore in a transmitted signal, consisting of many different frequencies, the *waveform* will suffer *distortion* due to this difference in attenuation. Attenuation distortion in the range of *voice frequencies* is measured by reference to 800 Hz, as specified by ITU-T, whilst in North America 1004 Hz is more commonly used. Figure A.22 illustrates distortion for a voice channel.

attenuator: A device which results in *attenuation* of the *signal* (which may be electrical or optical) without resulting in appreciable *distortion*. See also *optical attenuator*, *fixed attenuator* and *variable attenuator*.

attitude: Usually refers to the position of a *satellite* in its *orbit*, relative to the Earth and Sun.

attitude stabilisation: The control of a *satellite's attitude* in order to obtain the most effective use from the *gain* of its directive *antennas*.

attribute: The characteristics or properties of a system. For example the attribute of a display system could be its brightness, colour, size of displayed dots on the screen, etc.

ATUG: *Australian Telecommunications User Group*.

ATV: *Advanced Television*.

AU: *Administrative Unit*.

AUC: *Authentication Centre*.

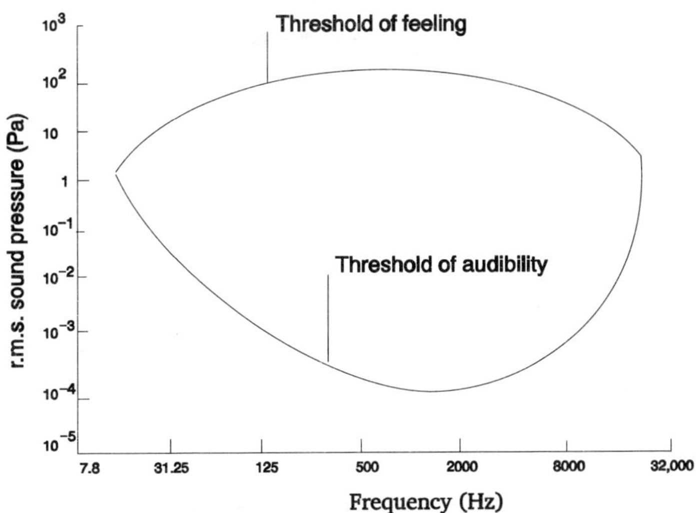

Figure A.23 Audibility

audibility: Ability of the human *ear* to hear a generated sound. This will vary depending on the level of sound as well as its *frequency*, and it is given by curves such as Figure A.23. The maximum sound level which can be comfortably tolerated is known as the *threshold of feeling*, and the minimum which can be heard is the *threshold of audibility* or *threshold of hearing*. The ear can detect sound in the 30 Hz to 20 kHz region, although the most sensitive band covers 1 kHz to 2 kHz.

audible alarm: An *alarm signal* which can be heard by the human *ear*.

audible ringing tone: The audible *signal* which the caller hears in the earpiece when the called *telephone* is *ringing*. It stops as soon as the *call* is answered and the receiver goes *off-hook*.

audio amplifier: An *amplifier* which can magnify *signals* in the *frequency band* which is audible to the human *ear*.

audio circuit: A component which is used in a *broadcast* radio or television service to carry the audio component of the *signal*. See also *speech circuit* and *voice circuit*.

audio coding: The conversion of audio *analogue signals* into a different format for *transmission*. Examples are *PCM*, *ADPCM* and *CELP*.

audio digital companding: *Companding* applied to *digital signals* in the *audio frequency band*.

Audio Frequency (AF): Frequencies in the range 30 Hz to 20 kHz which, when transmitted as sound waves, can be heard by the human *ear*.

audio frequency band: The *frequency spectrum* in the range 30 Hz to 20 kHz which the human *ear* can detect.

Audio Messaging Interchange Specification (AMIS): Standards for *voice mail systems*. Example is the digital standard based on the *ITU-T Recommendation* X.400.

audiometer: An instrument which can produce a *sound wave* of variable *frequency* and *amplitude* and is usually used in tests on hearing.

Audio Response Unit (ARU): Equipment which generates synthesised or pre-recorded speech in response to an input, which is usually made via a *dual tone multifrequency signalling* system. The *message* can vary depending on various conditions, such as the number on the *handset* pressed by the caller, the time of day, etc.

audio signal: An electrical *signal* which corresponds directly to an audible signal. For example the *microphone* in a *telephone* produces an electrical signal corresponding to the sound input and the earpiece produces an audible signal corresponding to the electrical input.

audiotext: The translation of *messages* stored in a *database* into spoken words. Audiotext is commonly used to provide information to users on topics such as weather forecasts, stock market prices, sports results. The user interacts with the system by means such as a *DTMF telephone* and by *voice recognition*. See also *premium rate service*.

audiovisual systems: Systems which handle audio and visual images. There are several *ITU-T Recommendations* covering this, such as Recommendation H.320 for audiovisual services at *transmission rates* in the band 64 kbit/s to 1920 kbit/s and Recommendation H.324 for *bit rates* below 64 kbit/s.

audit trail: A record of a sequence of events, which allows the events to be studied and recreated at a later date, if required.

aurora: Spontaneous, naturally occurring, *radiated emissions* which occur in the Earth's upper atmosphere. Those which occur in the northern hemisphere are know as aurora borealis and those in the southern hemisphere as the aurora australis. Aurora emissions interfere with *radio communications*.

AUSTEL: *Australian Telecommunications Authority*.

Australian Communications Authority: Australian body responsible for functions relating to *telephone numbering*, licensing and consumer protection.

Australian CCITT Committee (ACC): Now called the *Australian ITU-T Committee* it was formed as part of Australia's membership of the *ITU*. It coordinates all the technical activities and monitors the work done by other standards bodies in the region. The ACC is chaired by the *Australian Telecommunications Authority (AUSTEL)* and has representatives from the Australian Electronics Industry Association (AEIA), Australian carriers, AUSTEL, the Department of Trade and Communications (DOTAC), *Standards Australia*, and the Australian Information Industry

Association (AIIA). Associate members are also included from other Australian organisations or from abroad. The ACC sets up national *Study Groups* for each international *ITU-T* Study Group.

Australian Competition and Consumer Commission: Organisation which took over, on 1 July 1997, responsibility for endorsing *access* and *interconnection* arrangements within Australia.

Australian ITU-T Committee: See *Australian CCITT Committee.*

Australian Telecommunications Authority (AUSTEL): Formed in July 1989 to supervise the introduction of competition in the Australian telecommunications industry. Primarily a *telecommunications watchdog* with powers and functions obtained from the Telecommunications Act of 1991. Empowered by law to set mandatory Australian national standards relating to *Customer Premises Equipment (CPE)* and cabling.

Australian Telecommunications User Group (ATUG): A non-profit making user group whose executive is based in Sydney and members include individuals, organisations and government departments. It aims to improve the services available to its members, including the range of services and its quality, and it is also involved in pricing issues. It carries out political lobbying on behalf of its members and seeks to influence those organisations who are responsible for setting telecommunication regulations or legislation. Its executive or task force prepares papers on selected topics, which are submitted to relevant bodies, in order to influence them.

authentication: The process of verifying that the *message* received is genuine and comes from the expected source. Authentication is commonly used to prevent telecommunications fraud and often makes use of *encryption*

Authentication Centre (AUC): The central control point, for example of a *cellular radio system*, which carries out *authentication* of a *call.*

authorisation: The process of granting a user special privileges to carry out specific functions on a system.

authority: Usually refers to a *telecommunications authority*.

autoanswer facility: The facility available with some equipment, such as *modems*, which allow them to automatically answer an *incoming call* over the *PSTN. Attendant* operation is when manual intervention is needed to answer a call, such as changing over from *voice* to *data* in a modem.

autobaud rate facility: The ability of a system to detect the *baud rate* of the *incoming signal* and to adjust itself to match this reception rate. This enables the system to receive *data* from many different devices, operating at different *data rates*. It is also known as *automatic baud rate detection*.

autodialler: A communications system which either: (a) allows frequently dialled numbers to be stored and *calls* made to these numbers by one or two key strokes; or (b) automatically redials a busy number until it becomes free. See also *attack dialling*.

automated attendant: Usually refers to a *voice processing* application which supplements the switchboard operator by allowing some *calls* to be handled automatically, e.g. querying a *database* to obtain a user's extension number by using a *DTMF telephone*.

Automated Maritime Telecommunications System (AMTS): A USA maritime system which is used for communications with ship stations in specific areas along the coast.

automated teller machine: Equipment which allows users to communicate with a bank's *database* to carry out financial transactions, without human intervention.

automatic alternative routeing: In an *exchange hierarchy*, when a direct route is busy, *traffic* may be automatically routed using an alternative *line*, such as via a higher level *tandem* route. Automatic alternative routeing is also applicable to *private networks*, where traffic on a congested *private line* between two *PABXs* may be automatically routed via an overflow *PSTN* line.

automatic announcement: *Network* feature which allows pre-recorded messages to be delivered automatically to users, e.g. wake-up calls.

automatic baud rate detection: See *autobaud rate facility*.

automatic callback: *Network* feature which allows the user to instruct the system to retry calling a busy number when it becomes free. See also *attack dialling*.

Automatic Call Distribution (ACD): A feature of a *PABX* or *Centrex* which allows a large number of *incoming calls* to be distributed amongst a designated group of users (called agents) according to a pattern which is programmed into the system. The ACD system can also provide ancillary management information such as the number of calls received, the average waiting time before a call is answered, the number of incoming calls which hang up whilst still in a waiting queue, etc.

Automatic Calling Unit: A unit which allows *data terminals*, such as computers and *card diallers*, to automatically initiate *calls* over a telecommunications *network*.

Automatic Data Processing Equipment (ADPE): Collection of equipment which results in *data* being automatically processed. The amount of processing can vary.

automatic date and time display: A facility which allows the date and time of a communications to be recorded and displayed. This could be the time the *message* is sent, when it is received, or both.

automatic dialler: A system which automatically dials a *telephone number*. This may be a selection from a pre-set selection of numbers or it may be randomly selected from a *database*, as in *telemarketing*.

Automatic Digital Distribution Frame (ADDF): System used to replace manual *distribution frames* in *transmission networks* used for the *Plesiochronous Digital Hierarchy (PDH)*.

automatic exchange: A *telephone exchange* which can automatically route *calls* between *subscribers*, without human intervention.

Automatic Gain Control (AGC): The concept of automatically controlling the *gain* of a device, such as an *amplifier*, to compensate for variations of the input *signal* strength so that the output signal is maintain at a constant level.

Automatic Identified Outward Dialling (AIOD): Term used to describe a service in which the *subscriber* receives an *itemised bill* for all *calls* made.

Automatic Intercept System (AIS): A system which is programmed to provide information automatically to a telephone caller who has been *intercepted* and routed to it.

Automatic Link Establishment (ALE): The capability of radio *stations* to establish radio *links* between themselves automatically, usually under control of a processor.

Automatic Message Accounting (AMA): System which automatically records *data* on *network* usage in order to generate *billing* information.

Automatic Number Identification (ANI): Facility, available in the USA, where a *Central Office (CO)* switch can identify the number of the *calling party* and pass this on to the *called party*. The called party can usually see this number before answering the *call*. In the UK this facility is known as *Calling Line Identification (CLI)*.

Automatic Protection Switching (APS): In *transmission* the capability of the system to automatically switch the *traffic* to another route should a fault occur on the original route.

Automatic Repeat Request (ARQ): A method for error control in *transmission* in which the *receiver* checks the *message* for errors and automatically informs the *sender* if an error is found, which results in the sender re-transmitting the faulty part of the message. This is also known as Automatic Retransmissions Request.

Automatic Route Selection (ARS): The ability of switches, such as a *PABX*, to automatically select a *transmission path* depending on specified criteria, such as the shortest distance or least congestion.

Automatic Send and Receive (ASR): Transmission equipment which has storage facilities which allow it to store, receive and send *messages* unattended.

automatic sequential connection: A facility which allows a *data terminal* to be connected in sequence, over a *private network* or the *PSTN*, to a number of other data terminals.

automatic switching equipment: Equipment, usually housed in an *automatic exchange* which carries out all the *switching* functions automatically.

Automatic Voice Network (AUTOVON): Non-secure voice *network* developed for use by the US *Department of Defence*, and now replaced by the Defence Switched Network.

Autonomous System (AS): A group of *networks* which are under a single *span of control*, usually that of one a single *telecommunications authority*. *Routeing* within an Autonomous System is known as internal and that beyond its boundary is called external.

AUTOVON: *Automatic Voice Network.*

availability of service: A commonly used measure of the *Quality of Service (QoS)* provided by a *network*. It is measured as the ratio of time the network is available for use to the total time in the period being considered. It is stated as a percentage or as a decimal (e.g. 60% or 0.6). Availability is low on a *congested system* or an unreliable system, measured in terms of its *Bit Error Ratio (BER)*, *MTBF* and *MTTR*.

Available Bit Rate (ABR): An *ATM Transfer Capability* defined by the *ITU-T* for *connection management*. It makes use of any available *bandwidth* subject to a minimum value specified by the user.

available line: The telecommunication *circuit* between two points which is operational but is not currently in use.

available time: The time that a system is available for use. See *availability of service*.

Avalanche Photodiode (APD): A *photodiode* in which the hole-electron pairs released by the action of light are accelerated such that they gain sufficient energy to generate secondary hole-electron pairs, resulting in a rapid increase of photocurrent.

Average Busy Season Busy Hour (ABSBH): Measure of *traffic* intensity. It denotes the three months in the year having the highest average traffic during the *busy hour*. Special days, such as Christmas, are excluded from the calculation.

average call duration: A parameter used in *teletraffic theory* which measures the average amount of time which a *call* occupies. It is calculated by measuring the total amount of time occupied by all calls during a period and dividing this by the number of calls in the same period.

average traffic: The mean value of *traffic* over a period of time. If N are the number of *calls* made over a period of time P, and the mean *holding time* of a call is H, then the average traffic is given by NH/P.

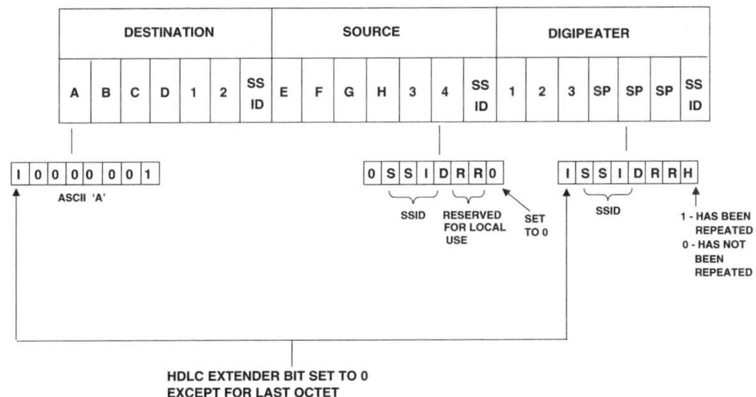

Figure A.24 AX.25 packet frame format

avoidance routeing: *Routeing* in which certain parts of the *network* are avoided, for example because they are known to be congested or defective.

AWG: *American Wire Gauge.*

AX.25: A derivative of the *X.25 protocol* used for *packet* based communications. Figure A.24 shows the *frame* format. Start and stop flags are used to define the *message* area. The information field contains a *network* protocol identifier (PID) and the frame includes a *frame check sequence (FCS)*

axial propagation coefficient: The *propagation coefficient* of a system measured along the axis of the *transmission medium.*

axial ray: A ray which travels along the axis of the *transmission medium.* Therefore for a *fibre optic* communications system an axial ray is the ray of light which travels along the axis parallel to the length of the fibre.

B

babble: Meaningless *transmission*, generated by systems on the *network*, which result in interference with other *signals* on the line.

BABT: *British Approvals Board for Telecommunications.*

backbone: That part of the *network* which carries all the high density *traffic*. The backbone for a switched network usually consists of main *trunk routes* and switches, and in a *data network* it would include the main links between local rings.

backbone routeing: *Routeing* which includes only the *backbone* of a *network*.

Background Block Error Ratio (BBER): A measure of *transmission* performance, specified within *ITU-T Recommendation* G.826, relating to digital paths at or above *E1* and *T1*. It applies to both *PDH* and *SDH* systems and is given by the ratio of *errored blocks* to total *blocks* transmitted during a specified time interval. All blocks occurring during the *Severely Errored Seconds* and unavailable time are not considered.

background noise: *Noise* which is present in a system even when there is no information *signal* being transmitted. This can apply to different systems, such as electrical, optical and acoustic.

backhauling: The practice of taking *traffic* back to a common point for *switching*. This can increase congestion on intermediate lines since the traffic may pass the same point twice. For example, much of the *Internet* traffic is backhauled from other countries back to the USA for switching, before being routed back to the countries concerned.

back-off: Usually applies to the process used for *multiple access* in *Local Area Networks (LAN)*, such as *CSMA/CD*, where one of the transmitting *terminals* backs-off (stops *transmission*) for a short time to prevent *collisions* with another terminals.

backplane: Usually refers to the *hardware* of a piece of equipment into which the individual cards plug, and which carries the *signals* between the cards.

backscatter: *Scattering* of an incident *electromagnetic wave* where the scattered component is in the opposite direction of propagation to the main wave.

Back Space (BS): A *Format Effector (EF)* which is used to move the print mechanism on a printer, or the cursor on a *Visual Display Unit (VDU)*, back one position from its current point.

backward channel: The *transmission channel* which is in the apposite direction to the main or *forward channel*. The main channel usually carries the user's information whilst the backward channel, which is of lower *bandwidth*, carries auxiliary information, such as *acknowledgements*, *error control signals* and *supervisory signals*. Also called the *reverse channel*.

Backward Explicit Congestion Notification (BECN): A method used in *frame relay* networks for informing the transmitting *node* that the route, in a direction opposite to that being taken, has become *congested*. This is done by setting a *bit* in the *header* of the *frame* to a logical 1.

Backward Indicator Bit (BIB): Method used, in *Signalling System No. 7*, developed by the *IUT-T*, for a receiving *terminal* to inform the sender that an error has occurred. The value of the *bit* is inverted.

balanced amplitude modulation: A *suppressed carrier modulation* system in which the *carrier signal* is suppressed by a *balanced circuit*. The output may be either a *single sideband* or a *double sideband* signal which is *amplitude modulated*. All *information* is contained in the sidebands and not in the carrier.

balanced circuit: A circuit whose impedance matches that of the *transmission line* so that *return loss* is minimised.

balanced code: A *line code* used in *PCM* systems in which the *frequency spectrum* of the coded *signal* has no *direct current* component.

balanced double current interchange circuit: A circuit used to interface a *DTE* to a *modem* to enable *data* to be transmitted over the *PSTN*. The characteristics of the circuit are specified in *ITU-T Recommendation* V.11.

balanced differential signalling: *Differential signalling* in which the current in the two wires are of equal and opposite strength, so they balance each other and their sum is equal to zero.

balanced line: A *transmission line* having two conductors and carrying *signals* which are of equal and opposite value so they cancel each other.

balanced modulator: A *modulator* used for *amplitude modulation* in which the output consists of *sidebands* only. The *carrier* is suppressed and *noise* associated with the carrier is balanced and removed.

balancing network: A *circuit* used to interface a *balanced line* to an *unbalanced line* in order to match their characteristics.

balun: Derived from the combination of the words BALanced and UNbalanced, a balun is primarily an impedance matching and *circuit* balancing device. It can interface one type of *transmission medium* to another, such as *twisted pair wire* to *coaxial cable*. Baluns must be used at either end of the original *transmission medium* and present its *characteristic impedance* to the transmitting and receiving *terminals*.

band limited signal: The *signal* which results when the *transmission medium* has low *bandwidth* so that *frequencies* within the signal are attenuated by different amounts (Figure B.1). Also referred to as bandwidth limited signal.

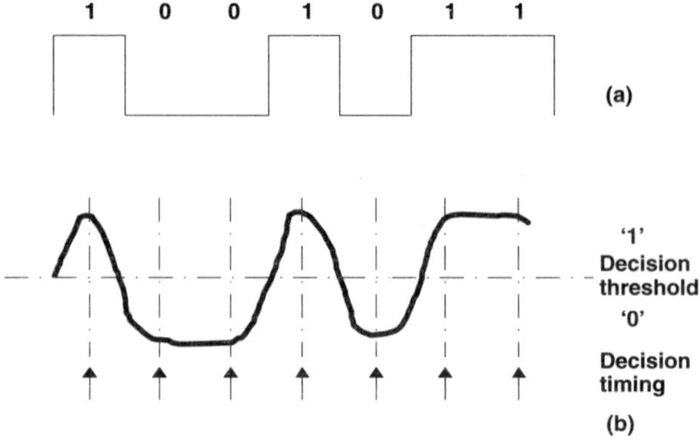

Figure B.1 Baseband waveform: (a) transmitted signal; (b) band limited signal.

band limited white noise: The output correlated *white noise* produced when uncorrelated white noise is passed through a circuit with low *bandwidth* handling capabilities.

band pass filter: A *filter* which lets through a *frequency range* or *frequency band* whilst blocking all others above and below this range. See also *low pass filter*, *high pass filter* and *band stop filter*.

band stop filter: A *filter* which stops or severely attenuates a *range* or band of *frequencies* but lets through other frequencies which are above or below this range. See also *band pass filter*.

bandwidth: The bandwidth of a device is usually a measure of the *frequency range* or *frequency band* over which it meets its operating characteristics. Therefore the bandwidth of a *transmission medium* is the band of frequencies which it can carry without significant loss or *distortion*. The bandwidth of a light emitter is the range of frequencies which it emits. See also *narrowband* and *wideband*.

bandwidth compression: A technique used for reducing the *bandwidth* of a *signal* without affecting the *information* carried by the signal. This means that the same amount of information can be transmitted using lower grade, lower bandwidth *channels*.

bandwidth-distance product: The product of the *bandwidth* of a *signal* and the distance over which it can be transmitted in a medium, such as *optical fibre*, whilst still maintaining its performance within specified values. Usually the greater the bandwidth the lower the distance it can be transmitted.

bandwidth on demand: A technique which allocates parts of the total system *bandwidth* to different users according to their requirements at any time. This ensure more efficient use of the whole bandwidth, with less unused *capacity*.

Bandwidth on Demand Interoperability Group (BONDING): An industrial group of companies formed to further the development and use of *bandwidth on demand* systems by ensuring that those provided by different vendors can interwork.

barrage jamming: *Jamming* of a transmitter by using a *jam signal* having wide *frequency range* which interfere with it. Because of the range of frequencies used it is possible that several transmitters will be jammed simultaneously and it is less likely that the jammed source will be able to avoid jamming by shifting its frequency. However, the wider frequency range used means that the power level will be lower so that a strong transmitting source may not be effectively jammed.

barrel distortion: *Distortion* of an image which results in straight sided objects or lines being bowed outwards, like the sides of a barrel.

barrier code: The *code*, consisting of a series of *digits*, which a user may need to dial prior to the *authorisation* code, to gain access to a *PABX*.

baseband Local Area Network: A *Local Area Network (LAN)* in which the *signals* are transmitted in their original form, without any *modulation* of a *carrier signal*.

baseband modem: The baseband modem takes a *digital signal* and transmits it in its existing, or slightly modified, form. This is not a true *modem* since it does not *modulate* a signal, in the accepted sense. It uses frequencies outside the *voice frequency* band, down to *direct current*. Therefore it is not suitable for use over the *PSTN* which has been enhanced with *loading* to carry voice speech. Its limited performance means it is only suitable for use over a relatively short distance, hence it is also often referred to as a *short-haul modem*.

baseband signal: Usually refers to the original generated *signal* before it has been used to *modulate* a *carrier signal*. Baseband signals are usually lower down in the *frequency spectrum*, and the lowest frequency could be zero (*direct current*).

base Earth station: A *base station* used within a *satellite* based mobile communications system. See also *Earth station*.

base standard: Usually refers to a standard produced by an international organisation, such as the *ITU-T*. It may contain permissible variants or alternative methods, which are implementor defined facilities. Adopting any of these variants would mean that the implementor was compliant with the standard, but there would be no guarantee of *interworking* between equipment based on these separate variants. Regional and national standards bodies usually adapt these base standards as *functional standards* or *profiles*, which contain only a limited subset of the permissible variants. Agreed test specifications and methods are also developed, to ensure that equipment designed to the different variants, permitted within the functional standard, will be able to interwork.

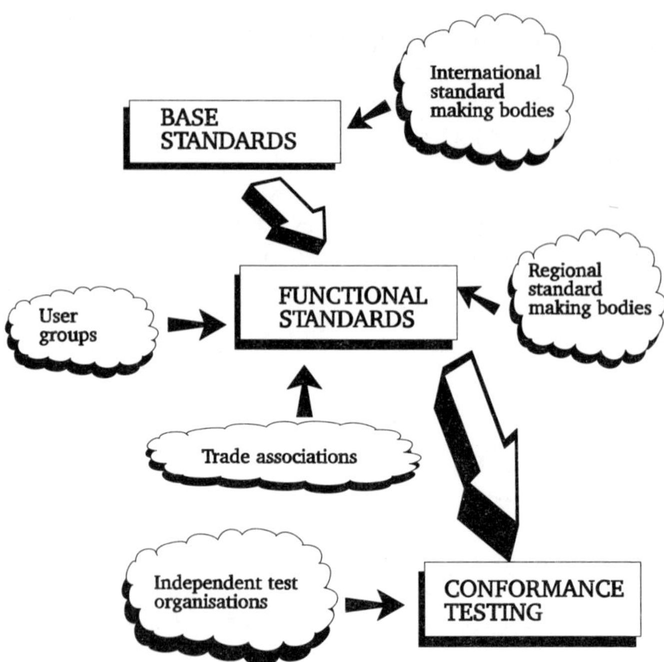

Figure B.2 Base standards, functional standards and conformance testing

Independent test houses carry out *conformance testing* against the selected profile and certify products which meet this requirement. Figure B.2 illustrates this standards making process.

base station: Radio station used within mobile communications, such as *cellular radio system*, for *transmission* to mobiles or to other base stations.

Base Station Controller (BSC): Part of the *Base Station Subsystem*.

Base Station Subsystem (BSS): Part of a *cellular radio system*, such as *GSM*. It contains the *BTS* and the *BSC* modules which are responsible for carrying out all the radio channel *management* functions, such as *configuration control*, power control, control of *frequency hopping*, and allocation of appropriate *channels* for *speech signals*, *data* and *signalling*.

Base Transceiver Station (BTS): Part of the *Base Station Subsystem*.

BASIC: BASIC stands for Beginners All-purpose Symbolic Instruction Code. It is a *high level language* used for programming computers and other processor based equipment.

Basic Exchange Radio (BEXR): System operating in the 450 MHz band and designed for providing *rural communications*.

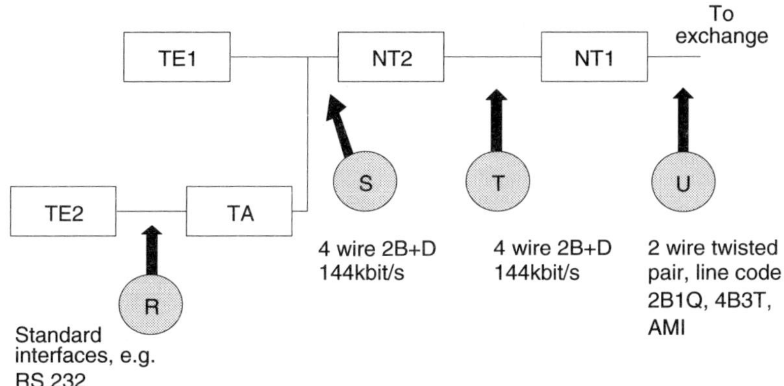

Figure B.3 ISDN Basic Rate Access

Basic Exchange Telecommunications Radio Service (BETRS): Basic *telephony* service for *rural communications*, authorised by the *FCC* and operating in the *UHF* and *VHF frequency bands*. *Digital* radio signals are transmitted to a local *antenna* and converted to *analogue signals* before being taken into the customer's premises.

Basic Mode Link Control (BMLC): *ISO* standard 646-1983 and *ITU-T Recommendation* V.3 for control of *data links* at the *data link layer* of the *OSI Basic Reference Model*.

Basic Rate Access (BRA): One of the access mechanisms specified in the *I Series Recommendations* issued by the *ITU-T* for *ISDN*. This is illustrated in Figure B.3, where R, S, T and U are the *reference points*, TE is *terminal equipment*, NT is a *network termination*, and TA is a *terminal adapter*. See also *Basic Rate Interface*, *two wire transmission* and *four wire transmission*.

Basic Rate Interface (BRI): An *ISDN* service which offers two *bearer channels*, or *B channels*, at 64 kbit/s each and a *signalling* channel, at *16 kbit/s*, making a total transmission rated of 144 kbit/s. This is also referred to as 2B+D.

basic services: Usually refers to the basic telecommunications services provided to *subscribers* to meet their ordinary requirements. See also *supplementary services*.

batched transmission: The *transmission* of *messages* from one *terminal* to another without any transmissions taking place in the reverse direction, in-between the messages.

batch processing: *Data processing* technique in which the *data* is collected and presented to the processor as a batch and the jobs are then processed without any interaction from the users.

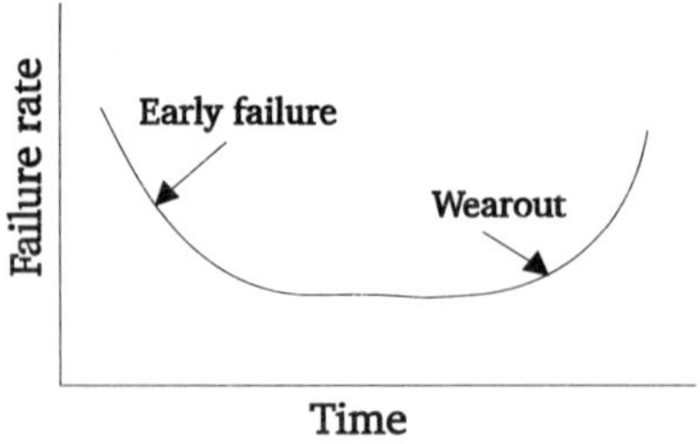

Figure B.4 Bathtub curve

bathtub curve: A curve which has a shape resembling a bathtub, as in Figure B.4. This is usually used to illustrate the failure rate of electronic equipment over time. There are a large number of early failures, but failure rate then decreases and remains fairly constant until the equipment reaches the end of its life when the failure rate again increases.

baud rate: It is a measure of the speed of a *transmission* or the *capacity* of a *channel*. It is given by the number of events in a second. If the *signal* consisted of a string of *bits* then the baud rate is equal to the speed of the signal measured in bits per second. If the signal goes through *modulation* and has several bits per modulation then the baud rate is less than the *bit rate*. Therefore in *dibit* which has four phase changes for modulation of a bit, a bit rate of 12000 bits/s would result in a baud rate of 600 bauds.

Baudot code: *Code*, developed by Emile Baudot, for *asynchronous data transmission*. It uses five *bits* to represent individual alphabetic *characters* and special functions such as carriage returns, line feeds, and space. Used mainly in *teleprinter* systems, it has now been replaced by the *International Alphabet No. 2*.

Bayes' theorem: A technique used in *probability* theory. If P(A) is the probability of an event occurring; $P(\overline{A})$ is the probability of it not occurring; P(B|A) is the probability of event B occurring assuming event A has already occurred; and P(A|B) is the probability of event A occurring when event B has already occurred, then Bayes' theorem states that: $P(A|B) = [P(A)P(B|A)]/[P(A)P(B|A) + P(\overline{A})P(B|\overline{A})]$. For example, suppose a company discovers that 80% of those who bought its products in a year had been on a company training course, and 30% of those who bought a competitor's equivalent product had also been on the same training course. During the year the company had 20% of the market share for this product and the company wishes to know what percentage of buyers actually went on its training course, in order to gauge its effectiveness. If B denotes a person who bought the company's product and T that he went on the training course, then the problem is to find P(B|T). From the data P(B) = 0.2, $P(\overline{B})$= 0.8, so from Bayes' theorem P(B|T) is equal to 0.4.

BBER: *Background Block Error Ratio.*
BBS: *Bulletin Board Service.*
BCC: *Block Check Character* or *Blocked Calls Cleared.*
BCCH: *Broadcast Control Channel.*
BCD: *Binary Coded Decimal.*
B channel: See *bearer channel.*
BCH: *Bids per Circuit per Hour.*
BCH code: *Bose, Chaudhuri, Hocquencgham code.*
BDF *Building Distribution Frame.*
BDT: The Telecommunications Development Bureau, which formed part of the previous *ITU* organisation. On 1 July 1994 the ITU restructured and the BDT combined with the *CTD* to form the new Telecommunications Development Sector.
beam: Usually refers to a group of *electromagnetic radiation* which is parallel, converging or diverging. Examples are a beam of light and a beam of *microwave* radio.
beam splitter: A device which is used to split a *beam*, such as that of light, into two or more separate beams. Generally the strength of these beams is lower than that of the combined beam unless they go through an *amplifier*.
beam steering: Changing the direction of radiation from the beam by altering that from its major *lobe*. In optical systems this may be achieved by mirrors and lenses and in radio systems by controlling the phase of signals fed into the *antenna*.
beam tilt: The tilt of the main *lobe* of the *beam* measured in degrees from the horizon.
beamwidth: The width of a *beam* measured as an angle. In radio systems it is usually taken to the *half power points*, i.e. 3 db points. For a *satellite* it is given by the area covered.
bearer channel: A 64kbit/s *channel* used for the *transmission* of user data (not *signalling*) within *ISDN*.
bearer services: These provide *subscribers* with *transmission capacity* which can be used for various purposes. In *ISDN* they involve only the first three layers of the *OSI Basic Reference Model*. Several bearer services have been defined by the *ITU-T*, such as: 64 kbit/s unrestricted; 2 × 64 kbits/s unrestricted; 384 kbit/s unrestricted; 1536 kbit/s unrestricted; and 1920 kbit/s unrestricted.
beat frequency: *Signal* frequency obtained by combining signals with two other *frequencies*. Usually these frequencies are either added or subtracted to obtain the beat frequency.
beat frequency oscillator: An oscillator which produces a *beat frequency*. Usually this is the result of combining an internally generated *signal* with

an incoming wave, which has not gone through *modulation*, to produce an audible signal.

BECN: *Backward Explicit Congestion Notification.*

B8ZS: *Bipolar with eight Zeros Substitution.*

bel: A measure of comparative magnitude. It is expressed as a logarithm to the base 10 of their ratios. For example, if a *signal* of strength P_1 is transmitted and it is received at strength P_2, then the loss (or gain) in the signal is equal to $\log_{10}(P_2/P_1)$ bels. One bel is therefore a factor of 10, two bels of 100, etc. In practice it is usually more convenient to express this in *decibels* which is equal to one tenth of a bel. Therefore 10 decibels is a factor of 10, 3 decibels is a factor of 2, etc.

BEL: A control character used in *transmission* to call for attention. It activates an alarm at the receiving end.

Bellcore: Bellcore (Bell Communications Research) was founded in 1984 following *divesture* and the forming of the *Bell Operating Companies (BOC)*. Its was owned by the BOCs and was tasked with developing standards to enable interoperability between the regional operators. It published Technical Advisories (TAs) and Technical References (TRs) which resulted in *ANSI* standards. It also determined and managed the North American numbering plan. With the move towards international standardisation and the growth of alliances and mergers between the Bell Operating Companies, the roll of Bellcore changed and in 1996 it was sold to Science Applications International Corp. (SAIC), a research and engineering services company with headquarters in San Diego.

Bell Operating Company (BOC): The twenty-two operating companies, owned by AT&T, which provided local telephone services to the USA. Following *divesture* these twenty-two companies were split from AT&T and formed into seven independent *Regional Bell Operating Companies (RBOC)*.

Bellman-Ford algorithm: An *algorithm* used in data *routeing* in which all *routers* on the *network* send their *routeing tables* by *broadcast* to all other routers so a data *packet* can choose the shortest route through the network.

bending loss: *Attenuation* in *optical fibre* due to bends (including *micro-bends*) which cause loss of transmitted light at the curvatures.

Bending radius: The maximum amount that a *optical fibre* can be bent without loss of any transmitted light.

BEP: *Block Error Probability.*

BER: *Bit Error Ratio.*

Berkley Internet Domain (BIND): The *Internet domain* which provides many *DNS* services and was first implemented as part of the Berkley *UNIX* suite.

BERT: *Bit Error Ratio Tester.*

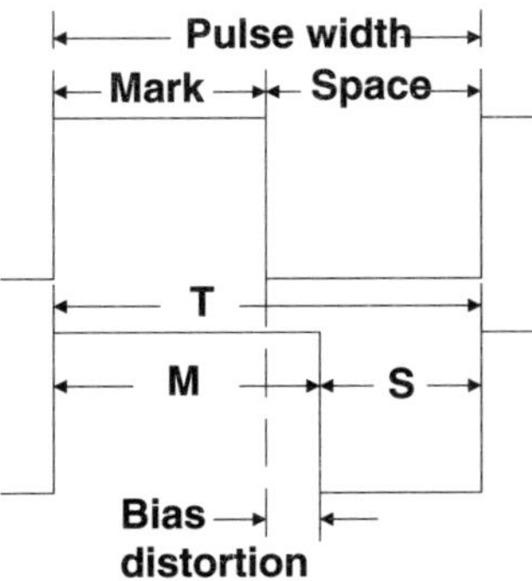

Figure B.5 Bias distortion

BETRS: *Basic Exchange Telecommunications Radio Service.*
BEX: *Broadband Exchange.*
BEXR: *Basic Exchange Radio.*
BGP: *Border Gateway Protocol.*
BHCA: *Busy Hour Call Attempts.*
bias distortion: The *distortion* caused in a *binary signal* due to a shift in the *mark-space* ratio, as shown in Figure B.5. Bias distortion, for this example, is given by (M-S)/T and it is usually expressed as a per cent by multiplying this by 100.
BIB: *Backward Indicator Bit.*
bid: An attempt made by a user to secure a system, such as a *circuit* or a *circuit group.*
bidirectional asymmetry: The conditions which exists in a *bidirectional communications* system when the *transmission* characteristics are different in the two directions. See *bidirectional symmetry.*
bidirectional communications: A communications system in which users can transmit and receive *messages* over the same *network.* There are several techniques for doing this, such as *full duplex* and *half duplex.*
bidirectional coupler: A coupling device, used in *fibre optic* communications system, which allows *signals* to travel in either direction, and has facilities for *sampling* signals in both directions.
bidirectional symmetry: The condition which exists in a *bidirectional communications* system in which the *transmission* characteristics are the same in the two directions. See *bidirectional asymmetry.*
Bids per Circuit per Hour (BCH): A measure of the ability of a *switched circuit* to handle the *traffic* demand. It is given by the ratio of the number of *bids* made in an hour to the number of *circuits* which are available.
bilateral control: Control technique between two *exchanges* in which each exchange controls *synchronisation* of the other exchange. This can be

done by each exchange directly controlling the *clock* in the other exchange. Alternatively each exchange can use the incoming *data* from the other exchange for deriving its *timing*.

billboard antenna: An *antenna* system with parallel *dipoles* having flat reflectors arranged in a straight line. The spacing between the dipoles is *wavelength* dependent. The phase of the *signals* in the dipoles also controls the main *lobe* direction.

billed telephone number: The *telephone number* against which a bill is tendered, irrespective of the number of lines connected with that number.

billing: The process of determining the amount of use *subscribers* have made of the telecommunications service and then charging them appropriately for this use.

binary: Refers to a system which can only occupy one of two states or have one of two values. These two states are normally referred to as *binary digits* or *bits*.

binary channel: A *channel* used for the *transmission* of *binary data.*

binary code: *Code* made from selecting and arranging *binary* units. For example computers can operate on words and numbers which are made from two state, logical 1s and logical 0s, only.

binary coded: *Information* which is represented in *binary code*.

Binary Coded Decimal (BCD): A coding system in which the decimal numerals, 0 to 9, are represented by four *binary digits*, as in Table B.1.

binary digit (bit): A digit in a *binary* system. Usually refers to a logical 1 or a logical 0 state.

binary digit interval (bit interval): The time between corresponding points on two successive *bits*. In a *transmission* system also refers to the time needed to transmit a bit.

binary notation: The use of *binary digits* to represent different notations or numbers. There are several such systems, such as *Binary Coded Decimal* and *Gray code*.

binary number: A number in which each position is represented by a *binary digit* and is to a radix of two. Therefore in Table B.2 the bit positions represent 2^0 (=1), 2^1 (=2), 2^2 (=4), 2^3 (=8) and 2^4 (=16) so the binary numbers which have the logical 1 *bits* in the positions shown have the digital number given alongside.

Table B.1 Binary Coded Decimal

Number	0	1	2	3	4	5	6	7	8	9
BCD	0000	0001	0010	0011	0100	0101	0110	0111	1000	1001

Binary Phase Shift Keying (BPSK): A *Phase Shift Keying* system in which only two phases exist, e.g. in-phase and 180° out of phase.

binary signalling: *Signalling* using only two states, such as a voltage present or absent, and a switch open or closed.

binary string: A group of *binary digits* which occurs in succession and together represent *binary coded* information.

binary symmetric channel: A *binary channel* in which the probability of errors, measured as the *Bit Error Ratio*, is the same for logical 0 or logical 1 *signals*.

Binary Synchronous Communications (BSC or BiSync): A *protocol* used for *synchronous data transmission* of *binary coded* information.

binary tariff: A *tariff* which is made up of two parts. For example there could be a fixed charge for access to a telecommunications service with a second variable charge depending on the time for which the service is used.

BIND: *Berkley Internet Domain (BIND).*

binomial distribution: A *probability distribution* used for discrete events (i.e. the characteristic can only take discrete values, such as 0, 1, 2, etc.). If p is the *probability* that an event will occur, q (= 1–p) is the probability that it will not occur, and n is the number of selections, then, if a binomial distribution is followed, the probability of an even occurring m successive times is given by $^nC_mp^mq^{n-m}$. In this expression ^nC_m is the *combination* of m out of n.

biphase coding: A coding scheme in which two phases are used. For example a logical one can be represented by a $+90^o$ shift about zero and a logical zero by a -90° shift about zero, so that there is a 180° phase shift between the two.

biphase signalling: A *signalling* system in which *binary digits* are represented by *sine waves* which are shifted by 180°.

bipolar coding: A coding method used for *bipolar signals* in which a zero *bit* is represented by a zero value (e.g. zero voltage) and a one bit is represented by a pulse (e.g. of voltage) which has opposite positive and

Table B.2 Binary number

Bit position	2^0	2^1	2^2	2^3	2^4	Decimal number
Binary number 1	1	0	1	1	0	13
Binary number 2	0	1	1	0	1	22
Binary number 3	1	1	0	1	1	27

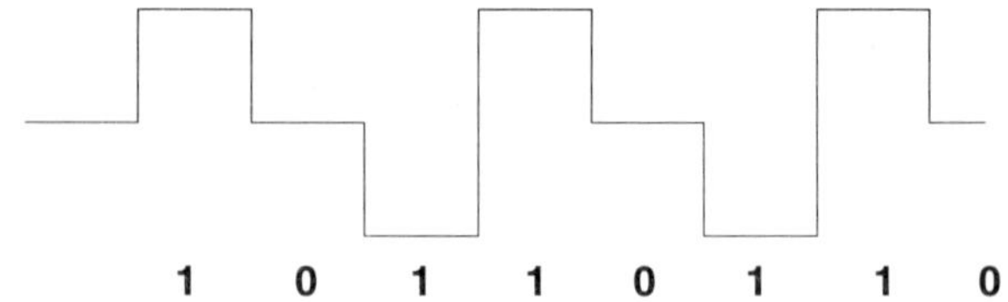

Figure B.6 Bipolar non-return to zero signal

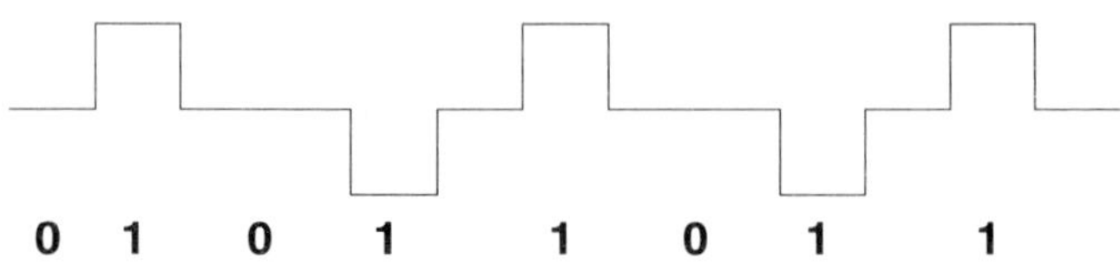

Figure B.7 Bipolar return to zero signal

negative values. This results in a *signal* which has an average value of zero.

bipolar non-return to zero signal: A *bipolar signal* in which a logical zero is represented by a zero level and a logical one by alternate positive and negative values, and the signal does not stay at a zero level in between two consecutive logical one states (Figure B.6). See *bipolar return to zero signal.*

bipolar return to zero signal: A *bipolar signal* in which a logical zero is represented by a zero level and a logical one by alternative positive and negative values, but the signal returns to zero in between two consecutive logical one states (Figure B.7).

bipolar signal: A *signal* which has positive and negative values. For example a logical zero may be represented by a negative voltage and a logical one by a positive voltage. Figure B.8 shows a bipolar signal $v_S(t)$ in which the decision level between a logical 1 and logical 0 is at the zero voltage point.

bipolar violation: The violation of the sequence of alternate positive and negative signal state, used in *bipolar transmission*, caused by the occurrence of two logical one bits having the same polarity (Figure B.9).

bipolar violation rate: A measure of the *transmission* accuracy in *bipolar signals*. It is given by the number of *bipolar violations* in a unit of time, usually a second.

Bipolar with eight Zeros Substitution (B8ZS): A *line code*, which aids in *synchronisation* and *signalling* between the sending and receiving *terminals*. *Bipolar violations* can be inserted if the *signal* contains eight or

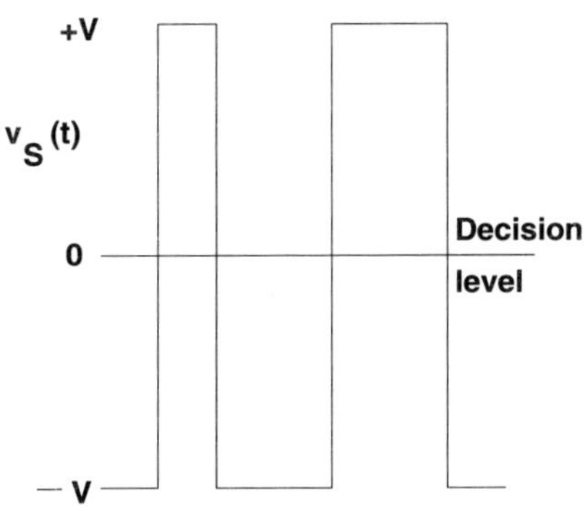

Figure B.8 Bipolar signal

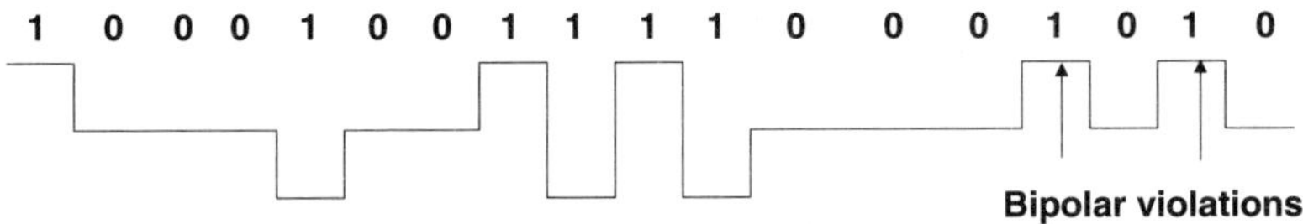

Figure B.9 Bipolar violation

more consecutive logical zeros *bits*, which are replaced by the original zero bits at the receiving end. Used in *T1* transmission (1.544 Mbit/s).

Bipolar with six Zero Substitution (B6ZS): A *line code* in which *bipolar violations* are inserted if the *signal* contains six or more consecutive logical zeros. Used in *T2* transmission (6.312 Mbit/s).

Bipolar with three Zero Substitution (B3ZS): A *line code* in which *bipolar violations* are inserted if the *signal* contains three or more consecutive logical zeros. Used in *T3* transmission (44 Mbit/s).

biquinary code: A coding method in which a *decimal digit* D is represented by two digits, d_1 and d_2, which satisfies the equation $D = 5d_1 + d_2$, where d_1 may be 0 or 1 and d_2 can be 0, 1, 2, 3 or 4, as in Table B.3.

birth-death model: A mathematical model, used for the analysis of queues such as in *teletraffic theory*, in which arriving customers to the queue are regarded as births and customers leaving the queue as deaths.

B-ISDN: *Broadband Integrated Services Digital Network.*

B-ISDN Protocol Reference Model: This is defined in *ITU-T Recommendation* I.321 and is shown in Figure B.10. Horizontal planes through the cube cover the *physical layer*, the *ATM* layer, the *ATM adaptation layer*, and the higher layers. The three vertical planes are the user plane, the control plane and the management plane.

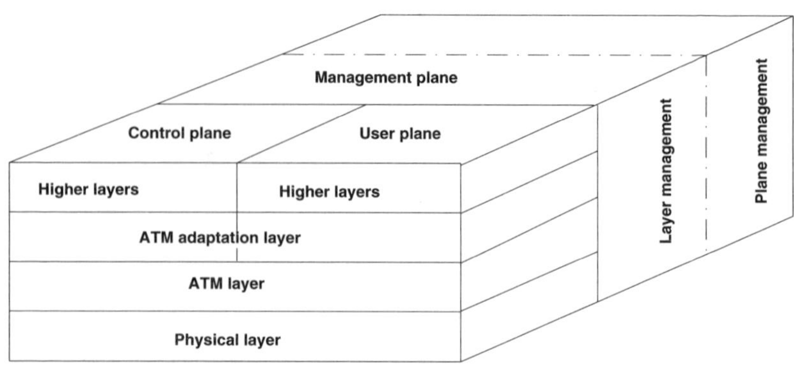

Figure B.10 BISDN protocol reference model

Table B.3 Biquinary code

Decimal digit	*Biquinary code*	
	d_1	d_2
0	0	0
1	0	1
2	0	2
3	0	3
4	0	4
5	1	0
6	1	1
7	1	2
8	1	3
9	1	4

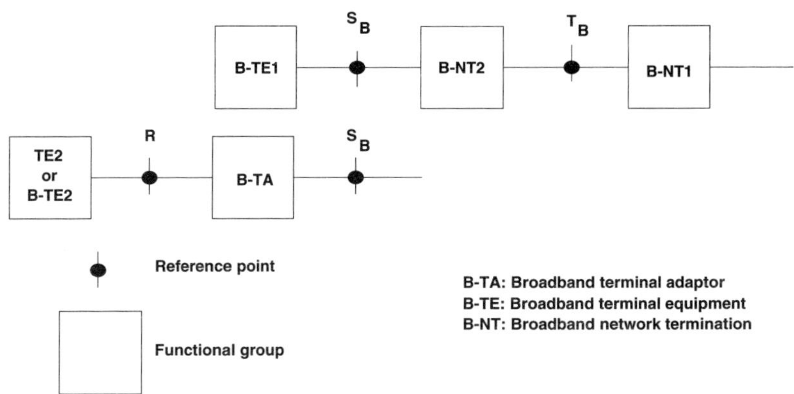

Figure B.11 BISDN User Network Interface

B-ISDN User-Network Interface (B-UNI): Specified in *ITU-T Recommendation* I.413 this defines the functional groups and reference points for *B-ISDN*, as shown in Figure B.11. This is similar to the *narrowband ISDN user-network interface*, the *reference points* being suffixed by the letter B. The characteristics of the T_B and S_B reference points are described, along with a list of the functions which may be included in each of the functional groups (B-TN1, B-NT2, B-TE1, B-TE2 and B-TA. See *terminal adaptor*, *terminal equipment* and *network termination.*

bistatic scatter: The *scattering* of *electromagnetic waves* by atmospheric conditions such as rain and ice clouds.

BiSync: *Binary Synchronous communications.*

bit: See *binary digit.*

bit by bit code: A variety of *line code*, examples of which are *Manchester code*, *Code Mark Inversion (CMI)*, and *Alternate Mark Inversion (AMI).*

bit count integrity: The process for ensuring that the number of *bits* generated remains unchanged between the sender and the receiver.

biternary signal: A *signal* which is formed by combining two *binary* signals. It has three possible states, +1, 0 and –1, and allows two binary signals to be transmitted at reduced *bandwidth*. (Figure B.12.)

bit error: Errors which occur during the *transmission* of a *bit* of *information* which result in the sent and received bits being different, i.e. a logical 1 instead of a logical 0 and vice versa.

Bit Error Ratio (BER): A measure of *transmission* quality, it is given by the ratio of the number of *bits* incorrectly received to the total number of bits sent by the sender in a given time interval. It is also referred to as the Bit Error Rate and is normally written as a power to 10. Therefore if 2 incorrect bits are received out of 100000 transmitted, the BER is equal

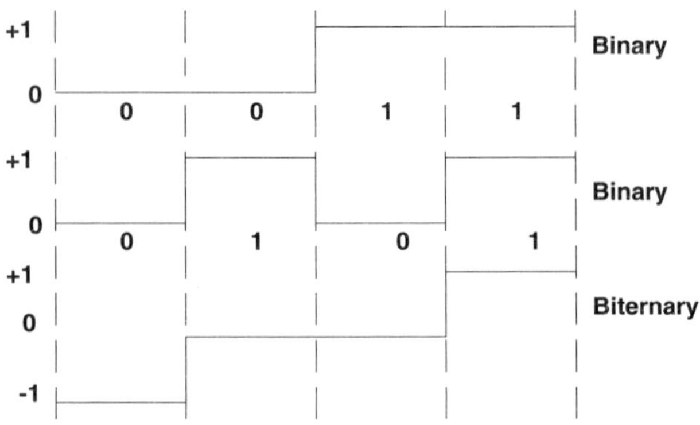

Figure B.12 Biternary signal

to 2×10^{-5}. Table B.4 gives some examples of target BER for various types of transmission. Long term BER must usually be supplemented by other measures of quality, such as *burst errors*.

Bit Error Ratio Tester: A test instrument which has facilities for generating a known *pseudo-random* pattern *signal* and comparing this with the signal received back after *transmission* to calculate the *BER* of the *transmission medium*. Also called a Bit Error Rate Tester.

bit interleaving: A technique use in *PDH multiplexing* systems in which *bits* from different *bit streams* are combined by mixing them one bit at a time. For example if three bit streams A, B and C are combined the bit interleaved signal would have bit 1 from the three streams, then bit 2, etc, i.e. a patter of A_1, B_1, C_1, A_2, B_2, C_2, etc.

bit interval: See *binary digit interval*.

Table B.4 Bit Error Ratio

Digital service	*Transmission rates (approximate)*	*Long term mean BER*
Voice: log-law PCM	64 kbit/s	2×10^{-5}
Voice: ADPCM	32 kbit/s	10^{-4}
Video: Linear PCM	60 Mbit/s	2×10^{-7}
Video: Inter-frame coding	2 Mbit/s	10^{-10} (approx.)
Data	16 kbit/s to 600 Mbit/s	10^{-7} (approx.)

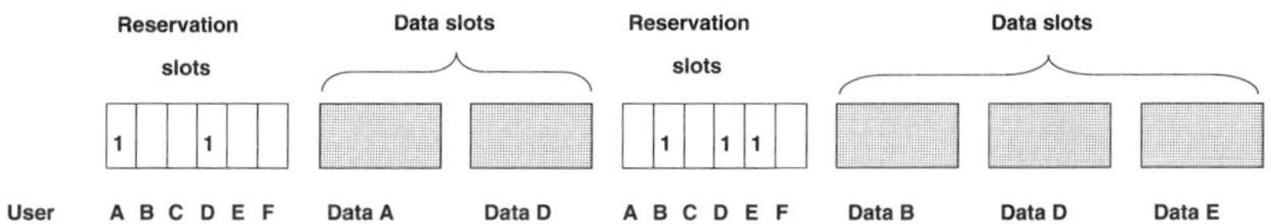

Figure B.13 Bit-map protocol

bit inversion: The changing of the value which represents the *bit* to that corresponding to the value which represents the bit of the opposite polarity. For example if a logical 1 is represented by +5 volts and a logical 0 by 0 volts then inverting the logical 1 bit puts its value at 0 volts.

bit-map protocol: A *multiple access* technique which uses a structure similar to *R-ALOHA*, with explicit *reservations* except that no *contention* is used to access the reservation slots. Figure B.13 illustrates its operation for six users, A to F. If there are N users, then initially there will be N reservation slots with each user allocated exclusive use of one of these reservation slots. If a user wishes to transmit it puts a logical '1' into its allocated slot and if it does not wish to transmit then it leaves this slot empty. The first station who puts a '1' in its slot transmit in the first data slot following the end of the reservation slots. In Figure B.13 this is user A. The next slot is allocated to the next user in sequence who marked its reservation slot, i.e. user D. When all the users who reserved slots have transmitted the reservation slots begin again.

bit masking: The facility to ignore the value of the *bit* in certain positions of a *binary string* and not in others. These bits are then said to be masked.

bit oriented protocol: A *protocol* in which the information can be coded into a single *bit*, such as for the *Synchronous Data Link Control protocol (SDLC)*.

bit pairing: An arrangement of *bits* within a *code* which are identical except for a few bit positions, and which are used to represent pairs of *information*. For example, in *ASCII* code, Table A.2, the upper and lower case letters have the same *bit patterns* for the bit positions 0 to 5 but have opposite polarities in bit position 6.

bit parallel transmission: See *parallel transmission*.

bit pattern: A sequence of *bits* within a *binary string*, used to represent information.

bit rate: It is defined as the number of *binary digits* passing through a *transmission medium* per unit of time, usually a second.

bit rate length product: The maximum product of the *bit rate* and the length of the *transmission medium* which can be tolerated with acceptable performance, measured, for example, by the *Bit Error Ratio* or the *attenuation*.

Bit Rate Reduction (BRR): Techniques used to reduce the *bit rate* needed to convey *information*. Reducing the bit rate allows more information to be carried in a *bandwidth* limited *transmission channel*. See also *compression*.

bit robbing: A technique used for conveying supplementary *signals* by using slots in a normal *voice channel*. For example, the original Bell D1 system used bit robbing of the 8th. *bit* in every *frame* of the voice channel for *timing*. The new D2 system also allowed bit robbing in every 6th. frame to carry control and *signalling* information.

bit serial transmission: See *serial transmission*.

bit slip: The addition or deletion of *bits* in a *bit stream* so as to vary the position of a bit, usually to accommodate a change in *timing* between the sender and the receiver.

bit stream: A continuous sequence of *bits* flowing through a *transmission medium*.

bit string: A sequence of *binary digits*, usually part of a *bit stream*, with associated stop and start indicators, so that they form a unit.

bit stripping: A technique, commonly used in *statistical multiplexing*, in which the *stop bit* and the *start bit*, associated with *asynchronous transmission*, are removed. The information is then handled as in *synchronous* transmission.

bit stuffing: The addition of extra, non-information carrying, *bits* into a *bit stream*. This information is passed to the receiving end and the stuffed bits are removed to return to the original *signal*. Bit stuffing is used for several reasons, such as to maintain *synchronisation* between the sender and the receiver, and to ensure that the *data* sent is not confused with other special *bit strings* such as used for *flags*.

bit synchronisation: The *synchronisation* of *signals*, often by *clock extraction* or *timing recovery*, from the *bit stream*.

black box: Usually refers to any item which is specified in terms of its external characteristics and behaviour, and whose internal construction and operations are not relevant.

black noise: *Noise* which has zero value over the whole *noise bandwidth* except for a few high level values at selected *narrowbands* of frequencies.

blackout: The complete disruption of communications, usually due to abnormal conditions in the *transmission medium*. Examples are the loss of radio communications due to *sun spots* and a nuclear burst.

blanketing area: A area surrounding an *antenna* in which its *transmissions* interfere with *signals* being received by other antennas.

BLER: *Block Error Ratio*.

BLERT: *Block Error Rate Test*.

blind area: Usually refers to an area of space in which objects cannot be detected by *radar*.

blind spot: Refers to the area of the human *eye* where the *signals* from the retina are brought out by the optic nerve. This point contains no light receptors and is therefore insensitive to light.

blind transmission: *Transmission* which occurs without receipt of any *acknowledgement* from the receiving *terminal*. The failure to receive this acknowledgement may be unintentional, e.g. due to a defect, or intentional, e.g. for security reasons.

block: A group of *characters* which are transmitted as a single unit and which may contain other *bits* for *synchronisation*, *error control*, etc.

block acknowledgement counter: A *counter* in a transmitting *terminal* which keeps a total of the number of *acknowledgements* sent by the receiving terminal. See *block completed counter*.

block cancel character: A *character* which is used to indicate that the last transmitted *block* is to be ignored.

block check: The check performed on a *block* of *data* to ensure that it is correctly structured and does not contain errors. See *Block Check Character*.

Block Check Character (BCC): A *character* which is added to a *block* of transmitted *data* in order to check that no errors have occurred during *transmission*. Examples are *longitudinal redundancy check* and *cyclic redundancy check*. The block check character is calculated and added to each *message* block before transmission and the receiver goes through the same calculation to generate a second BCC to check if an error has occurred during transmission.

block cipher: Data *encryption* method in which the *message* is divided into *blocks* which are then encrypted.

block code: *Line code* consisting of information *blocks* with *error detection* and *error correction* capability. Blocks are coded and decoded independently from other blocks. Examples are *nB1C*, *mBnB* and *linear block codes*.

block completed counter: A *counter*, in the transmitting *terminal* which keeps a record of the total number of *blocks* transmitted. See *block acknowledgement counter*.

blocked call: A *call* which cannot be completed because the *transmission channel* between the *calling party* and the *called party* is saturated with *traffic*.

Blocked Calls Cleared (BCC): A technique used to prevent long *queues* by rejecting, without service, any *call requests* which have been blocked and therefore cannot be serviced within a reasonable time.

block encryption: The *encryption* of *blocks* of *data* independent of other blocks.

block efficiency: A measure of the overheads within a data *block*. It is given by the ratio of the number of information *bits* in the block to the total number of bits. The block overhead is the number of bits left after the information bits have been removed.

Block Error Probability (BEP): The probability of obtaining a value of *Block Error Ratio*.

Block Error Rate Test (BLERT): A test of *Block Error Ratio* in which *blocks* with a known pattern are transmitted and the ratio of the total number of received block errors to the total number of blocks received is calculated. This is the BLERT and the smaller the value the greater the quality of the *transmission channel*.

Block Error Ratio (BLER): A measure of the quality of *transmission*. It is given as the ratio of the number of *blocks* of *data* incorrectly received to the total number of blocks transmitted by the sender. This is also called Block Error Rate.

blocking: Grouping data into *blocks* for *transmission*. Also refers to the process of denying access to a system.

blocking signal: A *signal* which is sent to stop other users from *seizing* a *line*. It can also refer to a signal, usually sent to an *exchange*, to indicate that a specified line should not be used for any *outgoing calls*, but it can be used for *incoming calls*. This is usually done for maintenance purposes.

blocking signal acknowledgement: A *signal*, sent back by an *exchange* in response to a *blocking signal*, as an *acknowledgement* that the requested line has been blocked.

block loss ratio: The ratio, measured over a period of time, of the number of *blocks* of data which have been lost, primarily due to errors in *transmission*, to the total number of blocks sent.

block mode traffic: Refers to the situation where a large amount of *traffic*, in the form of *blocks* of *data*, is being transmitted. Usually this traffic is travelling in one direction with very little in the opposite direction, as occurs during file transfer.

block parity: A method for checking errors in the *transmission* of *data blocks* by allocating one or more *bits* in the block as *parity bits* and checking at the receiving end that the correct *parity* has occurred.

block separator: The *character*, used during *transmission* of *blocks* of *data*, to differentiate between the individual blocks.

block transfer time: The time between the start of *transmission* of a *block* of *data* and when it is successfully received at the other end.

BMLC: *Basic Mode Link Control*.

blue books: The collection of books, produced by the *CCITT* (now *ITU-T*), which contained all its published standards for the four year study period 1985–1988.

Bluetooth: A *Wireless Local Area Network (WLAN)* standard proposed by a consortium of manufacturers. It operates in the *ISM frequency band* of 2.4 GHz at a *bit rate* of about 1 Mbit/s and a *range* of 10 metres. It is primarily intended to be used for linking equipment together, such as laptop computers and mobiles.

B-MAC: *Broadcast* method of *transmission*, used for *satellites*, in which the analogue *MAC* signal is *multiplexed* with a *digital signal* containing auxiliary *information* such as video *sychronisation*, digital sound, and *authorisation* facilities.

BNC connector: A bayonet type of *connector*, used for *coaxial cable*. BNC stands for bayonet-Neill-Concelman.

BOC: *Bell Operating Company*.

BOND: A join between two surfaces, usually metallic, which is done in such a way as to provide a low electrical resistance.

BONDING: *Bandwidth on Demand Interoperability Group*.

boolean algebra: A form of algebra which uses variables with two states only and is used extensively in the analysis of logic circuits. It uses basic functions such as AND, OR, NAND, NOR, etc.

bootstrap: The process of achieving a desired state by means of one's own action. For example a bootstrap programme in a processing device will bring it into a ready to use state by means of its own instructions.

Border Gateway Protocol (BGP): A *gateway protocol* which enables a *router* on the *Internet* to act as a *gateway* between *Autonomous Systems*. It developed from the *Exterior Gateway Protocol*.

BORSCHT: The acronym describes the functions which a *Central Office (CO)* must provide over the *exchange line* to the *subscriber*. They stand for battery feed, overvoltage control, ringing, supervision, coding, *hybrid*, and testing.

Bose, Chaudhuri, Hochquengham code (BCH code): A powerful family of *cyclic codes* which are used for *error correction*. It is named after its three inventors.

Bouger's law: It specifies the strength of an electromagnetic field during *transmission* of *electromagnetic radiation*. If S_d is the strength at a distance d from the source, which has a strength of S_s, then $S_d = S_s e^{-ad}$, where a is the constant coefficient of the *transmission medium* and is dependent on its *absorption* and *scattering* properties.

bounded transmission medium: *Transmission medium* in which the *signal* travels wholly within the physical bounds of the medium. Examples are *optical fibre* and *coaxial cable*. See *unbounded medium*.

BPDU: *Bridge Protocol Data Unit*.

bps: A abbreviation used to denote bits per second. It is a measure of the data *transmission rate* given as the number of *binary digits* transmitted in a second. Also written as bits/s.

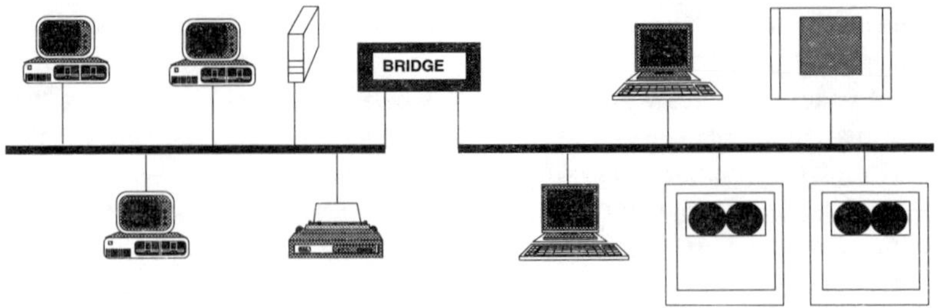

Figure B.14 LAN interconnection using a local bridge

BPSK: *Binary Phase Shift Keying.*

BRA: *Basic Rate Access.*

BRAN: *Broadband Radio Access Networks.*

branch exchange: Usually refers to a *Private Branch Exchange (PBX).*

branch feeder cable: A cable which carries *signals* to a branch of a telecommunications *network.*

BRAP: *Broadcast Recognition with Alternate Priorities.*

break in: A procedure used by a receiving *terminal* to interrupt the *transmission* from the sending terminal to request it to perform a special function, such as to pause, shift its transmission *frequency*, repeat a *message*, etc.

breakout box: A test equipment which connects to the pins of a cable *connector*, and allows the individual *signals* to be monitored, switched or otherwise controlled.

BRI: *Basic Rate Interface.*

bridge: A device which interconnects two *Local Area Networks (LAN)*, allowing them to operate as one unit. The advantage of using two separate LANs is that the traffic can be segregated (Figure B.14) so as to optimise the performance of devices connected to each LAN. The bridge operates at the *Data Link Layer* of the *OSI Basic Reference Model*, as shown in Figure B.15. On receiving a data *packet* the bridge examines

DATA LINK LAYER		DATA LINK LAYER
PHYSICAL LAYER		PHYSICAL LAYER
MEDIUM		

Figure B.15 OSI model of a bridge

its *address* information. If the sender and receiver are on the same LAN the packet is prevented from leaving the LAN. If they are on different LANs the bridge forwards it on to the corresponding LAN. This bridge operates in the same geographical area and is sometimes referred to as a *local bridge* to differentiate it from a *remote bridge* or a *half bridge*.

Bridge Protocol Data Unit (BPDU): A *protocol* in which *packets* are sent at periodic intervals by a *bridge* in order to monitor their *network*.

bridge tap: An electrical connection made across a cable pair to bring it into a customer's location. Also known as a bridging connection.

bridging loss: The loss which results due to an impedance being connected across the *transmission line*. It is measured as the ratio of *signal* power, at a point beyond the *bridge*, before and after the bridge is applied.

British Approvals Board for Telecommunications (BABT): UK body responsible for testing and approvals of telecommunications equipment connected to the *public switched network*.

British Standards Institution (BSI): The official standards body within the UK, formed in 1901 as the Engineering Standards Committee. It was granted a royal charter in 1929 for the coordination and preparation of standards and for making and registering goods which complied with these standards. It took its present name in 1931. Its activities cover a wide area, such as building material, domestic electric goods, and telecommunications. Its telecommunications standards responsibilities date back to the *British Telecommunications Act* of 1981, when the BSI agreed to assist the *DTI* in the *liberalisation* process within telecommunications, by producing British Standards. The BSI represents the UK in the *IEC*, *ISO*, and *CEN/CENLEC*, which have national standards bodies at meetings. It operates a series of technical committees which obtain the views of BSI members for presentation to these bodies. The BSI also has a Technical Committee (TCT/101) whose purpose is to coordinate the views of UK members to ETSI, and in this it works closely with the *DTI*.

British Telecommunications Act: Passed in July 1981 the Act was the start of *liberalisation* in the UK and it separated the telecommunications arm from the Post Office and created British Telecom (which was later renamed to BT). The Act was superseded by the *Telecommunications Act*.

broadband: A generic term for a system which covers a wide *frequency band*, usually much greater than that of a *voice channel*. Also known as *wideband*.

Broadband Exchange (BEX): A public *exchange* which is capable of handling *broadband signals*.

Broadband Integrated Services Digital Network (B-ISDN): A series of *ITU-T Recommendations* for *broadband* services, known as the *I Series Recommendations*. It uses *ATM* as the transfer mechanism.

broadband multiplexing channel: A *channel* which has a high *bandwidth* so that several *signals* can be transmitted through it using *multiplexing* techniques.

Broadband Radio Access Networks (BRAN): *ETSI* project, established in April 1997, to develop standards for a new generation of *broadband* radio access *networks*. A family of standards are being defined: HIPERLAN/2, for short range systems, with limited cordless mobility; HIPERACCESS, to provide wide area fixed radio access; and HIPERLINK for short range but very high speed point-to-point *links*.

broadcast: The distribution of a *signal* to all receivers, irrespective of whether it is intended for them. For example a television programme could be broadcast to all sites but only those with the required *authorisation* can receive it. Similarly *messages* on a *LAN* are broadcast to all other stations on the LAN but only the stations with the correct *address* accept it. See also *multicast* and *unicast*.

Broadcast Control Channel (BCCH): A control *channel* specified within the *GSM* specification.

broadcast frequency: The *frequency* used for a *broadcast*. This normally refers to a radio or television broadcast and the frequencies are tightly regulated to prevent interference between transmitting stations.

broadcasting organisation: An organisation which is licensed to *broadcast*, usually radio or television.

Broadcasting Satellite Service (BSS): A service which uses a *satellite* to *broadcast* the *signals*. Usually refers to radio or television broadcasts which are made *Direct To Home (DTH)*. See also *Direct Broadcast Satellite (DBS)*.

broadcasting service: A terrestrial based *broadcast* service which is intended for use by the general public.

Broadcast Recognition with Alternating Priorities (BRAP): A *multiple access* protocol, which is an extension of the *bit-map protocol* and is illustrated in Figure B.16. As for the bit-map protocol reservation slots are allocated for exclusive use of a user. However, as soon as a user

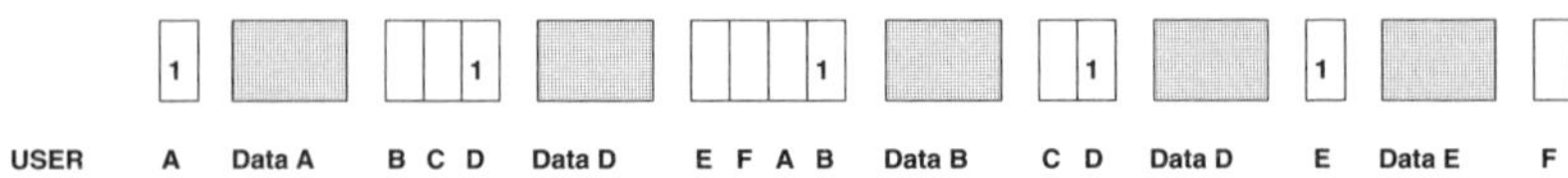

Figure B.16 BRAP protocol

makes a reservation it completes its *transmission* in a data slot. Also, the reservation slot sequence now does not start from the beginning but from the last user who transmitted. It overcomes the two limitations of the bit-map protocol: delay under light loads, caused by ready users who have made a reservation but have to wait until the end of the reservation cycle before they can transmit; and the asymmetry in the protocol, where some users get preferential treatment over others, depending on their position within the reservation frame. BRAP is very similar in operation to *MSAP*.

broadcast storm: A phenomena, often found in *Local Area Networks (LAN)* with looped data paths, in which *packets* of data continuously circulate around the network so that it becomes *congested* and performance deteriorates.

Broadcast Television Systems Committee/Multichannel Television Sound (BTSC/MTS): An American system for the *transmission* of high quality sound with television *broadcasts*. It contains a 15 kHz stereo quadrature *channel* along with a 10 kHz monophonic second audio channel and an *engineer orderwire* channel.

broker's call: A facility available on a telephone service in which a user can hold two simultaneous calls, and be able to move between them with secrecy on each.

brouter: A hybrid product which performs some of the functions of a *bridge* and a *router*. It is also referred to as a router bridge, or a bridge/router, or a *hybrid router bridge*. It can operate in both the *Data Link Layer* and the *Network Layer* of the *OSI Basic Reference Model*. It will route, using the Network Layer, if it supports the *protocol* of the transmitted data, and it will bridge, using the Data Link Layer for protocols it does not recognise.

Brownian noise: It is the noise associated with the random movement of electrons at temperatures above absolute zero. See *thermal noise* and *Gaussian noise*.

BRR: *Bit Rate Reduction.*

BSC: *Base Station Controller* or *Binary Synchronous Communications.*

BSI: *British Standards Institution.*

B6ZS: *Bipolar with six Zeros Substitution.*

BSS: *Base Station Subsystem* or *Broadcasting Satellite Service.*

B3ZS: *Bipolar with three Zeros Substitution.*

BTS: *Base Transceiver Station.*

BTSC/MTS: *Broadcast Television Systems Committee/Multichannel Television Sound.*

buffer: In communications systems refers to a physical storage device, or a *software* routine, which compensates for differences in *timing* or *data transmission* between two *terminals* by storing *information* from one terminal before sending it on to the second terminal. *Buffering* can be in one direction of data flow or in both directions. See *buffer storage*.

buffer storage: The physical device which acts as a *buffer* between two terminals operating at different *transmission rates*.

buffered repeater: A *repeater* which performs amplification and *regeneration* of the *signal* and also compensates for differences in data handling capability between the sender and the receiver.

buffered terminal: A *terminal* which has the capability to act as a *buffer*, usually by incorporating *buffer storage* which can store incoming and outgoing *messages*.

buffering: The process used to *buffer* data in communications systems.

Building Distribution Frame (BDF): A *distribution frame* which acts as a termination point for cable within the building and provides an interface between the *PABX* and the wiring in the building. It also gives a local cross-connect facility.

bulk absorption: The *absorption* of *electromagnetic waves* which occurs in a *transmission medium* over a unit length or unit volume.

bulk encryption: *Encryption* which is simultaneously applied to multiple *channels* of *data* in a multichannel *trunk*.

bulk scattering: The *scattering* of *electromagnetic waves* in a *transmission medium*, measured over a unit length or unit volume.

bulk storage: See *mass storage*.

bulkhead connector: A *connector* which enables two cables to be connected together. It is usually used to pass through a solid device, such as the side of a cabinet.

Bulletin Board Service (BBS): A messaging system, usually based on *electronic mail* which allows a sender to leave *messages* on a central processor and which can then be picked up by the addressees at their convenience.

bundling: The practice of combining several items under one price. For example *PTOs* frequently give a single price for a basket of services which they offer, so that *subscribers* cannot pick and choose those services which meet their requirements more closely. *Regulatory bodies* consider bundling to be uncompetitive and often force the operator to unbundle their services.

B-UNI: *B-ISDN User Network Interface*

buried facet SOA: It is important in *Semiconductor Optical Amplifiers (SOA)* to control the facet reflectivity, since this suppresses the *lasing* action of the *amplifier*. In the buried facet method of control the *waveguide* ends before the facet, as in Figure B.17, so the optical output can defract to a larger extent before *reflection* starts, reducing the amount of power coupled back into the amplifier. See also *angled facet SOA*.

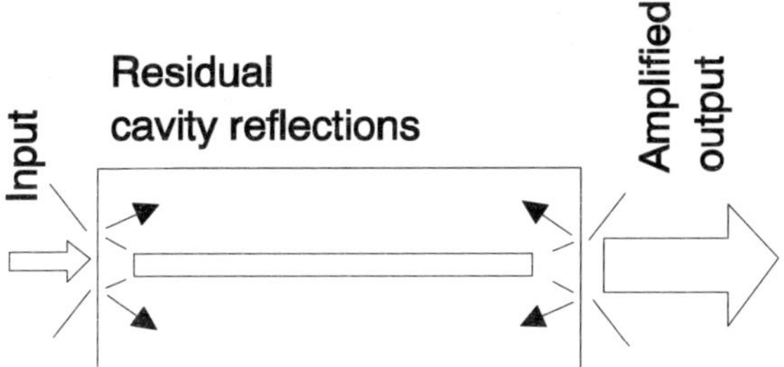

Figure B.17 Buried facet SOA

burst error: *Transmission error* in which several *digits* in sequence are affected, the error frequency being such that the number of correct *bits* between corrupted bits is less than a specified amount. As seen from Figure B.18 burst errors can result in the *BER* being much higher than the long term mean value during these intervals. A typical cause of burst errors is interference from other *signals*. See also *random error*.

burst error correcting code: *Error correcting code* designed to correct errors in *burst mode data transmission*, usually by the addition of *parity bits*. One type of *cyclic code*, for example, requires (3n-1) parity bits to correct bursts of up to n bits in length.

burst isochronous transmission: *Transmission* of *data* in *burst mode*, the *bits* within the burst being synchronised to a *clock*. In between the bursts there is no transmission. Also known as *interrupted isochronous transmission*.

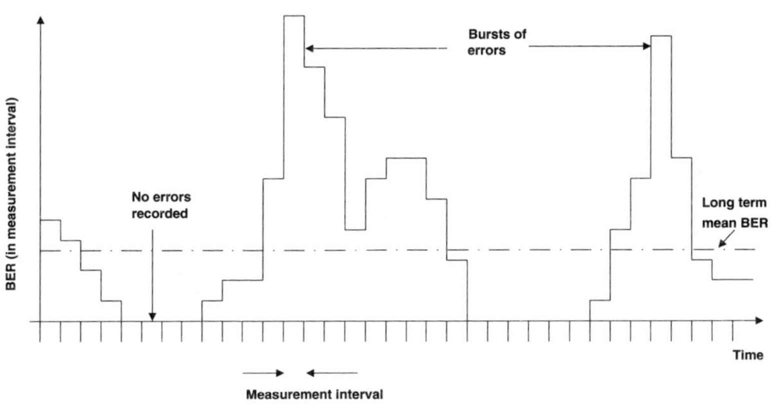

Figure B.18 Variation of BER in a burst error environment

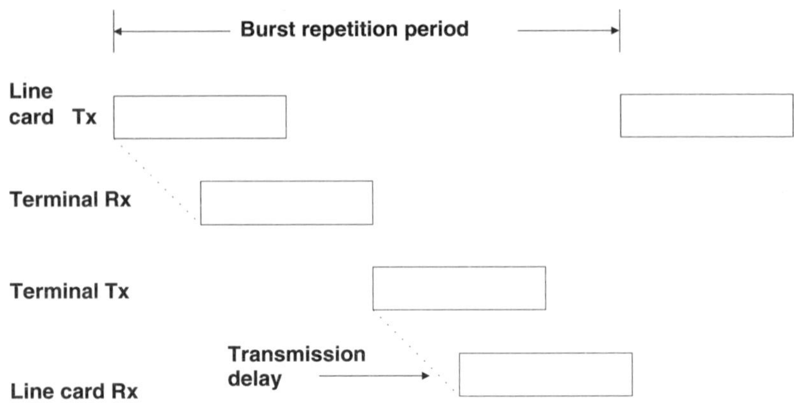

Figure B.19 Burst mode transmission

burst mode: The *transmission* of data in large continuous *blocks* (bursts), the transmitting *terminal* occupying the whole *channel* during the period of the transmission. For example, in Figure B.19, Terminal 1 transmit a burst of data and after a *delay time* Terminal 2 receives it. Terminal 2 then sends a burst of data, and so on. The time between successive bursts of data from the same terminal is the burst repetition period.

bursty traffic: *Traffic* which flows in bursts across the *network*. This is usually the case when there are a large number of users sharing a network and each has a relatively low amount of *data* to transmit and this is done in *burst mode*.

bus: A data *channel* which is shared by several users. This is normally a physical *link* connecting several devices and it may be made from copper or *optical fibre*. A bus can connect different cards within a piece of equipment (e.g. backplane bus) or it can connect several pieces of equipment on a *network* (e.g. *LAN*).

bus arbitration: The technique used for ensuring that different users on the bus get to use it in an equitable manner.

bus driver: An *amplifier* connected to a *bus* in order to increase the strength of the *signals* on it to ensure that it is received at the far end.

business terminal: A *terminal* used in a commercial rather than a residential environment. See *residential terminal*.

bus segment: A section of the *bus* which is electrically or optically continuous, e.g. not separated by *repeaters*, *bridges*, etc.

bus topology: See *network topology*.

busy: A communications state in which a device is fully occupied and cannot service any other users.

busy hour: The continuous sixty minute period during a 24 hour period when the *traffic* on the system is the greatest.

Busy Hour Call Attempts (BHCA): It is a measure of the *traffic* handling capability of an *exchange* and is given by the number of *call attempts* it will need to deal with during the *busy hour*.

busy signal: The audible or visual *signal* sent to the *calling party* to indicate that the *line* or the *terminal* being accessed is not available (*busy*), probably because it is occupied on another *call*.

busy verification: The facility available on a *PSTN* to enable an attendant to break into the *circuit* to check whether the system is *busy* with a *call* or is defective. Generally a *signal* is sent to both the *called party* and the *calling party* to indicate that the attendant is present.

butt joint Usually refers to the join made between two *optical fibres* in which their ends are pressed close together so that light can travel from one fibre to the other with minimum loss of power.

bypass: Usually refers to the use of an alternative *network* or alternative service from that provided by the *local exchange carrier*. See also *bypass carrier*, *private circuit* and *overlay network*.

bypass carrier: A local *PTO* who provides services, usually to large businesses, by linking them to long distance operators at competitive rate, so as to *bypass* the *local exchange carrier*.

byte: A collection of *bits* which are treated as a single unit. Although bytes can have any number of bits the most common is an 8 bit byte.

byte order of transmission: The *transmission* of *data* in which the order is determined by the characteristics of the *bytes* which compose the data. For example it can be transmitted by sending the least significant bytes first, etc.

byte serial transmission: *Transmission* in which the *bytes* making up the *data* are sent serially. The *bits* within each byte may be sent serially or in parallel.

byte timing: An interchange circuit specified in *ITU-T Recommendation* X.21 which provides information on the grouping of *bits* within a *character*.

C

CA: *Conditional Access.*

cable: **(1)** A group of electrical conductors (wires) or *optical fibres* which are insulated from each other and combined in a protective enclosure. Each of the conductors or fibres can be used individually. Figure C.1 shows an example of a cable with fibre and copper. **(2)** A cable can also refer to a *message* which is sent using *telegraphy*.

Cable Act: Act passed in the USA in 1992 which allowed telephone companies to enter the cable business. This led to alliances between groups, such as cable and telephone companies, and cable and *Internet Service Providers (ISP)*.

cable assembly: A *cable* which is terminated by *connectors* and is ready for installation.

cable casting: The *broadcasting* of *information* over *cable* so that it can be simultaneously received by several *terminals*. Usually the broadcasts consist of *video* and audio direct to the public.

cable chart: A layout plan of the *cables* in the *network*, such as in a LAN, which includes *patch panels* components.

cable code: A type of *Morse code* primarily used in communications systems using *submarine cables*.

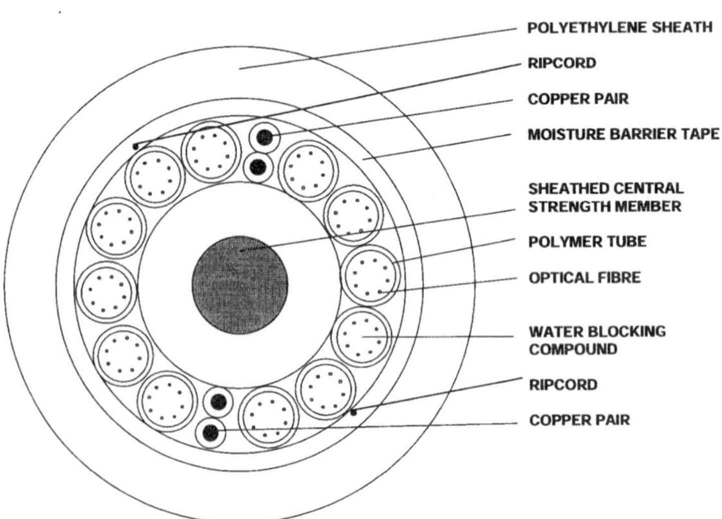

Figure C.1 Cable for duct use

Cable Communications Association (CCA): UK based association of *Cable Television (CATV)* operators.

Cable Communications Policy Act: USA legislation, passed in 1984, which prohibited local telephone companies from providing *video* programmes. The *MFJ* also restricted the line of business in which the *ROBOCs* could engage. See also *Cable Act.*

core: In a *optical fibre* the portion of the fibre which carries the light waves. It is surrounded by the *cladding*. See also *cable core.*

cable core: Usually refers to the central component of the *cable* which provides it with strength and does not carry any *signals*. See Figure C.1.

cable distribution system: A system for distributing *signals* over (usually) *coaxial cable*. The signals also often consist of *video* and sound entertainment programmes and are sent to subscribers. See also *cable casting* and *cable network.*

cable franchise: The licence or right granted to a telecommunications operator to provide *cable* based services in a fixed geographical area. Usually the operator has an exclusive franchise in that area.

cable jacket: The *cable* outer covering which protects the interior. The cable jacket does not carry any *signals* and it may have markings to identify the cable's contents, e.g. number of conductors in the cable.

cable loop: The arrangement of the *transmission cable* in a *Local Area Network (LAN)* in which all the *terminals* on the *network* are connected.

cable loss: The loss of *signal* within a *cable*. This may be due to several causes such as *attenuation*, *scattering*, etc.

cable modem: *Modem* used in conjunction with a *cable distribution system.*

cable network: Usually refers to a *network* made from *coaxial cable* and used for *broadcasting* television and radio programmes. There are many arrangements for cable networks, such as *tree and branch*, *switched star* and *Hybrid Fibre Coax (HFC)*. See also *cable distribution system.*

cable pressurisation: Technique, used in *cables* which need to operate in wet conditions, of enclosing them in dry air or inert gas under pressure. This prevents the ingress of moisture and at the same time loss of pressure indicates the presence of a fault condition.

cable splice: Joining two *optical fibre cables* together by *splicing* so as to get a good optical join, with low *signal* loss.

Cable Television (CATV): The *transmission* of television over *cable*. Sometimes this also covers other elements of the transmission system, such as the *microwave* link, if the *signals* are first transmitted via microwave radio to a local *antenna* and then carried over cable to individual houses. In fact cable television was first used to transmit television to remote communities. A community antenna was used to

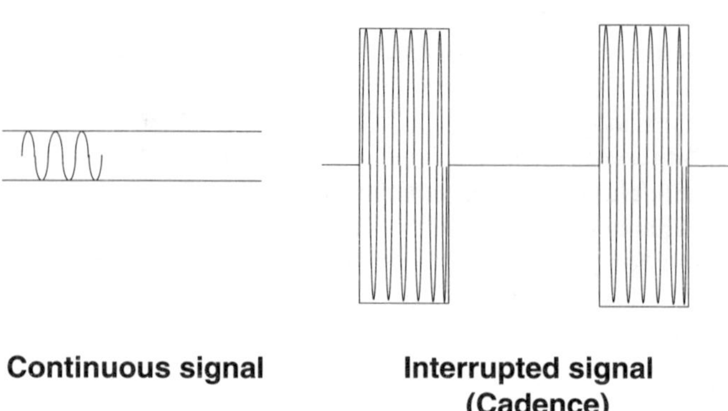

Figure C.2 Cadence ringing signal

receive the radio signals which were then carried by cable to individual houses. CATV originally stood for *Community Antenna Television.*

Cable Television ready set: There are several differences between *Cable Television (CATV)* and off-air systems. For example the *frequency range* is larger in *cable* systems so conventional TV tuners may not be able to cover the full range. Also *channel spacing* could be closer so that the tuner in the TV set may not be able to cope. *Television receivers* which have been designed for use with *cable distribution systems* do not have these limitations and are known as Cable Television ready sets.

CAC: *Commercial Action Committee* or *Connection Admission Control.*

cache memory: Small, fast, temporary memory store for instructions which are to be used next by a processing unit. Usually the cache memory gets the information from the processor's main memory.

cadenced signal: A *signal* which is not continuous, i.e. it is off for part of the time, as in Figure C.2. This is normally the case for a *ringing* signal, which has a relatively high magnitude during the on periods.

CAD: *Computer Aided Design.*

CAI: *Common Air Interface* or *Computer Based Instruction.*

calendar time: A measure of time taken on the *Coordinated Universal Time (CUT)* scale.

call: The request to set up a connection between a *calling terminal* and a *called terminal*, usually over the *PSTN*. It also refers to the overall process, i.e. *call setup* (or *call establishment*), the *call* itself (when information exchange takes place) and *call clearing* (or *call clear down* or *call disestablishment*).

call accepted packet: A *packet* sent to the *network* in a *packet switching* system, by the *called terminal*, to indicate that it is ready to accept the *call request* from the *calling terminal.*

call accepted signal: A *signal* sent by the *called terminal* to the *calling terminal* to indicate that it has accepted the *call request*. See also *call answered signal*.

Call Accounting System (CAS): The *hardware* and/or *software*, normally residing in an *exchange* or *switching centre*, which keeps an account of the *calls* made by *subscribers* for charging purposes.

call add on: The facility, usually part of a *PABX* or *switching centre*, which allows a second *call* to be added on to the one already taking place.

call answered signal: The *signal* which is sent by the *called terminal* to the *calling terminal* when the *call* is answered and, in the case of a *telephone* call, the receiver goes *off-hook*. See also *call accepted signal*.

call attempt: The request made by the *calling terminal* to the *called terminal* to set up a *call*. The request may not succeed and call attempts are used as a measure of performance in *teletraffic theory*.

callback: The process in which, once a *call attempt* is made from the *calling terminal* to the *called terminal*, the *call* is immediately abandoned and the called terminal immediately calls the calling terminal. This may be done, for example, for *security* reasons, to ensure positive identification of the calling terminal. It is also used when there are large difference in *accounting rates* between countries, when callback is used to originate calls from the cheaper country. See *callback modem*.

callback modem: A *modem* which performs a *security* function. It is programmed with *personal identification codes*, *passwords* and authorised *telephone numbers*. When users call and correctly enter their identification code and password the modem will disconnect them and then call them back using the stored telephone number corresponding to the identification code. See also *dialback modem*.

callback on busy: Feature, usually available on a *PABX*, which allows the *calling terminal* to *call* the *called terminal* until *call establishment* occurs, if the called terminal is busy when a *call attempt* is made.

call barring: The facility, available on a *terminal*, which prevents it from either making *outgoing calls* or receiving *incoming calls*.

call blocking: The inability to complete a *call* primarily because of heavy *traffic* resulting in *blocking*. This will usually result in a *busy tone* being returned to the *called terminal*.

call capacity: The *call handling* capability of a system which nables it to meet a certain *Quality of Service (QoS)*.

call centre: A central location where many *calls* are answered or made. The call centre would typically have many agents who are linked to a *CTI* system and are assisted by features such as *Automatic Call Distribution (ADC)* and *predictive dialling*.

call clear down time: The time taken to terminate a *call*, measured from the instant that the decision is taken to terminate the call to the instance when a free condition is obtained.

call clearing: The process used to end the *call* in an orderly manor, after the two parties have finished communicating. Also known as *call disestablishment.*

call clear packet: A *packet* which is sent in a *packet switched* system to end the *session*. It is equivalent to an *on-hook* signal at the end of a *telephony call*.

call collision: The *collision* which can occur during *call setup* either because a *terminal* is *signalling* a *call request* at the same time that a *call* is coming in, using the same *channel*, or because of simultaneous *seizing* of both ends of a *trunk*.

call completion ratio: A measure of the effectiveness of the system, it is given by the ratio of the number of *calls* which are successfully completed to the total number of *call attempts*.

call connected signal: The *signal* sent by the *called terminal* to the *calling terminal* to indicate that the *call* has been successful in establishing a connection.

call control: Generally refers to the whole process involved in connecting, maintaining and disconnecting a *call*.

call control signals: The *signals* used during the process of *call control*.

call delay: The delay involved in *call setup*, measured as the time from when a *call attempt* begins to when *ringing* starts at the *called terminal*.

Call Detail Recording (CDR): A feature, usually of the *exchange*, which enables it to note several particulars of the *call* for *billing* purposes.

call disestablishment: See *call clearing*.

call distribution: See *Automatic Call Distribution (ACD)*.

call diversion: The automatic switching or diversion of a *call* from the *address* which is being called to another address which has been preselected by the *subscriber* being called. Also known as *call forwarding*. Call diversion is normally used if users are away from their normal *terminal* and wish to receive calls at other locations which they are visiting. Call diversion can also be set to occur if the called number is busy or does not answer within a preset period of time.

call duration: The length of time of the *call* measured as the time from when the *called terminal* goes *off-hook* to when either the *calling terminal* or the called terminal goes back *on-hook* at the end of the call.

called line identification: A *network* feature which informs the *calling terminal* the *address* of the *called terminal* to which it is connected. See *Calling Line Identification (CLI)*.

called location: The location where the *called terminal*, which is being *addressed* by the *calling terminal*, is based. Frequently used to indicate the called terminal itself.

called party: Strictly the person who is to receive the *call* at the *called location*, although it is frequently used to mean the *called terminal*. Also referred to as *called subscriber*.

called subscriber: See *called party*.

called terminal: The *terminal* which is being *addressed* by the *calling terminal* and is to receive the *call*.

called terminal alerted state: The situation which exists when the *calling terminal* has completed the *call request* and *ringing* is occurring at the *called terminal*.

called terminal answered signal: The *signal* sent by the *called terminal* to the *calling terminal* to *acknowledge* that the *call* has been successfully answered and *call establishment* is in process.

called terminal engaged signal: The *signal* sent by the *called terminal* to the *calling terminal* to indicate that the *call request* has been unsuccessful since the *terminal* is occupied on another call.

called terminal free signal: The *signal* sent by the *called terminal* to the *calling terminal* to indicate that the called terminal is free to answer the *call* but has not yet done so. In a *telephone* system this would normally be a *ringing tone* transmitted to the calling terminal.

caller identification: See *Calling Line Identification (CLI)*.

call establishment: The initial steps which the *network* carries out during the *call setup* process between a *calling terminal* and a *called terminal*.

call failure signal: The *signal* sent by the *network* to the *calling terminal* to indicate that its *call request* could not be carried out successfully due to a failure in the *call establishment* process.

call forwarding: See *call diversion*. Call forwarding can also be used to implement *number portability* in which the donor network (the network to which the number was originally assigned) sends the call on to the recipient network (the network to which the number has been ported).

call forwarding operator line: The feature, normally available on a *PABX*, where a *call* is transferred to an operator if the *called terminal* is not available, e.g. busy or does not answer within a certain period of time. If *call forwarding* is in operation then transfer to an operator would occur only if the call forward number is also busy or does not answer.

call handling: The term normally used to cover a number of *value added services*, such as *call barring*, *call forwarding*, *conference calls*, *closed user groups*, *Calling Line Identification (CLI)*, and the *personal number service*.

call holding: The process of putting a *call* which is being answered into a waiting state whilst some other operation, such as answering another call, is carried out.

call hour: A measure of *traffic* intensity. One call hour is the accumulation of calls totalling one hour, e.g. one call for one hour, two calls for half hour duration each, etc. One call hour equals 60 *call minutes*, 3600 *call seconds* and 36 *Cent Call Seconds (CCC)*.

calling: The process of setting up a *call*, usually over a *switched network*.

calling card services: A form of credit card which provides *value added services* to its *subscribers*. Examples are: access from any country back to the home country of the subscriber, with all *traffic* billed to the card owner and international traffic being billed at the home country rates; *conference calls*; *call screening*; etc.

Calling Line Identification (CLI or CLID): A *network* feature which provides the *called terminal* with the identity (e.g. number) of the *calling terminal* after *ringing* has occurred but before the *call* is answered. This enables the calling terminal to be selective in the calls which it accepts. CLI is also used in *CTI* systems, operating in *call centres*, to provide agents with information on the calling *subscriber* (via a computer screen) before they answer the call. Also called *Calling Line Identification Presentation (CLIP)*. See also *Calling Line Identification Restriction (CLIR)*.

Calling Line Identification Presentation (CLIP): See *Calling Line Identification (CLI)*.

Calling Line Identification Restriction (CLIR): Facility which prevents the *called party* from knowing the identity of the *calling party*. See also *Calling Line Identification (CLI)*.

calling line identification signal: The *signal* which is sent from the *calling terminal* to the *called terminal* at the time of the *call* and containing the identity of the calling terminal.

calling location: Strictly the location of the *calling terminal* although it usually refers to the calling terminal itself.

calling party: The person making the *call*, although it frequently refers to the *calling terminal*. Also referred to as the *calling subscriber*.

calling party camp-on: A *network* service which allows a *call* to be completed even if the network *transmission* or *switching* facilities are temporarily busy. The system monitors these facilities and completes the call when they become free.

calling party clearing: A method of *call clearing* in which the *call* is not terminated until the *calling party* has put down the *handset*. If only the *called party* puts down the handset then this can be picked up again and the call continued without interruption. See also *first part clear*.

Calling Party Pays (CPP): The normal method for *billing* in a telecommunications system whereby the person who initiates the *call* is charged for it. It is now also used in *paging* systems where the *pager* is bought outright and the person initiating a paging request pays a premium which covers the cost of sending the *message*.

calling rate: A measure of the utilisation of a *terminal*. The calling rate is the amount of time a terminal is initiating *calls*, measured over a period of time, expressed as a fraction or in *erlangs*.

calling signal: The *signal* which is sent by a *calling terminal* to request the *network* to set up a *call*. It will include the *address* of the *called terminal*.

calling subscriber: See *calling party*.

calling terminal: The *terminal* that has initiated the *call*. See also *calling party* and *calling subscriber*.

call management: Generally used in the context of managing *calls* within a *call centre* which use *CTI* techniques and provide facilities such as *Automatic Call Distribution (ACD)* and *Calling Line Identification (CLI)* to give information regarding the caller to agents handling *incoming calls*.

call management system: Systems which provide *call management*. They are also used to provide information on certain aspects of *calls*, primarily for accounting purposes, such as the total cost of calls over a period, the usage on different lines, etc.

call minute: A measure of *traffic* intensity, measured by the accumulated amount of call time in minutes. For example one call minute equals one call made for one minute, or two calls for thirty seconds each, etc. See also *call hour*.

call not accepted signal: A *signal* sent by the *called terminal* to indicate that it cannot accept the *call* and therefore *call establishment* has not occurred.

call packet: A *packet* of *data* which carries information required for *call establishment*, such as the *address* of the *called terminal*. Used in *packet switching* systems.

call pending signal: A *signal* sent to the *called terminal*, if it is busy, to inform it that another *call* is waiting for *call establishment*. This signal is usually presented to the user as a visual indicator on the *terminal* or as an audio tone.

call pickup: The facility, available on a *PABX* and *Centrex*, which allows users, within a group, to answer *calls* which are *ringing* on other *terminals* within the group by *dialling* a short code on their own terminal.

call processing system: The system, usually an *exchange*, which processes the *call* from the instance that a *call request* is initiated through to when the call been completed and both the *calling terminal* and the *called terminal* are back *on-hook*.

call progress signal: The *signal* which is sent by the *network* to the *calling terminal* to inform it of the progress of the *call*. Examples are *dial tone* and *busy tone*.

call record: *Data* on a single *call*, such as the *call duration*, *address* of the *called terminal*, time call made, etc.

call release time: See *call clear down time*.

call request: The *signal* sent to the *network* by the *calling terminal* to request that it set up a connection to the *called terminal*. The request would contain the *address* of the called terminal and, in the case of a *telephone* call, it would be initiated by the *handset* going *off-hook* and *dialling* taking place.

call request packet: A *packet*, sent by the *calling terminal* in a *packet switched network*, requesting that a connection be set up to the *calling terminal*. It would contain the *address* of the *called terminal*.

call request time: During *call establishment* it is the time from when a *calling terminal* initiates the *call* to when it receives a *signal* back from the *network* asking it to proceed. In a *telephony* system this would take the form of a *dial tone*.

call restriction: A feature of a *PABX* or *Centrex* which places restriction on the use of certain features by some of the *terminals* within its *network*. This could be, for example, to restrict these terminals from placing *long-distance calls*.

call routeing: The *routeing* of *calls* in an *exchange* between the *calling terminal* and the *called terminal*. A few examples are shown in Figure C.3. A proportion of *traffic* will be between terminals on the same *exchange*. Most will be *incoming calls* or *outgoing calls* to other exchanges. The transit or *tandem* traffic will vary from a very small percentage on rural exchanges to 100% on a *tandem exchange*.

call screening: *Security* feature which allows *authentication* of all *call requests*, such as required when users access a *database* of confidential information.

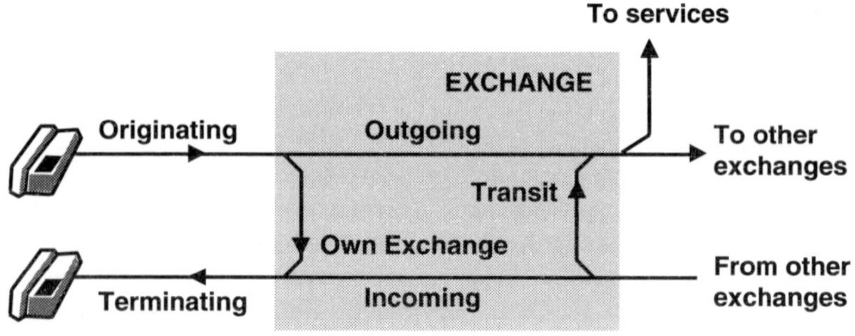

Figure C.3 Call routeing

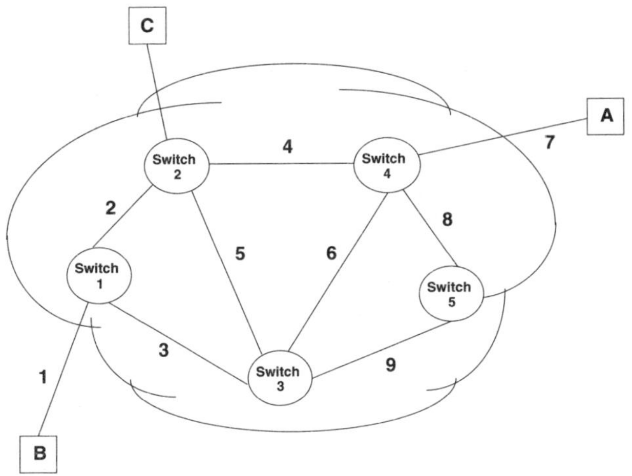

Figure C.4 Call setup

call second: A unit of *traffic* intensity. One call second is equal to one call of one second duration. 50 call seconds could equal one call of 50 second duration, or two calls of 25 second duration, etc. See *call hour* and *call minute*.

call setup: The process of setting up a *call* between two *terminals*. It takes place in all types of *networks*, such as mobile, *circuit switched* and *packet switched*. In the packet switched example of Figure C.4 call setup between terminal A and B would occur with A sending a *call request packet* to the *packet switching exchange* 4. This contains the address of terminal B and may contain the *address* of terminal A, so that it can be identified by the receiver. Switch 4 checks its *routeing table* for the most appropriate path, which could be to switch 2. This switch then carries out a similar process, and eventually sends the request for call setup to terminal B. If terminal B accepts the call it sends a *call accepted packet* back to A along the route which was set up by the original call request packet. The call is now set up correctly and information transfer can take place between the terminals.

call setup time: Time needed for *call setup*, measured from the time a *call request* is initiated to the time that information exchange commences. In the example of Figure C.4 this would be the time from the start of *transmission* of the *call request packet* to the time the *call accepted packet* is received back by the *calling terminal*.

call sign: The *address* used to identify communications systems, and usually consists of a combination of *characters*. It is mainly used for *call setup* and maintenance.

call spillover: Terminology used to describe the effect which occurs in a *common channel signalling* system when a *call control signal*, belonging to a previous *call* which has been grossly delayed, arrives at the *switching centre* at the same time that the switch is performing *call setup* for the next call.

call splitting: The feature of the *network* which allows a *telephone* operator to talk privately to either party making the *call*. It can also be used in a *conference call* to allow one of the callers to talk privately to one or more of the parties in the conference.

call store: Generally refers to the temporary store which is used to hold records of *calls* which are in progress, in a *Stored Programme Controller (SPC)* system. Once the calls have been completed this *data* is transferred to permanent memory within the system.

call ticket: A record of a *call*, containing information on the call, such as the *addresses* of the *calling terminal* and the *called terminal*, the date and time of day of the call, the *call duration*, the call charges, etc.

call tracing: (1) *Network* feature which allows *subscribers* to obtain the *telephone number* of the last *call* which was made to them, unless this was specifically withheld by the user. **(2)** The *network* feature which allows the whole route taken by a call to be identified.

call transfer: The *network* feature which allows a user to redirect the *call* received to another *terminal* by providing the *address* of the new terminal.

Call User Data (CUD): In a *packet switching* system, the information which is contained in the *call request packet* and is transmitted from the *calling terminal* to the *called terminal*.

call waiting: A *network* provided visual and/or audio *signal* which informs the person who is currently on a *call* that another call is waiting to be answered.

CAM: *Computer Aided Manufacture.*

CAMA: *Centralised Automatic Message Accounting.*

Cambridge Ring: A *Local Area Network (LAN)* operating on the *slotted ring* principle for *multiple access*. It can handle up to 255 *nodes* with *data rates* of 10 Mbit/s.

CAMEL: *Customised Applications for Mobile networks Enhanced Logic.*

camp on: A *network* feature which allows *calls* to wait in a queue if the facility they wish to use is busy, the call automatically being connected once the facility becomes free. This feature can also be used with *call waiting*, so that a *signal* is sent to the *called terminal* to indicate that a call is waiting. See also *calling party camp on*.

Campus Area Network (CAN): A *network* which extends over a campus, connecting a group of buildings together, which are situated geographically relatively close to each other. It is in range between a *Local Area Network (LAN)* and a *Wide Area Network (WAN)*.

campus network: See *Campus Area Network (CAN)*.

CAN: *Cancel* or *Campus Area Network*.

Canadian Radio, television and Telecommunications Commission (CRTC): Canadian government telecommunications regulatory body.

Canadian Standards Association (CSA): Canadian standards making organisation which also tests and certifies products for compliance to standards. It is a private organisation and operates on a non-profit making basis. See also *Standards Council of Canada*.

Cancel (CAN): A *character* used in *binary code* to indicate that the previous *message* is to be ignored, probably because it has an error in it.

candela (cd): A unit of measure of light emission from a source.

candlepower: A unit of measure of the illuminating power of a light source. See also *lumen*.

CANTO: *Caribbean Association of National Telecommunication Organisations*.

CAP: *Carrierless Amplitude Phase modulation* or *Competitive Access Provider*.

Capability Set 1 (CS1): An *ITU-T* standard for *Intelligent Networks*.

capacity: Refers to the amount of *data* which the system can handle. For a *switching* system it would be measured by the amount of data that can be switched in a given time. For a *transmission* system it is given by the amount of data transmitted in a period of time, e.g. its *bandwidth*. Also referred to as *channel capacity*.

capture effect: The effect, seen in *frequency modulation*, in which if two *signals* are present at approximately the same *frequency* but different strengths, then only the stronger of the two will be selected and will appear in the output, the other being suppressed completely.

carbon block: A voltage surge protection device, which consists of two electrodes spaced apart so that a voltage above a specified level will jump the gap and cause the line to be earthed.

carbon granule transmitter: Device which was used in a *telephone* to convert *sound waves* into electrical signals, as shown in Figure C.5. Sound pressure on the diaphragm compresses the carbon granules and varies their resistance. This will give an electrical signal between the carbon electrodes, which varies according to the sound pressure.

card dialler: A device, associated with a *terminal*, which accepts a card containing the *telephone number* in coded form, often punched holes, and carries out automatic *dialling* of the number.

Caribbean Association of National Telecommunication Organisations (CANTO): Association of state owned or private telecommunications *carriers* providing domestic and inernational services in the Caribbean.

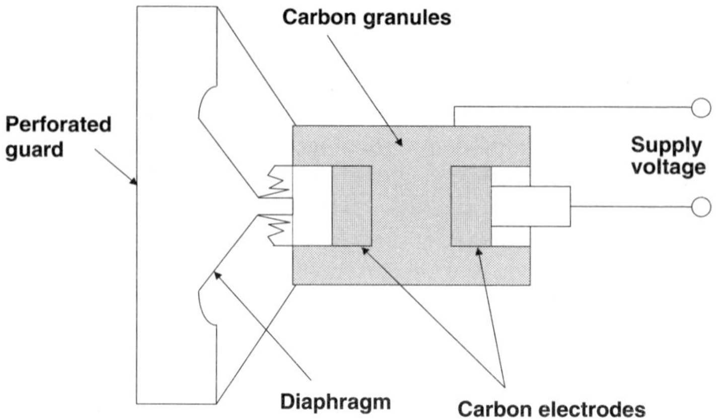

Figure C.5 Carbon granule transmitter

carriage control character: A *character*, used in devices such as a *teleprinter*, a *teletypewriter*, and a *Visual Display Unit (VDU)*, which causes a line to be skipped, i.e. a blank line is produced.

carriage return character (CR): A *character*, used in a device such as a *teleprinter*, a *teletypewriter*, and a *Visual Display Unit (VDU)*, which causes the print head to move to the start of the next line. The carriage return character is specified in *ASCII* and *EBCDIC*.

carried traffic: The *traffic* which is accepted by the system for *transmission* or *switching*. See also *offered traffic*.

carrier: **(1)** An *electromagnetic wave*, such as a *sine wave* or a *pulse train*, which has not as yet been through *modulation* and therefore does not contain any information. Often called the *carrier wave* or *carrier signal*. **(2)** Shortened term for *common carrier*.

carrier band: The continuous *frequency band* that forms the *carrier*.

carrier detector signal: The *signal* sent by a *modem* to its associated *terminal* to indicate that it is on and ready to accept *data* from a *calling terminal*. This prevents the *called terminal* from accepting *noise* on the *line*, mistaking it for data.

carrier frequency: The *frequency* of the *carrier wave*.

Carrier Identification Code (CIC): *Code* which allows a *user* to originate an *inter-LATA call* via a specific *Interexhange Carrier (IEC)*.

Carrierless Amplitude Phase modulation (CAP): One of two signal *modulation* techniques used with *ADSL*, the other being *DMT*. CAP uses incoming *data* to modulate a single *carrier* and this is then transmitted along the wire. The carrier has no *information* content so it is suppressed before *transmission* and reconstructed at the other end, hence the term 'carrierless'.

carrier modulation: See *modulation.*

carrier noise level: The variation of the level of the *carrier wave* in the absence of a *modulating signal*. This variation is usually caused by *noise*.

carrier power: The average power provided to an *antenna feed* when no information is being transmitted, i.e. the *carrier* without *modulation.*

carrier preselection: The facility available to *subscribers* connected to a *local network* to have all their long-distance and international *calls* routed through another *carrier* without the need to dial any extra *digits*. The alternative carrier becomes the default carrier and special digits need only be entered if this is to change. See also *carrier selection.*

carrier selection: The facility available to *subscribers* connected to a *local network* to be able to choose another *carrier* for their long-distance or international calls by *dialling* a special code before the call is made. See also *carrier preselection.*

Carrier Sense Multiple Access (CSMA): A *multiple access* method which relies on the sender of a *message* sensing the state of the *transmission channel* and basing its actions on this. It can therefore only be used in channels which have short *propagation delays*, since for channels with long delays (e.g. satellite) the sensed data is considerably out of date. In a perfect channel (zero transmission delay) the sender can listen to the state of the channel when it is ready to transmit and only send when the channel is free, so avoiding any *collisions*. If t is the time from the start of a transmission from one user to when all users sense the presence of the signal on the line (Figure C.6), i.e. the *propagation delay* on the network plus the sense time, then the vulnerable period, when collisions can occur, is 2t, i.e. t after the end of one transmission and t to complete

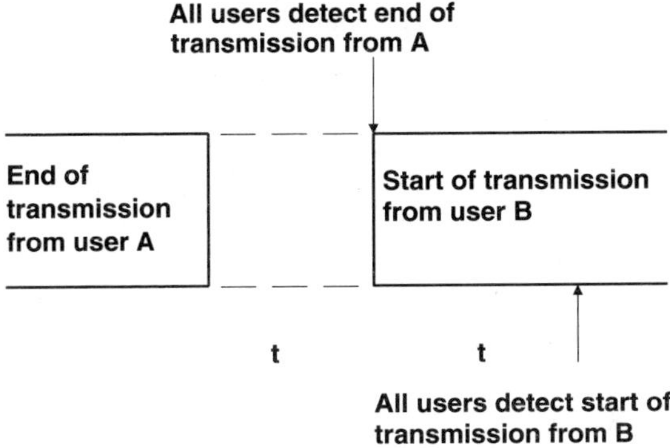

Figure C.6 Vulnerable period in CSMA

a transmission. If the channels are slotted into t second *timeslots*, then the vulnerable period is reduced to t since transmission must start at the start point of each slot. t is small for low propagation delays on the medium. CSMA can operate in three modes: *1-persistent CSMA*, *non-persistent CSMA*and *p-persistent CSMA*. In addition there is a variant called *CSMA/CD*.

Carrier Sense Multiple Access with Collision Detection (CSMA/CD): In a *CSMA* system once *transmission* commences it is completed even if a *collision* occurs on the first *bit*. This is clearly wasteful of *bandwidth*. In CSMA/CD a user monitors the *line* even when it commences transmission and stops once a collision is detected. After that the user waits a random time before sensing the line again. This *multiple access* method has been standardised by the *IEEE* as 802.3. Is is not always possible for a sender to effectively sense the line during its transmission since the strength of the transmitted signal may be so strong as to swamp any signal returned back. Often CSMA/CD *algorithms* require each user who detects a collision to transmit a short *jamming signal* to immediately inform all other users that a collision has occurred on the line.

carrier signal: See *carrier*.

carrier system: Usually refers to the systems which are used for *multiplexing* several different *data channels* onto the *transmission medium*. Examples are the *E1* and *T1* systems.

Carrier to Noise Ratio (CNR): The ratio, measured in *decibels*, of the level of the *carrier* to that of the *noise* before any *modulation* has occurred. Also shortened to C/N.

carrier wave: See *carrier*.

Carsons rule: It defines the *bandwidth* B required for *transmission* in *angle modulation*. For a maximum *baseband frequency* of Q_m it is given by $B = 2Q_m(1+ \beta)$ where if ΔQ is the maximum frequency deviation of the carrier then β is the ratio of $\Delta Q/Q_m$.

Carterphone decision: The ruling by the *FCC*, in 1968, that equipment could be connected to the national *telephone network* (the Bell system) even though it was not produced by the network operator concerned (AT&T in this case). This was on the assumption that the equipment did not do any harm to the public network.

CAS: *Call Accounting System* or *Channel Associated Signalling* or *Centralised Attendant Service*.

cascaded systems: The connection of entities together in such a way that the output from one goes into the input of the next. For example, in cascaded *amplifiers* a series of amplifiers are connected together such that the output form one goes to the input of the next, the *signal* being amplified further at each stage.

CASE: *Computer Aided Software Engineering*.

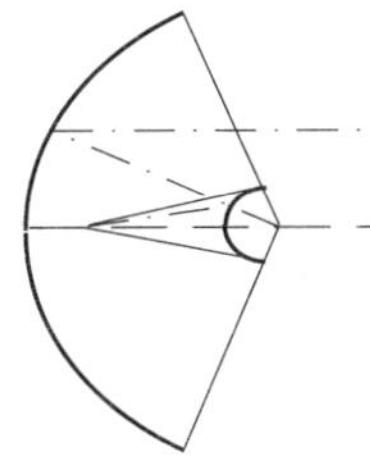

Figure C.7 Cassegrain dual reflector antenna

Cassegrain antenna: A development of the *reflector antenna* in which a convex hyperbolic sub-reflector is inserted at the focus point of the main parabolic reflector. This reflects signals back into the *antenna feed*. (Figure C.7.) It provides a larger depth of focus and field of view, and higher performance feeds can be used with more convenient locations.

category n cable: *Cable* specification by *ANSI*, *ISO* and *EIA/TIA* (568) for 100 ohm *unshielded twisted pair* and its associated *hardware* such as *connectors*. The letter n can take several values such as 1 and 2 for *voice transmission*, and 3, 4 and 5 for *data transmissions* up to 16 Mbit/s, 20 Mbit/s and 100 Mbit/s respectively.

Cathode Ray Tube (CRT): A vacuum tube which is widely used as a display device in a variety of equipment, such as *television receivers*, *Visual Display Units (VDU)*, *terminals*, etc. It works on the principle of deflecting an *electron beam*, which scans across a *phosphorescent* coated screen, so as to produce the required pattern.

CATV: *Cable Television*. Previously known as *Community Antenna Television*.

cat whisker: A flexible piece of wire which was adjusted on a crystal and used for tuning very early radio sets.

CB board: *Central Battery board.*

CBCS: *Cordless Business Communications System.*

CBDS: *Connectionless Broadband Data Service.*

CBEMA: *Computer and Business Equipment Manufacturers Association.*

CBR: *Constant Bit Rate* or *Current Bit Rate.*

CBT: *Computer Based Training.*

CBX: *Computerised Branch Exchange.*

CC: *Cluster Controller* or *Clear Channel.*

CCA: *Cable Communications Association.*

CCB: *Coin Collection Box.*

CCC: *Clear Channel Capacity.*

CCCH: *Common Control Channel.*

CCD: *Charge Coupled Device.*
CCD scanner: *Charge Coupled Device scanner.*
CCH: *Connections per Circuit Hour.*
CCIA: *Computer and Communications Industry Association.*
CCIR: Comite Consultatif des Radiocommunications. (International Consultative Committee for Radio). Now renamed the *ITU-R.*
CCIS: Common Channel Interoffice Signalling.
CCITT: Comite Consultatif International de Telegraphique et Telephonique. (Consultative Committee for International Telegraph and Telephone.) Now renamed *ITU-T.*
CCL: *Cordless Class License.*
CCR: *Commitment, Concurrency and Recovery.*
CCS: *Cent Call Seconds* or *Common Channel Signalling.*
CCSA: *Common Control Switching Arrangement.*
CCSC: *Common Channel System Codeword.*
CCSS: Common Channel Signalling System. See *Common Channel Signalling.*
CCTV: *Closed Circuit Television.*
CDDI: *Copper Distributed Data Interface.*
CDF: *Combined Distribution Frame.*
CDMA: *Code Division Multiple Access.*
CDO: *Community Dial Office.*
CDPD: *Cellular Digital Packet Radio.*
CDPSK: *Coherent Differential Phase Shift Keying.*
CDR: *Call Detail Recording.*
CDV: *Cell Delay Variation.*
CEC: *Commission of the European Communities.*
Ceefax: A *teletext* system, introduced in the UK by the BBC in the 1970s, for transmitting *data* over the normal *television transmission signal.*
CEI: *Comparably Efficient Interconnection.*
cell: **(1)** In a *cellular radio system* it is the geographical area covered by a *base station* and using the same *frequency.* **(2)** In a *transmission system,* such as *packet switching* or *ATM,* it is the group of *bits* which contains user information, and is usually made up of a *payload,* a *header* and a trailer, as in Figure C.8.
Cell Delay Variation (CDV): A measure of *Quality of Service (QoS),* used in *ATM* systems, which defines the variation in delay of a transmitted *cell.*
Cell Insertion Ratio (CIR): A *Quality of Service (QoS)* performance measure in *ATM* systems. It is caused by *bit errors* in the header *address* field and is measured as the ratio of the inserted *cells* to the total number of cells entering a *Virtual Circuit (VC).*

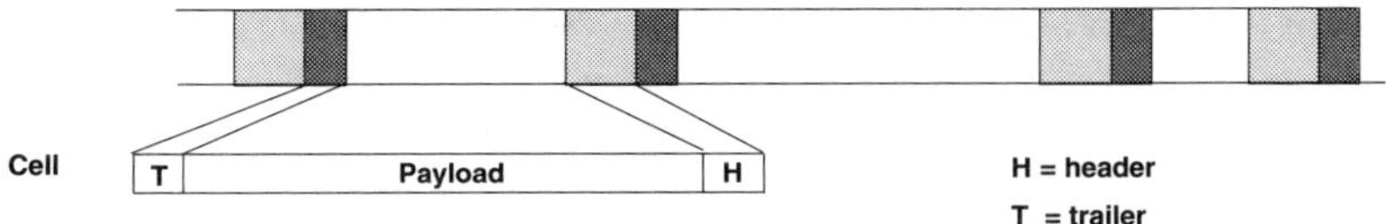

Figure C.8 A cell arrangement

Cell Loss Priority (CLP): An indicator *bit* in an *ATM frame* which informs the *network* which *cells* can be discarded in a *congested system*, whilst still meeting the *Quality of Service (QoS)* requirements. The definition of CLP is given in *ITU-T Recommendation* I.150.

Cell Loss Ratio (CLR): This is a key *Quality of Service (QoS)* measure in *ATM* systems. It is measured as the ratio of the *cells* lost in a given *Virtual Circuit (VC)* to the total number of cells entering that Virtual Circuit. Several factors can lead to cell loss, such as errors in the cell *header* or overflowing of *buffers*.

cell relay: A *fast packet switching* technique which uses fixed length *frames*, unlike *frame relay* and *X.25* (*packet switching*) which use variable length frames. *ATM* is an example of cell relay and uses a fixed frame of 53 *octets*, of which 5 octets form the *header*. Figure C.9 shows the transport of different types of *traffic*, *voice*, *data* and *video*, using fixed length cells.

cell repeat pattern: In a *cellular radio system* each *cell* has a number of *radio channels* which can be used. To prevent *co-channel interference*, cells with the same *frequency* use must be spaced apart, and this results

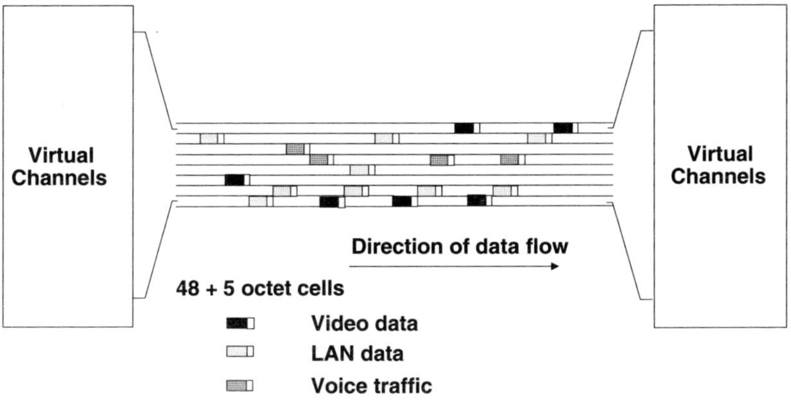

Figure C.9 Fixed length cells for the transport of data

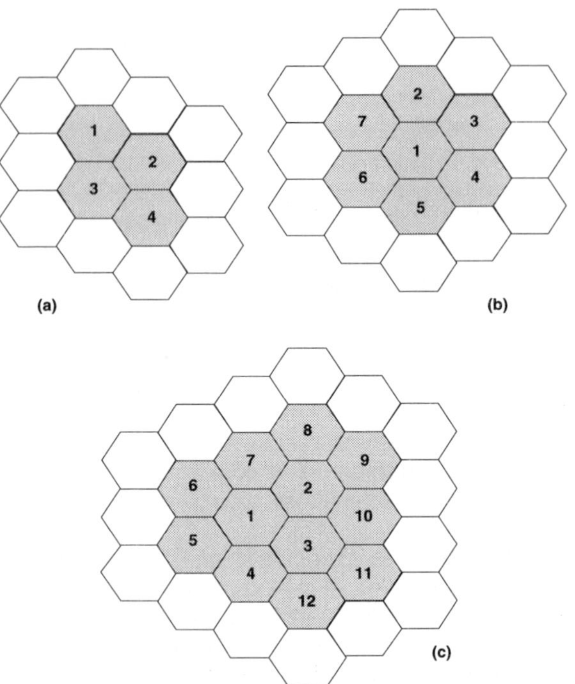

Figure C.10 Cell repeat pattern: (a) four cell repeat; (b) seven cell repeat; (c) twelve cell repeat

in a cell pattern, the most common being the four, seven and twelve cell repeat patterns, as shown in Figure C.10.

cell splitting: Technique used in *cellular radio systems* to increase the *capacity* by splitting an existing *cell* into a smaller geographical areas, as in Figure C. 11. This shows different cell sizes for rural, suburban and urban areas. The problem with cell splitting is that a *base station* is needed for each of the new cells, which can get expensive, and for small cell sizes *co-channel interference* can be a problem even though the same *cell repeat pattern* is maintained.

Cell Stream: The UK's first commercial ATM service, launched by BT in 1997.

Cellular Digital Packet Data (CDPD): *Packet data* service over *cellular radio systems*, used primarily in the USA, and operating at 19.2 kbit/s.

Cellular Geographic Service Area (CGSA): The area covered by a *cellular radio system* licence.

cellular mobile radio: See *cellular radio system.*

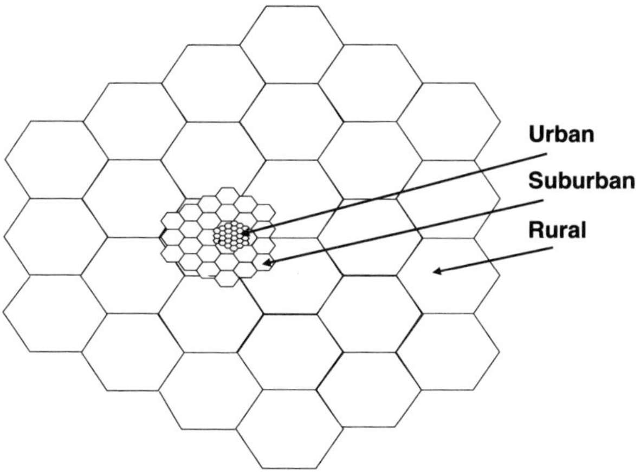

Figure C.11 Cell splitting

cellular radio system: A radio system for mobile users which is based on *cells*, each with its own *base station*, as shown in Figure C.12. Each base station connects to a *Mobile Switching Centre (MSC)*, which in turn is connected to other MSCs and to the *PSTN* so that calls can be made between mobiles and to/from the public telephone network (*PSTN*) to a mobile. *Handoff* capability exists between cells so that *calls* can be continued if the user is on the move and crosses from one cell to another.

CELP: *Code Excited Linear Prediction.*

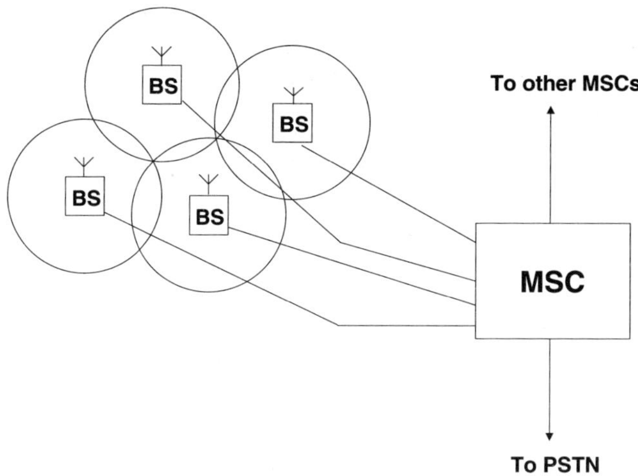

Figure C.12 Cellular radio system

CEN: European standards writing body, like *ETSI*, which works closely with *CENELEC*. CENELEC is primarily responsible for electro-technical matters and CEN for other areas. In this respect CENELEC is the European equivalent of the *IEC* and CEN of *ISO*. The two bodies function separately except in the area of information technology where they combine as the *Joint European Standards Institute (JESI)*. The secretariat for CEN/CENELEC is in Brussels and their membership is drawn from the European national standards organisations. The output from CEN/CENELEC are standards, referred to as ENs (Europaische Norm) or European Standards. When in draft form they are called European pre-Standards, or ENVs. There is usually a two-year conversion period from an ENV to an EV, to give users time to adopt them. Whenever possible these are based on international standards, such as from the *ITU-T* and ISO.

CENELEC: See *CEN*.

Cent Call Second (CCS): A unit of *traffic* which is defined as hundreds of *call seconds* per hour. It is also related to the *erlang*, where 36 CCS equals one erlang. Also known as *hundred call seconds*.

centimetric wave: An *electromagnetic wave* with a *wavelength* in the 1 cm to 10 cm range, corresponding to a *frequency* of between 3 GHz and 30 GHz.

Central Battery board: *Signalling* method used in *manual exchanges* where gravity switch contacts are used at the *telephone* to light a lamp on the switchboard at the *exchange* end, using relay contacts.

Centralised Attendant Service (CAS): Facility in which *calls* to several remote locations are attended to by one or more operators located in one central area serving these remote sites.

Centralised Automatic Message Accounting (CAMA): An *Automatic Message Accounting (AMA)* system in which the *billing* information for several *Central Offices (CO)* is collected centrally.

Central Office (CO): Refers to the location, belonging to a *common carrier* (*Public Telephone Operator (PTO)*), which houses the *switching equipment* for terminating and interconnecting the different *subscriber lines*. Also know as an *exchange*, a *telephone exchange* or a *switching centre*.

central office trunk: The *trunk line* which connects a *Central Office (CO)* to a *PABX* belonging to a *subscriber*. These lines do not run between COs. Also known as central office connecting facility.

central station: Usually refers to a *local exchange* serving several *outstations*, as in a *rural communications* system. A *microwave* based system is shown in Figure C.13, with three outstations, a central station and a *repeater* used to access an outstation which is out of *range* of the central station.

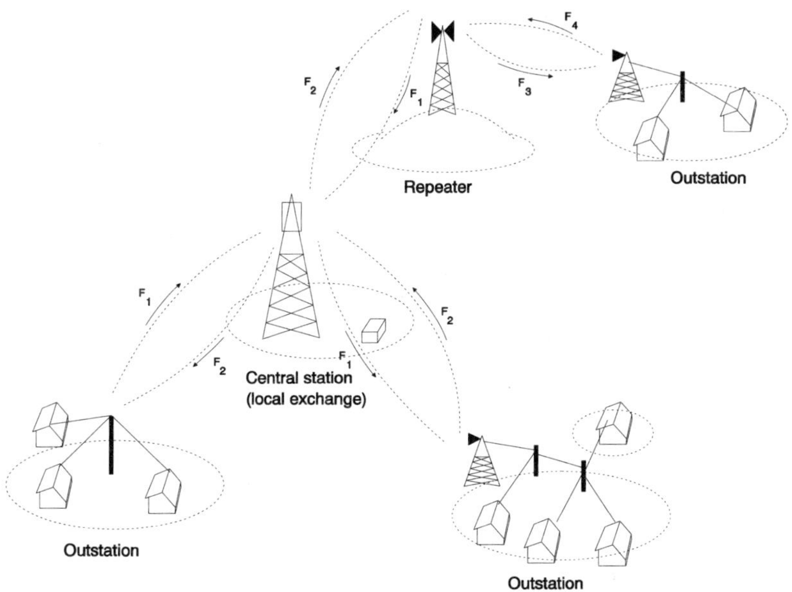

Figure C.13 Central station in a rural multipoint system

central wavelength: In a communications system based on an *electromagnetic wave* (such as optical communications) it is the *wavelength* at which the power is primarily transmitted.

Centre for Telecommunication Development (CTD): A body of the *ITU* which was set up in 1986, following the *Plenipotentiary* of that year, in Geneva, to encourage the growth of telecommunications in developing countries. When the ITU reorganised in 1994 the CTD became part of the *ITU Development Sector.*

centre frequency: The *frequency* of the *carrier signal* used in *Frequency Modulation (FM).*

centrex: A service offered by a *PTO* in which the *switching equipment*, located in the *Central Office (CO)*, is used to provide the facilities to a *subscriber* which would normally be obtained from a *PABX* located on the subscriber's premises. Examples of these services are *Direct Inward Dialling (DID)*, *call pickup*, *voice mail*, etc. CENTREX stands for Central Office Exchange service.

CEPT: Conference Europeen des Administrations des Postes et des Telecommunications. (Conference of European Posts and Telecommunications.) Formed in 1958 by the *PTTs* with the aim of harmonising standards. It presently consist of over 30 members, covering all the countries of the *European Community (EC)* and the *European Free Trade Association (EFTA)*, plus PTTs from other European countries. CEPT is a sister organisation to *CEN/CENELEC* and participates in many of its work programmes. In January 1998 CEPT set up an independent body, *ETSI*, to carry out all the standards activities on its behalf. CEPT still maintained the *Technical Recommendations Application Committee (TRAC)*, formed in 1986, to approve standards for connection of equipment to public networks. In February 1990 CEPT reorganised to form several new groups: *ETNO*, *ECTRA*, *ERC*, *ETIS*, and *EURESCOM*.

cesium standard: A *frequency* standard based on the transition of cesium-133 atoms. It is a primary standard in that it does not need calibration.

CFM: *Composite Fade Margin.*

C450: An analogue *cellular radio system* developed by Siemens during the early 1980s with commercial service starting in Germany in 1985. It has a *channel spacing* of 20 kHz at 450 MHz, and *Frequency Modulation (FM)* is used for speech encoding. *Signalling*, for *call control*, is *FSK* at 5.28 kbit/s.

CFSP: Common Foreign and Security Policy. One of the three pillars for the further unification of the *European Community (EC)*, created by the *Maastricht Treaty*.

CGSA: *Cellular Geographic Service Area.*

chain code: *Code* obtained from a sequence of *binary digits* in such a way that the codes are based on a sequence and are connected to each other. For example the codes derived from the binary sequence 101101 would be 101, 011, 110, etc.

challenge/reply system: A procedure used for *authentication* of the communications between two systems. The *calling terminal* identifies itself and issues a challenge to the *called terminal* to do the same. The called terminal does so using a special *code* sequence which matches that used in the challenge.

changeback: The process of returning the *traffic* to the main *path* after it has been temporarily diverted to a secondary path, to allow maintenance actions to take place on the main path. See also *changeover*.

changeover: The process of moving *traffic* from the main *path* to a secondary path because the main path is faulty or is to be used for another purpose. See also *changeback*.

channel: (1) A *path* along which a *signal* is transmitted from one user to another. It may consist of a physical path or a logical one, e.g. an *E1* system contains 32 channels and a *T1* system has 24 channels. **(2)** A *frequency* used in radio or television which identifies different *transmitting stations*.

Channel Associated Signalling (CAS): A *signalling* system in which the signalling information needed to control a *data channel* is either carried

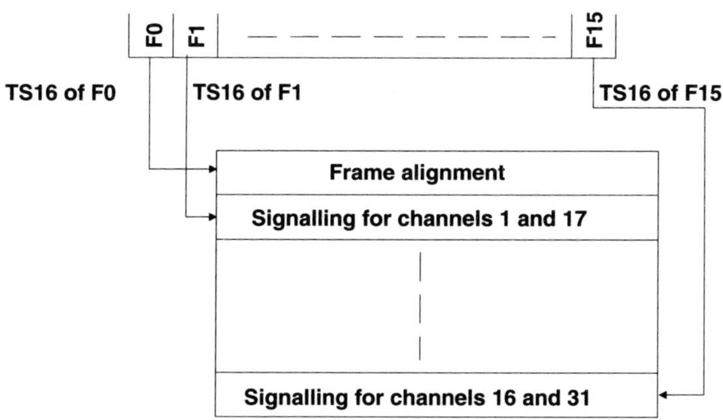

Figure C.14 Channel Associated Signalling multiframe structure

within the channel itself or in a separate channel associated with the data channel. For example the 24 channel *T1* system uses *bit robbing* for signalling, where some of the eight bits within a channel are used for signalling. The *ETSI* 30 channel *E1* system uses a separate channel, *timeslot* 16, to carry the signalling information for the remainder of the 30 channels. This is illustrated in Figure C.14 where a *multiframe* arrangement is used, channel 16 within each frame providing signalling for two other specific channels.

channel bank: Terminology used, usually in the North America, for a *multiplexer* which combines several lower speed *channels* into a higher speed one, often at *T1*. It also has the ability to handle *signalling* and *framing* information.

channel bypass: Similar technique to *drop and insert* except that those *channels* which are to pass through the intermediate *node* are not dropped off and reinserted (as in the *drop and insert* technique) but remain in the main *trunk*.

channel capacity: The maximum *transmission rate*, often specified as the maximum *bit rate* that the *channel* can carry.

channel coding: The process of varying the information being carried by the *channel* to suit the conditions of the channel. For example *redun-*

dancy bits may be added to the *data* to provide *error detection* and *error correction* capabilities if the channel is noisy and prone to errors.

channel down-speeding: The facility of being able to run a *channel* at less than its rated speed in order to maintain a service, although at reduced performance, when the available resources become limited.

channel group: The combining of several *channels* to form a larger set of channels. For example in *Frequency Division Multiplexing (FDM)* 12 *voice frequency* channels (or 16 if 3 kHz channelling) are combined into a single group in the *range* 60 kHz to 108 kHz.

channelisation: The subdivision of a *wideband channel* into a number of lower *frequency* channels so that different types of *information* can be carried by these individual channels.

channel mastergroup: The grouping of *channel supergroups* into a larger unit. Various schemes are use for this. The UK used 15 supergroups to form the mastergroup in the band 312 kHz to 4028 kHz. Europe used 5 supergroups, which spanned 812 KHz to 2044 kHz, and the US Bell system used 10 supergroups in different supergroup *carrier frequencies*, depending on system usage, which resulted in different mastergroup frequency allocations. The Bell mastergroup contained 600 channels.

channel noise line weighting: *Weightings* applied to different *frequencies* in a *noise frequency spectrum*, to allow for the fact that these cause different levels of annoyance to the listener.

channel packing: The technique used for improving the utilisation of *voice frequency channels* for *data transmission* by *multiplexing* several onto a single *voice channel*.

channel priority: The setting of priorities to different *channels* within a communications system so that the higher priority channels can be maintained if system faults occur and only limited service is available.

channel rerouteing: The facility for moving *channels* from a failed *link* to a working link so that service is maintained in the advent of a fault.

Channel Service Unit (CSU): A device, located on the customer's premises, which terminates a digital line from the *Central Office (CO)*. It is used to perform certain line conditioning and testing functions in response to signals from the Central Office.

channel spacing: The difference in *frequency* between adjacent *channels*, such as used in the formation of *channel groups*, or *transmission* of radio *signals*.

channel supergroup: The combining of several *channel groups* into a single unit, called the channel supergroup. For *Frequency Division Multiplexing (FDM)* 5 groups are combined using *carriers* spaced 48 kHz apart, giving a supergroup in the *frequency range* 312 kHz to 552 kHz. A supergroup contains 60 channels.

channel supermaster group: The unit formed by combining several *channel mastergroups*. In the North American Bell system 6 mastergroups were combined to give a 3600 *channel* supermaster group.

character: A letter, a number or a symbol which is used to represent *information*. The character may be coded, such as in *ASCII* code or *binary code*.

character check: The process used to verify that the *character* is valid, i.e. conforms to the rules of the *code* which was used in its creation. It is part of an *error detection* system.

character code: The *code* used to represent the *character*. Examples are *ASCII*, *EBCDIC*, and *BCD*.

character count integrity: The process of ensuring that the number of *characters* received are the same as the number sent.

character error ratio: A measure of the quality of a *transmission* system. It is the ratio of the number of *characters* which have errors in them when received to the total number of characters sent.

character framing: A method of *transmission* in which each *character* is preceded and followed by a special start and stop *code*, so that the sending *terminal* and receiving terminal are synchronised to the character and there is no *synchronisation* between characters.

character generator: The unit which creates a graphical representation of a *character*, from its *character code*, for displaying, e.g. onto the screen of a *Visual Display Unit (VDU)*.

characteristic distortion: *Signal distortion* which has been caused by transients in the *transmission channel*, often due to the effects of *modulation*.

characteristic impedance: The characteristic impedance at any point of a *transmission line* is given by the ratio of the applied voltage to the resultant current at the same point. It varies with *frequency* due to the variation of the inductance and capacitive effects. If R is the resistance in ohms per unit length, G the leakage in siemens per unit length, L the inductance in henries per unit length, and C the capacitance in farads per unit length, then the characteristic impedance Z_0 is given by $Z_0 = [(R+j\omega L)/(G+j\omega C)]^{1/2}$. Changes in characteristic impedance along the length of the cable will result in *signal distortion* and *reflections*. If the cable is terminated in its characteristic impedance then it will have no reflections from its end and there will be a constant ratio of voltage to current along its entire length.

characteristic wave: Usually refers to the *electromagnetic wave* that propagates in the *ionosphere*. There are two characteristic waves, called the ordinary wave and the extraordinary wave, which are generally elliptically polarised. Any wave launched into the ionosphere is split into the ordinary and extraordinary waves. The fraction of total power imparted to these two wave components depends on the match between their *polarisations* and that of the launched wave, the latter being a function of *antenna* geometry, frequency and direction of propagation.

Character Mode Terminal (CMT): A slow speed *terminal* which uses *asynchronous transmission*, each *character* being framed by a *start bit* and a *stop bit*, i.e. in a *stop-start transmission* mode. It cannot display graphical information.

character order of transmission: *Transmission* in which the *characters* are transmitted in a particular order, such as, for example, in ascending order of significance.

character set: All the *characters* which make up a group which is used for a specific purpose. Usually this also includes the coding format. Examples of character sets are the letters of the alphabet, the codes used in the *American Standard Code for Information Interchange (ASCII)*, etc.

character string: A sequence of *characters*, each with its own *start bit* and *stop bit*, which are treated as a single unit.

chargeable duration: The time which will be used to determine the *billing* information for the *call* made by a *subscriber*.

Charge Coupled Device (CCD): A semiconductor device which works on the principle of storing and moving electron charge between different internal cells. It is used in several applications, such as *data* and image storage.

Charge Coupled Device scanner: A *scanner* which uses a *Charge Coupled Device (CCD)* for storing the scanned image and is typically used in applications such as *facsimile* machines.

charging plan: The plan used in the *network* to determine the *billing* information. Usually this is related to the *numbering plan*, the charges being dependent on the distance the *call* traverses, the number of *switching centres* which handle the call, and the duration of the call.

check bit: A *bit* which is added to a block of *data* in order to check for errors in processing or *transmission*. An example of a check bit is a *parity bit*.

check character: A *character* which is generated at the transmitting end, using a special *algorithm*. This character is added to the *block* of *data* being transmitted. It is then recreated at the receiving end, using the same algorithm, and is compared with the transmitted check character. If the two do not agree then an error in *transmission* is suspected and corrective action is taken. See also *Block Check Character (BCC)*.

check digit: A *digit* which is generated from the *data* and appended to the *message*, in a similar way to a *check character*, and is used for error checking.

check loop: A technique for connecting the transmit and receive *channels* together for testing purposes. See also *loop*.

checksum: A technique for error detection during *data transmission*. In it a number of *characters* are grouped into a *frame* and *bits* are combined according to an *algorithm* to generate a checksum. This is then transmitted in the trailer of the frame. At the receiving end the checksum is again computed using the same algorithm and the new checksum is compared to the transmitted one. If there are any differences than an error in transmission is assumed and the receiver takes the appropriate action, such as requesting a re-transmission.

Chemical Vapour Deposition (CVD): A process used in the preparation of very pure materials, such as the glass used in *optical fibres*, in which the material is deposited at high temperature from its compounds, which are in gaseous form.

CHILL: CCITT High Level Language. A high level computer programming language developed by the *CCITT* (now *ITU-T*). Not widely used.

chip frequency: One of the *frequencies* generated by a *frequency hopping* generator in a *spread spectrum* system.

chipping rate: The *rate* at which the sender of a *transmission* clocks the pseudorandom *noise code* sequence in a *Direct Sequence Spread Spectrum (DSSS)* system. This clock *pulse rate* is much higher than that of the information being transmitted.

chirped Bragg grating: A fibre grating, formed in the core of the optical fibre, which is used for *dispersion* compensation. Chirp indicates that the frequency of the grating changes along the fibre.

chirping: **(1)** Rapid change in the *frequency* of an *electromagnetic wave*, often occurring in pulses. **(2)** The rapid shifting of spectral lines in emitted *wavelengths*.

Chi-square distribution: Used in statistical analysis (*sampling*) to determine the significance of a difference between an observed result (O) and an expected result (E). It is given by $\chi^2 = \sum (O - E)^2/E$.

choke: A term sometimes used for an *inductance*.

choke packet: In a *packet switching* system it is the *packet* which is sent by a *node* in a *congested system*, which causes the *data rate*, from the node which is causing the congestion, to be reduced.

chroma: The quality of colour, often used to describe the control of the colour image in a *video* receiver.

chromatic aberration: The defects in a video image caused by the *chromatic dispersion* effects.

chromatic dispersion: See *intramodal dispersion*.

chromatic dispersion coefficient: It is the derivative of the *group delay* with respect to *wavelength* (λ) for an *electromagnetic wave* travelling in an optical *transmission medium*. It is given by $dD/d\lambda$ for a given frequency.

chrominance: A measure of the quality of colour in a colour *television receiver*. It is the difference in *wavelength* between the colour reproduced on the receiver to that of a standard reference colour with the same *radiance*.

chrominance components: The parts of the picture on a colour television which provide the colour information.

chrominance signal: The colour television *signal* which carries the colour information.

churn: The term used to describe the phenomena where *subscribers* move, relatively frequently, from one *PTO* or service provider to another. Churn costs the telecommunications operators a considerable amount of money since it has been estimated that it costs between $300 and $500 to acquire a new subscriber.

CI: *Common Interface.*

CIC: *Carrier Identification Code.*

CICC: *Contactless Integrated Circuit Card.*

CIDR: *Classless Inter-Domain Routeing.*

CIF: *Common Intermediate Format.*

cifax: The *encryption* of a *facsimile signal* to provide confidentiality during *transmission.*

cipher: An *encryption* technique in which plain text is represented by arbitrary symbols or groups of symbols.

ciphony: The use of *encryption* in *telephony* systems.

CIR: *Cell Insertion Ratio* or *Committed Information Rate.*

C/I ratio: Refers to the ratio of the wanted signal, or *carrier* (C), to the interference signal (I), usually in *cellular radio systems*. These systems can tolerate a certain amount of interference, but if too large then speech quality is severely degraded. Analogue *TACS* systems, for example, will work with C/I ratios of about 17 dB.

circuit: In telecommunications refers to the physical *path* which connects two or more *terminals* together and carrying several *transmission channels*. The communications can be either *simplex* or *duplex*. See also *two wire circuit* and *four wire circuit*.

circuit access point: Points on a *circuit* which provide access to service engineers for the purpose of testing and maintenance.

circuit breaker: A device which has the ability to disconnect a *circuit*, either under operator control or automatically under certain fault conditions, such as a current or voltage overload.

circuit designator: *Characters* assigned to a *circuit* which provide information regarding its characteristics, such as the number of *channels*, operating *frequency*, etc.

circuit group: Several *circuits* which have been combined for a special purpose, such as to carry heavy *traffic* between two locations.

circuit group congestion signal: A *signal* which is sent to the *calling terminal* to indicate that *call establishment* could not be carried out due to *congestion* on the *circuit group* which is to be used.

circuit hour: A measure of *traffic volume*. One circuit hour is equivalent to one circuit being used for one hour, or two circuits being used for half an hour each, etc.

circuit release: Freeing up a circuit so that it is available to another user. In *telephony* this occurs when a *subscriber* goes *on-hook*.

circuit restoration: The re-establishment of connection between two users after an *outage*, usually by using another *circuit*.

circuit segment: A *circuit* which provides a point-to-point connection between two *terminals*. See *link*.

circuit switched connection: The *traffic path* which is established, on demand, between two *terminals* using a *switching centre* and in which each user has exclusive use of the *circuit* until the *call* has been completed.

circuit switched exchange: An *exchange* which contains *switching equipment* working on the *circuit switching* principle. Also called a *circuit switching centre*.

Circuit Switched Public Data Network (CSPDN): A *Public Data Network (PDN)* which uses *circuit switching* techniques.

circuit switched network: A *network* which operates on the *circuit switching* principle for *call setup* between *terminals*. Usually the calls go through a *switching centre*. The *PSTN* is a circuit switching network. See also *packet switching network* and *message switched network*.

circuit switching: The principle used to connect two *terminals* together in which a dedicated *circuit* is set up between the two for the duration of the *call*. This circuit is set up on demand and will normally be established through a *switching centre*. *Billing* for a circuit switched call is usually on a time and distance basis. Circuit switching is implemented using space/time switches that move *timeslots* from one *frame* to another. Access to the timeslots is via *multiplexers* operating in *drop and insert* mode. (Figure C.15.) See also *packet switching* and *message switching*.

circuit switching centre: See *circuit switched exchange*.

Circuit Transfer Mode (CTM): The *transfer mode* used in *ISDN* in which a number of *channels*, or a fixed amount of *bandwidth*, is allocated to a connection.

circular polarisation: *Polarisation* of an electrical or magnetic field in which the field vector rotates in the plane normal to the direction of propagation, such that the locus of the extremity of the vector describes a circle. See also *linear polarisation* and *elliptical polarisation*.

CIS: *Contact Image Sensor*.

CISPR: *Comite International Special des Pertubations Radioelectriques*.

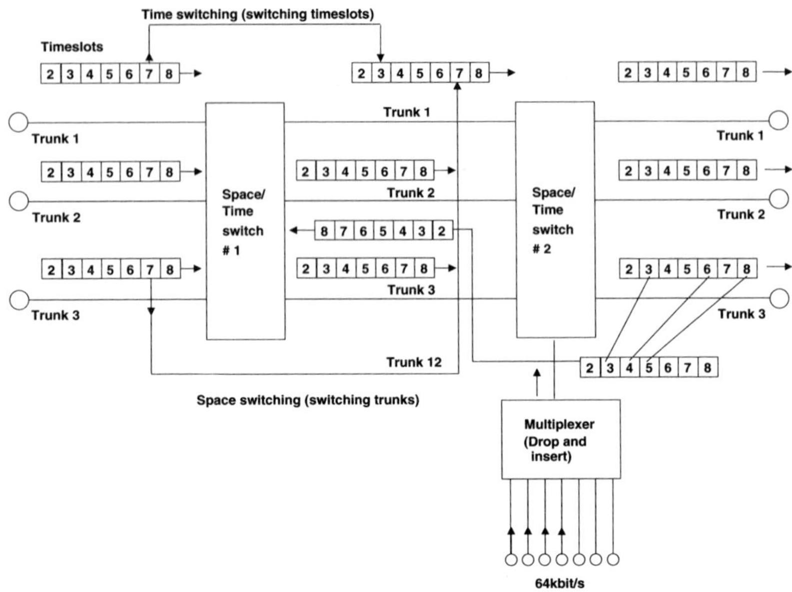

Figure C.15 Circuit switching used to route frames

CIT: *Computer Integrated Telephony.*

Citizen Band (CB) radio: A radio *transmission* system which is available to the public and uses a limited *frequency band* and low power. For example in the USA there are two CB bands, 26.965 MHz to 27.225 MHz, and 462.55 MHz to 469.95 MHz. The power output allowed is 4 watts giving the system a range of a few miles.

civision: The *encryption* of *video signals*, such as in television *transmission.*

cladding: **(1)** The part of the *optical fibre* which has a lower refractive index than its *core* material, and surrounds it. This ensures that the light travelling down the length of the fibre is contained within it due to *total internal reflection*. **(2)** The metal covering over the length of a metallic cable, such as a *coaxial cable*. The metal is attached to the cable by means of extrusion, pressure rolling, drawing, etc., to achieve a good bond.

cladding mode: A *transmission* mode in which the light propagation is within the *cladding* of the *optical fibre*.

cladding ray: The ray of light which propagates in *cladding mode*.

clamping circuit: A device which limits the voltage to a predetermined fixed level. Usually used to protect telecommunications equipment from damage due to an overvoltage surge.

CLAN: *Cordless Local Area Network.*

CLASS: *Custom Local Area Signalling Services.*

Classless Inter-Domain Routeing: Routeing protocol, specified in RFC-1654, which overcomes the shortages of Class B addresses in the *Internet* by allowing subnetting.

classmark: An indication of the characteristics, such as the privileges and restrictions, assigned to a *line* or *trunk* accessing a switch.

class of exchange: The hierarchy of *exchanges* which make up a national telecommunications *network*. An example is shown in Figure C.16 which illustrates how the *subscriber lines* are connected through the various levels of exchanges to the *international gateway exchange* and international lines.

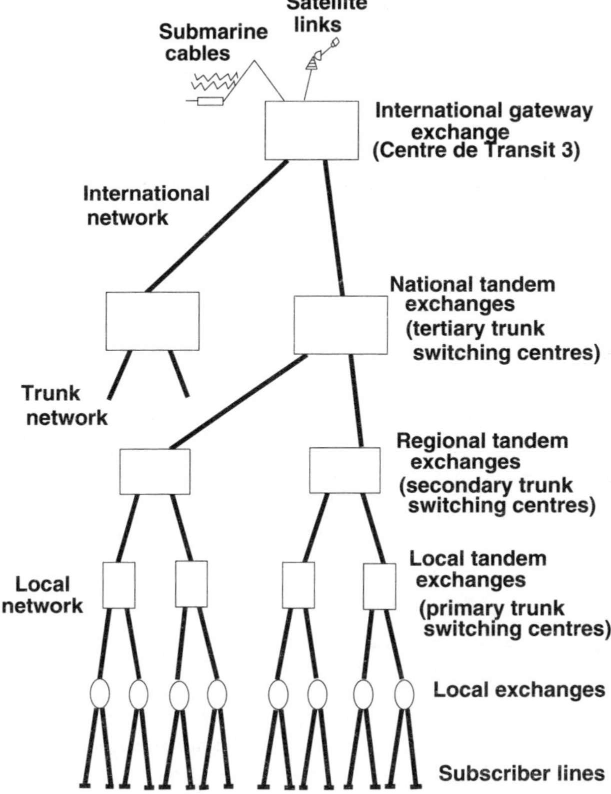

Figure C.16 Class of exchange

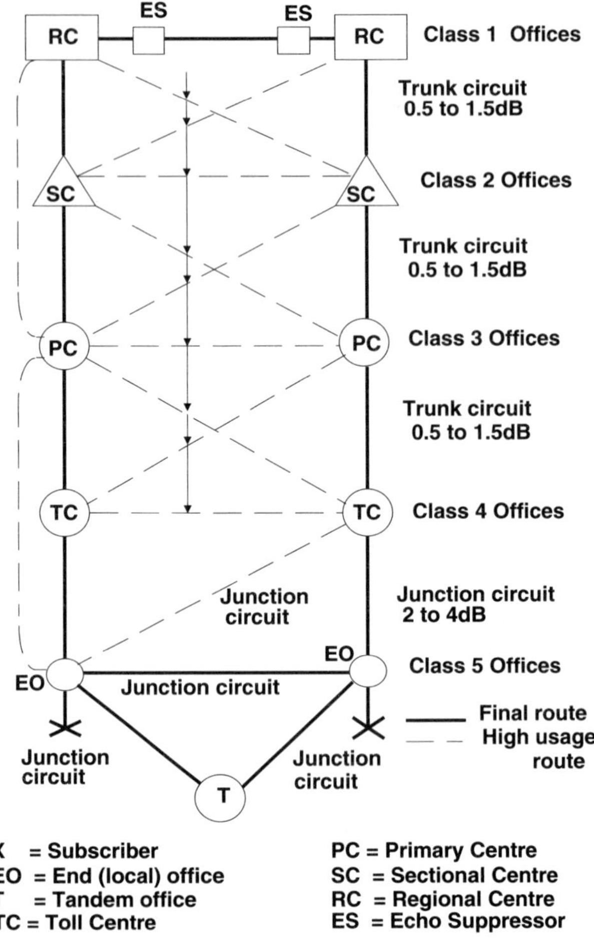

Figure C.17 Class of office

class of office: Terminology used in the USA to describe the hierarchy of *exchanges* or *switching centres* used on the telecommunications *network*. It interconnects various levels of exchanges, referred to as class of offices, as shown in Figure C.17 for the former North American analogue network. The Class 5 office connects to the *subscriber* and the Class 1 office is the highest in the hierarchy.

Class 1 protection: A level of protection provided to the user in the event of a fault. For example a *subscriber* using an instrument with basic

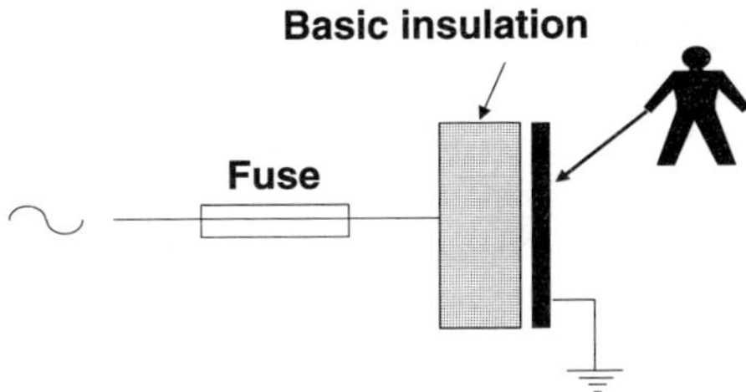

Figure C.18 Class 1 protection

insulation can be protected from a fault on the line (Figure C.18) by a fuse as well as by earthing the metal outside of the instrument.

Class 2 protection: Class 2 protection supplements the protection provided by *Class 1 protection* for example by adding supplementary insulation to the basic insulation, such as by forming the body of the *subscriber's* instrument out of insulating material (Figure C.19).

clear: The process of terminating a *call* and of freeing up *circuits*. Also known as *clearing* or *clear down*. See also *clear-back signal* and *clear-forward signal*.

clear-back signal: The signal sent by the *called terminal* to the *calling terminal* to indicate that the *called party* has terminated the *call*. See also *clear-forward signal*.

Clear Channel (CC): A communications *channel* in which the full *bandwidth* is available to the user. Any control or *signalling* information needed for this channel are transmitted over a separate channel.

Clear Channel Capacity (CCC): The *capacity* of a *Clear Channel (CC)*.

clear confirmation: A *signal* sent during *call control* to indicate that the *message* to *clear* has been received.

clear down: See *clear*.

clear-forward signal: The signal sent by the *calling terminal* to the *called terminal* to indicate that the *calling party* has terminated the *call*. See also *clear-back signal*.

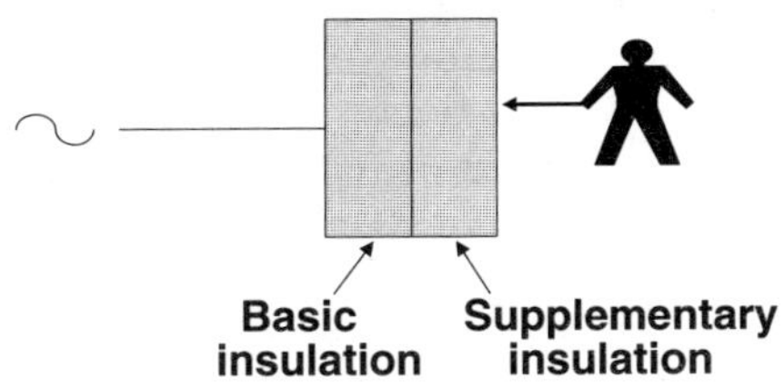

Figure C.19 Class 2 protection

clearing: See *clear.*

clear request: A *signal* sent by a *terminal* to the *network* requesting that the *call* be terminated.

clear request packet: In a *packet switching* system, the *packet* sent by the *terminal* to the *network* requesting that the *call* is terminated. See also *call request packet.*

clear traffic: *Traffic* which has not been through *encryption* or been changed in any similar way.

Clear To Send (CTS): The signal sent by a *modem* to the *Data Terminal Equipment (DTE)*, following the DTE's *Request to Send (RTS)* signal, to indicate that the DTE can begin *transmission.*

CLEC: *Competitive Local Exchange Carrier.*

CLI: *Calling Line Identification.*

CLID: *Calling Line Identification.*

client: Usually refers to *software*, resident on a *terminal*, which allows the terminal to access a *server* on the *network*, to obtain services.

client layer: Usually refers to the layers within the *OSI Basic Reference Model* which are not provided by the *network* and have to be provided by the user.

client-server architecture: A networking architecture in which a central *server* (large computer) is accessed by many *clients* (smaller computers). Processing takes place on both the server and client machines, resulting in distributed computing.

CLIP: *Calling Line Identification Presentation.*

clipping: **(1)** The loss of parts of speech, usually consisting of syllables at the beginning or end of a conversation, due to network conditions. **(2)** The process of limiting the peak *amplitude* of a *signal* by external means.

CLIR: *Calling Line Identification Restriction.*

CLNS: *Connectionless Network Service.*

clock: Refers to a *signal* which has an accurate *frequency* and is use for several applications, such as *synchronisation*, *timing*, etc. It also refers to the physical device which produces this signal.

clock extraction: The process whereby a system extracts a *clock signal*, from another source which contains the clock and other *data*, and uses this to maintain its own operation. For example a system can extract a clock from an incoming signal and use this for internal *timing* and *synchronisation.* Also known as *timing recovery* and *clock recovery.*

clock frequency: The *frequency* at which the *clock* operates. Also sometimes referred to as the clock pulse rate.

clock recovery: See *clock extraction.*

clock tolerance: The maximum amount that the *clock frequency* can deviate from the specified value, usually measured against a time standard such as the *Coordinated Universal Time (CUT).*

closed circuit: Usually refers to the *transmission* of radio or television *signals* such that it reaches a limited number of *terminals*, rather than being *broadcast* to a much larger audience.

Closed Circuit Television (CCTV): Television in which *transmission* is over a *closed circuit.*

closed numbering plan: A *numbering plan* which has a fixed number of *digits* for *dialling* on any *call*, regardless of the geographical location within the *network* of the *calling terminal* and the *called terminal.*

Closed User Group (CUG): A group of *subscribers* on a *network* who can communicate with each other and share a common service, but cannot communicate with any other subscribers on the network. A user can belong to more than one user group. There are also variations of the basic Closed User Group, such as those with outgoing access in which a member of the group can send *messages* to others outside the group but cannot receive any messages from outside the group. See also *International Closed User Group (ICUG).*

Closed User Group indicator: Information contained in the preamble to a *call* which indicates whether the *calling terminal* belongs to a *Closed User Group (CUG).*

cloud attenuation: *Attenuation* of *radio waves* in clouds due to effects such as *scattering* and *absorption* by water droplets or ice particles.

CLP: *Cell Loss Priority.*

CLR: *Cell Loss Ratio.*

CLTP: *Connectionless Transport Protocol.*

cluster: A collection of two or more devices, such as *terminals*, at a single location.

Cluster Controller (CC): A device which provides remote control of a *cluster*, managing their communications over the *network.* Figure C.20 shows an example of a Cluster Controller operating over an *analogue network.*

clutter: Usually refers to the effect of *noise* on a radar system which causes parts of the *display* screen to be obscured.

CM: *Connection Management.*

C-message line weighing: A method of *noise* measurement, using *line noise weighting* to take into account the effect that *noise* at different *frequencies* has on the human *ear*. It is used in the USA and operates at a reference frequency of 1 kHz, with a reference power level of –90 dBm. The unit used for this noise measurement is referred to as decibels above reference noise C-message (dBrnC).

CMI: *Code Mark Inversion.*

CMIP: *Common Management Information Protocol.*

CMIP Over Logical Link Control (CMOL): A *network management* standard proposed by several manufacturers for the transport of *CMIP*

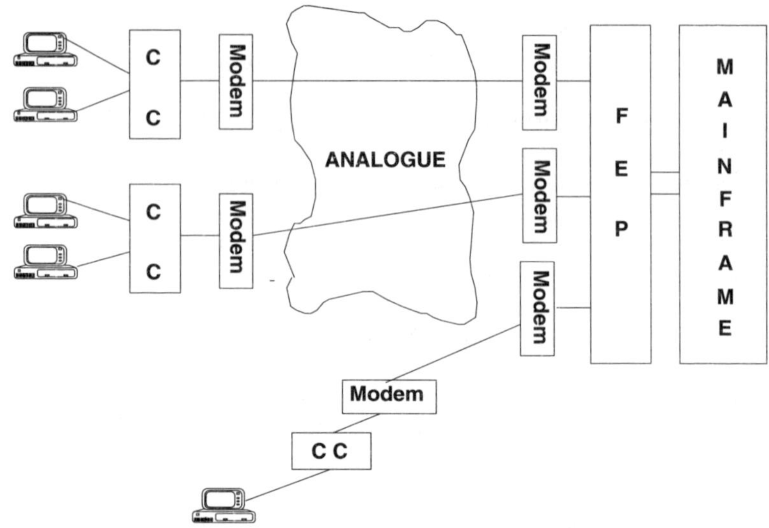

Figure C.20 Distributed processing. (CC = Cluster Controller; FEP = Front End Processor)

over *Logical Link Control*. The idea is to save memory by transporting *OSI* based messages over a shortened *protocol* stack.

CMIS: *Common Management Information Service.*

CMISE: *Common Management Information Service Element.*

CMOT: *Common Management information protocol Over TCP/IP.*

CMOL: *CMIP Over Logical Link Control.*

CMMR: *Common Mode Rejection Ratio.*

CMT: *Character Mode Terminal.*

C/N: See *Carrier to Noise Ratio (CNR).*

CNM: *Customer Network Management.*

CNR: *Carrier to Noise Ratio.*

CO: *Central Office.*

COAM: *Customer Owned And Maintained equipment.*

coax: *coaxial cable.*

coaxial cable (coax): A metallic *cable* which is constructed with a central insulated metal core, surrounded by a metal tube, the whole assembly being protected by an outer sheath. The coaxial cable has a high *bandwidth* and low susceptibility to interference. Also referred to as a coaxial pair cable. Several coaxial cables may be combined into a larger cable assembly.

COBOL: Common Business Oriented Language. A high level computer programming language used mainly for business applications, such as record and file handling.

co-channel interference: Interference which results between two *signals* which occupy the same *transmission channel*. See also *adjacent channel interference*.

cocktail party effect: Usually refers to the ability of the human brain to concentrate on sounds which are coming from one direction and to ignore sounds from other directions, which constitute *background noise*.

COCOT: *Consumer Owned Coin Operated Telephone*.

code: The representation of *data* following a set of rules. The code can take the form of characters or signals. For example the *American Standard Code for Information Interchange (ASCII)* or the various codes used in computer programmes, such as COBOL.

codec: Codec stands for *coder-decoder*. It converts *analogue signals* into *digital signals* and then back again from digital to analogue. For example a codec is used in a *telephony* to convert *voice* (analogue signal) into a digital signal for *transmission* and *switching* and then changing the received digital signal back into analogue for reception by the *ear*.

code compression: The process of reducing the space required for information storage or *transmission* by applying a special *code* to it.

code conversion: The transformation of *data* which has been formatted in one *code* into that of another code.

coded character set: A *character set* which has been formed using fixed rules i.e. *code*.

code dependent system: A system in which the correct *codes* must be used by the *terminals* connect to it for it to operate. See also *code independent system*.

Code Division Multiple Access (CDMA): A *multiple access* and *multiplexing* technique which is an application of *Spread Spectrum Multiple Access (SSMA)* and the two terms are often used synonymously. CDMA allows the *transmission* of *information* from various users to overlap in *frequency* and time within the same *channel* (unlike *FDMA* and *TDMA*, where one of these parameters is used for separation) but segregation between users is obtained by using different *codes* which are matched between corresponding *transmitting terminal* and *receiving terminal*.

coded modulation: The introduction of redundant *digits* into a *digital signal*, before *modulation*, for *error detection* and *error correction* purposes. See, for example, *block coding* and *convolution coding*.

Coded Orthogonal Frequency Division Multiplexing (COFDM): A *transmission* method for a *bit stream* having a high *data rate*, such as is found in *video* systems, by *modulation* of many *carriers* in parallel, each at a comparatively low *data rate*.

Code Excited Linear Prediction (CELP): A *speech compression* and coding technique which allows good quality voice to be obtained at *data rates* down to 8 kbit/s. See also *Linear Predictive Coding (LPC)*.

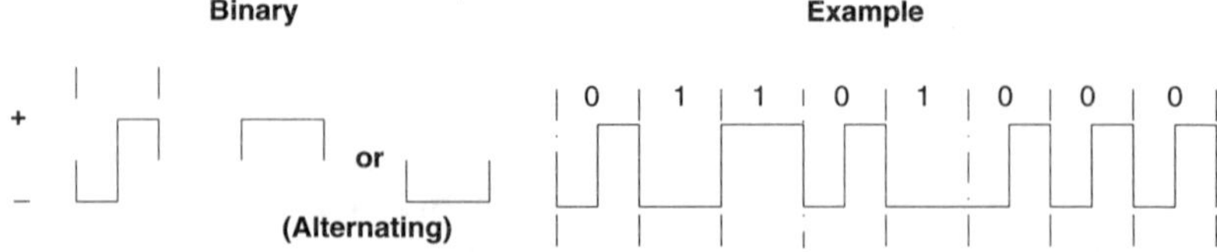

Figure C.21 Code Mark Inversion

code independent system: A system whose correct operation is not dependent on the *terminals* connected to it having to use a particular *code*. See also *code dependent system*.

Code Mark Inversion (CMI): *Transmission code*, specified in *ITU-T Recommendation* G.703, for use over short distances. It is shown in Figure C.21 and uses alternating positive and negative pulses to represent a *binary* 1, in order to achieve a polarity balance over a long term.

coder-decoder: See *codec*.

code word: **(1)** In a coded system it is the sequence of symbols which have been assembled according to a *code*. **(2)** A word which has been assigned a special meaning, for example MAYDAY is the international distress word.

coding gain: A measure of the effectiveness of an *error correcting code*. It is given by the saving in energy per source *bit* of *information* for the coded system, relative to an uncoded system delivering the same *BER*.

codress: A *code* in which the entire *address* is contained within the coded *message*.

coexistence interface: A specification within the *DECT* standard which allows equipment which does not conform to the *Common Interface (CI)* to coexist within the same spectrum.

COFDM: *Coded Orthogonal Frequency Division Multiplexing*.

coherence plane: Usually refers to the plane which is perpendicular to the direction of propagation, at which there is *coherent light*.

coherent demodulator: A system used for *demodulation* of signals, such as *DSBSC*, which uses a *balanced modulator* supplied with a locally generated *carrier*.

Coherent Differential Phase Shift Keying (CDPSK): A method of *modulation* using *Phase Shift Keying (PSK)* in which *demodulation* occurs by comparing the *phase* between the *carrier* and successive pulses.

coherent light: Light whose parameters are predictable at all points on its *waveform*. Generally a *LASER* emits coherent light, which has a narrow *frequency band* and all the light waves are in phase.

Coin Collection Box (CCB): Unit used to collect the *tariff* (coins) for calls made from public payphones.

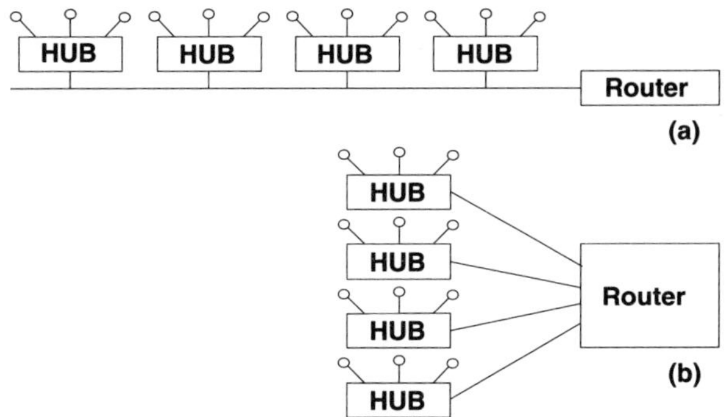

Figure C.22 Backbones: (a) traditional; (b) collapsed

cold standby: Equipment which is connected to a *network* but has not been powered up, so that it can be brought into service at short notice, if required. See also *hot standby*.

cold start: Often referred to as cold restart. A method of starting a system in which all previous *information*, such as stored data and *traffic* queues, are cleared prior to starting. Usually done to clear a fault and the power to the system is usually turned off before being turned on again.

collapsed backbone: The connection of *Local Area Network (LAN)* components to a central router via individual connections, as in Figure C.22(b), rather than via a traditional *backbone* as in Figure C.22(a).

collect call: A *call* in which the *calling party* requests the *called party* to accept *billing* for the call.

collective routing: The *network* facility which enables a *switching centre* to automatically route *messages* to a group of recipients, rather than have the sender *address* each of the recipients individually within the message.

collimated light: A beam of light in which all the rays are parallel, i.e. no convergent or divergent rays.

collision: The effect of two *terminals* trying to transmit *data* on the same *channel* at the same time. This can occur in a *Local Area Network (LAN)* when a terminal transmits a *packet* while another terminal is still in the process of transmitting its packet. In a *half-duplex* transmission system it can occur if the terminals at both ends are trying to transmit data at the same time.

collision detection: The process by with *terminals* on a *network* determine if a *collision* has occurred. See *CSMA/CD*.

collision resolution: The process which is followed by *terminals* on a *network* following a *collision detection*, in order to recover from the situation and to transmit their *data*.

colocation: Usually refers to the installation of another operator's equipment on the premise of a *PTO*. This was mandated by the FCC, whereby competing companies could locate their switches within each other's *Central Office (CO)*, subject to certain safeguards.

colour: The effect produced by light of different *wavelengths*, in the visible region of the *electromagnetic spectrum*. Any colour is defined by three characteristics: its brightness (*luminance*), its *hue* (i.e. the dominant electromagnetic wavelength or *frequency*) and its *intensity* (*saturation*).

colour temperature: It is the standard adopted to define the whiteness of a source. This determines the amount of blue or green in peak white. It is measured as the temperature of a blackbody which emits the same amount of light, of the same colour, as the device whose colour temperature is being determined. It is measured in Kelvins. In Europe the standard is Illuminant D corresponding to 6500 K.

combinations: It is the number of ways in which a proportion can be chosen from a group. Therefore the number of ways in which two letters can be chosen from a group of four letters A, B, C, D is equal to 6, i.e. AB, AC, AD, BC, BD, CD. This is expressed as ${}^4C_2 = 6$. The number of combinations of r times from a group of n is given by ${}^nC_r = (n!)/(r!(n-r)!)$ where ! represents the factorial of the number.

Combined Distribution Frame (CDF): A *distribution frame* which combines the facilities provided by several distribution frames, such as a main distribution frame and an intermediate distribution frame.

Comite International Special des Perturbations Radioelectriques (CISPR): *IEC* International Special Committee on Radio Interference.

Commercial Action Committee (CAC): A group within *CEPT* which has the responsibility for making the organisation more open in its outlook. In April 1990 it organised CEPT's first meeting in which user groups and user associations were invited and consulted on various topics, such as *Electronic Data Interchange (EDI)*, international *packet switching*, *videotext*, and *ISDN*.

Commission of the European Communities (CEC): One of the four main bodies of the *European Community (EC)*. It may be considered to be the civil service of the Community as well as its think-tank and its referee. It is charged with guarding European treaties, and defending the Community's interests. It has twenty members, appointed by Member States and is divided into twenty-three Directorate Generals, each with specific responsibilities (Table C.1). Members of the Commission perform functions similar to those of ministers within their own countries, each being responsible for specific portfolios within the Community's activities. The Commission has the power to take Member States to the *European Court of Justice (ECJ)* if they fail to carry out a (*Treaty of*

Table C.1 Directorates Generals within the Commission of the European Communities

Directorates General	*Services*
DG I	External Relations
DG II	Economic and Financial Affairs
DG III	Industry
DG IV	Competition
DG V	Employment
DG VI	Agriculture
DG VII	Transport
DG VIII	Development
DG IX	Personnel and Administration
DG X	Information
DG XI	Environment
DG XII	Science
DG XIII	Telecommunications
DG XIV	Fisheries
DG XV	Internal Market and Financial Services
DG XVI	Regional Policies and Cohesion
DG XVII	Energy
DG XIX	Budgets
DG XX	Financial Control
DG XXI	Customs and Indirect Taxation
DG XXII	Education
DG XXIII	Enterprise Policy
DG XXIV	Consumer Policy

Rome) or legal obligation. The Commission meets once a week in Brussels.

Commitment, Concurrency and Recovery (CCR): One of the standards within the common *Application Layer* of the *OSI Basic Reference Model*. It ensures that multi-step operations, in which each successive step is dependent on the previous one, are carried out correctly. CCR is used to ensure data consistency between different distributed open systems.

Committed Information Rate (CIR): The *bandwidth* which is reserved for exclusive use by a *Virtual Circuit (VC)* in *Frame Relay* systems. This ensures that *frames* from any one user cannot monopolise the *transmission channel*, so preventing other users from transmitting their *data*.

Common Air Interface (CAI): An *ETSI* standard, part of the second generation *Cordless Telephony (CT)* systems, such as *CT2* and *DECT*. It ensures that these systems can provide a public access service using *handsets* and *base stations* from different manufacturers, provide they conform to the CAI standard.

common battery exchange: An *exchange* in which a source such as a battery provides power needed for systems such as user calling and *supervisory signals*.

common battery signalling: The provision of power from a *switching centre* to a *telephone* on the user's premises to support its calling and *supervisory signals*.

common carrier: Term used, primarily in the USA, to describe a private or public organisation which supplies telecommunications services to the public. Common carriers are primarily concerned with the telecommunications infrastructure rather than the content of the information which is carried over it. In Europe a common carrier is often referred to as a *PTO* or *PTT*.

Common Channel Interoffice Signalling (CCIS): A variation of the *Common Channel Signalling (CCS)* technique in which one separate *channel* carries the *signalling* information for several *trunks* between *Central Offices (COs)*.

Common Channel Signalling (CCS): A *signalling* system in which a common *channel* is used to carry the signalling information for a large number of different *data channels*, so that a much greater number of different users and services can be supported than with other forms of signalling, such as *CAS*. CCS is widely used for *trunk lines*, *PABX* tie lines, *ISDN*, and *Intelligent Networks (INs)*. Examples of CCS are *Signalling System No. 6* and *Signalling System No. 7*.

Common Channel System Codeword (CCSC): A control *channel*, operating on the *slotted ALOHA* principle, used within the *MPT1327 trunked radio* system.

common control: The use of a common system for *call control* of all *channels* coming into a *switching centre*. Older systems used a separate call control device for each *line*, which was expensive to implement. *Signalling* and other control information is also carried on separate *lines* to the *traffic*.

Common Control Channel (CCCH): One of the *GSM Control Channels*.

Common Control Switching Arrangement (CCSA): The use of *switching equipment*, located in a *Central Office (CO)*, for carrying out all *switching*, including that for a *private network*. See also *centrex* and *Virtual Private Network (VPN)*.

common equipment: Equipment on the *network* which is used by more than one *subscriber*, for example the *switching equipment* in the *Central Office (CO)*.

Common Interface (CI): *DECT* specification that enables conforming equipment from different manufacturers to interwork when operated in a public environment. Also covered by *ETSI* in the *Generic Access Profile (GAP)*.

Common Intermediate Format (CIF): A *video signal* coding technique, specified in *ITU-T Recommendation* H.261, which aims to overcome the differences between the 625/50 and 525/60 video *scanning* standards. The *luminance* component of CIF has 288 lines of 352 coded *pels*, repeated 29.97 times per second. See also *Quarter Common Intermediate Format (QCIF)*.

Common Management Information Protocol (CMIP): A part of the *network management protocol*, developed by the *ISO*. CMIP is the protocol used to carry the *Common Management Information Service (CMIS)*. (See Figure M.1.)

Common Management information protocol Over TCP/IP (CMOT): *Protocol* which enables *CMIP* to be transported over *TCP/IP* as illustrated in Figure C.23. The structure of CMOT is shown in Figure C.24. The *application layer* is *OSI*, the *transport layer* is *TCP/UDP* and the *network layer* is *IP*. The *presentation layer* consists of a *Lightweight Presentation Protocol (LLP)*, as defined in RFC 1085, which provides a means for carrying OSI applications over the TCP/IP environment.

Common Management Information Service (CMIS): Part of the *OSI network management protocol* it provides the necessary service needed to support *Common Management Information Service Element (CMISE)* applications. The manager interacts with remote agents (see *agent process*) using CMIS, which is transmitted by *CMIP* (Figure M.1). The manager can request information from the agent manager, for example by using various services provided by CMIS, such as GET (which fetches the attribute values from the *MIB*) SET (which replaces the attribute values specified by the new values contained within the SET

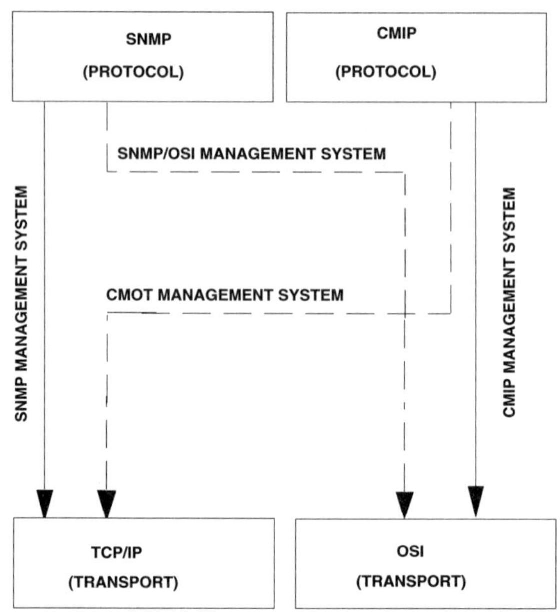

Figure C.23 Alternatives for mixed SNMP/CMIP systems

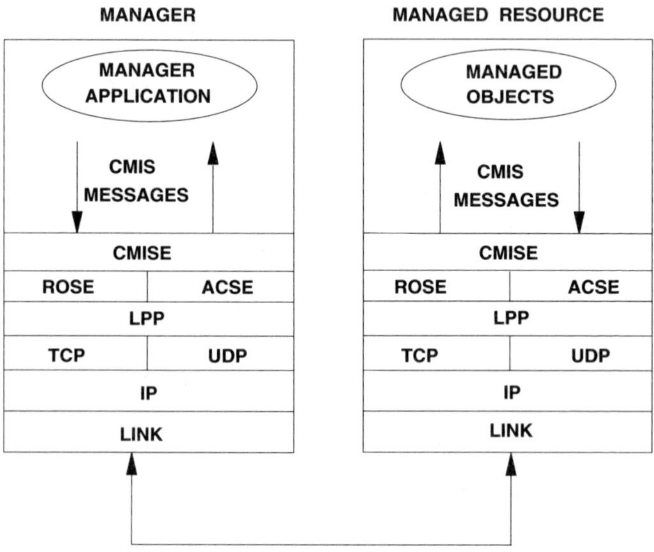

Figure C.24 Communications within CMOT

command) CREATE (which adds a new instant of a *managed object*) DELETE (which deletes the instance of a managed object) and ACTION (which provides a general facility for requesting actions against existing managed objects).

Common Management Information Service Element (CMISE): Part of the *OSI network management protocol* it provides functions for transmitting management information within an *open system architecture*, and it is therefore of prime importance in network management systems. It is a *transaction* oriented standard which uses the services of common standards, such as *ACSE* and *ROSE*. Contained within CMISE are *CMIS* and *CMIP*.

common mode interference: Refers to interference which appears between two *signal* leads or between a signal lead and ground. Also refers to interference which affects two or more components in a *network* in a similar way, rather than affecting a single component.

Common Mode Rejection Ratio (CMMR): The ratio of the *common mode interference* which appears at the input to a system to that appearing at the output from that system.

common return: A *path* which returns current to its source and is common to two or more circuits.

Common Technical Regulation (CTR): At the end of 1991 the procedure for the development of a *NET* was revised. Some of the NETs were now known as CTRs and a new body was set up to be responsible for approvals, called the *Approvals Committee for Terminal Equipment (ACTE)*. A CTR consist of subsets of an *ETS* and the choice of which parts of the ETS are made mandatory is done by ACTE. For example, *PSTN* requirements for all the *European Union (EU)* member states are specified in standard NET4 which runs to about 1000 pages. This covers all the national variants along with tests used to verify them. By contrast, the equivalent CTR (CTR21) specifies a common approach for the whole union and it is under 50 pages in length. The process for development of a CTR is shown in Figure C.25. The *European Commission* initiates the proposal for a new CTR and this is directed to *ETSI* by the *Senior Officials Group on Telecommunications (SOGT)*. The ETS produced by ETSI goes to ACTE for approval and conversion to a CTR, and then to SOGT for enforcement.

communal base station: Generally refers to a *base station* which is shared between users in a *PMR* system, usually to provide coverage in remote areas. To ensure privacy between users selective *signalling* is used and time-outs are normally incorporated to prevent users from hogging the system.

communication channel: See *channel*.

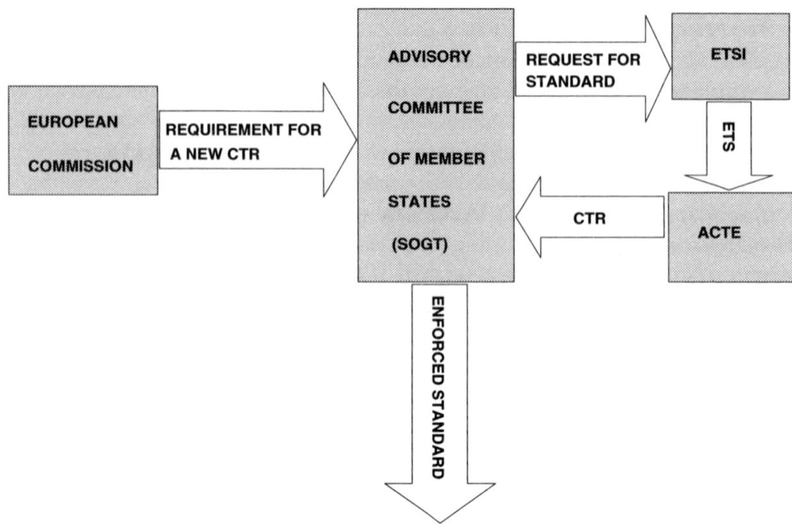

Figure C.25 Process for development of a CTR

communication interface standard: Standards which have been designed to ensure interworking between different *terminals* which follow these standards, at the *logical interface*, *mechanical interface* and *electrical interface* levels. Examples are the *V-Series standards* and the *X-Series standards*.

Communications Act: Passed by Congress in 1934, it set the regulatory framework of US telecommunications. It also set up the *Federal Communications Commission (FCC)* to administer telecommunications on behalf of the US government. Telecommunications competition, however, did not start in the US until 1959, when the FCC authorised the use of private *microwave* systems.

communications controller: A control device, situated on a *Local Area Network (LAN)*, which manages the communications between a *host* processor (*server*) and distributed *terminals*.

communications link: See *link*.

communications network: See *network*.

communications protocol: See *protocol*.

communications satellite: See *satellite*.

communications silence: The process of avoiding any form of communications, whether it be *transmission*, radiation or emission. The system can listen but only if this does not result in any of the above.

communications sink: A device which receives *data*, *control signals*, *timing signals*, etc. from other *communications sources*. Examples of communications sinks are a *television receiver* and a *Visual Display Unit (VDU)*.

communications source: A device which transmits *data*, *control signals*, *timing signals*, etc. to other *communications sinks*. Examples of communications sources are a *radio transmitter* and a *transmitting terminal*.

Community Antenna Television (CATV): See *Cable Television*.

Community Dial Office (CDO): Usually refers to an unattended *switching centre* which serves a small community and is controlled from a larger *Central Office (CO)*.

community of interest: A group of *subscribers* who have a common interest and tend to generate most of their *traffic* in communications with each other. Examples could be users in the same organisation, or members of a club, etc. See also *Closed User Group (CUG)*.

compander: A device which carries out *companding*.

companding: The word derives from 'compressing' and 'expanding'. It is the process used to reduce the *dynamic range* of an *analogue signal* by compression so that it occupies less *bandwidth*. This means that it requires a lower grade *transmission line* during *transmission*. The original signal is recreated by expansion at the receiving end.

Comparably Efficient Interconnection (CEI): An *FCC* requirement that when an operator introduces an enhanced service it should also provide *network interconnection* opportunities to other operators that are comparably efficient to the interconnection that its enhanced service enjoys.

compatible sideband transmission: *Transmission* of the *sideband* in which the *carrier* which has been suppressed is reintroduced at a lower level, so that reception by *Amplitude Modulation (AM)* receivers is facilitated.

compelled signalling: A method of *signalling* in which the *transmitting terminal* does not send any further *data* until it has received back an *acknowledgement* from the *receiving terminal* that the previous *transmission* was received satisfactorily.

compensated optical fibre: *Optical fibre* which has been so designed that the light rays travelling through its core and its *cladding* arrive at the far end at the almost the same time, so that *modal dispersion* is greatly reduced.

Competitive Access Provider (CAP): An operator who has been licensed, in the USA, to provide *local area services* in competition with the incumbent *PTO*.

Competitive Local Exchange Carrier (CLEC): A telecommunications company providing services in competition with a *Local Exchange Carrier (LEC)* in the USA.

compiler: *Software* which converts *code* written in a *programming language*, which a human can understand, into *machine code*, which a computer can understand.

completed call: A *call* in which the preliminary *call setup* has been completed and *information* exchange is still taking place. *Call disestablishment* has not yet occurred.

composite: The output from a *multiplexer* or *data concentrator* which contains all the *data* from the input *multiplexed channels*.

composite code: A *code* which is obtained from two or more other codes.

composited circuit: A *circuit* which is used for simultaneous *transmission* of either *telephony* and *telegraphy* or for *signalling*. *Frequency Division Multiplexing (FDM)* is used to separate the two.

Composite Fade Margin (CFM): The *fading effect* of a *microwave* signal, measured in terms of its *fade margin*. The Composite Fade Margin (CFM) combines the *Dispersive Fade Margin (DFM)* and the *Flat Fade Margin (FFM)* according to the equation: $CFM = -10\log(10^{-FFM/10} + 10^{-DFM/10})$. A plot of CFM versus FFM for DFM = 30 dB and 40 dB is shown in Figure C.26.

composite signalling: A *direct current signalling* system in which the signalling *information* is sent on the same *lines* as the *speech signals*, a transformer being used at both ends of the line to inject and detect the signalling voltages. In the UK BT developed a DC signalling protocol known as DC10.

composite video signal: A *video signal* which combines the picture information (including colour) and the *synchronisation* information.

compressed dialling: Method of *dialling* using shortened *telephone numbers* used, for example, for those numbers which are frequently called.

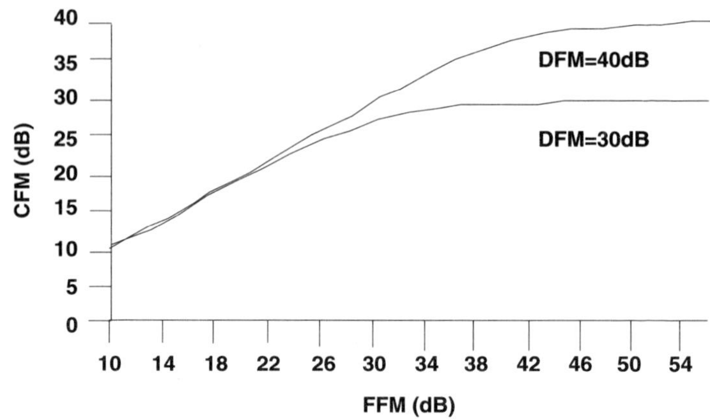

Figure C.26 Composite Fade Margin

compressed video: The reduction in the *bandwidth* occupied by a *video signal*, by use of *compression* techniques.

compressed voice: The reduction in the *bandwidth* occupied by a *voice signal*, by use of *compression* techniques.

compression: (1) The process of reducing the number of *bits* needed to represent a piece of *data*. This will also reduce the time needed to transmit the information and the amount of storage space needed to save it. This is known as data compression. **(2)** The process of reducing the amount of *bandwidth* needed to transmit an *analogue signal*. This is also known as *companding*.

compression ratio: The ratio of the *dynamic range* of a *signal* before and after *compression*.

Computer Aided Design (CAD): The use of a processor (computer) to automate the design process. This could be the design of a telecommunications circuit, the design of the physical layout of cables in a *structured cabling system*, etc.

Computer Aided Manufacture (CAM): The use of a processor (computer) to automate the manufacturing process.

Computer Aided Software Engineering (CASE): The use of a processor (computer) to automate the *software* development process.

Computer and Business Equipment Manufacturers' Association (CBEMA): A trade association of manufacturers and suppliers of *hardware*, *software* and services to the computer, business and telecommunications industries. It is active in many areas, including standards formulation, where it aims at ensuring consensus amongst its members on the long term direction of the industry. The standards activities within CBEMA are administered by its Standards Programme Management Committee (SPMC). The Committee does not make its own standards, but instead participates in other groups developing standards. CBEMA serves as the secretariat to the *X.3 Committee* and has been appointed by *ANSI* to be the technical administrator of the *US Technical Advisory Group*.

Computer and Communications Industry Association (CCIA): A trade association representing the interest of the computer, data communications and telecommunications industry, with headquarters in Arlington VA. See also *CBEMA*.

Computer Based Instruction (CBI): The use of a processor (computer) for teaching. Usually the instructions are programmed into the computer and no human teacher is needed. Also referred to as *Computer Based Training (CBT)*.

Computer Based Training (CBT): See *Computer Based Instruction (CBI)*.

Computer Inquiry: A series of studies carried out by the *FCC* into the *data communications* industry in the USA. Computer Inquiry I, issued in 1971, stated that *data processing systems* were unregulated services, which in effect meant that the then Bell companies could not offer them. Computer Inquiry II, issued in 1980, distinguished between *basic services* and *enhanced services* and stated that the latter was unregulated, so opening it up to competition. Computer Inquiry III, issued in 1986, removed the separation between basic and enhanced services and stated that FFC regulation of a *common carrier's* services would be in line with the *open network architecture*.

Computer Integrated Telephony (CIT): Older name for *Computer Telephony Integration (CIT)*.

Computerised Branch Exchange (CBX): A computer incorporated *PABX* and *switching network*.

computer network: A *network* connecting computers and in which *data processing* plays an important role.

Computer Supported Telephony Application (CSTA): A standard, developed by the *European Computer Manufacturers Association (ECMA)*, for the integration of telephone systems with computers. See also *Computer Telephony Integration (CTI)*.

Computer Telephony Integration (CTI): The integration of computers and *telephony* systems, either directly or via some other device, such as a *PABX*, to provide features which aid the user. Examples are *call handling* and *call management*. CTI is widely used in *call centres* and Figure C.27 shows the interfaces in a typical system.

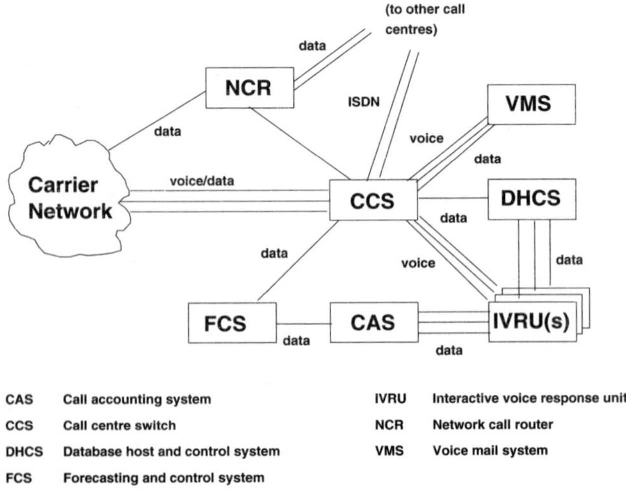

Figure C.27 CTI interfaces in a call centre

COMSAT: Communications Satellite Corporation, a satellite service operator.

concatenated coding: The application of *error correcting codes* in *tandem*, i.e. one error correcting code is applied to the *data* and then a second error correcting code is applied to the output from the first, and so on. Concatenated coding is used when the error correction needed is greater than that which can be obtained from a single code.

concatenation: The connection of systems in *tandem* so that the output from one stage is fed into the input of the next. Examples are connection of *amplifiers* such that the output from the first stage feeds into the input of the second, so giving greater amplification, and the connection of two or more *optical fibres* together to make one long continuous length.

concentrator: A device which concentrates *data* from many *lines* onto a fewer number of lines. Concentrators are effective when linking many users to a resource, where each user is only using the resource for a small amount of time, such as the many *subscriber lines* connected to a smaller number of *trunk lines* in a *switching centre*.

Conditional Access: Term used to describe the ability for users to receive (usually) television programmes depending on the subscription paid. The signals go through an *encryption* process and can only be decoded by subscribers who have been granted access.

conditioned voice grade circuit a *voice grade circuit* whose characteristics have been modified (usually by the addition of *conditioning equipment*) to meet certain performance parameters or *Quality of Service (QoS)* requirements.

conditioning: See *line conditioning*.

conditioning equipment: Equipment which has been added to a *transmission line* in order to improve its characteristics or quality. For example a *voice grade circuit* can be conditioned to carry *data*. There are several types of conditioning. For example C conditioning applies to *frequency response* and *delay distortion* characteristics.

conducted emission: *RFI* emissions which are conducted along a *transmission medium* such as the mains leads of the equipment. Regulatory bodies have set legal limits for conducted emissions, as shown in Figure C.28. See also *radiated emission*.

conducted interference: Interference which results from *conducted emissions*.

cone of silence: The cone shaped region originating from an *antenna* in which there is *signal attenuation*.

cone vision: The parts of the *human eye* which dominate when light levels are high, such as in bright sunshine. See also *rod vision* and Figure R.21.

conference call: A *call* which is between three or more *locations*, each having equal access to all other locations involved in the call. This is

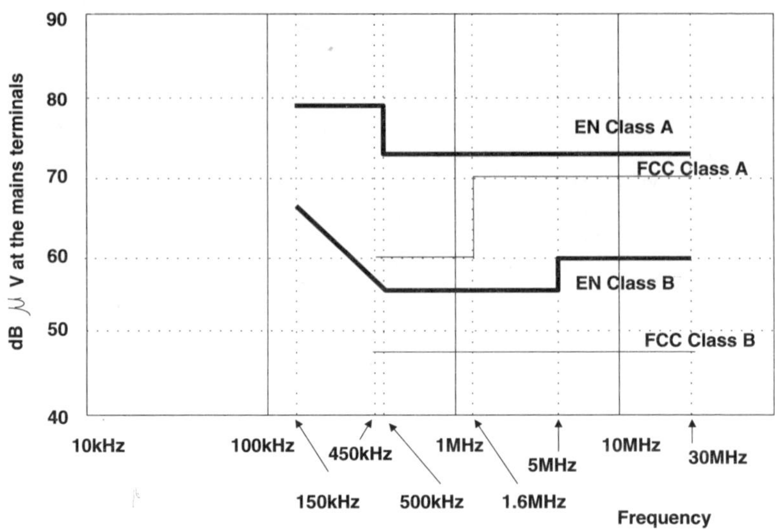

Figure C.28 Conducted emission limits

usually done by setting up a conference bridge linking all the sites concerned.

Conference of European Posts and Telecommunications (CEPT): See *CEPT*.

confidentiality: One of the three types of *security* available with *EDI* systems, the other two being *authentications* and *data integrity*. Confidentiality is normally achieved by *encryption* of the *data* being transmitted.

configuration: The set up of the *network* components, both *hardware* and *software*, to meet an application requirement.

configuration control: The control of the *configuration* to ensure that it continues to meet the requirements of the application for which it was set up.

configuration file: The file which keeps the *configuration* information. Also called a configuration table.

configuration management: One of the five *System Management Functional Areas (SMFA)* defined by *OSI* for *network management*. The other areas are *accounting management*, *security management*, *performance management* and *fault management*. Configuration management provides a mechanism for managing the network elements, often called *managed objects*, which are under control of the management system. The system should have the facility for changing the *configuration*, initialising management objects, shutting them down and removing them from service, collecting state information on a regular and on-demand basis, and *provisioning* services and resources to meet demand. Examples are: connecting end-to-end service; setting up *alternate path routeing* options under fault conditions; configuring gain on cards; providing alternative configurations depending on time of day; and downline loading of information.

conformance: Usually refers to a *network* equipment meeting a specified requirement, such as that defined in published standards.

conformance testing: Tests carried out to ensure that the equipment meets the requirement specified, such as in a published standard. Usually these tests are carried out by an independent testing laboratory and the product is then certified as having met the stated performance standards.

Conformance Test Service (CTS): Programme, supported by the *European Commission* for setting up test services throughout Europe so that vendors need only test their equipment once and receive approval for sale throughout the *EC*. This is covered by Memorandum M-IT-03. Figure C.29 illustrates the CTS process. The rules are formulated by the *European Committee for IT Testing and Certification (ECITC)*. For any particular area a Technical Support Service (TSS) is set up which provides advice to the ECITC and national members, and is responsible for definition and support of the test method, and for cooperating with other standard bodies in this area. The national certification coordinating member organises certification at the national level and confirms the certificates issued by other participating countries. It also accredits the

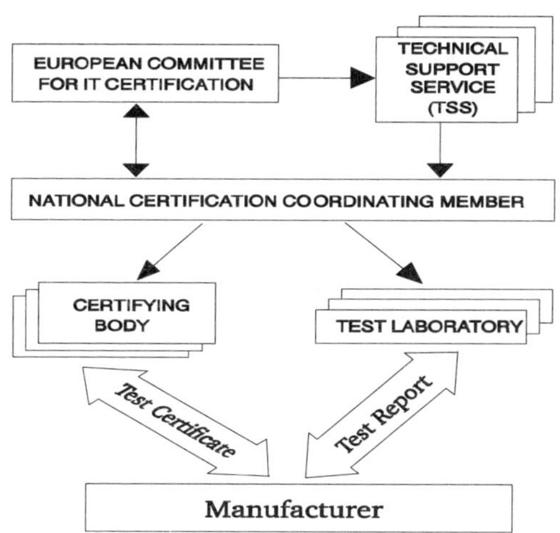

Figure C.29 The CTS process

test laboratories and the certifying bodies in its country. The test laboratory performs tests and issues harmonised test reports. These reports are then passed by the equipment manufacturer to the certifying body, who issues harmonised European certificates for the IT or telecommunication product. Conforming products are able to carry the CE mark.

congested signal: A *signal* sent in the backward *transmission path*, to indicate that *call setup* could not be accomplished due unavailability of a *line* caused by excess *traffic* or failure of a part of the *network*.

congested system: A communications system which cannot handle all the *traffic* demands placed on it due to lack of resources. Under these conditions some of the traffic will be delayed or lost.

congestion avoidance: Measures taken by the *network* to prevent the occurrence of a *congested system* such as, for example, requesting *transmitting terminals* to cut down on the amount of *messages* sent if the system looks like it is becoming overloaded.

Connection Admission Control (CAC): That function, part of the *ATM* based *B-ISDN*, which decides on the *capacity* to be allocated to new *traffic* in order to provide the required *Quality of Service (QoS)* to new and existing users.

connection charge: The one-off charge which is levied on a new *subscriber* to cover the operators costs of connecting the subscriber to the *network* or service.

Connectionless Broadband Data Service (CBDS): An *ETSI* standard, which is a variation of *Bellcore's* standard on *SMDS*, based on *IEEE 802.6*. The aim is to enable suppliers to speed up the introduction of *MAN* Interconnection across Europe.

connectionless mode transmission: *Transmission* in which no *path* is set up prior to or during the transmission. Usually used in *packet switching* systems in which the *packet* carries the *address* information and this is routed through the *network* according to the best route at the time, based on *algorithms* at the individual *switching centres*. See *connection mode transmission.*

Connectionless Network Service (CLNS): A *packet switching* transport mechanism in which individual *packets* have their own *address* information so they can be routed through the *network* using different *paths*. See *Connection Network Service (CONS).*

Connectionless Transport Protocol (CLTP): *Protocol* used within the *OSI Basic Reference Model* with facilities for transport and *error control* of *data* between two *terminals*, without any guarantee of *Quality of Service (QoS)* or *flow control.*

Connection Management (CM): The sublayer protocol, part of the *GSM signalling* system, which is concerned with *call control*, *call estab-*

lishment, *call clearing*, management of *supplementary services* and the *short message service.*

connection mode transmission: *Transmission* in which a *path* is initially established through the *network* and this is maintained for the duration of the *call.* In *packet switching* the first *packet* would normally contain the *address* and other information needed to set up the path. Subsequent packets would then not contain any address information but would follow the path set up by the first packet. Each packet is numbered and counted so that the end of transmission can be determined and also the incidence of lost packets. See also *Connectionless mode transmission.*

Connection Network Service (CONS): A *packet switching* transport mechanism in which the first *packet* contains the *address* information and it goes through the *network* and sets up a *virtual path.* All subsequent packets follow this path and do not have their own address information. See *Connectionless Network Service (CLNS).*

Connections per Circuit Hour (CCH): A measure of *traffic* handling capability. It gives the rate at which connections can be set up by *switching equipment.*

connectivity: Refers to the level of integration within the *network* where *subscribers* can communicate using a variety of devices, such as *facsimile*, *modems*, *data terminals*, *telephones*, etc.

connectivity exchange: The exchange of *information*, in *High Frequency (HF)* radio *networks*, concerning *transmission* parameters, such as the quality of the route, path availability, etc.

connector: The device which provides the physical interface to a cable, such as *coaxial cable* and *optical fibre*. There are many types of connectors, to meet the different specifications, such as RJ-45, D-type, FC/PC, SC, etc.

connector insertion loss: The additional loss which occurs in the *transmission medium* due to the introduction of a *connector*. It is therefore a measure of the loss which occurs in the system, with connector present, compared to that of a continuous line without any connectors. The loss is measured in *decibels.*

connect signal: The *signal* which is sent at the *call setup* stage to request the *network* to reserve a switch for the *call.*

CONS: *Connection Network Service.*

Constant Bit Rate (CBR): Services which require a constant transfer of *information.*

constant ratio code: A *cyclic code* which has a *block* length of n and each *code word* has a set of m bits. Also known as m-out-of-n code, it is a non-linear code and is not widely used.

Consumer Owned Coin Operated Telephone (COCOT): Privately owned coin boxes linking to the *Public Switched Telephone Network (PSTN)*.

Contact Image Sensor (CIS): A method of *scanning* used, for example, in *facsimile*, in which a single line scanner, occupying the whole length of the paper, is placed in close proximity to the paper at the document scanning point.

Contactless Integrated Circuit Card (CICC): *ISO* standard for a *smart card* in which *information* is stored and retrieved without use of conductive contacts.

content addressable storage: A storage system in which the locations are determined by their content rather than their position in the overall system. Also referred to as *associative storage*.

contention: The situation which arises when two or more users attempt to use the same *network* resource at the same time. Contention most commonly occurs in *multiple access* systems when several users attempt to use the same *transmission channel*.

contention avoidance: Multiple access techniques in which *contention* is avoided completely. Examples are *Reservation TDMA (R-TDMA)* and *Priority-Oriented Demand Assignment (PODA)* protocols. See also *contention minimisation*.

contention delay: The delay in a *contention* system caused, for example, by time spent in waiting to transmit data or in re-transmission following a *collision*.

contention minimisation: Techniques which are use to reduce the amount of *contention* which can occur in *multiple access* systems, but not completely avoid it. Examples are *Reservation ALOHA (R-ALOHA)* and *Carrier Sense Multiple Access (CSMA). See also contention avoidance.*

Contention Priority-Oriented Demand Assignment protocol (CPODA): A *PODA* method for *multiple access* in which *contention* is permitted for the *reservation slots*. A central station still manages the allocations based on reservations.

contention ring: A *ring network* architecture in which *contention* is used to secure network resources. If a user wishes to transmit in this system it checks to see if a *bit* is passing through it. If not it commences transmission and at the end of this it puts a *token* onto the ring. When the token again comes round to the originator it removes it from the ring. If there is traffic on the ring when a user wishes to commence transmission it waits until the token passes through it. This is then converted to a 'connector', usually by changing the polarity of the last bit, and then it commences transmission. Since there is now no token on the line no other station can transmit. If two stations initially start transmission simultaneously, because they both checked the line and did not find a bit passing through them, then they will read each other's *data* as it goes around the ring and will know that a *collision* has occurred. Both then back off for a random time and start again. When the ring is full, i.e. many users have data to transmit, the system operates basically as a *token passing* ring, each user transmitting in turn. On light loads it operates as a contention ring, with low *contention delay* since a ready station does not need to wait for a token to arrive before it can begin its transmission.

content provider: Refers to the telecommunications service provider who is responsible for management of the content of the service rather than the infrastructure which delivers it to the *subscriber*. Examples are the services provided to obtain the latest weather forecasts, or stock market prices, over the telephone.

contiguous ports: *Ports* which occur sequentially, in an unbroken numerical order.

continuity check: The test done to verify the a *transmission path* has not been broken at any point. This is often done by sending a continuous single *frequency tone*, which is looped back at the receiving end. A continuous path exists if the tone is received back. This test signal is referred to as the continuity check tone.

continuity failure signal: A signal received by a *transmitting terminal* to indicate that the *call* cannot be completed because the *continuity check* has indicated a failed *path*.

continuous DTMF: A *telephone* feature which allows *DTMF* signals to be sent as long as a key on the telephone pad is depressed. This allows facilities to be accessed which require a *tone* of long duration.

continuous phase modulation: *Phase modulation* in which the transition between phases of the *carrier* occurs smoothly, without discontinuities.

continuous receiver: A *receiving terminal* which can receive and record *messages* without operator intervention.

Continuous Tone Controlled Squelch System (CTCSS): A selective *signalling* system, widely used in mobile communications systems, such as *PMR* and *paging* (Figure C.30). Typically a single sub-audible *tone frequency* is transmitted along with the *speech signal*, the CTCSS subtones which are used internationally being given in Table C.2. At the receiving end only the audio output, above 300 Hz, is passed to the user so that the tone is suppressed.

Continuous Variable Slope Delta Modulation (CVSD): *Speech compression* technique which is used at rates of between 20 kbit/s and 32 kbit/s, i.e. *compression ratios* of 16:5 to 2:1. Its performance is not as good as *ADPCM* but is less complex to implement. It is based on *delta modulation* in which the periods used to approximate the *signal* are progressively changed to match the input *analogue waveform*.

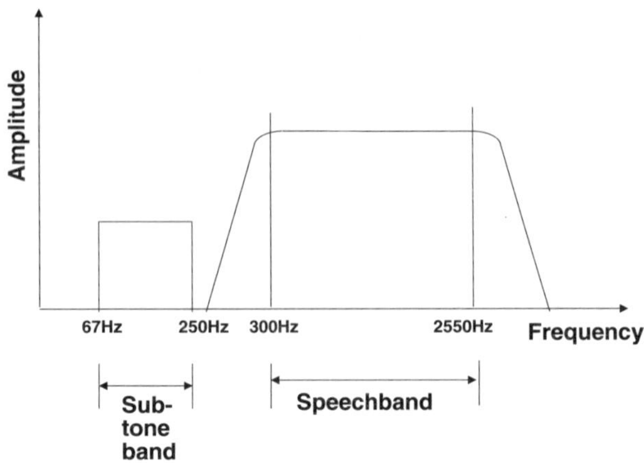

Figure C.30 Continuous Tone Controlled Squelch System

Continuous Wave (CW): *Signal* having a *waveform* with constant *amplitude* and constant *frequency*.

contrast ratio: A measure of the reflectivity of the image, such as a *facsimile* document or a display on a *Visual Display Unit (VDU)*. It is given as the ratio of the maximum reflectivity of a spot on the image being considered to the reflectivity at the point of interest.

control centre: A central location from which the operation of the *network* is controlled and managed. Also known as a *network management centre*.

Control Channel System Codeword (CCSC): A *signal* which is continuously transmitted in the *MPT1327 trunked radio* system. It provides details of the system's *network*, area, zone, control category, etc. CCSC operates on a *Slotted ALOHA (S-ALOHA)* basis.

control character: A *character* which takes the form of a *signal* (non-printing) which, when it occurs, causes an operation to be stopped, started or modified in some other way.

Table C.2 CTCSS frequencies (in Hz)

67.0	71.9	74.4	77.0	79.7	82.5	85.4	88.5	91.5	94.8
97.4	100	103.5	107.2	110.9	114.8	118.8	123.0	127.3	131.8
136.5	141.3	146.2	151.4	156.7	162.2	167.9	173.8	179.9	186.2
192.8	203.5	210.7	218.1	225.7	233.6	241.8	250.3		

control circuit: A *circuit* used to carry supervisory and *signalling* information in order to control the operations of other circuits. See also *control line*.

control field: The *field*, part of a *frame*, which carries *control signals* for the system.

control field extension: The extension of the *control field*, in protocols such as *HDLC*, to carry additional control information.

controlled access: The process of allowing access to certain network resources only to those with the correct *authorisation*.

controlled ALOHA: The method of controlling the *traffic* in an *ALOHA multiple access* system to reduce *congestion*. Techniques include reducing the probability of retransmissions of *packets* which have suffered *collisions*, and denying certain users the right to transmit over a period of time.

controlled maintenance: The process of maintaining the *network* using supporting data on equipment reliability, failure rates, etc., in order to get the maximum benefit from *preventative maintenance* and to reduce the amount of *corrective maintenance* needed.

controller: A device, part of the *network*, which acts as its organisation centre, relaying information and coordinating the activities of the other resources on the network.

control line: The *transmission line* between devices which is used to carry *control signals* and does not carry any *data*.

controlling exchange: Terminology used to describe the *exchange* which has overall control of an international *call*. Usually this is the exchange through which the *calling party* is operating.

control signal: The *signal* used to control the system, such as to start, stop or modify its functions in any other way.

control station: The *station* on a *network* which is responsible for overall supervision of the network, such as polling and selecting other stations, and which is responsible for ensuring an orderly recovery of the network following a fault.

conversational mode: A method of two-way communications between *terminals* in which they exchange information freely, following an agreed *protocol*. The communication is similar to that of a spoken conversation between two people.

conversational services: Specified in *ITU-T Recommendation I.211, for Broadband Integrated Services Digital Network (B-ISDN)*, it defines services such as *video telephony*, *videoconferencing* and high speed *data transmission*.

conversion filter: The use of a counter to convert from one *code* to another. For example if a *Delta Modulation (DM)* has been generated with a *sampling* rate of n times the *Nyquist rate* and the output is then fed into a *binary* counter, then the output from this counter, at intervals of n digits, will be *PCM* at the Nyquist rate.

convolutional coding: An *error correction* coding system in which the coder input and output are continuous streams of *digits*. Each m *bits* of input *data* are encoded into k output bits, where conversion is a function of the last n bits of the input. The code is described as 'a rate m/k code' and n is the constraint length of the code.

cooperation factor: In *facsimile* systems it refers to the product of the density of the scanning line and the scanning length.

cooperation index: In *facsimile* systems it is the value of the *cooperation factor* divided by π.

Cooperation in the field of Scientific and Technical research (COST): A programme of research established in 1971 as an open and flexible framework for R&D cooperation in Europe. It involved the 15 *EU* Member States as well as others countries. COST Actions exist in over 15 research domains, the largest of these is COST Telecommunications. Any group of scientists can propose new action on a research activity within COST and once five COST countries have signed a Memorandum of Understanding (MoU), the Action is operational. COST Telecommunications Actions can be informally grouped as: Optical Technology; Development and Simulation of Networks and Integrated *Satellite*/Terrestrial *Networks; Multimedia and Image Communications; Radio Systems; User Requirements and Telecommunications for the Disabled and Elderly; Speech* Technology; *Software* Verification and *Validation.* COST complements other *EC* funded research programmes, such as *ACTS*. The Fifth Framework Programme runs from 1999 to 2002 and its budget is in the region of ECUs 10 billion.

Coordinated Universal Time (CUT): Time which is measured on a scale of which the fundamental unit is one second. It is defined by *ITU-R Recommendation* 4604.

Copper Distributed Data Interface (CDDI): A *data transmission* system, providing *data rates* of 100 Mbit/s, which is the equivalent of *FDDI* but runs on *unshielded twisted pair* copper wire rather than on *optical fibre*.

copper pair: Refers to the construction of cable used to connect *subscribers* to the *local exchange*, i.e. in the *local loop*. It consists of two copper wires which are insulated from each other and then twisted together along their length. Several such copper pairs (25, 50 or 100 off) are then combined into a cable, with protective covering over them. This arrangement minimises *crosstalk* between the copper pairs.

Cordless Business Communications System (CBCS): Private, switched, relatively large, *Cordless Telephony (CT)* system which is operated using multi-cells, with roaming and in-call handover capability between *cells*.

Cordless Class Licence (CCL): A class licence provision under the UK *Telecommunications Act* which allows operators to offer digital public access cordless services to others without applying for a licence or paying a fee.

Cordless Local Area Network (CLAN): *Local Area Network* which uses radio *links* between *nodes.*

cordless PABX: A *PABX* in which some or all of the *terminals* are connected to the main switching unit without the use of wires. Several techniques can be used for the connection, such as radio, ultrasonic or *infrared light.* Also known as *Wireless PABX (W-PABX).*

cordless switchboard: Generally refers to an early operator controlled switchboard in which the cord used to make connections between *subscribers* is replaced by manually operated keys.

Cordless Telephony (CT): Although sometimes this term is used to describe the whole area of mobile communications, such as *cellular radio systems* and *Personal Mobile Radio (PMR),* in reality it covers the original application in which a telephone was used on the customer's premises and the *handset* was connected to the base via a radio signal. This was know as *CT1* and it has been extended to systems which also operate outside the home, such as *CT2*, *DECT* and *PHS*.

correlation ranging: *Ranging* technique used in a *Passive Optical Network (PON)* arrangement for determining the *round trip delay* of a new *Optical Network Unit (ONU)* which is to be added to the *network*. This enables the new ONU to be attached to a live system without disruption of the existing system. Also known as *photon ranging*.

core technology: Refers to technologies on which applications can be built. Examples of core technologies are microelectronics, *software*, optical, and assembly technologies, such as multichip modules. See also *application technologies*.

Corporation for Open Systems (COS): Formed in North America in 1985 primarily out of an initiative from the *Computer and Communications Industry Association (CCIA).* About forty major US corporations took part in establishing COS. It aims to speed up the penetration of *OSI* by working closely with other standards making bodies. COS is actively involved in standards setting and in *conformance testing*. It has members on many US technical committees, such as *IEEE* and *EIA*, and provides an input to *ITU-T* Study Groups and to *ISO* on OSI standards.

corrective maintenance: Work done to repair a device which has failed.

correlation: The technique for establishing the strength between variables. Correlation is said to be high if there is a connection between the items or events and it is low if there is no connection. The strength of the correlation is measured by the correlation coefficient, whose value varies between 1 (good correlation) to 0 (poor correlation).

correlative code: A *line code* which works on the principle of deliberately introducing a controlled amount of *Intersymbol Interference (ISI)* into the received signal, which produces a multi-level received signal which, with suitable pre-transmit coding, can be designed to provide the requisite line code features.

COS: *Corporation for Open Systems.*

cosine law: See *Lambert's cosine law.*

COST: *Cooperation in the field of Scientific and Technical research.*

Council of Ministers: One of the four main bodies of the *European Community (EC)*. It consists of ministers from the Member States and meets to take decisions on selected topics. There are specific councils, such as the Agriculture Council in which agricultural ministers sit. The Council also contains very small working groups of staff (mainly civil servants seconded from the Member States) who hammer out details of legislation. The Council is the final decision making body of the Community and the only one who can adopt legislation. Three forms of voting can take place in adopting proposals: unanimity, where all Member States must agree; qualified majority voting, where each Member State uses its votes, number of votes having been allocated according to the size of the country; and simple majority, where each state has one vote, irrespective of its size. The Council of Ministers meets in Brussels, and twice a year the heads of states, their foreign and finance ministers, along with the President of the Commission and two other Commissioners, convene in what is known as the European Council. This European Council does not have legislative power, but it is the main political authority of the Community and sets its political direction.

counter: A device which keeps tracks of events. Counters can be implemented in *hardware* or *software*.

counter-rotating ring: A *network* having a *ring topology* in which *signals* flow in both directions around the ring.

country code: The one, two or three digit number, placed in front of a *telephone number*, to indicate the country in which the *subscriber* is calling from. Country codes are needed for *international trunk dialling*.

country network identity: A *signal*, sent along a *backward channel*, towards the *calling terminal*, to indicate the country to which the *call* is being routed.

coupler: A device which connects two other devices together to enable energy to be transferred between them.

coupler transmittance: The *transmittance* measured between the input and output of a *coupler*.

coupling: The process of joining two systems together so that energy can pass between them.

coupling coefficient: A measure of the effectiveness of the *coupling* process. The coupling coefficient C is given by $C = Z_m/(Z_a Z_b)^{1/2}$, where Z_m is the mutual impedance between the two devices being coupled and Z_a and Z_b are their self impedances. If the impedances are capacitive then it is referred to as a capacitive coupling coefficient; if inductive it is an inductive coupling coefficient.

coupling loss: The power loss which occurs during *coupling*. It may be stated as an absolute value (watts or *decibels*) or as a fraction of the input power.

coverage area: Usually refers to the area served by a telecommunications system, such as a *cellular radio system.*

CPE: *Customer Premises Equipment.*

CPODA: *Contention Priority-Oriented Demand Assigned protocol.*

CPP: *Calling Party Pays.*

cps: A measure of the *data transmission* rate, given in *characters* per second.

cradle switch: The switch which is operated when a *telephone* goes *off-hook* and *on-hook*. Also referred to as a *switch hook.*

CRC: *Cyclic Redundancy Check.*

CRC-4: See *Cyclic Redundancy Check (CRC).*

CRC-6: See *Cyclic Redundancy Check (CRC).*

CRC-8: See *Cyclic Redundancy Check (CRC).*

critical angle: For light passing between two *transmission media* of different refractive indices, it is the smallest angle of incidence at which *total internal reflection* occurs.

critical frequency: The *frequency* below which the *signal* will be reflected back from the *ionosphere*. Frequencies above this value will be transmitted through.

critical radius: The limiting radius to which an *optical fibre* can be bent without any of the light, which is propagating along its length, leaking from its surface. Also known as the critical bend radius.

crossbar: A mechanical switch arrangement for making electrical connection between two points, used in the early *analogue switches*. It consists of horizontal and vertical bars which are moved by energising their associated electromagnets. The crosspoint contact, at the junction point of the two bars, only operates when the horizontal bar is selected before the vertical bar operates, and once operated the contact remains held by the vertical bar. The horizontal bar can then be used again to select another crosspoint in the row without disturbing those already in use.

crossbar exchange: An *exchange* which uses *crossbar* equipment for *switching*.

crossbar matrix: A connection of switches in an arrangement resembling a *crossbar*, as shown in Figure C.31. The switching elements at the

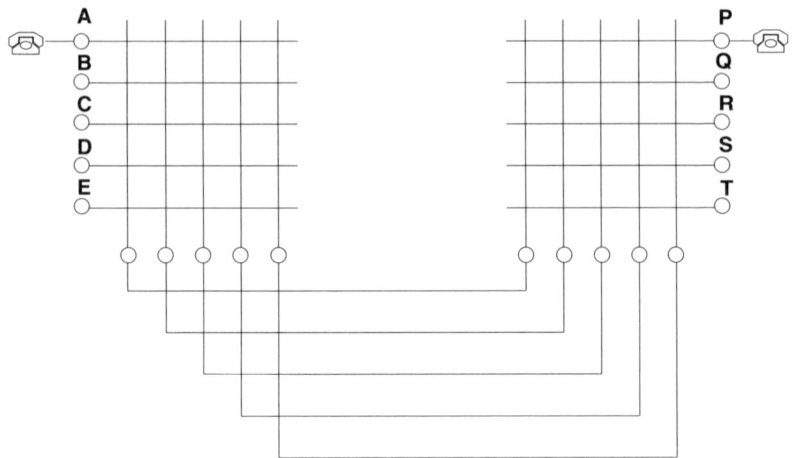

Figure C.31 A simple crossbar matrix

crosspoints can be obtained by a crossbar, as in Figure C.31 or by other means, such as relays. By closing the appropriate relays a connection can be made between any of the lines, A to E and P to T.

crossbar switch: See *crossbar*.

cross compiler: A *compiler* which runs on one type of computer but whose output compiled programme can run successfully on another type of computer.

cross-connect: Equipment which distributes *signals* in a *network*. The cross-connect consists of many input and output lines, with the ability to interconnect any input line to any other output line, so that *signal* paths can be varied. This cross-connection can be carried out manually, as in a *patch panel* of a *distribution frame*, or electronically.

crosslink: The radio *link* between two *satellites* in *geostationary orbit*.

cross modulation: The *crosstalk* effect in which two *carrier signals*, which are undergoing *modulation*, interfere with each other.

cross-office check: A *continuity check* which passes through a *central office* and is done to verify that a continuous *transmission path* exists between the points being tested.

cross-phase modulation (XPM): The non-linearity introduced into an optical wave due to the summation of temporal intensity variations of several co-propagating channels, as can occur in a *Dense Wavelength Division Multiplexing (DWDM)* system.

crosspoint: The contact point in a matrix switch. See, for example, *crossbar*.

cross polarisation: The effect of operating two radio system with the same frequency, but in which one has vertical *polarisation* and the other horizontal polarisation.

cross-polarisation discrimination (XPD): The technique for reducing the mutual interference between systems sharing the same or adjacent *frequency channels.* In *radio transmission* systems this can be done by careful design and alignment of the transmitting and receiving *antennas.*

crosstalk: The unwanted coupling between two *signals* which are being carried in two *transmission medium* located close to each other. See also *far-end crosstalk, near-end crosstalk, intelligible crosstalk* and *unintelligible crosstalk.*

CRT: *Cathode Ray Tube.*

CRTC: *Canadian Radio, television and Telecommunications Commission.*

cryptography: The process of *encryption* of *messages* between two *terminals,* so that they cannot be understood by a third party.

CSMA: *Carrier Sense Multiple Access.*

CSMA/CD: *Carrier Sense Multiple Access with Collision Detection.*

CS1: *Capability Set 1.*

CSPDN: *Circuit Switched Public Data Network.*

CSTA: *Computer Supported Telephony Application.*

CSU: *Channel Service Unit.*

CT: *Cordless Telephony.*

CTCSS: *Continuous Tone Controlled Signalling System.*

CTD: *Centre for Telecommunications Development.*

CTI: *Computer Telephony Integration.*

CTM: *Circuit Transfer Mode.*

CT1: Early *Cordless Telephony (CT)* standard for *analogue signals.* See also *CT2.*

CTR: *Common Technical Regulation.*

CTS: *Clear To Send* or *Conformance Test Service.*

CT2: Second generation *Cordless Telephony (CT)* which has a *Common Air Interface (CAI).*

CUD: *Call User Data.*

CUG: *Closed User Group.*

Current Bit Rate (CBR): One of the four service classes defined for *ATM* by the *ATM Forum.* It roughly corresponds to the *Deterministic Bit Rate (DBR)* (specified in the *ITU-T Recommendation* I.371) and provides a *virtual circuit* with fixed *bandwidth,* similar to that provided by a *circuit switched connection.*

cursor: A movable symbol on a *Visual Display Unit (VDU)* screen, which indicates the position of the next print character or can be moved to any point of interest on the screen. Cursors can have various shapes, such as flashing bar, block of light, etc.

customer access network: See *access network.*

customer loop: See *access line.*

Customer Network Management (CNM): Facillity within *ATM*, *Frame Relay* and *SMDS* which provides *subscribers* with a degree of management over their own *networks.*

Customer Owned And Maintained equipment (COAM): Equipment which is owned and maintained by the customer. Usually refers to *Customer Premises Equipment (CPE).*

Customer Premises Equipment (CPE): Telecommunications equipment which is located on a *subscriber's* premises and is connected into the *PSTN.* It may belong to the subscriber or to the *PTO.* A common example is a *PABX.*

Customised Appplications for Mobile networks Enhanced Logic (CAMEL): An *ETSI* standard for mobile *Intelligent Network (IN)* services.

Custom Local Area Signalling Services (CLASS): A *signalling* system which enables enhanced features to be provided, such as *Caller Line Identification (CLI), call forwarding*, etc.

cutoff frequency: *Frequency* above or below which the system will not perform as expected. For example below a certain frequency *radio waves* will not be able to penetrate and be transmitted along the *ionosphere* layer. Also in *transmission* along *optical fibre* below a certain frequency the specific propagation mode does not exist.

cutover: The transfer of a *circuit* or *lines* from one system to another, usually within a relatively short time and in a way that minimises disruption to the *subscriber.*

CVD: *Chemical Vapour Deposition.*

CVSD: *Continuous Variable Slope Delta modulation.*

CW: *Continuous Wave.*

cycle: A sequence of events which repeats itself, such as in a *waveform.*

cyclic code: A class of *linear block code* which is used for *error detection* and *error correction.* It works on the principle of using an *algorithm* on a *block* of *data* and including the remainder to the *transmission* as a *check character.* At the receiving end the same *algorithm* is used to compare the check character and if it is different then an error in transmission is assumed. See also *Cyclic Redundancy Check (CRC).*

Cyclic Redundancy Check (CRC): A technique for *error detection* during *data transmission.* A number of *characters* are grouped together into a *frame* and bits of it are combined according to an *algorithm* to give a *checksum* called the CRC or CRC checksum. (In some applications this checksum can be calculated from a preceding frame.) This is appended to the transmission. The *receiving terminal* carries out the same algorithm on the received data and generates another checksum. This is compared with the original one and if it is different then an error in transmission is assumed and an *ARQ* (request for re-transmission) is send to the *transmitting terminal*. The number of *bits* used in CRC can vary, the most common being four bits for CRC-4, six bits for CRC-6 and eight bits for CRC-8.

cycle time: The time needed to complete a *cycle*. For example, see *poll cycle*.

D

DAA: *Data Access Arrangement.*
DAC: *Digital to Analogue Converter.*
DAB: *Digital Audio Broadcasting.*
DACS: *Digital Access and Cross-connect System.*
daisy chain: Connecting devices together on a *network* such that *signals* pass from one device to the next. See also *cascaded system.*
DAM: *DECT Authentication Module.*
DAMA: *Demand Assignment Multiple Access*
D-AMPS: *Digital Advanced Mobile Phone System.*
DAP: *Directory Access Protocol.*
DAR: *Distributed Adaptive Routeing* or *Dynamically Adaptive Routeing* or *Dynamic Alternate Routeing.*
dark current: The current which flows in a *photodetector* when bias is applied to it but no light is incident on it. Usually this current is small and it can vary with external conditions, such as the amount of bias, temperature, etc.
dark fibre: *Optical fibre* which has been installed but is not carrying any optical *signals.* Usually this fibre is installed and owned by a different company to the one which will be using it for optical *transmission.*
DARPA: *Defence Advanced Research Projects Agency.*
DASS: *Digital Access Signalling System.*
data: Any *characters*, numbers, symbols, etc. which are arranged according to some formal system and which is used for the *transmission*, storage, display or manipulation of information.
Data Access Arrangement (DAA): A device which is required (by the *FCC*) to interface privately owned equipment to the public *network*, owned by a *common carrier*. The functionality of this device is normally built into a *DCE* and *DTE.*
database: A collection of *data*, usually relating to a specific subject. Telecommunications databases are often stored electronically and can be accessed by several different users, often simultaneously. Sometimes referred to as a data bank.
Database Host and Control System (DHCS): The *database* system which provides the primary interface to display *terminals* or *Personal Computers (PC)* used for the service positions in a *call centre*. It allows the process for handling many of the *incoming calls* and *outgoing calls* to be automated and is therefore an integral part of the *CTI* system.
Database Management System (DBMS): A *software* programme which helps in the creation, manipulation and maintenance of a *database.*

data buffer: A device which can be used to store *data* for short periods. For example the buffer could be used to store *traffic* which is being generated at a slow *data rate* and then to send it out in short bursts at higher speeds.

data bus: The *bus* used for the *transmission* of *data*.

data capture: The process of recognising *data* and storing it for future use. Data capture is commonly done in *data terminals* and computers.

data channel: A *channel* which can be used for the *transmission* of *data*. Usually data channels are considered to carry data in one direction only. See *data circuit*.

data circuit: Two *data channels* which combine to provide two-way communications. See *data channel*.

Data Circuit-terminating Equipment (DCE): Equipment which is used to connect a *Data Terminal Equipment (DTE)* to the line. A DCE may be a separate unit or incorporated into the DTE. It provides functions needed for *call establishment*, *call management* and *call disestablishment*. A modem is an example of a DCE, since it connects equipment to an *analogue* line. Several standards exist for the DCE-DTE interface, such as *RS-232*, *X.21*, etc. If specialised interfaces are needed, such as digital signals for connection to a *Public Data Network (PDN)* then the interface unit is known as a *Network Terminating Unit (NTU)*.

data collection: The process of collecting and storing *data* from various sources and of carrying out other actions, such as *encoding*, prior to *transmission*,

Data Communication Function (DCF): The function which handles the *data communications* within the *TMN functional architecture*.

Data Communication Network (DCN): Part of the *TMN physical architecture*, it is the *network* which supports the Data Communication Function (DCF). It consists of dedicated communications *paths* such as *X.25* and *voice channels*.

data communications: The transfer of *data* within a *network*, including all the functions associated with it, such as *encoding*, *transmission*, *error control*, *routeing*, etc.

Data Communications Channel (DCC): *Channel*, part of the *SONET* standard, which carries *information* such as performance monitoring and alarms.

data communications code: *Code* used during *data communications* to represent *characters*. Examples are the *International Alphabet No. 5* and the *ASCII*.

data compression: Reducing the amount of *data* which needs to be transmitted to convey *information*, by the use of data compression *algorithms*. This allows the same information to be represented by fewer *bits*, so reducing the requirements on the *transmission channel*. Exam-

Table D.1 Data compression standards

Standard		*Description*
ITU-T	V.42bis	Standard for data compression using BTLZ for throughput up to 400%
MNP	Level 5	Data compression/Throughput up to 200%
	Level 6	Data compression with V.29 technology/ Throughput up to 200%
	Level 7	Predictive data compression/Throughput up to 300%
	Level 8	Predictive data compression with V.29/ Throughput up to 300%
	Level 9	Predictive data compression with V.32/ Throughput up to 300%
	Level 10	Enhanced Level 9/Throughput up to 300%

ples of data compression *protocols* are the *V.42bis* and *MNP* used with *modems* and shown in Table D.1.

data concentrator: A device which enables many *data* sources to share a common *transmission medium*, where the number of *channels* on the medium is less than the total number of channels of all the sources combined.

data corruption: The act of violation of the *data integrity* of the *information.*

Data Country Code (DCC): A three *digit* number, part of the *Data Network Identification Code (DNIC)*, which is used on international numbers for *Public Data Networks (PDN)*, specified by the *ITU-T*. This allows a country to have several different data networks. For example the UK has been allocated codes 234 to 237, which enables it to have up to 40 different Public Data Networks.

data directory: Part of the *database* which contains an inventory of the elements on the *network*, such as a list of the data elements, names, ownership, usage, etc.

data encryption key: A *key* used for *encryption* and *decryption* of transmitted *data*, for security reasons.

Data Encryption Standard (DES): First public cryptosystem, developed by the American *National Bureau of Standards (NBS)*, now the *National*

Institute for Science and Technology (NIST). It is a *block cipher* with a 64 bit block length, involving both substitution and transportation, under control of a 56 bit *data encryption key*. It is specified in the *Federal Information Processing Standard (FIPS)* Publication 46, published in 1977.

data entry terminal: See *data terminal*.

data flow control: See *flow control*.

datagram: One of the three services provided in the *X.25 packet switching service*, the other two being *Permanent Virtual Circuit (PVC)* and *Switched Virtual Circuit (SVC)*. In the datagram system each *packet* contains the source and destination *address* so no *call setup* is needed, the *Packet Switched Network (PSN)* is left to route the packet to its destination. If the data to be sent can be confined to a single packet this method of transfer can be efficient. However there is no formal *acknowledgement* mechanism and the responsibility for *error correction* is placed on the *receiving terminal*. The *ITU-T Recommendation* for X.25 dropped support for the connectionless datagram method in 1984.

datagram network: A *network* which supports the *datagram* mode of *transmission*.

data integrity: A measure of the performance of the communications system, in that there are no undetected errors and none of the information content is missing.

data link: The *link* between two terminals which is used for *data communications* and which, usually, does not have any intermediate *switching equipment*.

Data Link Control (DLC): The *hardware* and *software* which ensures that *data transmission* over a *network* is error free. Several techniques are used, such as the addition of *parity bits* to the transmitted *frame*. See also *High level Data Link Control (HDLC)*.

data link controller: That part of the communicating system which carries out *Data Link Control (DLC)*.

Data Link Escape character (DLE): A *character*, used in *transmission control*, which changes the significance of a specified number of other characters which follow.

Data Link Layer: Layer 2 in the *OSI Basic Reference Model*, between the *network layer* and the *physical layer*. It carries out *synchronisation* and management of the *bit stream* to and from the physical layer and also *error detection*, but not *error correction*. The *ISO* 8886 standard provides the Data Link Layer Service Definition for OSI.

data mode: The state of a *Data Circuit-terminating Equipment (DCE)* when it is connected to the *data network* and is ready for *data transmission*.

data network: A *network* consisting of *data circuits* which are connected together and used for *data communications* between *Data Terminal Equipments (DTE)*.

Data Network Identification Code (DNIC): Part of the *ITU-T* international *numbering plan* for *Public Data Networks (PDN)*. It is a four *digit* number which contains a three digit *Data Country Code (DCC)*, which is followed by a digit which denotes the network number within the country.

Data Over Voice (DOV): A technique for transmitting *data* over the same *transmission medium* as normally used to carry *voice*. Usually both the data and voice are carried as *analogue signals*, and this can be achieved by devices such as *modems*. Commonly use in *centrex* or *PABX* systems. Data Over Voice is also necessary to carry data services over *twisted pair wire*, to residential *subscribers*.

data PBX: A *PBX* which has the ability to set up and switch data lines and not voice. Only *digital signals* are supported and connections are usually between computers and peripherals.

Dataphone Digital Service (DDS): Approved by the *FCC* in December 1974 as *data transmission* service in the USA, provided by AT&T (Bell). Figure D.1 shows the basic arrangement. It is an all synchronous facility, and *signalling* used is modified *bipolar coding*. A wide range of *data rates* are supported, such as 2.4, 4.8, 9.6 and 56 kbit/s.

data processing system: Any system which is used to manipulate raw *data* to produce an ordered result, such as sorting, storing, transmitting.

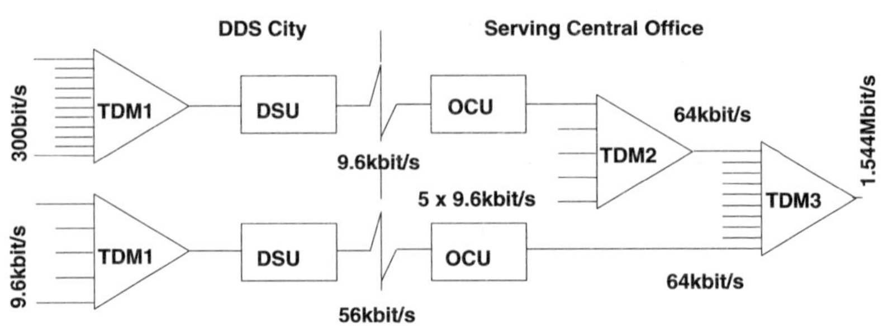

Figure D.1 DDS carrier structure and multiplexing arrangement

Usually it refers to those systems which operate automatically, such as computers.

data processing terminal: A *terminal* used to send or receive *data* in a *data network*.

data rate: A measure of the speed at which *data transmission* occurs. Normally specified in *bits* per second.

data security: Procedures which are used to ensure that the *data* is not, either intentionally or unintentionally, modified, deleted or disclosed to an unauthorised party.

data service: Services which are primarily concerned with data, rather than *voice* or *video*. Examples are *facsimile*, and *telegraphy*.

Data Service Unit (DSU): A device which connects *Data Terminal Equipment (DTE)* to the *digital transmission network*. It has facilities for *equalisation*, testing and *timing*. It may incorporate a *Channel Service Unit (CSU)*.

Data Set Ready (DSR): An *RS-232 signal* sent by the *DCE* to the *DTE* to indicate that the DCE is connected to the *network* and is ready to start the *handshake protocol* at the beginning of *transmission*. See also *Data Terminal Ready (DTR)*.

data signal: A *signal* which contains *data* and may include information for *error control*.

data signalling rate: The rate at which *data* signal elements can be generated, accepted or transmitted along a *transmission path*, measured in *bits* per second.

data signalling rate transparency: The ability of the *network* to interface *data terminals* which are operating at different *data signalling rates*.

data sink: A device which accepts *data signals* from a transmitting device. See also *data source*.

data source: A device which generates *data* for *transmission*. See *data sink*.

data stream: A sequence of *characters* and *bits*, representing *information*, transmitted through a *channel*.

Data Switching Exchange (DSE): *Switching equipment*, located in a *switching centre* and used to switch *data signals*. It can operate as a *circuit switched exchange* or a *packet switched exchange*.

data terminal: Any device which is used to transmit or receive *data* over a *network*. Examples are simple *RS-232* terminals and computers. See also *Data Terminal Equipment (DTE)*.

Data Terminal Equipment (DTE): A *data terminal* transmitting over a *data network* and operating according to a defined *protocol*. The DTE interfaces physically to the network through a *DCE* and it performs the higher level functions.

Data Terminal Ready (DTR): An *RS-232 signal* sent by the *DTE* to its *DCE* to indicate that it is ready for *transmission* and requesting that the DCE make connection to the *data network*. See also *Data Set Ready (DSR)*.

data traffic: *Traffic* which is composed of *data* rather than *speech signals*. It is also usually composed of *digital signals*.

data transfer: The movement of *data* from one location to another. In the *OSI Basic Reference Model* data transfer occurs within the *seven layers*, the unit of transfer being its *Protocol Data Unit (PDU)*. Data transfer can also occur by *connection mode transmission* or by *connectionless mode transmission*.

data transfer phase: The period, during a *call*, in which the *data transfer* occurs. It does not include other aspects of the call, such as *call establishment* and *call disestablishment*.

data transfer rate: The speed at which *data transfer* occurs, measured in *bits* per second, *blocks* per second, *characters* per second, etc. It only includes the main *information* part of the data, i.e. excluding *synchronisation*, *signalling*, *error control*, etc. It is also referred to as the *information rate*.

data transfer request signal: A *signal* sent by a *DCE* to its *DTE* to indicate that a distant *DTE* has requested *data transfer*.

data transmission: The general term used to cover all aspects of the *transmission* of *data* over a *data network*.

Data User Part (DUP): The level 4 *protocol*, part of the *ITU-T Signalling System No. 7*, which is used for *data transmission* and *switching*.

DAVIC: *Digital Audio-Visual Council*.

dB: *Decibel*.

dBa: A unit of *acoustic noise* measurement, referenced to approximately 3.16 picowatts (which is 0 dBa or -85 *dBm*.)

dBm: Unit of power, measured in *decibels (dB)*, which is referenced to 1mW of power. Therefore, for example, a level of +3dBm means that the *signal* under consideration is 3dB above 1mW (i.e. equal to 2mW).

dBm0: A unit of absolute power, measured at the point of zero relative transmission level. It is stated in *dBm*.

DBMS: *Database Management System*.

dBmV: A unit of absolute power in dB relative to 1mV across a 75 ohm resistor.

DBR: *Deterministic Bit Rate*.

DBS: *Direct Broadcast Satellite*.

DC: *Direct Current*.

DCA: *Dynamic Channel Allocation*.

DCC: *Data Communication Channel* or *Data Country Code*.

DCCE: *Digital Cell Centre Exchange*.

DCCH: *Digital Control Channel.*
DCDM: *Digitally Coded Delta Modulation.*
DCE: *Data Circuit-terminating Equipment*
DCF: *Data Communications Function.*
DC5 signalling: See *E&M signalling.*
D channel: *Channel* intended to carry *signalling* for *circuit switching* by the *ISDN.* In addition to signalling it may also be used to carry *packet switched information.* It uses a *layered protocol,* as specified in *ITU-T Recommendations* I.440, I.441, I.450 and I.451. For *Basic Rate Access (BRA)* the D channel occupies a *bandwidth* of 16 kbit/s and it carries signalling for the two *B channels.* For *Primary Rate Access (PRA)* the D channel occupies 64 kbit/s and carries signalling for 30 B channels (in the *E1* system) or 24 B channels (in the *T1* system).
DCM: *Digital Circuit Multiplication.*
DCN: *Data Communication Function.*
DCPSK: *Differentially Coherent Phase Shift Keying.*
DCR: Dynamically Controlled Routeing.
DCS: *Digital Code Squelch* or *Dynamic Channel Selection* or *Digital Cross-connect System* or *Digital Cellular System.*
DCT: *Discrete Cosine Transform.*
DC10 signalling: See *composite signalling.*
DDD: *Direct Distance Dialling.*
DDF: *Digital Distribution Frame.*
DDI: *Direct Dial Inward.*
DDN: *Defence Data Network.*
DDS: *Dataphone Digital Service.*
DDSN: *Digital Derived Service Network.*
DE: *Defect Event.*
decentralised control: Mechanism for controlling a *network* in which the control and *data processing systems* are distributed over several elements in the network rather than being concentrated in a few devices.
deception jamming: A *jammer* which uses a *jamming signal* which can simulate the characteristics of the system it is jamming.
decibel (dB): One tenth of a *bel.* It is the measure of relative power at two points. If P_1 is the power at one point and P_2 is that at another then dB = $10\log_{10}(P_1/P_2)$. If P_1 is less than P_2 then the value of dB is negative. The expression for the power can be stated in terms of voltage, current and resistance, if required. Table D.2 shows the value of decibels for different power and voltage ratios.
decimal digit: A *digit* which can have any one of ten values in the range 0–9. The decimal numbering system works to a base of 10.
decimetric wave: *Electromagnetic wave* having a *wavelength* of between 0.1 metre and 1.0 metre.

Table D.2 Power and voltage ratios at different decibels

Decibels	*Power ratio*	*Voltage ratio*
0	1	1
1	1.26	1.12
2	1.58	1.26
3	2.00	1.41
4	2.51	1.58
5	3.16	1.78
6	3.98	2.00
7	5.01	2.24
8	6.31	2.51
9	7.94	2.81
10	10.0	3.16
20	100.0	10.0
30	1000.0	31.6
40	10000	100.0

decimillimetric wave: *Electromagnetic wave* having a *wavelength* of between 0.0001 metre and 0.001 metre.

decipher: To convert text, which has been *enciphered*, back into *plain text*, by use of the appropriate cipher.

decision: One of the legal instruments available under the *Treaty of Rome* in order to enforce *European Community* regulations. Decisions are primarily issued by the *European Council* and are binding in their entirety on those to whom they are addressed.

decision feedback system: A system which provides *error control* and will generate an *Automatic Repeat Request (ARQ)*, if appropriate.

decision system: A system which can take autonomous action to produce an output based on the state of its input and on criteria which have been programmed into it.

decoder: **(1)** A device which converts a *signal* back into its original form, i.e. before it went through *encoding*. It is usually used to change a *digital* *signal* into its equivalent *analogue signal* or into another form of digital signal. **(2)** To convert text which has been through *encoding* back into *plain text*, using the appropriate *code*.

decollimation: Causing the rays within a beam of light to converge or diverge further, i.e. move further away from the parallel.

decryption: Converting text which has been through *encryption* back into *plain text* using the appropriate encryption key.

DECT: *Digital Enhanced Cordless Telecommunications*.

DECT Authentication Centre (DAM): A card used within *DECT handsets*, similar to the *Subscriber Identity Module (SIM)* used with *GSM*, for *user authentication*.

DECT protocol: *Protocol* used within *DECT*, as shown in Figure D.2, which indicates its relationship to the *OSI Basic Reference Model*. The *physical layer* carries out *modulation* and *demodulation*, *synchronisation*, etc. The *Medium Access Control (MAC)* layer selects *radio channels*, establishes and releases *links* and carries out *multiplexing* and *demultiplexing* of all *information* into packages. The *Data Link Control (DLC) layer provides reliable data links* to the *Network Layer* and it has two operational planes, C-plane and U-plane. The Network Layer is the prime *signalling* layer.

DED: *Dual Error Detection*

dedicated access: The connection of a *subscriber's* equipment, such as a *PABX*, to the *Interexchange Carrier's (IXC)* switch, without using a *Local Exchange Carrier (LEC)* facilities.

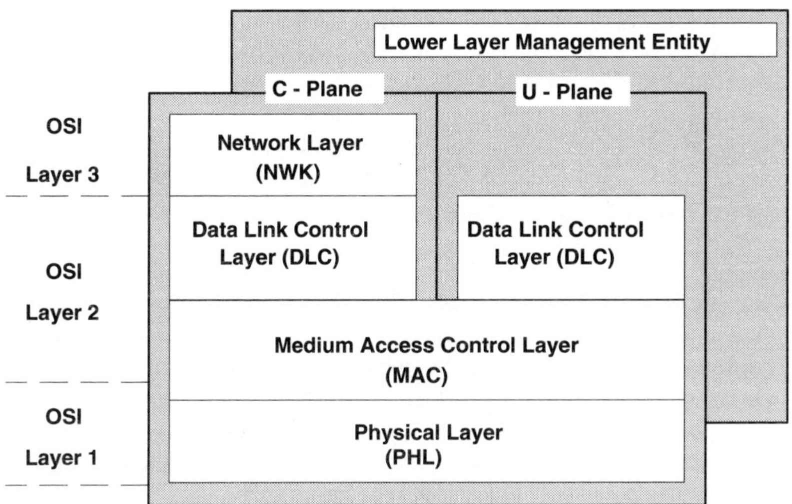

Figure D.2 DECT structure in relation to the OSI reference model

dedicated line: See *leased line*.

dedicated network: See *private network*.

dedicated service: A specific communications service which is provide to a group of users only, such as to the military, academic institutions, etc.

deemphasis: The process of modifying the characteristics of a received *signal*, such as its *frequency*, *amplitude*, *phase*, etc., so that it more closely resembles that of the signal when it was originally transmitted.

deep channel: *Channel* used for communications to and from deep space, i.e. space which is at a distance further from the Earth than the Moon.

de facto standard: A standard which is widely used and supported, but which has not been produced or adopted by any formal standard setting organisation.

default configuration: A configuration which the system, such as a *network*, will adopt when its parameters have not been specified.

Defect Event (DE): A fault, such as loss of *signal*, loss of *frame synchronisation*, etc. as specified in *ITU-T Recommendation* M.550 for digital circuit testing.

Defence Advanced Research Projects Agency (DARPA): US funding agency which devised TCP/IP in the late 1970s as the transport mechanism for use in the *ARPANET*.

Defence Data Network (DDN): A *packet switching network* used by the US *Department of Defence (DoD)* for worldwide secure and non-secure communications.

Defence Information Systems Network (DISN): The *network* interconnecting many of the US *Department of Defence (DoD)* networks in the USA, so as to form one coherent *Wide Area Network (WAN)*.

degradation: The condition which causes one or more of the system's parameters to fall outside of its rated values.

degraded minute: A measure of the quality of *transmission*. It is specified in *ITU-T Recommendation* G.821 which states that the objective is that fewer than 10% of one minute intervals are to have a *Bit Error Ratio (BER)* worse than 10^{-6}.

DEL: *Delete*.

delay distortion: The *distortion* which occurs when different *frequency* components of a *signal* propagate through a *transmission medium* at different speeds, and so arrive at the receiving end at different times.

delayed delivery: A *network* feature which allows some *data* to be temporarily stored so that it can be delivered at a later time. Usually delayed delivery would occur if the *called terminal* was *busy*, the temporarily stored data being delivered when it becomes free.

delay equaliser: A device used in a *transmission* system, to make its delays constant over a *frequency band* and so minimise *delay distortion*.

delay line: A circuit which is introduced into a *transmission medium* so as to deliberately delay a *signal* which is propagating through it. This is normally done for *timing* reasons.

delay modulation: A *signal modulation* technique which introduces different delays into the *signal.*

delay time: The time between a stimulus and its expected response. For example, during *call setup* it is the time following the end of the *calling signal* before the *called terminal* provides a response.

Delete (DEL): A *transmission character*, used in *codes* such as *ASCII*, to indicate that a transmitted character is to be deleted or that an operation is to be ended.

delimiter: Special *characters*, groups of characters, or *signals* which are used between *blocks* of *data* so as to separate them. Examples are the *End Of Text (EOT)* character and the *end of block signal*.

delivery confirmation: *Message* sent back to the *calling terminal* to confirm that the *data* has now been successfully delivered to the *called terminal.* This is often used if the *delayed delivery* feature has been applied.

Dellinger effect: The rapid loss or *fading* of the *radio wave transmission* in the *ionosphere* caused by solar storms. This effect can last from a few minutes to several hours.

Delta Modulation (DM): A *modulation* technique for converting *analogue signals* into *digital signals.* It is the limiting case of *Differential Pulse Code Modulation (DPCM)* in which the number of *bits* is reduced to one. The basic form of DM is illustrated in Figure D.3. (See also *Delta Sigma Modulation (DSM)* and *Adaptive Delta Modulation (ADM)*.) The analogue *baseband* input signal is fed into the comparator along with the signal from the integrator and this causes the pulse generator to provide

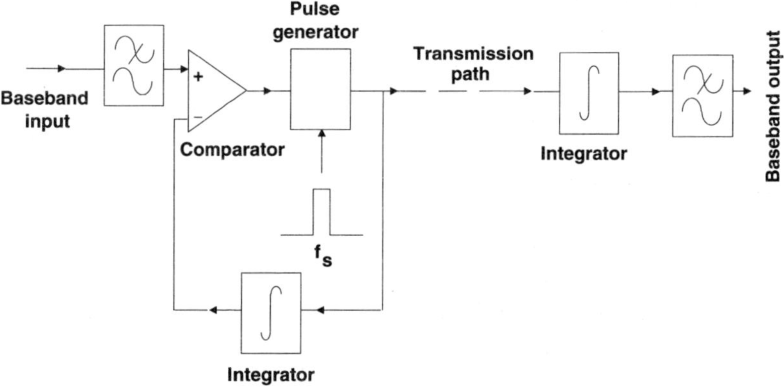

Figure D.3 Delta modulation system

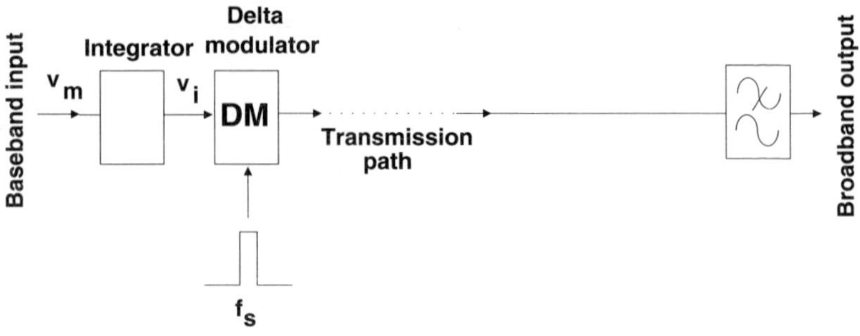

Figure D.4 Delta sigma modulation system

a positive (logical 1) output when the input voltage exceeds that from the integrator, and to transmit a negative, or zero, (logical 0) output when the input voltage is less. At the receiving end these positive and negative signals cause the integrator output to rise and fall by one step, so recreating the original input baseband signal.

Delta Sigma Modulation (DSM) An enhancement of the simple *Delta Modulation (DM)* technique in which the integral of the input *baseband signal* goes through *modulation* rather than the signal itself, as shown in Figure D.4. The integrator reduces the high frequency energy of the input signal and is therefore more suitable for use when the input baseband signal has a flat *frequency spectrum.*

Demand Assignment Multiple Access (DAMA): A *non-contention multiple access* method in which the communication *channel* is allocated to users as demanded, rather than on a fixed basis. There are several types of DAMA techniques, such as *polling*, *token passing*, *slotted ring*, and *register insertion ring*.

demarcation strip: The terminating point, usually a block, strip or board, which forms the interface between the equipment on a customer's premises and the lines entering the premises from the *PTO*.

democratic network: A *network* which several *clocks* are used but each clock has equal status, i.e. no one clock has control over the whole network. The network clock rate is the mean of the individual clocks involved. See also *despotic network*.

demodulation: The reverse process to *modulation* in which the output *signal* has the same characteristics as the original *modulating signal.*

demodulator: The device which carries out *demodulation*. It receives the *modulated carrier wave* and extracts from this the *modulating signal.*

demultiplexer (deMUX): A device which carries *demultiplexing*. Usually the same physical equipment can act as a *multiplexer* and a demultiplexer.

demultiplexing: The process of separating out the individual communication *channels*, which had been combined into one *circuit* during an earlier *multiplexing* process. It is therefore the reverse process to multiplexing.

deMUX:: *Demultiplexer.*

dense binary code: *Binary code* in which all the *bits* available are used in the code. For example if three bits are available then the equivalent dense binary code would have eight representations, i.e. 000, 001, 010, 011, 100, 101, 110, and 111.

Dense Wavelength Division Multiplexing (DWDM): *Wavelength Division Multiplexing (WDM)* technique which uses a large number of separate *wavelengths* for parallel communications along the same *optical fibre*. Usually the term DWDM is applied to systems having greater than four wavelengths although systems with over 64 wavelengths have been developed.

de-packetising: The process used in a *Packet Switched Network (PSN)* where the *codes* which have been added to the main *message* in the *packet*, to aid in *transmission*, are removed.

Department of Defence (DoD): A US government body which handles US military matters as well as its communications systems. It has been responsible for developing some commonly used *protocols*, such as *TCP/IP*.

Department of Defence master clock: The *clock* to which all time and *frequency* measurements for the US *Department of Defence (DoD)* are related.

Department of Trade and Industry (DTI): UK government department whose Telecommunications and Posts Division is responsible for many aspects of telecommunications. It acts as the UK Administration for Posts and Telecommunications matters, following the privatisation of its public operator (BT). Therefore the DTI represents the UK in *CEPT* and in the European Commission (Figure D.5). It also acts as the government department responsible for all UK legislation on telecommunications matters. The DTI represents the UK within *ITU-T*, especially at *Plenary Assemblies*, where it casts the UK vote approving *ITU-T Recommendations*. For each ITU-T Study Group the DTI organises a coordinating committee, which develops the UK view. (Figure D.6.) Individual members attending these Study Groups then support the UK view. A similar procedure is adopted for *ETSI*, except that individual members are more fully represented in ETSI committees. However the DTI uses coordinating committees to develop and sound out the UK view and then casts its national vote accordingly in the ETSI Technical and General Assemblies.

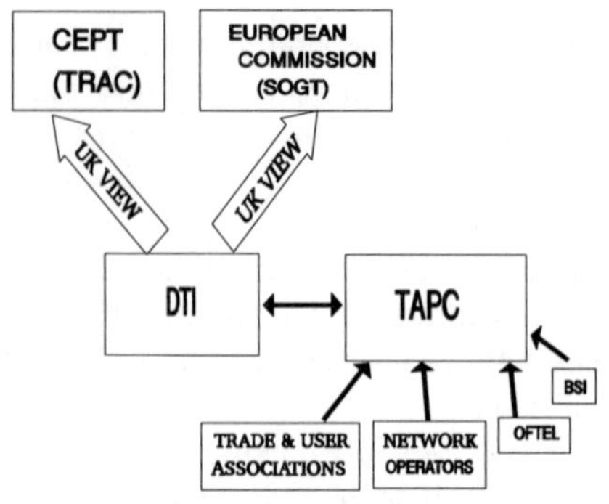

Figure D.5 DTI's coordination role for CEPT and EC

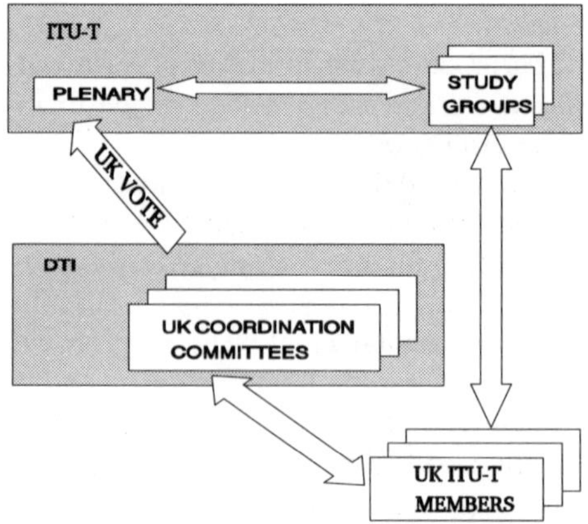

Figure D.6 DTI's coordination role for ITU-T representation

Department Of Transport And Communications (DOTAC): Australian government department which represents Australia in the *ITU*. It delegates coordination within *ITU-T* to the Australian ITU-T (formerly *CCITT*) Committee (*ACC*).

departure angle: The angle, in an *antenna* transmission pattern, between the horizontal and the main *lobe*.

depolarisation: The opposite of *polarisation*, i.e. the reduction of the polarisation of an *electromagnetic wave*.

deregulation: Government action which reduces the amount of regulatory control, for example, on the requirements for market entry. Deregulation is usually intended to promote competition and to result in *tariff* reductions, so benefiting the consumer.

DES: *Data Encryption Standard.*

descrambling: The opposite process to *scrambling*, i.e. restoring the *signal* back to the state it was in before it was scrambled.

despotic network: A *network* in which all the *timing* is derived from one *master clock*. See also *synchronised network* and *democratic network*.

despun antenna: In the case of an *antenna* mounted on a *satellite* which is rotating in its *orbit*, the position of the antenna is continuously adjusted, or despun, so that the *satellite footprint* on the Earth's surface remains fixed.

destination address: The *information* contained in a *message*, usually in its *header field*, which provides the complete *address* of the location where the message is to be delivered, e.g. the address of the *called terminal*.

destination address field: The *field* in the *message header* which contains the *destination address*.

destination terminal: The *terminal* which is to receive the *message*. Also referred to as the *called terminal*.

destuffing: The opposite process to *bit stuffing*, i.e. the removal of the stuffed bits from the original *signal*.

detection: **(1)** The process of sensing the presence of a *signal*, such as an optical sensor detecting the presence of light. **(2)** The process used in *demodulation*, involving the recovery of the *modulating signal* from the *modulated carrier wave*. See also *envelope detection* and *synchronous detection*.

Deterministic Bit Rate (DBR): One of the four *ATM Transfer Capabilities (ATC)* specified in *ITU-T Recommendation* I.371. It provides a *virtual circuit* with fixed *bandwidth*, emulating *circuit switching*.

deterministic routeing: *Routeing* within a *network* in which the routes between *nodes* are determined in advance, often by the use of *routeing tables*.

deterministic signal: A *signal* which can be predicted before it occurs, i.e. an explicit mathematical expression can be written for it. See also *random signal*.

Deutsche Institute for Normung (DIN): A private non-profit making association which was founded in 1917. It is the German authority for standards making and represents the country in *ISO*. DIN standards are voluntary but they are accepted by industry and government, and are usually made the requirement of purchasing agreements. All interested parties can participate in formulating DIN standards.

deviation: In a *Frequency Modulation (FM)* system it refers to the change in the *frequency* of the *carrier signal* when it goes through *modulation*. It is usually stated in kilohertz.

deviation ratio: In a *Frequency Modulation (FM)* system it is the ratio of the maximum *deviation* of the *carrier signal* to the maximum *frequency* of the *modulating signal.*

device control: *Characters* which are used to control remote equipment, such as turning it on or off, rather than carrying any *information.* Examples of device control codes are DC1, DC2, DC3, DC4, *X-on* and *X-off.*

device driver: *Software* and *hardware* components which allow a control and *data transfer* between *peripheral equipment* or other communications devices.

DFB: *Distributed Feedback LASER.*

DFM: *Dispersive Fade Margin.*

DFSK: *Double Frequency Shift Keying.*

DFT: *Discrete Fourier Transform.*

DGPS: *Differential Global Positioning System.*

DGPT: Direction Generale des Posts et Telecommunications. France's national regulatory authority, part of its Ministry of Post and Telecommunications.

DGT: Director General of Telecommunications. The UK's telecommunications regulatory authority.

DHCS: *Database Host and Control System.*

diagnostic test: Tests carried out to localise, detect, isolate and, sometimes, correct faults and malfunctions within *hardware*, *software* or systems.

diagnostic modem: A *modem* which is used to communicate *diagnostic test* information and other control *signals* with another modem.

dialback modem: A security feature in which a called *modem* will verify the identity of the calling modem (usually by use of a password) and will then disconnect it and dial it back.

dial backup: The term used to describe the use of the *PSTN* to back up a *private line* or a *leased line*, in the event of a failure of these lines.

dialling: The process of generating a *signal* to establish a connection to another *terminal.* Dialling is the first step in the *call setup* process.

dial mode: The state of a *DCE* in which its components, which are connected with *dialling*, are established to a communications *channel.*

dialogue: The interchange of *information* between two *terminals* which are connected to each other. This takes place according to a recognised *protocol*, which ensures orderly *call establishment*, *information* interchange, and *call disestablishment.*

dial pulse: A *pulse train* produced as a result of *dialling*. This is usually produced by breaking a *Direct Current (DC)* circuit, the number of pulses depending on the length of the *digit* which is dialled.

dial pulse signalling: *Signalling* using *dial pulses*. It takes its name from the older *rotary dial* telephones in which the dial is moved to the desired

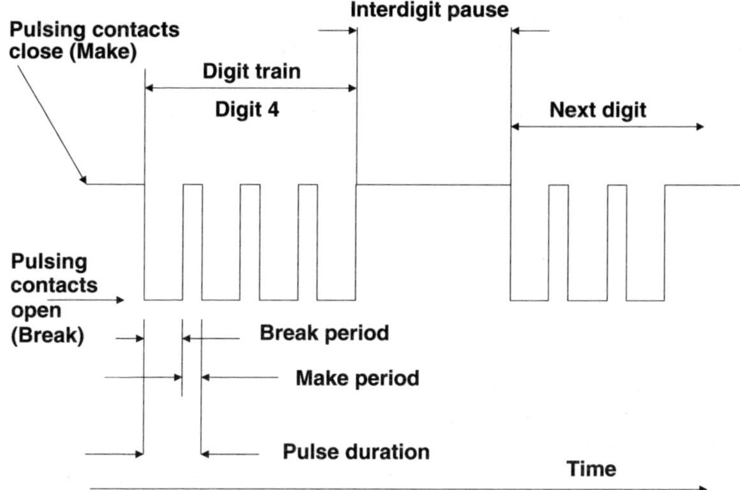

Figure D.7 Dial pulse signalling

number and on the return it breaks a DC circuit and generates the dial pulses, which signal the *address* of the *called terminal* to the *exchange*. The *pulse train*, shown in Figure D.7, operates at a rate of 10 pulses per second, with a tolerance of ±1 pulse. The make to break ratio is either 60% or 67%. The interdigital pause period at the end of pulsing, as the dial returns to rest, is greater than 400 ms and is usually between 600 ms and 800 ms. Also called *Loop Disconnect (LD) signalling*.

Dial Service Assistance (DSA): The *network* facility, connected with an *exchange*, in which an attendant is available to provide certain services, such as directory enquiries and random conferencing.

dial signalling: Applies to *dial pulse signalling* or to *Dual Tone Multifrequency signalling (DTMF)*.

dial through: A *network* feature which enables users of a *PABX* to make *outgoing calls* once the PABX has established the initial connection.

dial tone: A *signal* generated by the *exchange*, or a *PABX*, to indicate that it is ready to receive the *dial pulses* from the *calling terminal*. This tone can be heard in the earpiece of a *telephone*.

dial tone delay: The time interval between the *telephone* going *off-hook* and the *subscriber* hearing the *dial tone*.

dial tone first coin service: A payphone service in which *dial tones* are received as soon as the instrument goes *off-hook* and before any coins are inserted. This means that the operator can be contacted without inserting coins and coins must only be inserted when a *called terminal* answers.

Table D.3 Dibit operation

Dibits	*Phase*
00	+90
01	0
10	+180
11	+270

dial-up connection: Connection made between *terminals* using the *PSTN* so that a *line* is accessed via a *switching centre* for the duration of the *call* only and different lines are likely to be used for different calls. The alternative is a *leased line* connection which uses the same line for all calls.

dial-up network: Usually refers to the *Public Switched Telephone Network (PSTN).*

dial-up modem: A *modem* which is able to automatically make a *dial-up connection*. It can therefore dial other modems and use the *PSTN* for *data* transfer.

dibit: A set of two *bits*. If one *modulation* equals two bits, each modulation represents a dibit. An example of the dibit operation is given by *ITU-T Recommendation* V.22 which uses four phase changes for modulation as in Table D.3.

dictionary tree: Mechanism used in *modem* V.42bis *algorithms* for *data compression*. The algorithm identifies recurring *character* strings in the *data* and uses these *code* words to compile dictionaries, which are duplicated at both end of the *transmission line* (Figure D.8). As data is

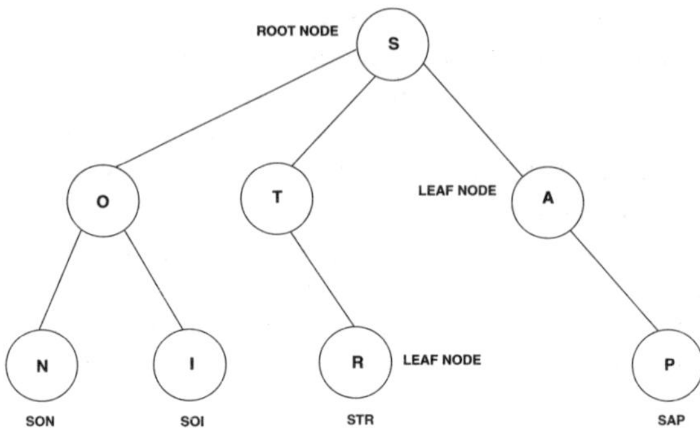

Figure D.8 V.42bis dictionary tree

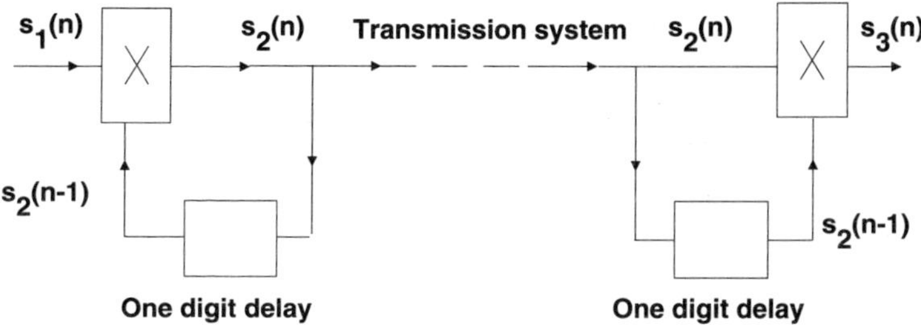

Figure D.9 Differential coding and decoding

transmitted the dictionary is searched using a tree structure. This shows a search on the code word 'STRING'. The initial character is read in the leaf node and then a search is started for the next character on the leaf nodes, and all the way down until a match is found.

DID: *Direct Inward Dialling*.

differential echo suppressor: An *echo suppressor* which operates by monitoring signals on both *channels* of a *four wire circuit.*

differential encoding: *Encoding* in which *signal* conditions are referred to changes from previous values rather than with reference to a fixed value. This is illustrated in Figure D.9. The output signal $s_2(n)$ is dependent on the input referred to the value of the previous signal, i.e. $s_2(n) = s_1(n)s_2(n-1)$. *Phase Shift Keying (PSK)* is an example of differential encoding since *information* is coded by the phase difference between successive symbols rather than the phase relative to some reference value.

Differential Global Positioning System (DGPS): An enhancement to the *Global Positioning System (GPS)* which uses error information of the *satellite* range to correct the positioning measurements.

Differentially Coherent Phase Shift Keying (DCPSK): See *Coherent Differential Phase Shift Keying (CDPSK).*

differential modulation: *Modulation* in which the condition of the *signal* at any time is dependent on the state of the previous signal element. *Delta Modulation (DM)* is an example of differential modulation.

Differential Phase Shift Keying (DPSK): A *Phase Shift Keying (PSK)* technique for signal *modulation* in which *differential encoding* is used, i.e. the *information* is conveyed by changes in phase between *digits* rather than by the phase deviation of each from a reference *carrier*. This is shown in Figure D.10. The *differential encoding* (coding and decoding) circuit of Figure D.9 is used and the output from this modulates the carrier signal.

Differential Pulse Code Modulation (DPCM): A *Pulse Code Modulation (PCM)* system in which *differential encoding* is used, i.e. the difference

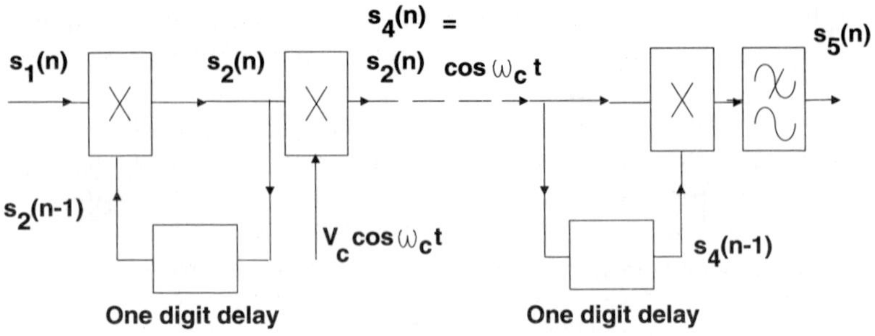

Figure D.10 Differential Phase Shift Keying

between each sample and the previous one is encoded and transmitted rather than encoding the actual value of each sample. In circumstances where a *signal* is not changing rapidly between samples this can result in each sample being represented by fewer bits (e.g. four instead of eight, as for PCM) resulting in greater efficiency. Figure D.11 shows the principle of DPCM. Each sample at the coder is the difference between the present value of the input *baseband signal* and the integrated value of the previous transmitted samples. At the receiver the output of the *decoder* is fed to the baseband output *channel* via a similar integrator.

differential signalling: *Signalling* in which the voltage levels in two wires are compared to derive the signalling state, rather than the voltage with reference to a fixed value. An example of differential signalling is RS-442.

Differential Trellis Coded Modulation (DTCM): *Trellis Coded Modulation (TCM)* in which differences in *phase* are used rather than the absolute phase value.

differentiating network: A *network* whose output *signal*, at any instant, is proportional to the rate at which the input signal is changing.

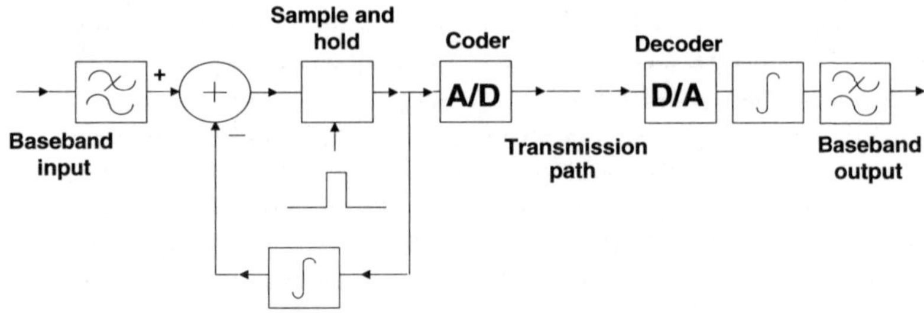

Figure D.11 Differential Pulse Code Modulation

diffraction: The bending or deflection of an *electromagnetic wave* when its path is restricted by a very small object. Usually these objects are no bigger than the *wavelength* of the electromagnetic wave itself.

digit: One of the elements selected from a set of whole numbers. For example any of the numbers 0 to 9 represents a digit, and either logical 0 or logical 1 represents a digit.

digital: Any item whose magnitude can vary discreetly only, rather than vary continuously (when it is known as analogue).

Digital Access and Crossconnect System (DACS): A switch, usually located in an *exchange*, which allows customer *circuits* at the *trunk* level to be routed. For example, these may be at the *E1* or *T1* level.

Digital Access Signalling System (DASS): A *signalling* system, used in the UK for *ISDN primary rate access*. It is a variant of *Channel Associated Signalling (CAS)* in which messaging is via *timeslot* 16 using *Link Access Procedure D (LAPD)*. Three types of *calls* are identified in DASS: category 1, which are calls that require end-to-end digital paths; category 2, which do not need this end-to-end *digital path*; and category 3, which can only be use for *telephony* and not *data*.

Digital Advanced Mobile Phone System (D-AMPS): A second generation version of *AMPS* which incorporates *digital* speech encoding with *Time Division Multiple Access (TDMA)*. This gives three digital voice *channels* for one 30 kHz radio channel, as used with AMPS. The D-AMP standard (IS-54) has enhanced the basic architecture and signalling protocol of AMPS, including a *short messaging service* and *authentication* for security control.

digital audio: The generation and *transmission* of *audio frequencies* in *digital* form.

Digital Audio Broadcasting (DAB): The *broadcasting* of *audio frequencies* in *digital* form, usually used in commercial radio or television.

Digital Audio-Visual Council (DAVIC): Established in August 1994 to increase the global compatibility between audio and *video transmission* systems. It is based in Geneva and has a membership of over 200 companies from more than 25 countries. It represents many aspects of the audio-visual industry, such as equipment manufacturers, service providers, research establishments and government agencies. The DAVIC aims to develop specifications for acceptance by formal standards bodies in the area of *multimedia* systems. Its goal is to provide interoperability across applications by defining 'tools' rather than full subsystems. DAVIC specifications consist of two parts: nominative and informative. Nominative parts need to be implemented in order for a subsystem to be able to claim conformity to the DAVIC specification, whilst the informative part primarily provides clarification of the nominative part and helps in its implementation. DAVIC members are represented in the General Assembly which elects the Board of Directors. It has three key Advisory Committees: Standardisation, Strategic Planning, and Management. All the technical work of the DAVIC is carried out by the Technical Committees which are supervised by the Management Committee.

Digital Cell Centre Exchange (DCCE): *Switching centre*, part of the BT *Integrated Digital Network (IDN)* used in the UK. It acts as a *local exchange* serving *subscribers* in a local area. Subscribers can connect to a DCCE either directly or via a *Remote Concentrator Unit (RCU)* if they are situated too far from the DCCE. Each DCCE has a *junction circuit* to at least two *Digital Main Switching Units (DMSU)*, to provide a level of protection in the event of a line fault.

Digital Cellular System (DCS): See *cellular radio system.*

Digital Circuit Multiplication (DCM): A variant of the analogue *TASI*, in which *digital speech encoding* is used at 64 kbit/s. It provides a four to five times increase in the *capacity* of a *E1* or *T1* system which uses normal *Pulse Code Modulation (PCM)*. DCM is used on several *transmission paths* where maximum use of capacity is important, such as on *submarine cables* like *TAT-8*.

Digital Code Squelch (DCS): *Signalling* system used in *Personal Mobile Radio (PMR)*. The DCS codeword is defined in MPT1318. It consists of a 223 bit *frame* which is transmitted continuously on the carrier frequency at 134.4 bit/s. The codeword is generated from a cyclic *Golay code* and is shown in Figure D.12.

digital connection: A *digital path* set up between two *terminals.*

Digital Control Channel (DCCH): The *signalling channel* used within *cellular mobile radio* and *Personal Communications Service (PCS)* systems.

Digital Cross-connect System (DCS): See *Digital Access and Cross-connect System (DACS).*

digital data channel: A *channel* which carries *data* in *digital* form.

digital data link: A *link* which carries *data* in *digital* form.

digital data transmission: *Data transmission* in which all the *data* is transmitted in *digital* form.

Digital Derived Services Network (DDSN): An early implementation, by BT in the UK, of an *Intelligent Network (IN)*, as shown in Figure D.13. It consists of 10 *Digital Derived Service Switching Centres (DDSSC)*

Figure D.12 Components of a DCS codeword

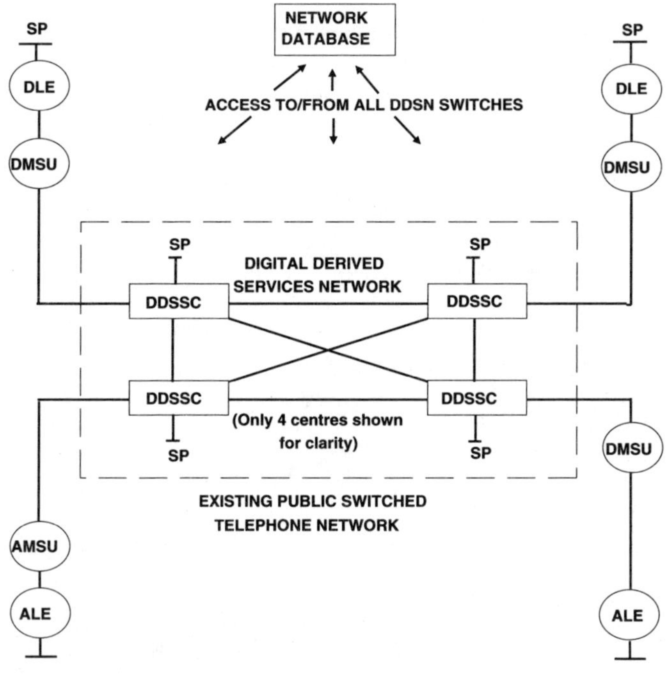

Figure D.13 Digital Derived Services Network

which perform the functions of a *Service Switching Point (SSP)* and are connected to the *Digital Main Switching Units (DMSU)* of the *trunk network*. This network provides IN facilities, such as *premium rate service* and *freephone service*.

Digital Distribution Frame (DDF) A *distribution frame* for connection of *digital circuits*. See also *Automatic Digital Distribution Frame (ADDF)*.

Digital Enhanced Cordless Telecommunications (DECT): A standard for a cordless telephony system, developed by *ETSI*. It was originally known as the Digital European Cordless Telecommunications system. The specification has two levels: the *Common Interface*, which allows conforming equipment to work with each other, and the coexistence interface, which allows equipment designed to proprietary standards to

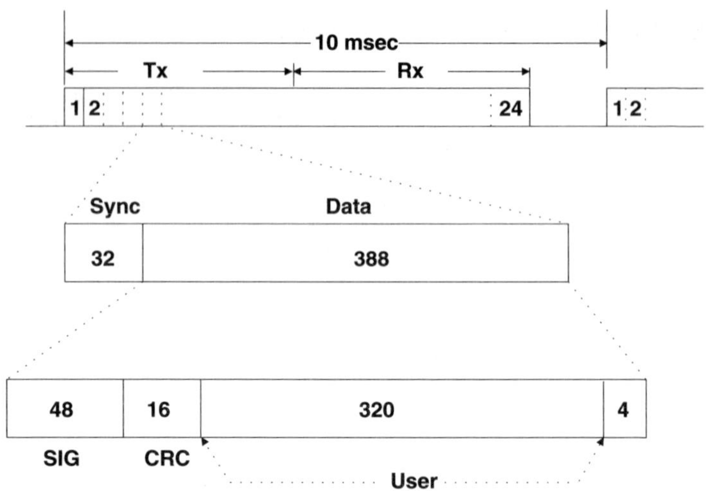

Figure D.14 DECT frame structure

coexist in the same *frequency spectrum*. DECT provides speech bearers by using 32 kbit/s *ADPCM* to *ITU-T Recommendation* G.726. It operates in the 1880 MHz to 1900 MHz band and provides 120 *duplex channels*. Figure D.14 shows the DECT *frame* structure. It is a multiple carrier, *Time Division Multiple Access (TDMA)* system which uses 10 channels at 1.728 MHz intervals, with 12 duplex channels multiplexed onto each carrier and a transmitted *bit rate* of 1152 kbit/s. See also *DECT protocol*.

digital exchange: An *exchange* which uses *digital switching equipment* and in which all *traffic* is in the form of *digital signals*.

digital filling: See *bit stuffing*.

digital hierarchy: Refers to the hierarchy of *transmission channels*. For the *Plesiochronous Digital Hierarchy (PDH)* this is shown in Table D.4 for both the North American (USA) and European systems. Table D.5 shows the values for the *Synchronous Digital Hierarchy (SDH)* up to the STM64 rate.

digital leased line: A *leased line* which is to carry *digital signals*.

digital loopback test: *Loopback test* applied to *digital* equipment carrying *digital signals*.

Digital Loop Carrier (DLC): *Network* equipment and services, forming part of the *local loop*, which is used for *multiplexing* several *telephone channels* on to a common *line*.

Digitally Coded Delta Modulation (DCDM): *Delta Modulation (DM)* technique in which the step size is controlled by the *bit* sequence produced by the *sampling* and *quantisation*.

Table D.4 European and American digital hierarchy

Order		*Number of telephone channels at 64 kbit/s*		*Total bit rate (Mbit/s)*	
Europe	*USA*	*Europe*	*USA*	*Europe*	*USA*
0	DS0	1	1	0.064	0.064
1	DS1	30	24	2.048	1.544
2	DS1-C	120	48	8.448	3.152
3	DS2	480	96	34.368	6.312
4	DS3	1920	672	139.264	44.736
5	DS4	7680	4032	565.148	274.176

Table D.5 Synchronous Digital Hierarchy

STM rate	*Transmission rate (Mbit/s)*
1	155.52
4	662.08
8	1244.16
12	1866.24
16	2488.32
64	9953.28

Digital Main Switching Unit (DMSU): The main *switching centre* used on the BT *Integrated Digital Network (IDN)* in the UK. It forms part of the *trunk network* which contains only one level of switching centres, which are fully interconnected for security. The DMSU connects to the *local network* through a *Digital Cell Centre Exchange (DCCE).*

digital microwave: A *microwave* system which uses *digital modulation* of its *carrier*, which is operating at microwave *frequencies.*

digital modulation: The term is often applied to either the *modulation* of an *analogue carrier* by a *digital baseband signal* or to the modulation of a digital carrier by an analogue baseband signal. Examples of the first case are *Amplitude Modulation (AM)*, *Frequency Modulation (FM)*, or *Phase Modulation (PM)*, and it is used in applications such as the *transmission* of digital *data* over analogue lines, using *modems*. An important example of the second case is *Pulse Code Modulation (PCM)*, and it is used for digital transmission in *telephony*.

digital multiplexer: A *multiplexer* which multiplexes *digital signals*, usually using *Time Division Multiplexing (TDM)*.

digital multiplex hierarchy: The *digital hierarchy*, built up by *multiplexing* signals together.

digital path: The *path* set up between two *terminals* and which is carrying *digital signals*.

Digital Private Network Signalling System (DPNSS): A system used for *signalling* between two *PABXs*, first developed for use in the UK in 1983. It is similar on *Common Channel Signalling (CCS)* with 2 Mbit/s links using *ITU-T Recommendation* G.703 interface with signalling information carried on a common 64 kbit/s *channel* in *timeslot* 16.

digital repeater: A *repeater* which regenerates *digital signals*.

digital section: That part of a *transmission medium* which has been optimised to carry *digital signals*.

Digital Sense Multiple Access (DSMA): A *multiple access* technique, similar to *Carrier Sense Multiple Access (CSMA)*, used for cellular data communications.

digital signal: A *signal* which is made up of discrete values, which are usually coded to carry the required *information*. See also *analogue signal*.

Digital Signal level: Term used in the USA to describe the different levels in the *digital hierarchy*. These are Digital Signal level 0 (*DS-0*), Digital Signal level 1 (*DS-1*), etc. See Table D.4.

Digital Signal Processing (DSP): The processing of *digital signals* which were originally *analogue signals* and have been converted into digital form. DSP is widely used for carrying out many tasks within telecommunications, such as *echo cancellation*, *voice compression*, etc.

Digital Speech Interpolation (DSI): A technique for increasing the number of simultaneous *voice* conversations which can be carried on a *transmission medium*, by *multiplexing* in which *information* from a different user is interleaved into the silent periods of a conversation. It is a form of *voice compression*. See also *TASI*.

Digital Subscriber Line (DSL): The collective name given to a number of techniques used for transmitting *digital data* over the *local loop* or *subscriber line*. These are also know collectively as xDSL. Examples are *ADSL*, *HDSL*, *VDSL*, *MDSL*, and *RDSL*.

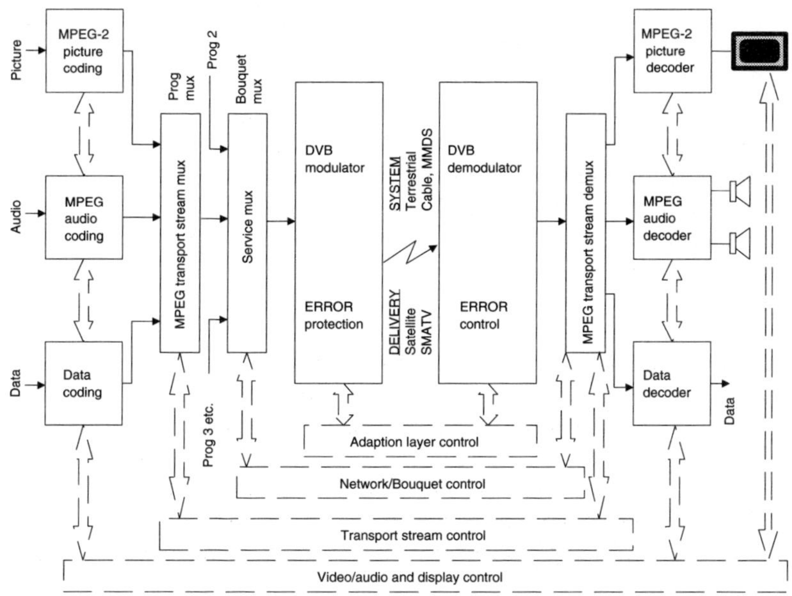

Figure D.15 Simple Digital to Analogue Converter

digital switch: A switch which is used to route *digital signals* and which normally operates in a *Time Division Multiplex (TDM)* mode.

digital switched circuit: A *switched circuit* which carries *digital signals* and uses *digital switches.*

digital telephone: A *telephone* which has a *codec* built into the *handset* so as to convert *speech signals* (*analogue signals*) to *digital signals* and vice versa.

digital terminal: A *terminal* which can transmit and receive *digital signals.*

Digital to Analogue Converter (DAC): A device which converts a *digital signal* into an *analogue signal* which has the equivalent information. Figure D.15 shows a simple circuit which converts a digital input into an analogue output. The resistors to the summing *amplifier* have a *weighting* which is binary. See also *Analogue to Digital Converter (ADC).*

digital traffic: *Traffic* which is in the form of *digital signals.*

digital transmission: *Transmission* in which *digital signals* are involved.

Digital Video Broadcasting (DVB): European project to develop standards for the *broadcasting* of *digital video signals.* The aim has been to

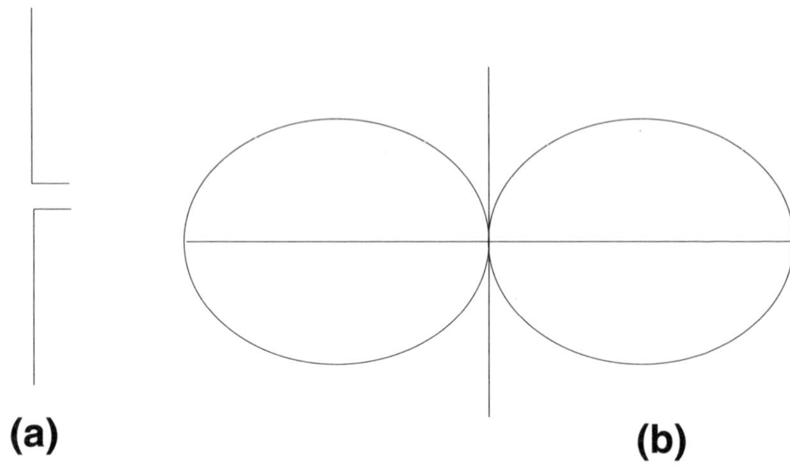

Figure D.16 Digital Video Broadcasting

provide specifications for all forms of programme delivery. The following are some of the specifications available: DVB-B Satellite Broadcasting (ETS 300 421); DVB-C Cable Transmission (ETS 300 429); DVB-CS Transmission via *SMATV* (ETS 300 473); DVB-T Terrestrial Broadcasting (ETS 300 744); DVB-MS *MVDS* Broadcasting; DVB- SI Service Information (ETS 300 468); DVB-TXT Fixed Format *Teletext* (ETS 300 472). Figure D.16 shows the overall concept of Digital Video Broadcasting, as developed by the DVB Project.

digital video coding: The coding of *video signals* for *transmission*. Several techniques are used, such as *PCM*, *DPCM*, *VLC*, *transform coding*, *JPEG*, Vector Quantisation (VQ), *MPEG*, etc.

digitise: To convert a *signal* into a *digital* form using the appropriate *code*, such as *PCM*.

digitised voice: *Voice frequency signals* which have been *digitised* and converted into *digital* form, usually ready for *transmission* or storage.

digit rate: The number of *digits* which are transmitted in a given time.

digroup: The basic digital multiplexing group. For North America and Japan it is *T1*, i.e. 1.544 Mbit/s supporting 24 *voice channels*, and for Europe it is *E1*, i.e. 2.048 Mbit/s supporting 32 voice channels, including overheads.

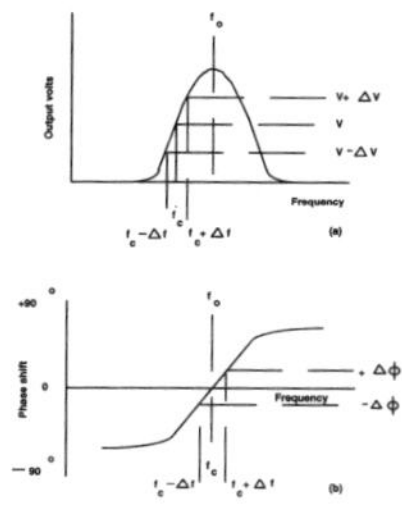

Figure D.17 Dipole: (a) construction; (b) polar pattern

DIN: *Deutsche Institute fur Normung.*

dipole: Usually refers to an *antenna* which consists of a straight section, fed in the centre, and having an electrical length equal to one half *wavelength* at the operating *frequency*. Figure D.17(a) shows the constructruction of a dipole and Figure D.17(b) its electrical field plotted as a *polar diagram.*

Direct Broadcast Satellite (DBS): The service which uses *satellites* to *broadcast* television and radio programmes down to Earth based *antennas*. These antenna may be centrally located, the signals then being conducted by cable to individual homes. Alternatively these can be small dish antennas located on individual homes and buildings. A relatively large number of *digital* broadcast channels can be supported, in the region of 150–200.

direct call: A *network* facility which allows a *call* to be established with a predetermined *address* so that the *call setup* time is considerably reduced.

Direct Current (DC): Current which has a single direction of flow. See also *Alternating Current (AC).*

Direct Current signalling: *Signalling* in which the *Direct Current (DC)* in the line if broken and reconnected as a consequence of *dialling*. See also *E&M signalling*, *loop disconnect signalling* and *composite signalling*.

Direct Dial Inward (DDI): A facility, available on *centrex*, or a *PABX*, which allows callers from outside a company to call an extension directly, without first going through an operator. Also referred to as *Direct Inward Dialling (DID).*

direct dialling PABX: A *PABX* which supports the *Direct Dial Inward (DDI)* feature.

direct dialling service: A *network* feature which allows a caller to make some special types of *calls*, such as collect calls and credit card calls, without operator intervention.

Direct Distance Dialling (DDD): A term used in the USA to describe the *network* feature which allows users to make *long-distance calls* without operator intervention. In the UK this is referred to as *Subscriber Trunk Dialling (STD)*.

Direct Inward Dialling (DID): See *Direct Dial Inward (DDI)*.

Direct Inward System Access (DISA): A feature of a *PABX* which allows external callers to dial into it and, after providing the appropriate password, make use of all the PABX facilities. The most common use of DISA is employees making *local calls* into the PABX, from home, and making use of the cheaper lines of the company to initiate *long-distance calls*.

directional antenna: An *antenna* in which the bulk of the radiation pattern is confined to one direction. It is used in point-to-point communications.

directional coupler: A device which can be used to sample either the forward or backward wave during *transmission*. A unidirectional coupler can only sample waves in one direction.

directional selectivity: For an *antenna* it is represented by its *radiation pattern*, and is a plot of the relative strength of the radiated field as a function of the angle.

direction finding: The process of locating an object, such as a source of *electromagnetic radiation*, by noting the strength of the *signals* from two or more directions.

Directive: Refers to one of the legal instruments available under the *Treaty of Rome* to enforce *European Community (EC)* regulations. The aims of all Directives are binding on Member States, but the method of attaining the aims are left to the legal system within the individual Member States.

direct line: A *line* which connects two switchboards without passing through any *switching centre*.

Directorate General: The *Commission of the European Community* is divided into twenty-three *Directorate Generals*, each with specific responsibilities, as given in Table C.1.

direct orbit: A *satellite orbit* in which the projection of the orbit on the plane of the equator moves in the same direction as the Earth's rotation.

Directory Access Protocol (DAP): Protocol used for communications between the Directory Service Agent (DSA) and the Directory User Agent (DUA), as defined in the *ITU-T Recommendation* X.500. This is used for *Electronic Data Interchange (EDI)*.

directory assistance: Service provided by *PTOs* for obtaining the *telephone numbers* of other *subscribers*.

directory number: The *telephone number* of a *subscriber*, often listed in a printed telephone directory.

directory service: **(1)** A standard for storage of information on *network* resources, as specified by the *ITU-T Recommendation* X.500. **(2)** Standards for the electronic supply and update of information, consisting, for example, of telephone and telex numbers. It is part of the *application layer* specified in the *OSI Basic Reference Model*.

Direct Outward Dialling (DOD): A facility of a *PABX* which allows users internal to the organisation to make outgoing *calls* without operator intervention. Usually a preliminary number has to be dialled, such as a 9, and call baring can be used on certain extensions to prohibit *outgoing calls*.

Direct Sequence Spread Spectrum (DSSS): A *spread spectrum* technique, which is used to combat interference, such as from a *jamming signal* and to make it more difficult to detect or demodulate the signal, for privacy. In this technique the sender uses a *pseudo-random* noise *code* sequence which is clocked at a rate which is much higher than that of the *information* being transmitted. This is called the chipping rate and the resultant chopped signal is used to modulate a fixed frequency *carrier*. This spreads the *bandwidth* of the original *data* over a much wider *frequency range*, for *transmission*. The receiver sees this wide frequency range, but uses an identical pseudo-random noise sequence to autocorrelate and determine the desired signal.

direct service circuit: A *circuit* which links two *terminals* and is reserved for their exclusive use.

direct signalling: *Signalling* in which the signalling conditions are applied directly to the speech (or associated) *lines*. Examples are *loop-disconnect signalling*, *MF signalling*, and *E&M signalling*.

Direct To Home (DTH): Term used to describe the delivery of multichannel sound and *video* programmes direct to *subscriber's* homes, usually either using *Cable Television (CATV)* or *Direct Broadcast Satellite (DBS)*.

DIS: *Draft International Standard*.

DISA: *Direct Inward System Access*.

disabling signal: A *signal* which is sent to stop an event from continuing or occurring.

disconnect signal: A *signal*, usually sent in a *switched network*, to indicate that the established connection is to be broken. This signal is usually sent from one end of a *circuit* to the other end.

Discrete Cosine Transform (DCT): A mathematical technique for the *compression* of *video signals*.

Discrete Fourier Transform (DFT): Mathematical technique which is widely used in *Digital Signal Processing (DSP)* applications, and for the detection of periodic signals.

Discrete Multitone (DMT): An *ADSL* modulation technique. It distributes the incoming *data* across many individual *carriers*, resulting in 256 discrete sub-channels between 26 kHz and 1.1 MHz. It has good performance since each channel can vary the *data rate* to avoid specific noise areas. See also *Carrierless Amplitude Phase modulation (CAP)*.

discriminating ringing: A feature of a *PABX* in which different *ringing tones* are used for different numbers. Usually this uses different tones for *internal calls* and *external calls*, so the *called party* can determine the type of call before answering it. Also known as *distinctive ringing*.

discriminator: A circuit used for *detection* in *Frequency Modulation (FM)*. It works by converting frequency variations in the input *signal* to variations of *amplitude* or *phase* and then putting these through *demodulation*. Figure D.18 shows the principle of a tuned circuit discriminator. In Figure D.18(a) the resonant frequency of the circuit is offset from the centre frequency of the incoming FM signal and a deviation of the input frequency varies the output voltage. This can then be detected by an *envelope demodulator* to give the *baseband* output voltage. For the

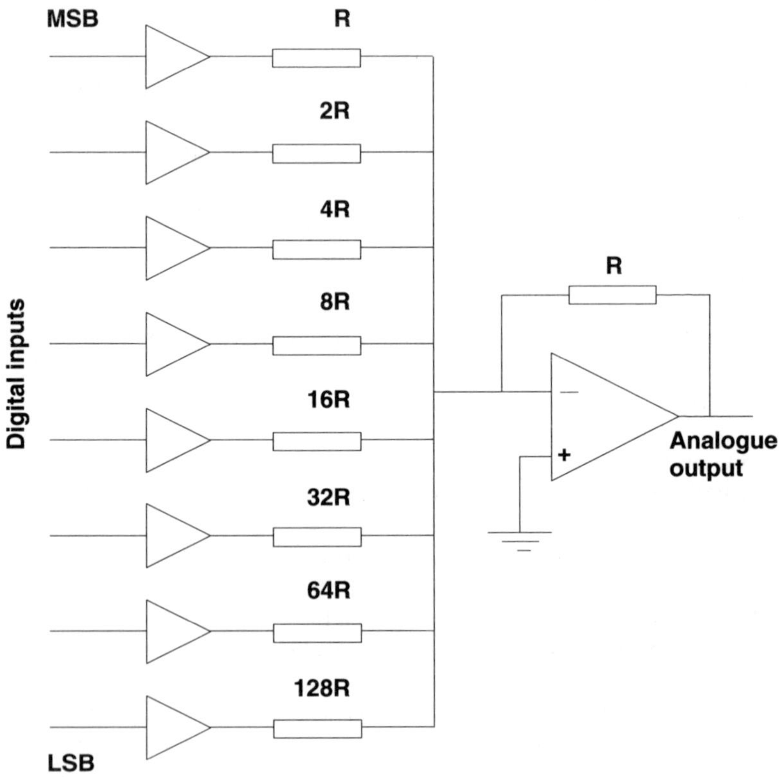

Figure D.18 Tuned circuit discriminator: (a) frequency to amplitude conversion; (b) frequency to phase conversion

frequency to phase conversion of Figure D.18(b) a phase shift is produced depending on the input frequency deviation, and the baseband can be recovered from this by using a *coherent demodulator.*

dish: Common name applied to a parabolic *antenna* used to receive *electromagnetic waves*, for example from a *satellite* transmitter or a *microwave* or *radar* system.

DISN: *Defence Information Systems Network.*

dispersion: The spreading effect of a pulse of light which results in *distortion* and reduces the rate at which the light can be transmitted in the *transmission medium.* Dispersion can be caused by several effects, in *optical fibre* they can be grouped as *intramodal dispersion* and *intermodal dispersion.*

dispersion coefficient: A measure of *dispersion.* It is given as a function of the *spectral width* and length along the *transmission medium*, and is usually expressed in picoseconds/(kilometre.nanometre).

dispersion compensator: A device which compensates for the *dispersion* which occurs in the *transmission medium.* An example of a fibre grating compensator is shown in Figure D.19. In travelling through the transmission medium the longer *wavelengths* are delayed more than the shorter wavelengths. When this pulse of light enters the dispersion compensator the longer wavelengths are reflected from the chirped grating earlier than the shorter wavelengths, which delays the shorter wavelengths and so recompresses the pulse, as shown.

Dispersion Shifted Fibre (DSF): An *optical fibre* in which the dispersion *wavelength* has been shifted, by appropriate design, to the region of minimum attenuation. Dispersion Shifted Fibre is widely used in *Wave-*

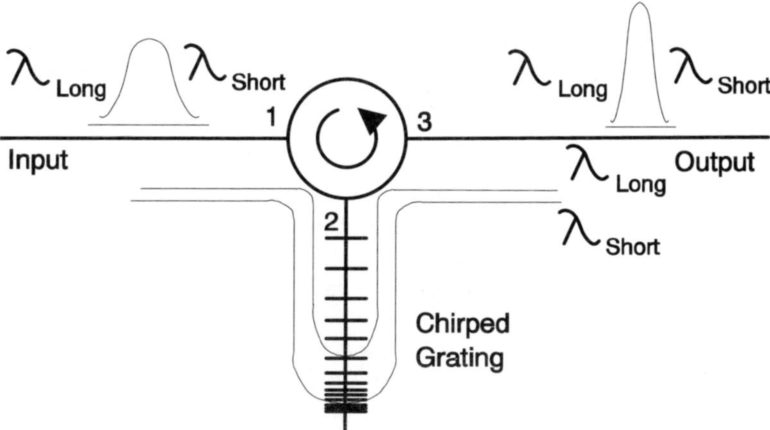

Figure D.19 Fibre grating dispersion compensator

length Division Multiplexing (WDM) systems. See also *Non-Dispersion Shifted Fibre (NDSF)*.

Dispersive Fade Margin (DFM): A concept used to allow the outage time in analogue radio systems, caused by *fading*, to be calculated. This can be done by substituting the Dispersive Fade Margin for *fade margin* in the relevant equations, such as given in *Bellocre* Technical Advisories.

dispersive fading: See *multipath fading*.

display: A method of representing *information* in visual form. Usually refers to the device which will show the display, such as a *Visual Display Unit (VDU)*.

distinctive ringing: See *discriminating ringing*.

Distinguishing Name (DN): In a *Management Information Tree (MIT)*, as defined in the *OSI Basic Reference Model*, each *managed object* can take its name relative to the object at the head, or root, of this tree. This name is called its Distinguishing Name. See also *Relative Distinguishing Name (RDN)*.

distortion: Any unwanted change in the characteristics of an entity. Usually it refers to the change in the value, shape or timing of a *signal* as it moves through a *transmission medium*, caused by unwanted changes in its *amplitude*, *frequency* or *phase*. See also *amplitude distortion*, *aperture distortion*, *barrel distortion*, *bias distortion*, *characteristic distortion*, *delay distortion*, *frequency distortion*, *phase distortion*, *pulse duration distortion*, *quantisation distortion*.

distress call: Internationally recognised *signals* which are sent before an actual message requesting help. The distress call includes the *address* of the station which is in distress and is usually *broadcast* over *frequency bands* reserved for this purpose. Examples of distress calls are SOS and MAYDAY.

Distributed Adaptive Routeing (DAR): A *dynamic routeing* technique in which the *routeing* through the *network* is based on *information* which is interchanged between *nodes* connected to the network. This information is updated as network conditions change. In operation each network condition is assigned a 'value' based on criteria such as the available *bandwidth* on the *link*, the transit delay, the number of links in the route, congestion at any of the nodes, etc. This information is stored at the nodes in *routeing tables* which are constantly updated as network conditions change. Routeing decisions are based on the information held in these tables.

distributed control: The state of a system or *network* where control of the elements occurs from several points, distributed around the system or network, rather than from one central point.

distributed data processing: A *data processing system* in which the key elements of the system, such as data storage, system control, and pro-

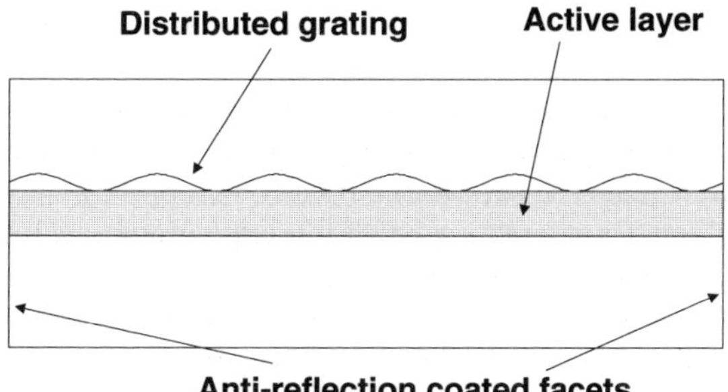

Figure D.20 Distributed Feedback Laser

gramme execution, are carried out within several processing units distributed across the *network*.

Distributed Feedback LASER (DFB): A *LASER* in which the optical feedback is not provided by mirrors, as in a conventional LASER, but by a grating built into the LASER which provides continuous feedback along the whole length of the device, as in Figure D.20. This results in a virtual singlemode LASER because the other modes are well below the fundamental.

distributed frame alignment signal: A *frame alignment signal* which is distributed over several *timeslots*.

distributed network: A network which has *distributed control* and in which the key elements, such as the *switching equipment*, are distributed at several points over the network, often being placed close to the *subscribers*.

Distributed Queue Dual Bus (DQDB): An *IEEE* standard (802.4) for a *Metropolitan Area Network (MAN)*. It comprises two contra-directional *data buses*, as shown in Figure D.21, which can be up to 150 km long

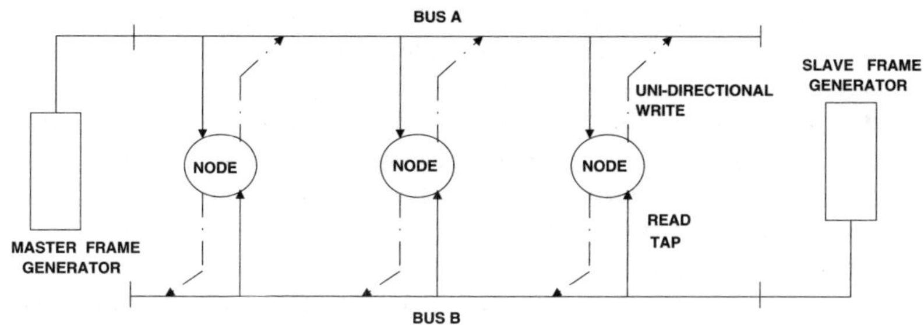

Figure D.21 Distributed Queue Dual Bus architecture

and operate at a *data rate* up to 140 Mbit/s. The bus can be configured into a *ring topology* for greater resilience. DQDB MAN is used in the *Switched Multimegabit Data Service (SMDS)*.

distributed switching: See *distributed network*.

Distributed Transaction processing (DTP): Part of the common application layer standards within *OSI*, it provides functions which enable distributed transaction-oriented processing between elements in the *open system architecture*. It uses *CCR* as an underlying service provider.

distribution frame: The device used to terminate *cables*, which also provides other facilities, such as inerconnection between sections of cable, test point, and overvoltage protection. See also *Main Distribution Frame (MDF)*, *voice distribution frame* and *Building Distribution Frame (BDF)*.

distribution service: Part of the *ITU-T Recommendation* I.211, which provides a *service classification*. Distribution services are subdivided into those with user control and those without user control. Examples of services without user control are *broadcast* television or radio. In distribution services with user control the user has presentation control, for example selecting the material for viewing from a broadcast. Examples are weather forecasts and stock price broadcasts.

distributor: A device which accepts *signals* from one *line* and places them on to several other lines. Distributors are commonly used in *switching systems* and forms one of the three key functions of *switching*. *Calls* first enter the *concentrator* where they are combined onto a smaller number of lines before being passed to a distributor. The distributor switches these onto the appropriate *trunks* and the switched calls then go to an *expander* for the final connection to the appropriate *subscriber* line.

District Switching Centre (DSC): An *exchange*, part of the BT *analogue network*, as illustrated in Figure A.15, which provided *switching* at the secondary level. It was also called a trunk transit exchange.

diurnal variations: Variations of events or *signal* conditions on a daily basis or on the time of day. For example, changes in the *ionosphere*.

diverse routing: See *route diversity*.

diversity: The use of different methods or systems. Usually diversity is used to improve reliability, such as in *route diversity*.

diversity gain: The *gain* obtained by *diversity*, measured as the ratio of the *field strength* of the *signal* obtained by diversity to that obtained without diversity, i.e. when the signal passes through a single *path* or *channel*.

diversity transmission: A *transmission* system in which two separate *carriers* are *modulated* by identical *modulating signals* and these are then transmitted using *route diversity* so that when recombined it provides an improved signal.

divestiture: AT&T in the USA previously owned the whole of the US telecommunications network, the Bell system. It also owned a manufacturing subsidiary (Western Electric) which produced equipment exclusively for the Bell system. In 1982 a ruling was made by the District Court for the District of Columbia that the regional Bell system had to be split off from AT&T. This became effective on 1 January 1984 and is known as the divestiture of AT&T. Following divestiture the twenty-two local operating companies of the Bell system were formed into seven regional holding companies, often referred to as the *Regional Bell Operating Companies (RBOC)* or 'Baby Bells'.

D layer: The layer of the *Earth's atmosphere*, closest to the Earth's surface, where ionisation has occurred due to the effects of solar radiation. It is therefore not present at night. The D Layer effects *radio communications* by reflecting *frequencies* below about 50 kHz. Higher frequencies are subject to *absorption* and *attenuation*.

DLC: *Data Link Control* or *Digital Loop Carrier*.

DM: *Delta Modulation*.

D-MAC: A variation of the *Multiplexed Analogue Component (MAC)* standard, it was adopted in the UK for 625 line television transmission.

DMSU: *Digital Main Switching Unit*.

DMT: *Discrete Multitone*.

DN: *Distinguishing Name*.

DNHR: *Dynamic Non-Hierarchical Routeing*.

DNIC: *Data Network Identification Code*.

DNS: *Domain Name Service* or *Domain Name Server* or *Domain Name System*.

document facsimile system: See *facsimile*.

document facsimile telegraphy: A *facsimile telegraphy* system in which documents are being transmitted rather than photographs. The system is able to recreate two levels of density rather than grey scales.

DoD: *Department of Defence*.

DOD: *Direct Outward Dialling*.

domain: A collection of elements, on a *distributed network*, such as the *Internet*, which are under control of a set of processors.

domain name: Name of a *domain*, usually on the *Internet*. This is represented by groups of names or words separated by dots.

Domain Name Server (DNS): A *server* which hosts the *database* containing the *Domain Name Service (DNS)*.

Domain Name Service (DNS): Distributed *database*, with an associated application, which can translate *Internet Protocol (IP)* addresses into human readable form (text). Also known as *Domain Name System (DNS)*.

Domain Name System (DNS): See *Domain Name Service*.

donor network: In considerations of *number portability* it is the *network* to which the *subscriber's* number was originally assigned. See also *recipient network.*

Doppler effect: This is the apparent change of *frequency* of a received *signal* when there is a relative movement between the source and recipient of the signal. This frequency change is known as the *Doppler shift.*

Doppler radar: *Radar* which uses the *Doppler effect* to determine the radial component of an object, relative to its *antenna.*

Doppler shift: See *Doppler effect.*

DOTAC: *Department Of Transport And Communications.*

dot matrix: A two-dimensional array of dots which can be used for the *transmission* or *display* of *characters* and images.

double current circuit: A *circuit* which uses current with positive and negative polarity to represent the two *binary* states. See also *single current circuit.*

double current signalling: *Signalling* which uses voltages or currents of equal and opposite magnitude to indicate *line* conditions. Figure D.22 shows a *call progress signal* using double current signalling. Generally the system has discrete send and receive *paths*, with a common return, and is inherently *full duplex* in operation. See also *single current signalling.*

double ended control: A method of *synchronisation* for two *exchanges* which are linked together, in which the *phase* of the incoming *signal* and the phase of the *clocks* in the two exchanges are compared and this is used to control the clock of one of the exchanges. This ensures synchronisation of the two exchanges.

Double Frequency Shift Keying (DFSK): A modification of *Frequency Shift Keying (FSK)* which uses two *signals* which are combined by

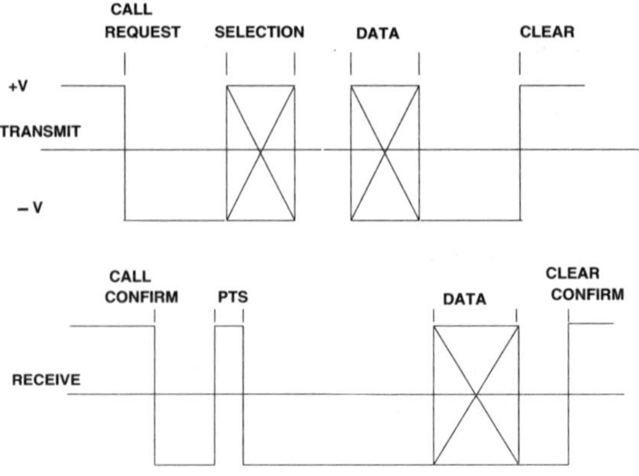

Figure D.22 Double current signalling

multiplexing and then *transmission* by *frequency shifting* among four *frequencies*.

double interlaced scanning: A technique, used in television *broadcasting*, in which alternate lines of the image are scanned first, to form a field, with the missing lines traced out subsequently as a second field, the two fields forming one complete frame. See also *sequential scanning*.

double modulation: A *modulation* technique in which the output from the first modulation is used as a *modulating signal* for a second *carrier* operating at a higher *frequency*.

Double Sideband transmission (DSB): *Transmission* in which both the upper and lower *sidebands* are equally spaced, following the process of *Amplitude Modulation*, and are transmitted, along with the *carrier signal*, which is at full power. See also *Double Sideband Suppressed Carrier transmission (BSBSC)*, *Double Sideband Reduced Carrier transmission (DSB-RC)* and *Independent Sideband transmission (ISB)*.

double sideband modulation: See *double sideband transmission*.

Double Sideband Reduced Carrier transmission (DSB-RC): *Double Sideband transmission (DSB)* in which the *carrier* is transmitted along with the *sidebands*, but at a strength which is less than that at *modulation*.

Double Sideband Suppressed Carrier transmission (DSBSC): A special case of *Double Sideband Reduced Carrier transmission (DSB-RC)* in which the *carrier* is completely suppressed, i.e. it is not transmitted with the *sidebands*.

down converter: A device which is used to lower the *frequency*, or *frequency band* of a *signal*.

downline loading: The process of sending programmes and operating instructions for storage in *nodes* connected to a *network* using the same *lines* as are used by the nodes for normal communications.

downlink: Usually refers to the *radio frequency* used by a *satellite* to send *signals* down to an *Earth station*. Set *frequency bands* have been allocated by the *ITU* for this. See also *uplink*.

DPCM: *Differential Pulse Code Modulation*.

DPNSS: *Digital Private Network Signalling System*.

DPSK: *Differential Phase Shift Keying*.

DQDB: *Distributed Queue Dual Bus*.

Draft International Standard (DIS): Standard, developed by the *International Standards Organisation (ISO)*, which has been issued for comment and before it has been adopted as a full standard.

DRCS: *Dynamically Redefinable Character Set*.

drift: The gradual change in the characteristics or value of a parameter. Usually it is undesirable and can lead to the performance of the system going outside its operating limits. See also *jitter* and *wander*, in which the variations occur over a shorter time.

drop and insert: The technique used, usually in *multiplexing*, of terminating one or more *channels* at a point and of inserting other channels into the *data stream.*

dropback: One of the techniques used to implement *number portability* in which the *donor network* releases the *call* once it has been established that the number has been ported. There are two forms of dropback: return to pivot and query on release. In the return to pivot dropback method the call is sent back to the previous switch for re-routeing to the new location. In the query on release dropback method the call is sent right back to the initiating switch for re-routeing. If only a few customers have been ported then the dropback method is efficient since it limits the number of *database* queries which would be needed. However, this method becomes inefficient as the quantity of ported numbers increases.

drop cable: (1) The short piece of *cable* connecting a *DTE* to the main *LAN network*. **(2)** The cable which connects a residential house to the main feed from a *Central Office (CO).*

dropped call: A *call* which is terminated unexpectedly. Usually this happens during wireless transmission, such as in a *cellular radio system*, due to an unexpected break in the *radio transmission link.*

dropped channel: *Channels* in the *transmission* which are terminated at a *node* and not continued as part of the rest of the *signal.*

DS-0, DS-1, etc.: See *Digital Signal level.*

DSA: *Dial Service Assistant.*

DSB: *Double Sideband transmission.*

DSB-RC: *Double Sideband Reduced Carrier transmission.*

DSBSC: *Double Sideband Suppressed Carrier transmission.*

DSC: *District Switching Centre.*

DSE: *Data Switching Exchange.*

DSF: *Dispersion Shifted Fibre.*

DSI: *Digital Speech Interpolation.*

DSL: *Digital Subscriber Line.*

DSM: *Delta Sigma Modulation.*

DSMA: *Digital Sense Multiple Access.*

DSP: *Digital Signal Processing*

DSR: *Data Set Ready*

DSSS: *Direct Sequence Spread Spectrum.*

DSU: *Data Service Unit.*

DTCM: *Differential Trellis Coded Modulation.*

DTE: *Data Terminal Equipment.*

DTE clear request: A *call control signal*, sent by a *DTE* to clear a call.

DTE waiting: A *call control* signal, sent by the *DTE/DCE* interface, to indicate that the DTE is waiting for further control *signals* from the DCE.

DTH: *Direct To Home.*

DTI: *Department of Trade and Industry.*
DTMF: *Dual Tone Multifrequency.*
DTP: *Distributed Transaction Processing.*
DTR: *Data Terminal Ready.*
dual bus: Two *buses* which are parallel, the *data* flowing in opposite directions in each bus for *transmission* and reception. Also called *dual cable*.
dual cable: See *dual bus*.
Dual Error Detection (DED): A *code* which can detect dual error which occur during *transmission*. Example is the *Hamming code*. See also *Single Error Detection (SED)*.
Dual Tone Multifrequency signalling (DTMF): A *signalling* method which uses a combination of two *tones*, having *frequencies* within the speech band. Each combination of two frequencies represents a single *digit*, as shown in Table D.6. The tones are in two groups, a high band and a low band, and they are geometrically spaced to ensure that any two frequencies of a valid combination are not harmonically related. The tones are present on the line for about 100 ms, followed by an intertone pause of about 100 ms before the next combination of tones is sent, so that *dialling* is much faster than in *loop-disconnect signalling*. Also DTMF has the advantage of end-to-end signalling so that signals can be transmitted in-band over the *link* to control distance equipment, such as *voice mail*, *answering machines*, etc.
ducting: The process of confining a *signal* within a *transmission medium*, such as a *cable*, air layer, etc. Can also refer to the mechanical structure through which several cables are passed. Ducting effects often occur in *microwave* transmission, where the *radio waves* from the transmitter are diverted, by an atmospheric duct, away from the receiver, as shown in Figure D.23.

Table D.6 DTMF tone assignments

Low	*High*			
	1209 HZ	*1336 Hz*	*1447 Hz*	*1633 Hz*
697 Hz	1	2	3	A
770 Hz	4	5	6	B
852 Hz	7	8	9	C
941 Hz	*	0	#	D

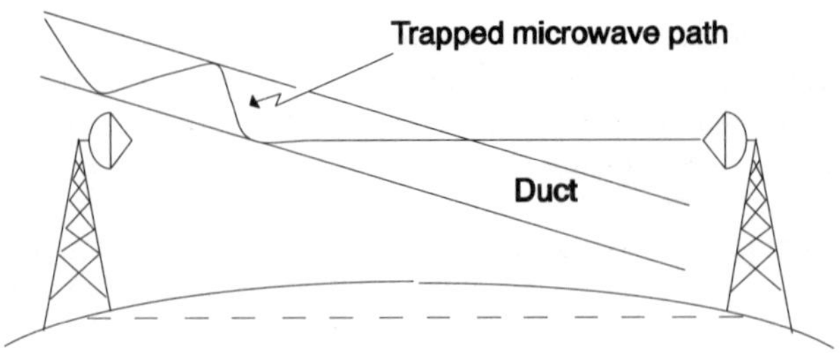

Figure D.23 Atmospheric ducting

dumb terminal: An *asynchronous terminal*, such as a *Visual Display Unit (VDU)*, which has very little intelligence of its own, deriving this from the processing system to which it is connected. It uses no communications *protocol* and can usually only handle one coded *character set*, such as *ASCII*.

dummy character: A *character* which is included in a *data* store or *data transmission* to achieve a specific characteristic; it does not carry the meaning of the *message*. Dummy characters are frequently used to fill a *frame* or a *buffer*.

dummy load: A device, usually added at the end of a *transmission line*, to give the line a specific characteristic. The energy in the dummy load is usually dissipated as heat and does not serve any useful purpose. The dummy load is usually chosen such that it adds to the actual load on the line and the two equal its *characteristic impedance*, so as to avoid reflections.

duobinary code: A *partial response code* in which the output is influenced by the previous input. In this a logical 0 in the input results in a logical 0 in the output. A logical 1 in the input will result in a change in the output only if the number of logical 0s since the last logical 1 in the input sequence is odd; otherwise no change occurs. Figure D.24 shows an example of duobinary coding.

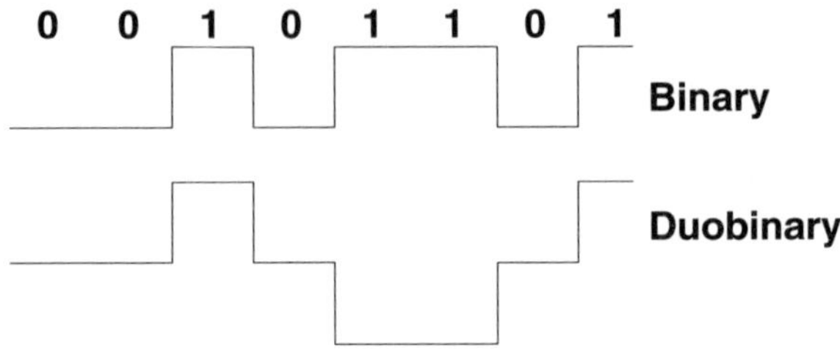

Figure D.24 Duobinary code

duopoly review: Review, conducted in the UK by the *DTI* in 1991, into the telecommunications environment, and the operation of its two major operators at that time, BT and Mercury (Cable and Wireless). It resulted in further *liberalisation* of the market with the granting of licences to a wide spectrum of infrastructure and service providers.

duplex: Literally means two. A duplex optical cable, for example, contains two fibres. However the term is usually used to mean *duplex transmission*, i.e. *transmission* in both direction simultaneously. See *full duplex* and *half duplex*.

duplex circuit: A *circuit* which can carry *duplex transmission*.

duplex signalling: A system which is able to send the *signalling* information over the same cable as the *voice signal*, without the use of filters.

duplex transmission: See *duplex*.

duty cycle: The duration of an event, measured as the ratio of the time over which the event occurs to the total time under consideration. Therefore if an event occurs for two hours in every 10 the duty cycle is 0.2 or 20%. The duty cycle of a repetitive waveform, such as a *pulse train* shown in Figure D.25, is the ratio of the on period to the total period, i.e. t/T for the pulse train.

DVB: *Digital Video Broadcasting*.

DWDM: *Dense Wavelength Division Multiplexing*.

Dynamically Adaptive Routeing (DAR): A *routeing* technique, used in *Packet Switched Networks (PSN)*, in which an *algorithm* is used to determine an alternative path through the *network* so as to avoid *congested* or defective *trunks* or *switches*.

Dynamically Controlled Routeing (DCR): A *routeing* strategy which is based on a central controller which monitors the state of all the *links* in the *network*, on a regular basis. This information is sent to the *nodes* on the network and is used to determine the most appropriate *route*.

Dynamically Redefinable Character Set (DRCS): The ability, in a *Visual Display Unit (VDU)* to alter the *character* set being displayed, for example between English, Chinese and Arabic. This is usually done by storing these characters in a *database*, located on the *network*, to which the VDU has access.

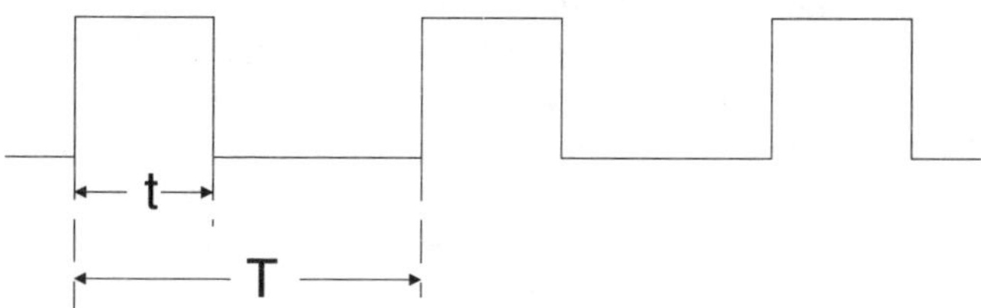

Figure D.25 Duty cycle of a pulse train

Dynamic Alternative Routeing (DAR): A *routeing* strategy, first used on the BT network in the UK, in which alternative routes are automatically selected through the *network* if the prime path becomes *congested* or in other ways unusable. If a *call* is successful through this new route then it is maintained. If, however, the call fails on this route then a new route is selected for the next call. This routeing technique is also referred to as 'sticky random routeing'.

dynamic bandwidth allocation: The facility of sharing *bandwidth*, available on a *transmission medium*, by allocating a part of this to *transmitting terminals* on an on-demand basis. This allocation is done automatically and is just sufficient to meet the terminal's requirements at the time. Once the terminal has completed its transmission the bandwidth is returned to the pool for use by other users.

Dynamic Channel Allocation (DCA): A system in which *transmission channels* are allocated to *users* on an on-demand basis, rather than by a planned assignment.

Dynamic Channel Selection (DCS): The facility, available with some radio communications standards, for the transmitting device, on *call setup*, to automatically scan the *transmission channels* available and to select that one which is free and has the least *co-channel interference*. This system improves the utilisation of the radio system and the quality of the *transmission*.

dynamic equaliser: An *equaliser* which can adapt automatically to changing *line* conditions.

Dynamic Non-Hierarchical Routeing (DNHR): *Dynamic routeing* strategy which has been in operation in the AT&T *network*, in the USA, since 1987. The strategy is to maintain fixed sequences of alternative routes for each source-destination *node* pair, but the sequences are changed to follow variations in the patterns of the *offered traffic*.

dynamic range: An indication of the spread of a parameter or system. It is usually measured as the ratio of the maximum value of a parameter (*frequency*, voltage, power, etc.) to its minimum value, expressed in *decibels (dB)*. For a *transmission medium* it indicates its *signal* carrying capability, measured as the ratio of the maximum signal power which can be handled without *distortion*, to the noise level in the system, stated in decibels.

dynamic routeing: *Routeing* in which the *route* through the *network* can be adjusted on an on-going basis depending on the network conditions at the time, so as to make the most effective use of network resources and to improve performance, such as making the system more robust and minimising delay. There are many systems in use, for example, *Dynamic Non-Hierarchical Routeing (DNHR)*, *Dynamic Alternative Routeing (DAR)* and *Dynamically Controlled Routeing (DCR)*.

E

EAMPS: *Expanded AMPS*.

E&M signalling: A form of *direct current signalling*, which was developed in the UK by BT as *DC5 signalling*, for use on the PSTN. It is also used for *tie line* signalling through a *PABX*. Separate wires are used for *voice* and signalling, the signalling taking place on two wires, the E (ear) wire and the M (mouth) wire. The PABX sends *signals* on the M wire and receives signals on the E wire of another PABX (Figure E.1). The E lead has an earth potential applied from a contact and the receiving equipment has a battery connected relay as the receiving circuit. The same circuit is used in the opposite direction on the M lead, but in reverse. The impedance between sending and receiving ends is limited to about 140 ohms. The relay contact is pulsed at about 10 pulses per second from the PABX control system.

ear: See *human ear*.

Ear and Mouth signalling: See *E&M signalling*.

Early Bird: The first international communications *satellite*, launched in 1965 to provide both fixed and mobile services. It was known as *INTELSAT* I, nicknamed Early Bird.

earth bulge effect: For a *microwave transmission* system, as shown in Figure E.2, the formula for calculating the earth surface curvature (h, in feet) is given by $h = (d_1 d_2)/(1.5K)$, where d_1 and d_2 are the distances in miles from the point being considered to either end of the microwave transmitters. K is a constant whose value depends on the propagation conditions. Usually K is one and propagation is in a straight line. Under

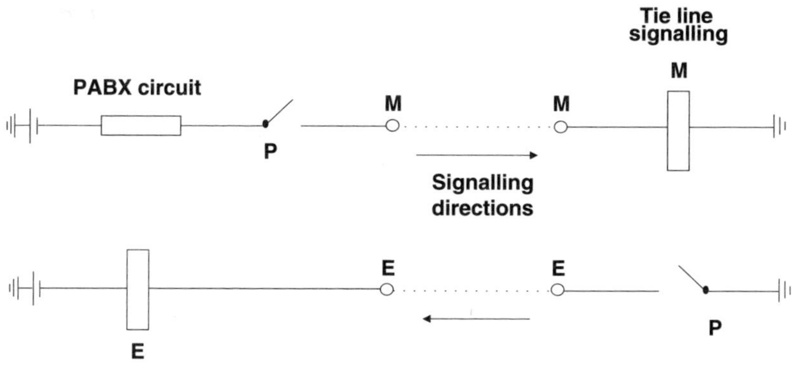

Figure E.1 E&M signalling

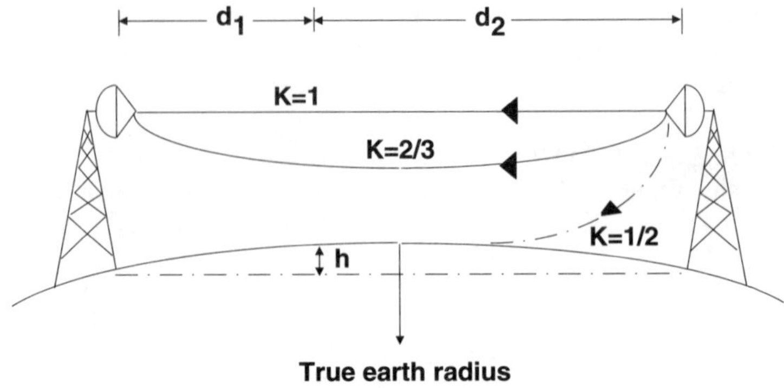

Figure E.2 Earth bulge effect

some atmospheric conditions an inversion layer can occur in which the atmospheric density increased with elevation. This will cause the microwave rays to bend in an opposite direction to the earth's surface and, if severe enough, it will strike the earth. This is termed the earth bulge effect and a figure of K of ⅔ is chosen to compensate for this, although very conservative designs may use a factor of ½.

earth calling: A *signalling* method, commonly used in a *PABX*, as shown in Figure E.3. The PABX applies a local earth via a relay contact to one leg of a pair from the *local exchange* when it wants service. The exchange detects the current in this wire and applies the exchange

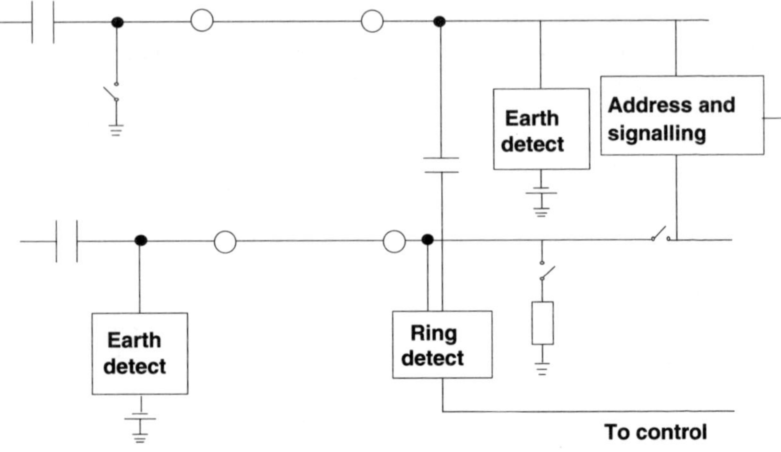

Figure E.3 Earth calling

battery across the pair with *dial tone* superimposed. The PABX then sends the *digits* to the exchange using techniques such as *loop-disconnect signalling* or *Dual Tone Multifrequecy signalling (DTMF)*. For an *incoming call* the exchange applies an earth potential to one leg of the pair (the opposite one to that used by the PABX to call the exchange) and the PABX detects this. Also known as *ground start* signalling.

earth calling PABX: A *PABX* which uses *earth calling* for its *signalling*. In North America it is known as a *ground start PABX*.

earthing: Connecting a circuit directly to earth so as to give a path of zero electrical resistance. Earthing is used for several reasons, such as: to conduct the high voltages generated by a lightning strike away from sensitive equipment; for safety reasons, to protect operators; to prevent interference being generated by spurious *signals* in adjoining equipment (see *RFI*). Also known as *grounding*. See also *ground*.

Earth satellite: A *satellite* which has an *orbit* around the Earth.

Earth's atmosphere: This is the atmospheric layers above the surface of the Earth, as shown in Figure E.4. The atmosphere is significantly influenced by the activity of the Sun, which causes ionised layers to form, such as the *D layer* and the *E layer*. These have an important impact on *radio transmission*. Meteorological activity and cloud formation occurs in the first 10 km.

Earth's ionised layers: These are the layers of the *Earth's atmosphere* in which ionisation has occurred (release of free electrons). They are primarily known as the *D layer*, *E layer* and the *F layer*, which is subdivided into the F_1 and F_2 layers.

Earth station: A radio transmitting and receiving *station* which is located on the surface of the Earth, or within the Earth's atmosphere, such as on the earth, on board a ship or on an aircraft. The station communicates with stations in *space*, such as on an *Earth satellite*. Also known as *ground station*.

Earth station antenna: An *antenna* which is mounted on an *Earth station*.

Earth terminal: A *terminal* mounted on an *Earth station* and used for communications with a station in *space*.

EAS: *Extended Area Service*.

EB: *Errored Block*.

EBCDIC: *Extended Binary Coded Decimal Interchange Code*.

EBIT: *European Broadband Interconnect Trial*.

EC: *European Community*.

ECC: *Embedded Communications Channel* or *Error Control Coding*.

echo: The *signal* which has been reflected back to its source with a characteristic which is very similar to that of a source but with a time delay. Sound echoes are common and for these there must be a delay of more than 30 ms to 40 ms between the first arrival of the sound and its

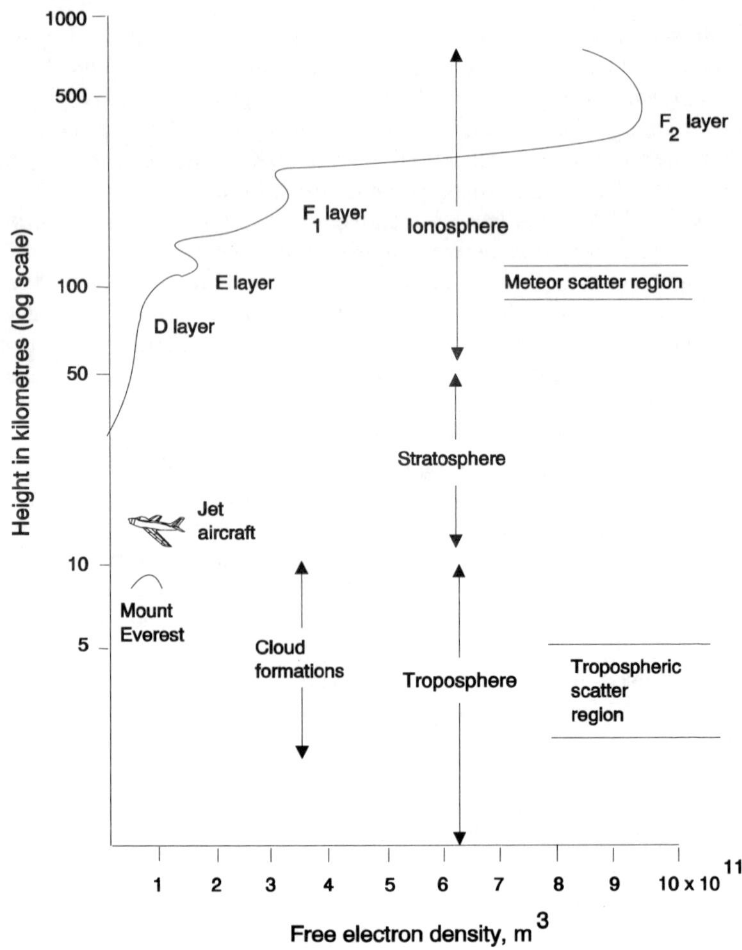

Figure E.4 The Earth's atmosphere

echo for the brain to be able to distinguish them as two different sounds. Echoes occur in *four wire circuits* due to poor impedance matching on the *line*, and this would cause speakers to hear their own voice after a delay. Echoes also occur in *radio transmission* due to *multipath effects*. See also *talker echo* and *listener echo*.

echo attenuation: The difference in strength between the original *signal* and its *echo*.

echo attenuation ratio: The ratio of the strength of the original *signal* and its *echo*, measured in *decibels*.

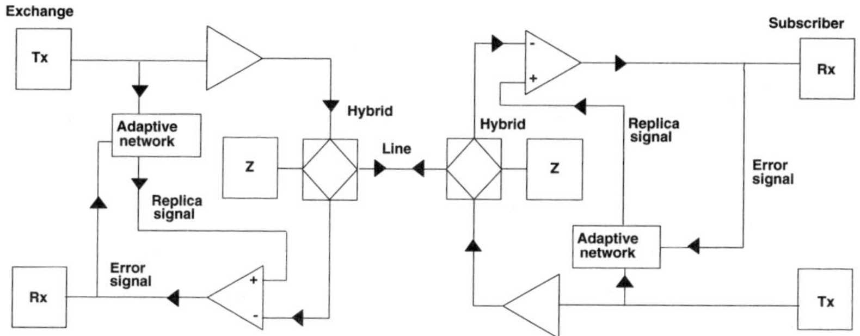

Figure E.5 Echo cancellation

echo cancellation: A technique for *duplex* operation in *two wire transmission.* In this both communicating *terminals* can transmit simultaneously over the same *line*, but each terminal estimates the echo which will be generated as a result of its own transmission and then substracts this from the received *signal* to arrive at the true signal from the other terminal. Figure E.5 shows an arrangement which can be used for this. The adaptive network models the echo path and this signal is subtracted from the received input to leave the true receive line signal.

echo suppressor: A device, used in *transmission lines*, to suppress *echoes*. It is usually voice activated so that it is not operative whilst the speaker is talking, but then suppresses any signals which are reflected back as echoes. In effect it therefore acts as a one-way system. The suppressor is turned off to permit high speed data transmission. It is also turned off by the high pitch *tone* generated by an answering *modem*.

ECITC: *European Committee for IT Testing and Certification.*

ECJ: *European Court of Justice.*

eclipse: The blockage of light from a source due to an obstruction coming in front of the source. In a *satellite* system it is the blockage of the light from a source (such as the Sun, Moon, etc.) due to another celestial body coming it front of it.

ECM: *Electronic Countermeasure* or *Error Correction Mode.*

ECMA: *European Computer Manufacturers Association.*

e-commerce: See *electronic commerce.*

ECSA: *Exchange Carriers Standards Association.*

ECSC: *European Coal and Steel Community.*

ECTRA: *European Committee for Telecommunications Regulatory Affairs*

ECTUA: *European Council of Telecommunication Users' Association.*

ECU: *European Currency Unit.*

EDFA: *Erbium Doped Fibre Amplifier.*

EDI: *Electronic Data Interchange.*
EDIFACT: *Electronic Data Interchange for Administration, Commerce and Transport.*
EDSL: *Extended Digital Subscriber Line.*
EDTV: *Enhanced Definition Television.*
EEC: *European Economic Community.*
EET: *Equipment Engaged Tone.*
Effective Isotropic Radiated Power (EIRP): A measure of the strength of a *signal* from an *antenna.* It is given by the product of the power supplied to the antenna and the *gain* of the antenna, relative to an *isotropic antenna.*
Effective Monopole Radiated Power (EMRP): A measure of *antenna signal* strength, given by the product of the power supplied to the antenna and the *gain* of the antenna, relative to a short vertical antenna.
effective noise bandwidth: It is defined (Figure E.6) as the width of a rectangular *frequency response* curve having a height equal to the maximum height of this curve and corresponding to the same total *noise power.*
Effective Radiated Power (ERP): A measure of the strength of the *signal* from an *antenna.* It is given by the product of the power supplied to the antenna and its gain relative to a half *wavelength dipole.* The numerical power gain of a half wave dipole relative to an *isotropic antenna* is 1.641 (2.15 *decibels*).
EFIS: *Electronic Flight Information System.*
EFS: *Error Free Seconds.*
EFT: *Electronic Funds Transfer.*
EFTA: *European Free Trade Association.*
EFTPOS: *Electronic Funds Transfer at Point Of Sale.*
EGP: *Exterior Gateway Protocol.*
EHF: *Extremely High Frequency.*

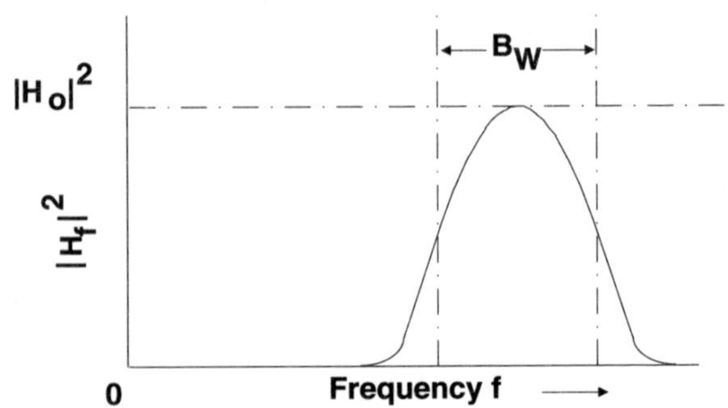

Figure E.6 Effective noise bandwidth

EIA: *Electronic Industries Association.*

800 service: Often written as 0800, it is a *billing* arrangement in which calls made to all numbers which start with 0800 are billed to the *called party* rather than the *calling party.*

80 column display: A standard used in *displays* on *terminals* in which *information* is formatted as 80 columns of *characters* on a page.

EIR: *Equipment Identity Register.*

EIRP: *Effective Isotropic Radiated Power.*

EKMS: *Electronic Key Management System.*

elastic buffer: A *buffer* in which some of the parameters can be varied, such as its *data* storage capacity and the value of the *signal* delay which it introduces.

E layer: One of the *Earth's ionised layers*, part of the *Earth's atmosphere.* It is a layer of high free electrons, caused by the ionisation effects of the Sun and extends from about 95 km to 160 km above the surface of the Earth, between the *D Layer* and the *F layer.* It has an important effect on *radio transmission* in the *High Frequency (HF)* region.

electrical interface: The specification of the *signals* which need to pass between two systems in order for them to interwork with each other. Examples are the *amplitude* of the signal, its *frequency*, the *connector* pins on which these signals appear, and any *codes* used for *encoding.* See also *mechanical interface.*

electrically despun antenna: An *antenna*, such as a *phased array*, which can electrically move its main *lobe* so that it maintains a fixed direction in spite of movements of the platform on which it is mounted. No mechanical movement is necessary.

electrical signal: See *signal.*

electrical telegraph: See *telegraph.*

electroluminescence: The process in which electrical energy is converted directly into light. This occurs in a device such as a *Light Emitting Diode (LED).*

electromagnetic: Refers to the phenomena of interaction between electrical and magnetic fields. See also *electromagnetic spectrum* and *electromagnetic waves.*

Electromagnetic Compatibility (EMC): The state which exists when two or more pieces of equipment are collocated or work with each other without generating harmful *electromagnetic interference* or being effected by electromagnetic interference generated by other devices. Many standards exist for EMC, some of them being legal requirements. A few of these standards are given in Table E.1.

Electromagnetic Interference (EMI): Interference which occurs as a result of *electromagnetic* phenomena. It may be as a result of *radiated emissions* or *conducted emissions* from other devices.

Electromagnetic Pulse (EMP): An *electromagnetic* energy pulse, of high intensity and short duration. This is usually produced by a nuclear explosion and it can cause many electronic devices in its vicinity to fail.

electromagnetic radiation: The radiation consisting of *electromagnetic waves*, which propagates in a *transmission medium* at the speed of light. It includes a wide range of *gnals*, such as radio, light, X rays, etc.

electromagnetic receiver: One of the earliest types of receivers used in a *telephone handsets*, as shown in Figure E.7. The *voice signal* in the coil

Table E.1 EMC Basic and Generic Standards. (Continued on next page)

Emission	*Immunity*	*Scope*
Basic standards		
EN 61000-3-2		Mains harmonics: all equipment < 16A per phase
EN 61000-3-3		Mains flicker: all equipment < 16A per phase
	EN 61000-4-2	Electrostatic discharge
	EN 61000-4-3	Radiated radio frequency field (ENV 50141)
	EN 61000-4-4	Electrical fast transient bursts
	EN 61000-4-5	Surge
	EN 61000-4-6	Conducted radio frequency (ENV 50140)
	EN 61000-4-8	Power frequency magnetic field
EN61000-4-11		Voltage dips, interruptions and variations
Generic standards		
EN 50081-1	EN 50082-1	Apparatus intended for use in residential, commercial and light industrial environments
EN 50081-2	EN 50082-2	Apparatus intended for use in industrial environments

Table E.1 (Continued from previous page) EMC product standards

Emissions	*Immunity*	*Scope*
EN 50065-1		Mains signalling apparatus
EN 55011		Industrial, scientific and medical apparatus
EN 55013	EN 55020	Broadcast receivers and associated equipment
EN 55014-1	EN 55014-2	Household appliances, tools and similar apparatus
EN 55015:1996	EN 61547	Electrical lighting and similar equipment
EN 55022		Information technology equipment
EN 50091-2	EN 50091-2	Uninterruptible power systems
EN 50083-2	EN 50083-2	Cable TV equipment
	EN 50130-4	Fire, intruder and social alarm systems
EN 60870-2-1	EN 60870-2-1	Telecontrol equipment and systems
	prEN 55105	

of the electromagnet energises it and this moves the iron diaphragm to produce the sound in the ear piece.

electromagnetic spectrum: Generally the whole spectrum of frequencies from zero to infinity. There are several well-defined portions of this spectrum, such as the *visible spectrum*, the *ultraviolet radiation* spectrum, and the *infrared light* spectrum.

electromagnetic waves: Waves caused by the *electromagnetic* phenomena, i.e. the interaction of electrical and magnetic waves. Examples are *light waves* and *radio waves*. Electromagnetic waves travel at the speed of light in vacuum and can be *broadcast* through air or space, or conducted along a *transmission medium*.

electron beam: A stream of electrons, such as that emitted from the cathode of a *Cathode Ray Tube (CRT)*.

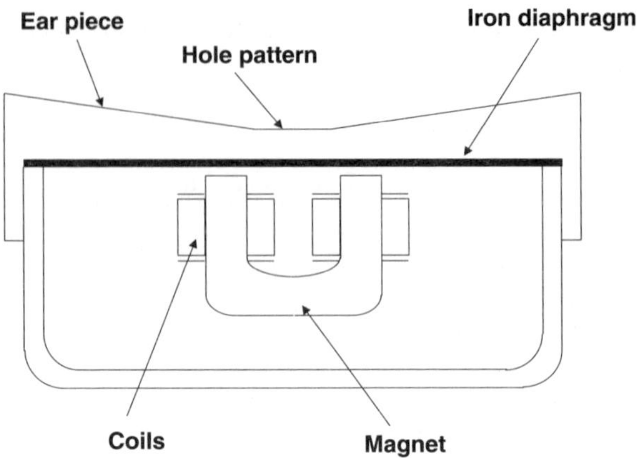

Figure E.7 Electromagnetic receiver

electron beam recording: Recording using an *electron beam*, such as when a beam of electrons, striking the screen of a *Cathode Ray Tube (CRT)*, is used to expose a film.

electron gun: The device which generates the *electron beam*.

electronic bulletin board: See *bulletin board service*.

electronic commerce: The generic term used to describe all forms of commercial activity which is conducted electronically. Often this is over a *Wide Area Network (WAN)* or the *Internet* and security is a key consideration. Several standards exist for this, such as X.400 and *EDI*.

Electronic Countermeasure (ECM): An action, taken as part of electronic warfare, to reduce the effective use of the *electromagnetic spectrum* by the enemy.

electronic crosspoint: A *crosspoint* which is formed by an electronic device.

Electronic Data Interchange (EDI): Standards which have been developed to allow the transfer of electronic *data* between computer based systems. Several standards exist for this, such as the *ANSI* X.12 and *EDIFACT*.

Electronic Data Interchange for Administration, Commerce and Transport (EDIFACT): An international standard for *EDI*, primarily intended to be used for *electronic commerce*. It was developed jointly by the *ANSI* and several European EDI groups, with the United Nations.

electronic exchange: An *exchange* which uses electronic switches rather than those based on mechanical or electromechanical components, such as *crossbar matrix* or *Stowger selectors*.

Electronic Flight Information System (EFIS): The instrument panel in the cockpit of an aircraft which provides information on navigation and the key performance functions of the aircraft, such as its engine and radio communications.

Electronic Funds Transfer (EFT): The process of transferring funds or accounts from one area to another, i.e. using electronic systems to debit the *database* in one area and to credit another database, with the same amount, in another area. The two databases may contain *data* relating to different people or organisations.

Electronic Funds Transfer at Point Of Sale (EFTPOS): An *Electronic Funds Transfer (EFT)* system in which the transfer of funds between the buyer and seller occurs at time of the sale, e.g. via the electronic tills in stores which transfer the information to a processing unit.

Electronic Industries Association (EIA): A trade organisation representing a large number of US electronic component and equipment manufacturers. It was founded in 1924 as the Radio Manufacturers Association. Over four thousand government and industrial representatives participate in its two hundred or so committees. The EIA is primarily involved in producing hardware oriented data communications standards. It has produced over six hundred standards, the best known being RS-232-C, now renamed EIA 232-C, which is used worldwide for computer interfaces. In 1988 the telecommunications sector of the EIA merged with the US Telecommunications Suppliers Association (USTSA) to form the Telecommunications Industry Association (TIA).

Electronic Key Management System (EKMS): A security system used, in applications such as *data communications* and *data processing*, for the generation, control and distribution of electronic *keys*.

electronic mail (e-mail): A method for sending *messages* electronically (such as by using the X.400 protocol) from one person to another. Messages can also be broadcast from one sender to many recipients. The messages primarily consist of text, although many different types of files, including graphics, can be attached to the e-mail. It is a store and forward service, the messages being stored waiting for collection by the recipient. E-mails can be sent over any *data communications network*, including the *Internet*. See also *electronic message system*.

electronic mail box: The store used for the electronic *messages* sent by *e-mail* where they are held pending collection by the addressee.

electronic message system: A modification of the *e-mail* in which the electronic *message* is delivered to the addressee automatically. In this, unlike e-mail, the addressee does not have to retrieve the message from the *electronic mail box*, e.g. by use of a 'get mail' command.

Electronic Programme Guide (EPG): Electronic based guides which enable viewers to select the most suitable programmes to view in advanced video based systems, such as with *Video On Demand (VOD)* and *Digital Video Broadcasting (DVB)*.

Electronic Switching System (ESS): *Switching equipment* which is controlled by a processor. Also known as a *Stored Programme Controller (SPC)*.

electro-optic: Term used to describe the phenomena in which an electrical field is used to modify the characteristics of light. For example to change the direction of its *polarisation* plane.

electro-optic switching: The use of the *electro-optic* phenomena to switch different *wavelengths* of light in devices such as *Optical Cross-connects (OXC)*.

Electrostatic Discharge (ESD): The creation of a high current, short duration, surge, usually to *ground*, due to the high static voltage built up in a person or a device. This can damage sensitive electronic equipment, such as those containing semiconductors.

Element Management System (EMS): A *network management* system which is primarily concerned with the management of individual elements, or *nodes*, on the *network* rather than the whole network.

ELF: *Extremely Low Frequency*.

elevation angle: The angle, measured in the vertical plane, between the horizontal plane and the line drawn between the observer and the object being observed. The angle may be considered to be negative if the object is below the horizontal plane. (Figure E.8.)

elliptical polarisation: A wave in which the electrical field vector rotates in the plane normal to the direction of propagation whilst its amplitude varies such that the locus of the extremity of the vector traces an ellipse, as in Figure E.9. The wave is defined by three parameters: the axial ratio, i.e. the ratio of the minor axis to major axis; the orientation angle, i.e. the angle between the major axis and the reference axis (β); the sense of

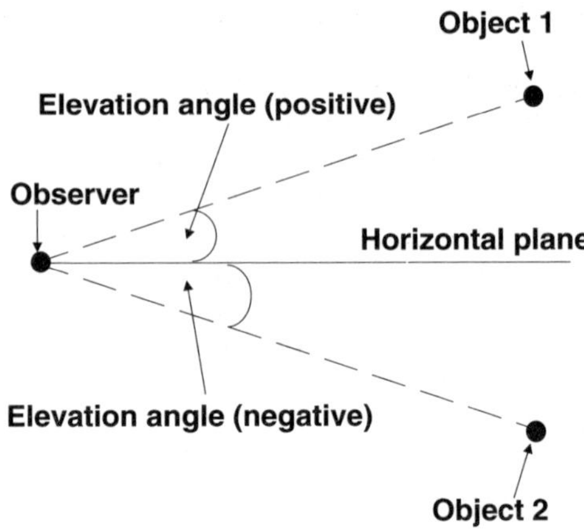

Figure E.8 Illustration of elevation angle

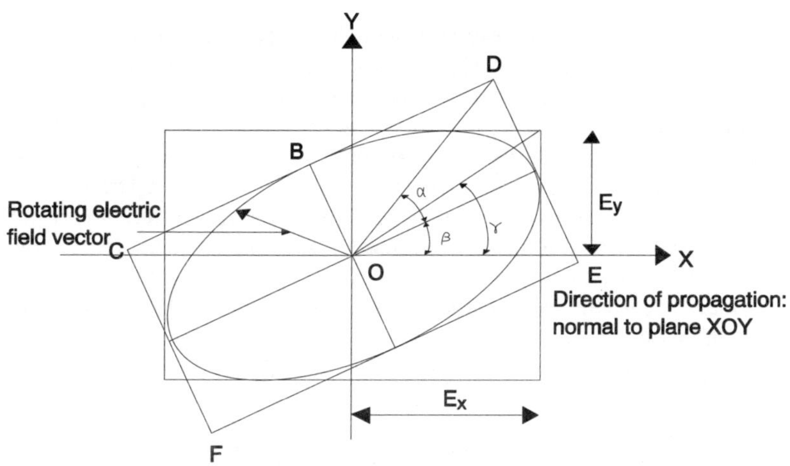

Figure E.9 Elliptical polarisation

rotation, i.e. clockwise or counter clockwise. An alternative way of defining the axial ratio is by the ellipticity angle (α). In Figure E.9 the angle γ is given by $\tan \gamma = E_y/E_x$. Elliptical polarisation is a special case of *linear polarisation* (axial ratio of zero) and *circular polarisation* (axial ratio of one).

EM: *Electronic Countermeasures* or *End of Medium.*

e-mail: See *electronic mail.*

Embedded Communications Channel (ECC): The *channel*, part of the *STM* section overhead, used within the *Synchronous Digital Hierarchy (SDH)* for carrying control information. It operates at rates of 192 kbit/s and 576 kbit/s.

Embedded Operations Channel (EOC): The *channel* used in *T1* systems for carrying maintenance and supervisory control information. It operates at 4 kbit/s and is also referred to as the *Facilities Data Link (FDL).*

EMC: *Electromagnetic Compatibility.*

EMC directive: Directive issued by the *European Commission* (89/336/EEC) with the aim of removing barriers to trade caused by different interpretations of *Electromagnetic Compatibility (EMC)* in European countries. It specifies the scope and coverage as well as the rules for compliance. Several standards are involved (see Table E.1).

EMC filtering: The process of reducing the amount of *conducted interference* by filtering out the unwanted high frequency components of a signal in the equipment's leads, for example its mains cable. This improves the equipment's *Electromagnetic Compatibility (EMC).* See also *EMC shielding.*

EMC shielding: The process of reducing the amount of *radiated interference* from an equipment by, for example, surrounding it with a conductive screen which absorbs and reflects the interfering *frequencies* radiating from it.

EMC standards: Standards which equipments need to meet in order to be regarded as having *Electromagnetic Compatibility (EMC)*. These standards have been developed by many bodies, and cover different types of equipment for both emission and immunity. Some of these are given in Table E.1, primarily developed by *CENELEC*. Other standards, developed by *ETSI* and relating to telecommunications equipment, are given in Table E.2.

emergency craft frequency: *Frequency*, allocated by the *ITU* for use in emergency situations, such as radio beacons for emergency craft position indication.

emergency frequency band: The *frequency band*, in the range 118 MHz and 132 MHz, which has been allocated for use in emergency situations, for sending *messages*.

emergency restart: The process of restarting a communications system, usually from its basic systems, following a major failure.

emergency routes: *Routeing plan* which is set up for use in emergency situations, when the existing *transmission paths* have failed.

EMI: *Electromagnetic Interference*.

EMP: *Electromagnetic Pulse*.

EMRP: *Effective Monopole Radiated Power*.

EMS: *Element Management System* or *European Monetary System*.

emulator: A *hardware*, *software* or *firmware* component which mimics or copies the characteristic of another component. It will provide the same

Table E.2 ETSI EMC standards

Standard	*Product sector*
ETS 300 127	Radiated emission testing of large systems
ETS 300 386-1	Public telecommunication network equipment
ETS 300 386-2-1	Switching equipment
ETS 300 386-2-2	Transmission equipment
ETS 300 386-2-3	Power supply equipment
ETS 300 386-2-4	Supervisory equipment

output for an input as the device it is copying although it may use a different internal process to arrive at the solution.

EN: *Europaische Norm* or *European standard.*

en-block signalling: *Signalling* in which the *address* information is contained in one or more *blocks*. This information is sufficient for *switching* and *transmission* from one stage to the next and does not occur until the full address information has been received.

encapsulation: The carrying of one *data* within another. For example the encapsulation of control *signals* within an *information block* or the *transmission* of one type of *protocol* by encapsulating it within another type of protocol.

encipher: To convert *plain text* into a secret from so that it cannot be recognised except by the intended recipient.

encoding: Converting *data* by using a designated *code*, so as achieve a given purpose, such as to hide its meaning or to improve the efficiency of *transmission*. The recipient then uses the same code to obtain the true data. See also *line code*.

encoding law: The *algorithm* used within *Pulse Code Modulation (PCM)* to define the *quantisation levels*. See also *A-law* and *μ-law*.

encryption: The process of converting *plain text* into a form so that the intended recipient can recover the original plain text but an unauthorised recipient cannot. The *algorithm* used to convert plain text into encrypted text and back again is called a *key*.

encryption algorithm: The *algorithm* used during *encryption*. These have two main components: substitution, in which *characters* which are used in the *plain text* are replaced by different characters in the encrypted text; and transportation, in which the locations of the characters in the encrypted text are altered from that in the plain text.

encryption key: See *key*.

end distortion: The movement in the end of a pulse, caused by speed related errors, as shown in Figure E.10. Often seen in start-stop *teletypewriters*.

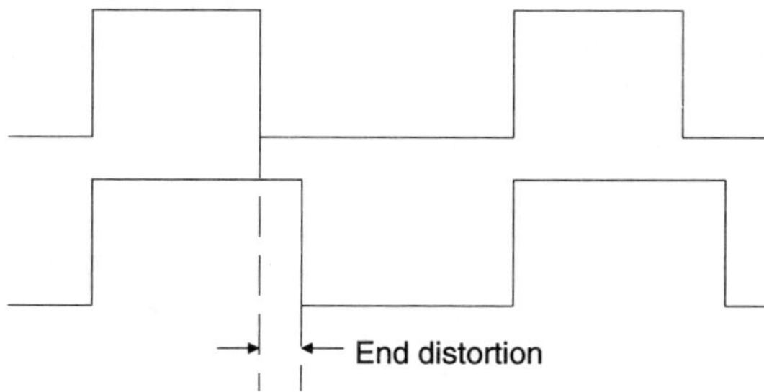

Figure E.10 Illustration of end distortion

end of address: A *signal* sent to indicate to the *network* that the previous information which has been transmitted is the complete *address* and it can be used for *routeing*. The main body of the *message* usually follows.

end of block signal: A *signal* which indicates the end of a *block* of *data*.

end office: A *Central Office (CO)* to which *subscriber* lines are connected. Also known as *local exchange*.

End Office Toll Trunking (EOTT): A US term for *trunks* between *end offices* situated in different toll areas.

End of Medium (EM): The last *character* on a medium containing *data* which indicates the end of the medium, or the end of the section of usable *data* on the medium. Examples are the last character on punched card, or paper tape or magnetic medium.

End Of Message (EOM): A *character* which indicates to the *network* that the *transmission* has been completed and the *information* received forms the complete *message*.

end of pulsing signal: A *signal* sent in a *telephone* system to indicate that the previous *dial pulses* constitute the full *address* and *routeing* can begin. Also known as *end of selection* signal.

end of selection signal: See *end of pulsing signal.*

End of Text (EXT): A control *character* used to indicate the end of transmitted text.

End Of transmission (EOT): An *ASCII* code which is sent to indicate the completion of *transmission*. This is used to initiate other actions, such as placing the *terminal* in a waiting mode, for example until an *acknowledgement* is received from the *receiving terminal*, or to initiate *call disestablishment* procedures.

End of Transmission Block (ETB): A control *character* which is used to indicate the end of a *block* of transmitted *data* in instances where the total *message* is divided into blocks for convenience.

end system: The *nodes* on the *network* which are the ultimate *data source* or *data sink* for the *data* being carried by the network. See also *relay system* and *OSI Basic Reference Model*.

end-to-end layer: The layers within the *OSI Basic Reference Model* (the *seven layer model*) which provide communications between *end systems* rather than via a *relay system*. These layers are the *Transport Layer*, the *Session Layer*, the *Presentation Layer* and the *Application Layer*. (See Figure O.4.)

end-to-end performance: Sytem performance measured from the *transmitting terminal* to the *receiving terminal* and including all the *transmission* and *switching* systems in between. Many factors affect this performance, such as *transmission loss*, *noise*, *frequency distortion*, *quantisation distortion*, *propagation delay*, *talker echo*, *listener echo*, *sidetones*, and *crosstalk*.

end-to-end protocol: A protocol which provides for the management of the total system, from the *transmitting terminal* to the *receiving terminal.* See *node-to-network protocol*

end user: The actual user of a system or service. For example a *telephone handset* can be sold by a manufacturer to a wholesaler, who then sells it on to a retailer. The person who ultimately buys this telephone, and installs and uses it, is the end user.

engaged tone: See *busy tone.*

engineering channel: A *channel* which is primarily used by installation and maintenance engineers and is used to carry *voice*, *data* and *video signals* concerned with maintenance of the *network* or individual *nodes* on the network.

engineering circuit: See *Engineering Orderwire (EOW)*

Engineering Orderwire (EOW): A *circuit* carrying *voice* or *data* which is to be used by engineers for the installation or maintenance of the *network* or *nodes* on the network.

Enhanced Definition Television (EDTV): An improvement in the standard television *signal,* which provides higher quality pictures and are compatible with existing receivers. However picture quality is not as good as in *High Definition Television (HDTV).* Also known as *Extended Definition Television (EDT).*

ENQ: *Enquiry.*

enquiry: A *transmission control character* which is used to determine if the remote *terminal* being accessed is ready to send or receive *data.*

enterprise network: A *network* serving a single organisation, and including all its local and remote locations. It can cover all its Local Area Networks (LANs) and *Wide Area Networks (WANs).*

entity: In a *network management* system it usually refers to the object on the *network,* usually defined in a *database*, which can be managed by the system.

entropy redundancy: One of the techniques for *compression* of a *digital signal,* for example as used within *MPEG-2*. It is based on the knowledge that in any non-random digitised *signal* there are some *code* values which occur more frequently than others. This allows shorter codes (which occupy less space) to be used for these more frequently occurring values. See also *spatial redundancy*, *temporal redundancy* and *psycho-visual redundancy.*

ENV: *European pre-standard.*

envelope delay distortion The *distortion* caused in a *signal* due to some of the *frequencies* in the transmitted *waveform* arriving at the sensing point before other frequencies. It can also occur in a system if the rate of change of the phase with frequency is not the same over the whole *frequency range.*

envelope demodulation: The process of recovering the original *signal*, following *modulation* of a *carrier*. See *envelope demodulator*. This process is also know as *envelope detection*.

envelope demodulator: A *demodulator* used for *waveforms* which have been through *Amplitude Modulation (AM)*. Figure E.11 shows a simple circuit which can be used for this and Figure E.12 shows its waveforms. The input voltage v_i is the modulated waveform, as in Figure E.12(a) and the voltage across the capacitor C is the output voltage demodulated

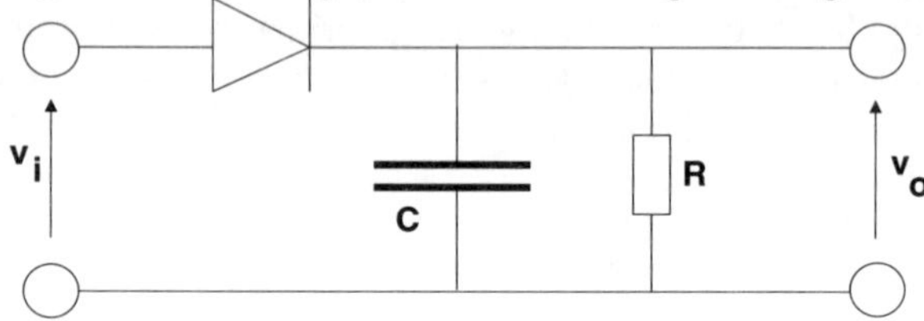

Figure E.11 Envelope demodulator circuit

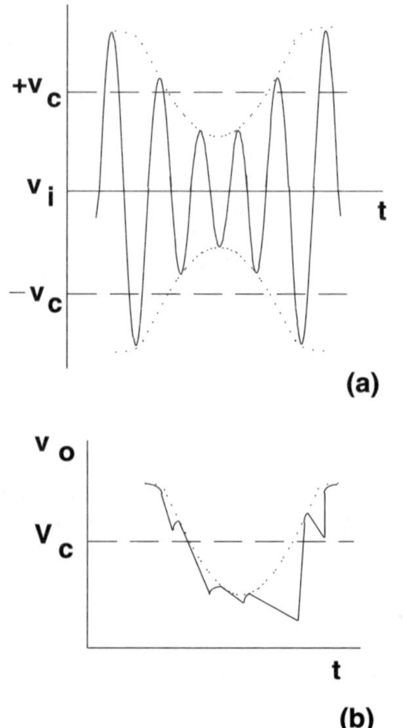

Figure E.12 Envelope demodulator waveforms: (a) input; (b) output

voltage signal v_0 as in Figure E.12(b). As seen the rectified voltage across the output resistor R follows the envelope of the modulated input signal. This is also known as *envelope detection.*

envelope detection: See *envelope demodulation.*

EOC: *Embedded Operations Channel.*

EOM: *End Of Message.*

E1: Refers to the European *transmission* standard for *signals* which have been through *Time Division Multiplexing (TDM).* Following *Pulse Code Modulation (PCM)* 64 kbit/s samples are combined into a *frame* of 32 *timeslots*, as shown in Figure E.13. This gives an overall transmission rate of 2.048 Mbit/s. Timeslot 0 carries control signals, such as *CRC*, and timeslot 16 is used for *signalling* for all the remaining 30 channels. These channels carry the user *information.* E1 is specified by *ITU-T Recommendation* G.732 and its main characteristics are summarised in Table E.3.

EOT: *End of Transmission.*

EOTT: *End Office Toll Trunking.*

EOTC: *European Organisation for Testing and Certification.*

EOW: *Engineering Orderwire.*

EP: *European Parliament.*

EPG: *Electronic Programme Guide.*

Table E.3 E1 main parameters

Parameter	*E1 (European System) ITU-T Recommendation G.732*
Sampling Frequency	f_s = 8kHz
Bit rate per channel	D_i = 64kbit/s
Number of time slots	32
Number of channels	30
Number of bits/frame	$32 \times 8 = 256$
Total bit rate	$256 \times$ 8kHz = 2.048kbit/s grouped word of 7 bits in the 0 channel of odd frames
Signalling	Out-of-octet, grouped in channel 16, consisting of 4 bits per channel, distributed over 16 frames (=1 multiframe)

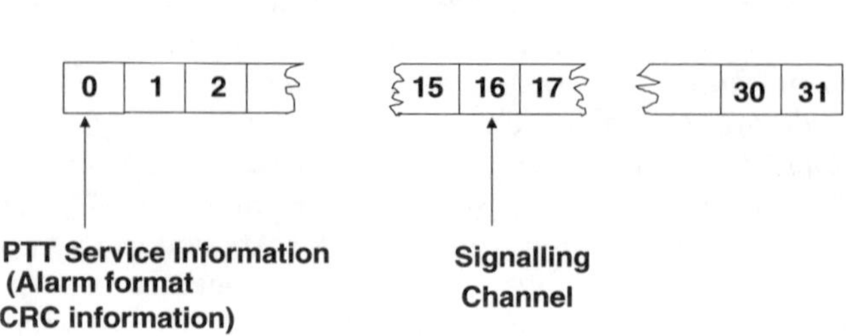

Figure E.13 E1 transmission frame

equal access: The requirement (first defined at the *divestiture* of AT&T) for all *subscribers* to be granted access to their operator of choice for *long-distance calls*. The ruling required the *Regional Bell Operating Companies (RBOCs) to provide equal access to their Central Office (CO)* switches to all *Interexchange Carriers (IXCs)*, including AT&T.

equalisation: Different *frequencies* in a *signal* travel at different rates in a *transmission line* and this can result in them arriving at different times at the receiving end, resulting in *distortion*, such as *delay distortion*. This can be compensated for by adding components, such as capacitors, resistors and inductors, to the line, a process known as equalisation.

equaliser: The system which is used for *equalisation* in a *transmission line*. It may be adjustable or fixed in value and if adjustable the values may be changed manually or they can change automatically in response to line conditions. These are called *adaptive equalisers*. Many different types of equalisers exist, such as the *time domain equaliser* (e.g. *Transversal Equaliser (TVE)*) and the *partial response equaliser*.

equalising repeaters: *Repeaters* which perform an *equalisation* function by compensating for effects of non-linearity in the *transmission line*.

Equipment Engaged Tone (EET): The tone a *calling terminal* receives if ther are no fee lines available to complete a *call*. Also called the *busy tone*.

Equipment Identity Register (EIR): A security feature on *GSM* which allows monitoring of the *International Mobile Equipment Identity (IMEI)*. This is used to validate mobile equipment, so that equipment which is not approved (e.g. faulty or stolen) cannot use the system.

equipotential: A surface on which every point is at the same voltage level, in relation to a voltage reference, usually the *ground*. This means that there is no current flow between any of the points on the surface and no build-up of charge.

equivalent bit rate: The number of *bits* transmitted over a *transmission medium*, in a given time, corresponding to the *information* content of a *signal*.

equivalent circuit: A *circuit* which, although different in construction from another circuit, provides the same response to stimuli. An equivalent circuit can therefore be studied to determine the behaviour of the actual circuit under different conditions. The equivalent circuit may be a theoretical (mathematical) circuit rather than an actual device. See also *equivalent network*.

equivalent four wire circuit: A *two wire circuit* which can provide *full duplex* operation (similar to a *four wire circuit*) by using *Frequency Division Multiplexing (FDM)*.

equivalent network: A *network* which has the same external characteristics as the actual network and can therefore replace it. It is the same as an *equivalent circuit*.

equivalent noise resistance: The equivalent resistance which produces, for calculation purposes, the same thermal *noise* as the noise source (e.g. *shot noise*).

equivalent noise temperature: The temperature of a resistance, at the input of a *noise* free component, which generates the same output noise.

Erbium Doped Fibre Amplifier (EDFA): An *optical amplifier* in which the *signal* passes through a section of erbium doped *optical fibre* and is amplified by a *LASER* pump diode.

ERC: *European Radio Committee*.

ERDF: *European Regional Development Fund*.

erlang: A measure of *traffic volume*, named after the Danish engineer (Agner Erlang) who did a considerable amount of work in *teletraffic theory*. It is defined as the number of *calls* per unit time, e.g. number of *call seconds* per second. It is a dimensionless unit and is numerically equal to the average number of simultaneous calls on a group of *circuits*. For a single circuit it is equivalent to the proportion of time for which it is occupied, and therefore the probability of finding that circuit *busy*. Traffic on a circuit cannot exceed one, and it is usually about 0.6 for a *junction circuit* and about 0.05 for a residential circuit to 0.5 for a business circuit. As an example, if S is the mean *call duration*, and N is the mean number of calls arriving in a time T, then the traffic (A) in erlangs is given by A = (NS)/T.

Erlang's loss formula: The formula calculates the probability that a *call* arriving at a *network* would find that the *link* is full and so it would be lost. Suppose that the link has a capacity of C and that n are the number of circuits occupied by the link, with n taking values between 0 and C, as shown in Figure E.14, with n following a *birth-death model*. Suppose that the call arrival rate is a *Poisson distribution* of rate λ and the *holding* *time* of the calls has an *exponential distribution* with a mean of 1 and is independent of the arrival process. Erlang's loss formula then states that the probability E(λ,C) of an arriving call finding the link full is given by

$$E(\lambda,C) = \frac{\lambda^C}{C!}\left[\sum_{i=0}^{C}\frac{\lambda^i}{i!}\right]^{-1}$$

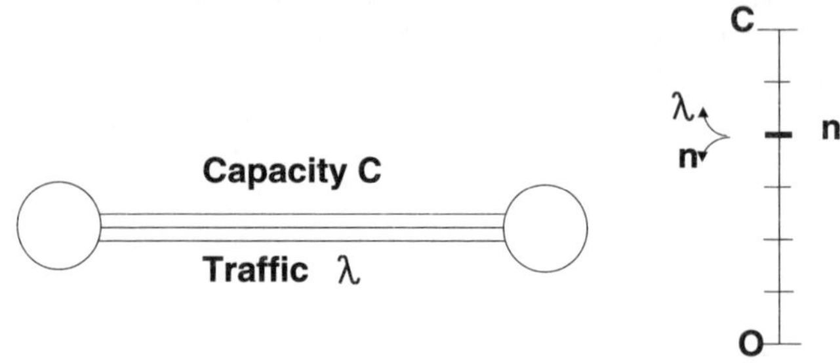

Figure E.14 Single link model for Erlang's loss formula

ERMES: *European Radio Messaging System.*

ERO: *European Radio Office.*

ERP: *Effective Radiated Power.*

error: Any difference between the true value and the value which has been computed, measured or estimated.

error bit: A *bit* (*binary digit*) which is in *error*, e.g. it has not been received correctly, a logical 1 being received as a logical 0 or vice versa.

error block: A *block* of *data* which is in *error*.

error budget: The allocation of the maximum errors which can occur on subsets of a communications system (such as *trunks*, *access lines*, *switching equipment*, etc.) so that the total errors of the overall system does not exceed the specified value.

error burst: See *burst error.*

error control: The process of controlling the occurrence of *errors*, usually by the use of *codes* which provide *error detection* and *error correction.*

Error Control Coding (ECC): The control of *errors* in transmitted *data*, by use of *codes*, usually by the addition of redundancy *bits* into the transmitted *bit stream*. Coding is such that it can enable *error detection* and *error correction* of errors which occur in the *transmission channel*. The received data must be passed through a *decoder* to arrive at the correct information. There are two main techniques for ECC, *convolutional coding* and *block codes*. See also *feedforward error correction* and *feedback error correction.*

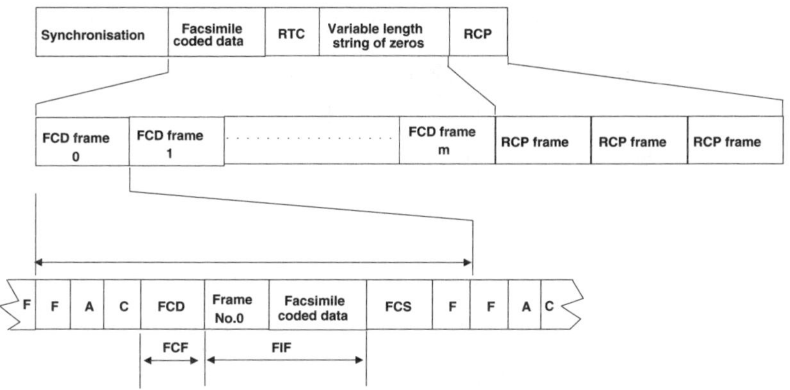

Figure E.15 ECM transmitted frame structure

error correcting code: See *Error Control Coding (ECC).*

Error Correcting Mode (ECM): An *ITU-T* enhancement for Group 3 *facsimile* (see *Group x facsimile*) in which the transmitted signal is separated into *HDLC frames*. This allows the *data* accuracy to be checked on the receiver by use of the *Frame Check Sequence (FCS)* bytes. Figure E.15 shows the EMC transmitted frame structure, in which each frame contains 256 *octets* of image plus its frame number.

error correction: The process of taking action to correct *errors* which have been determined to have occurred in received *data*. Error correction therefore usually follows *error detection* and the most commonly used method of correcting the error is for the *receiving terminal* to request the *transmitting terminal* to resend the data. This is known as *Automatic Repeat Request (ARQ)*. Some codes, such as *Feedforward Error Correction (FEC)*, however, can correct some errors without needing a retransmission.

error detection: The process of detecting *errors* which occur in the received *data*. Several *codes* exist for this, such as use of the *parity bit*, *Cyclic Redundancy Check (CRC)*, etc. Error detection is usually followed by *error correction*.

Errored Block (EB): An *error* event, defined in *ITU-T Recommendations*, such as G. 821 and G.826, which is applicable to both in-service and out-of-service testing for *PDH* and *SDH transmission* systems. It is defined as a *block* in which one or more *bits* are in error.

Errored Second Ratio (ESR): Specified in *ITU-T Recommendations*, such as G.821 and G.826, it is a performance parameter for *PDH* and *SDH* systems, used in *error* measurements. It is the ratio of the *Errored*

Seconds (ES) to the total seconds available during a fixed measurement period.

Errored Seconds (ES): It is an *error* event, specified in *ITU-T Recommendations*, such as G.821 and G.826 and it is a one second period with one or more *block errors*.

Error Free Seconds (EFS): A measure of the quality of a *transmission*. It is given by the proportion of one second intervals (expressed as a percentage or a ratio) in a specific time interval, in which the transmitted *data* is delivered error free.

error injection: The introduction of a known *error* pattern into the transmitted *bit stream* for a *Bit Error Rate Tester (BERT)*.

error performance: The performance of the system in relation to the number of *errors* which are generated. In *digital networks* the error performance can be specified and measured over a complete end-to-end connection, called a *path*, or over parts of the network called *lines* and sections, as in Figure E.16. Path error performance indicates the overall *Quality of Service (QoS)* to the *subscriber*. Line and section error performance measurements are made to check that transmission performance objectives are being met and for trouble-shooting, installation and maintenance. Table E.4 gives the values of some of the error performance objectives specified in *ITU-T Recommendation* G.821, and Figure E.17 shows their relationship to each other.

error rate: A measure of the amount of *errors* occurring in a system, it is given as the fraction of the *data bits* which have errors. For example, if on average one bit in every ten thousand bits is in error then the error rate is given by 1×10^{-4}.

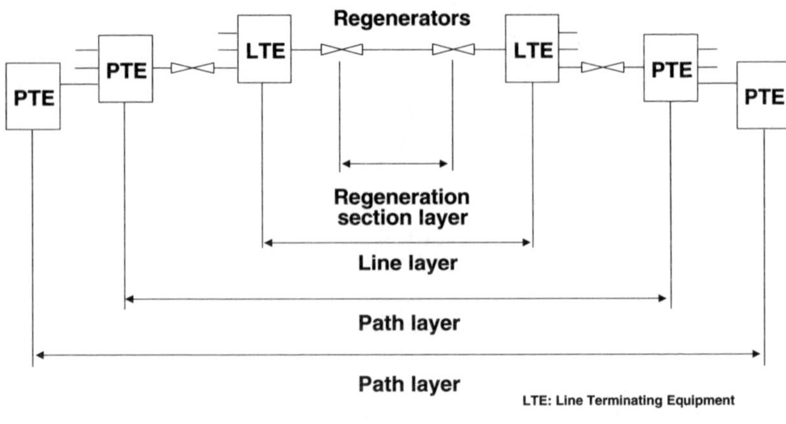

Figure E.16 Illustration of lines, paths and sections

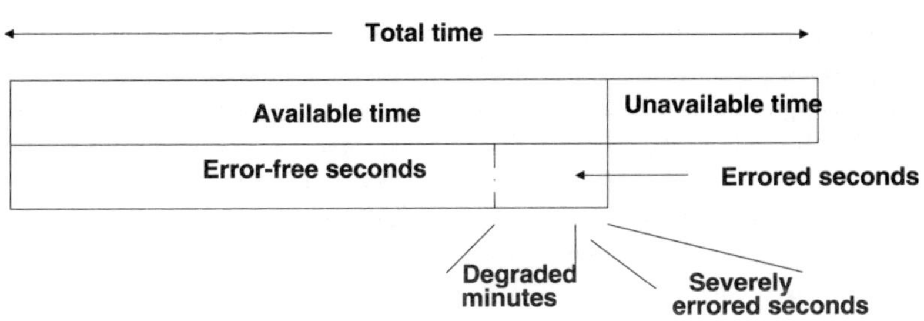

Figure E.17 Error performance parameters

error recovery: The process a system uses to recover from an *error* and to continue with its normal operation.

error signal: The *signal*, usually audio or visual, which indicates to the operator that an *error* has occurred in the system.

ES: *Errored Seconds.*

ESC: *Escape.*

Escape: A control *character* used to indicate that the characters which follow have a different meaning from normal. It can be used to terminate an operation or to extend the meaning of a character *code*.

ESD: *Electrostatic Discharge.*

ESF: *Extended Superframe Format.*

ESPRIT: *European Strategic Programme for Research and development in Information Technology.*

ESR: *Errored Second Ratio.*

ESS: *Electronic Switching System.*

ETACS: *Exended Total Access Communications System.*

ETB: *End of Transmission Block.*

Table E.4 ITU-T Recommendation G.821

Performance classification	*Objectives*
Degraded Minutes (DM)	Fewer than 10% of one minute intervals to have a Bit Error Ratio (BER) worse than 10^{-6}
Severely Errored Seconds (SES)	Fewer than 0.2% of one second intervals to have a Bit Error Ratio (BER) worse than 10^{-3}
Errored Seconds (ES)	Fewer than 8% of one second intervals to have any errors – equivalent to 92% error free seconds

Ethernet: One of the first *protocols* developed for *data transmission* over a *Local Area Network (LAN)*, and still widely used. It uses *CSMA/CD* as the *multiple access technique* and has been standardised by the *IEEE* as 802.3. It can operate at *data rates* of 10 Mbit/s over a shared *bus*. When used with *twisted pair wire*, Ethernet is referred to as *10BaseT*.

ETIS: *European Telecommunications Information Services foundation.*

ETNO: *European Telecommunications Network Operators.*

ETR: *ETSI Technical Report.*

ETS: *European Telecommunication Standard.*

ETSI: *European Telecommunications Standards Institute.*

ETSI Technical Report (ETR): Issued by *ETSI* to provide guidance on matters which are outside the scope of an *ETS* or *I-ETS*. They are published immediately following approval by the Technical Committee, without going through a Public Enquiry or Voting procedures.

ETX: *End of Text.*

EU: *European Union.*

EUCATEL: *European Conference of Associations of Telecommunication Industries.*

EURATOM: *European Atomic Energy Community*

EURESCOM: *European Institute of Research and Strategic Studies in Communications.*

Euromessage: *Radio paging* system, which started service in 1989 and is primarily used in Germany, UK, France and Italy. There are variations in these countries, as shown in Table E.5.

Europaische Norm (EN): European standard produced by *CEN* and *CENELEC*. Wherever possible these standards are based on international standards, such as from *ITU-T* and *ISO*. When in draft form they are called European pre-standards, or ENVs. There is usually a two year conversion period from an ENV to an EN, to give users time to adopt them.

European Atomic Energy Community (EURATOM): One of the three separate communities of the *European Community (EC)*, the other two being the *European Coal and Steel Community (ECSC)* and the *European Economic Community (EEC)*. EURATOM was established by the *Treaty of Rome* to increase the energy available within Member States, primarily by the adoption of nuclear power.

European Broadband Interconnect Trial (EBIT): An initiative, by several *PTOs* from *CEPT* countries, to support the *RACE* programme by developing a switched *broadband network* across Europe.

European Coal and Steel Community (ECSC): One of the three separate communities of the *European Community (EC)*, the other two being the *European Atomic Energy Community (EURATOM)* and the *European Economic Community (EEC)*. The ECSC was formed by the Treaty of

Table E.5 Methods used by Euromessage

Country	*Germany*	*UK*	*France*	*Italy*
Preamble length in bits	576	608	576	576
Synchronisation codeword	RPC1	RPC1	RPC1	RPC1
Idle codeword	RPC1	RPC1	RPC1	RPC125
Max batches between preambles	24 (or 42-96)	61	25	25
Transmission cycle time in seconds	84 (or 66-141)	30	11.8	12
Number of timeslots	3 or 1	1	1	1
Longest timeslot in seconds	28 (or 22-47)	30	11.8	12 or 24

Paris and signed by France, Germany, Italy, The Netherlands, Belgium and Luxembourg. It came into effect on 25 August 1952 and aimed to create a common market for coal and steel.

European Committee for IT Testing and Certification (ECITC): One of the first sectorial committees endorsed by the *EOTC* as part of the *CTS* process. It is usually a government or recognised agency which represents the relevant government on ECTIC. See also *EOTC*.

European Committee for Telecommunications Regulatory Affairs (ECTRA): Part of *CEPT*, it represents the telecommunications regulatory authorities from CEPT's member countries. It sets policies, for example on licensing procedures for private value added service providers, to ensure commonalty across Europe; on international accounting principles and whether changes should be proposed to any *ITU-T Recommendations*; on the role of regulators within the European equipment testing and certification field; on the harmonisation of regulatory policy on mobile communications across Europe; and on collaboration on proposals for *satellite* communications.

European Commission: The European Commission forms the executive arm (or civil service) of the *European Community (EC)*. It has commissioners drawn from the Member States, and is led by a president, who is also one of the commissioners. Larger countries, such as the UK, supply two representatives whilst smaller ones provide one. The European Commission meets once a week in Brussels. Members of the Commission perform functions similar to those of ministers within individual countries, each being responsible for one or more portfolios within the Community's activities.

European Community (EC): The European Community can be considered to be made from three separate communities: the *European Coal and Steel Community (ECSC)*, the *European Economic Community (EEC)* and the *European Atomic Energy Community (EURATOM)*. Generally, however, the term European Community is taken to mean the European Economic Community. The EC is a single market made from fifteen countries (Table E.6), with other countries applying to join. Eleven different official or working languages have been accepted (Table E.7) and the population is well over 350 million. In addition other minority languages are also recognised, such as Catalan, Frisian, Irish and Letzebuergesch.

European Computer Manufacturers Association (ECMA): Established in 1961 by European computer manufacturers to produce standards and interim guides for its members. It has a Secretariat in Geneva, and now includes several North American companies as well. There are two classes of members, Ordinary Members, who are European companies making and marketing data processing equipment, and Associated Members, who have an interest and experience in the European area, or other areas of interest to the Technical Committees. The organisation operates along *ISO* lines, with Technical Committees and Task Groups. It is dedicated to the production of *data processing* standards, which are not covered by other standards making bodies, and technical reports for its members. These standards are voluntary and are submitted to the standards bodies as technical contributions or to propose new work. Several ECMA standards have been endorsed by *ISO/IEC* and by *CENELEC* and published by them as international or European standards. ECMA is a liaison member of the *JTCI* set up by ISO/IEC.

European Conference of Associations of Telecommunication Industries (EUCATEL): An organisation of European national trade associations of telecommunication equipment manufacturers. It obtains and coordinates the views of its members on standards which are developed by other standards making bodies. It does not produce any standards of its own.

European Co-operation in the field of Scientific and Technical research (COST): See *COST*.

European Council: Summit meetings, held twice a year, at which heads of governments from the *European Community (EC)* Member States attend, accompanied by their foreign affairs and finance ministers. The

Table E.6 European country groupings

European Community *(Population: 350 million)*	*Eastern Europe* *(Population: 120 million)*	*European Free Trade Association* *(Population: 30 million)*
France	Czechoslovakia	Norway
Germany	Bulgaria	Iceland
Italy	Albania	Switzerland
Spain	Hungary	Liechenstein
Greece	Poland	
Belgium	Romania	
Denmark	Yugoslavia	
Portugal		
Ireland		
The Netherlands		
Luxembourg		
United Kingdom		
Finland		
Sweden		
Austria		

European Council does not have legislative powers, but it is the main political authority of the European Community and sets its political direction.

European Council of Telecommunications Users' Association (ECTUA): Set up in 1986 with the aim of providing a forum for its members to coordinate their views on standards being developed by other standards making bodies. It therefore has a very similar aim to the *EUCATEL* but its members are primarily national telecommunications users groups, although individual companies operating within Europe,

Table E.7 Official and working languages within the EEC

Language	*Country*	*Approximate population (millions)*
German	Germany, Austria, Belgium, Italy, Luxembourg	89
French	France, Belgium, Luxembourg, Italy	63
English	UK, Republic of Ireland	60
Italian	Italy	56
Spanish	Spain	39
Dutch	Netherlands, Belgium	21
Greek	Greece	10
Portuguese	Portugal	10
Swedish	Sweden, Finland	9
Danish	Denmark	5
Finnish	Finland	5

who are telecommunications users, can also join as associate members. ECTUA organises round table meetings of experts, at which all members can participate. This formulates the view of the organisation, which is then put to standards bodies and to governments.

European Court of Justice (ECJ): One of the four main bodies of the *European Community (EC)*. The ECJ consists of thirteen judges and sits in Luxembourg. It hears cases which are concerned with the implementation and interpretation of Community law. Rulings made by the ECJ are binding on Members and take precedence over national courts. Other three main bodies are the *Commission of the European Communities (CEC)*, the *Council of Ministers* and the *European Parliament (EP)*.

European Currency Unit (ECU): A unit of currency, created in 1981, as a common currency for all the *European Union* members. It has a value approximately equal to that of the US dollar.

European Economic Community (EEC): The EEC is one of the three communities of the *European Community (EC)*, the other two being the

European Coal and Steel Community (ECSC) and the *European Atomic Energy Community (EURATOM)*. The EEC officially came into being on 31 December 1992, although this date is largely symbolic; changes began long before this time and continued after it. On this date all *tariffs*, custom regulations and other trade barriers between Member States were removed. See also *European Community (EC)*.

European Free Trade Association (EFTA): Countries (Norway, Iceland, Switzerland and Liechenstein) which have agreed to have a special trading arrangement with each other. These cover a population of some 30 million. EFTA participates in standards making through *ETSI*. Representatives from EFTA have a special status as counsellors on the General Assembly, which is the governing body of ETSI, but have no right to vote.

European institute for Research and Strategic Studies in Communications (EURESCOM): Organisation which is part of *CEPT* and carries out R&D on behalf of CEPT members. It has laboratories in Heidelberg, Germany. It therefore has a function similar to that held by Bell Communications Research Inc., which did research for the UK *Regional Bell Operating Companies (RBOC)*.

European Interconnection Directive: Adopted by the *European Parliament (EP)* on 30 June 1997, it aims to harmonise conditions for open and effective interconnection of public *networks* across Europe. It enables negotiations to occur between operators but one of the parties can ask the regulator to investigate if it considers interconnect charges to be anti-competitive. The regulator can also 'name' operators who have 'Significant Market Power' which places an obligation on these operators to follow certain conditions in dealing with other operators. It is also the intention that, by the year 2000, operators in Europe who have 'Significant Market Power' will have to implement *carrier preselection*, in which subscribers can choose to have all their *long-distance calls* re-routed over a specified competitive carrier. Carrier preselection is commonly used in the USA.

European Monetary System (EMS): A monetary system which was agreed in 1974 between the *European Economic Community (EEC)* members. It came into effect on 9 March 1979 and has four elements: a European Unit of Account; a mechanism for exchange and information; transfer arrangements; and credit arrangements.

European Organisation for Testing and Certification (EOTC): Established in April 1990 by representatives from the *EC*, *EFTA*, *CEN* and *CENELEC* to act as a focus for all European matters concerned with conformity assessment of products and services. EOTC provides the framework and basic rules to enable European certification, based on the technical criteria defined within EN29000 and EN45000 standards. It is structured into sectors, each of which consists of 'agreement groups' made up of bodies from different countries interested in a particular area.

European Parliament (EP): One of the four main bodies of the *European Community (EC)*. It consists of five hundred and seventy six Members of the European Parliament (MEPs), which are directly elected by the citizens of the Member States, the number of representatives from each country varying according to its size. The MEPs belong to European political groupings and sit accordingly in the Parliament, which meets in Strasbourg. The Parliament has a consultative and advisory role, rather than a legislative one. It makes recommendations to the *Council of Ministers*. The other three main bodies of the EC are the *Commission of the European Communities*, the *Council of Ministers* and the *European Court of Justice (ECJ)*.

European pre-standard (ENV): See *Europaische Norm (EN)*.

European Radio Committee (ERC): Part of *CEPT*. It has a permanent group of experts to carry out studies and propose strategies for the management of radio *frequencies*. It ensures that it confers with all interested parties in this field.

European Radio Messaging System (ERMES): *Protocol* developed by a consortium of manufacturers, working with *ETSI*, for international *paging* systems. Issued as ETSI standard ETS 300-133 the first commercial ERMES service was launched in 1994. A variety of services are provided, such as *tone*, *numeric* and *alphanumeric*. Table E.8 provides brief details on the specification for ERMES and Figure E.18 shows the

Table E.8 ERMES specification

Parameter	*Value*
Frequency range	169.4125MHz to 169.8125MHz
Channels	16 channels 25kHz spacing
Modulation	4 level (4-PAM/FM) ±4687.5 and ±1562.5Hz
Data rate	6.25Kbit/s
Symbol rate	3.125kbaud
Error correction	2 bits (30, 18) shortened cyclic code
Interleaving to provide burst error correction	Message only to a depth of 9 code words

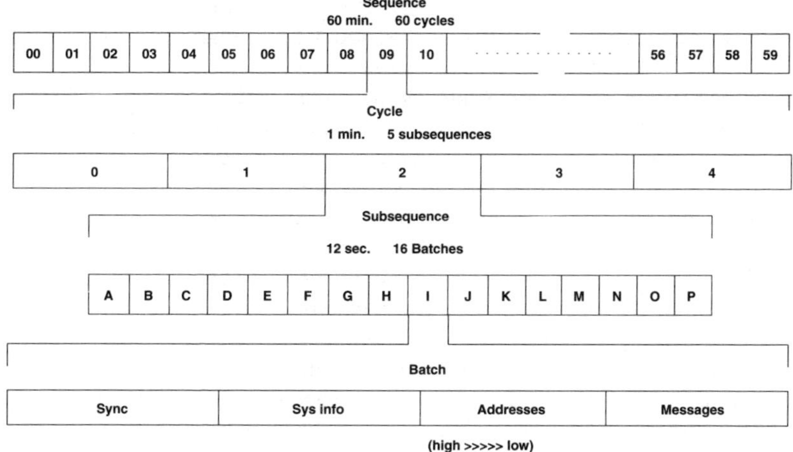

Figure E.18 ERMES protocol timing

construction of the transmitted *code*. This shows a sixty-minute sequence, which is then broken down into a 1 minute cycle and a 12 second subsequence. Batch I is then expanded into its *synchronisation*, system *information*, *pager address block* and pager *message* block of *code words*.

European Radio Office (ERO): An organ of the *European Radio Committee (ERC).*

European Regional Development Fund (ERDF): Fund operated by the *European Community (EC)* to improve the infrastructure in the less advanced regions of the Community. It provided close to 1 billion *ECUs* for the *STAR* project.

European Satellite Communications System (EUTELSAT): A series of European *satellites* launched for communications purposes. EUTELSAT II was launched in 1991 and uses a 14 GHz *frequency band* for *uplink* transmission and the 11 GHz and 12 GHz bands for *downlink* transmission.

European standard: See *Europaische Norm (EN).*

European Strategic Programme for Research and development in Information Technology (ESPRIT): A research and development programme, funded by the *European Community (EC)*, into Information Technology (IT).

European Telecommunications Information Services foundation (ETIS): Part of *CEPT*, it has as its members the information systems managers from the public telecommunication companies. Its aim is to allow CEPT members to share some of the development costs which they need to spend on *software* required for carrying out their internal *network* administration.

European Telecommunications Network Operators group (ETNO): Set up in May 1992 as a general policy association representing European operators. The aim of ETNO is to obtain maximum benefit for its members by influencing the development of European policy on telecommunications. All telecommunications operators who are active in countries falling within the geographical area of Europe can become members. There are two classes of members: full members, who provide public voice telephony service over fixed or mobile systems; and associated members who provide other value added services. The structure of ETNO consists of the General Assembly, Executive Board, Administrative Board, Secretariat and several Working Groups and Rapporteurs. The General Assembly is the highest decision making body of the association and only full members can vote at its meetings.

European Telecommunication Standard (ETS): A telecommunications standard issued by the *European Telecommunications Standards Institute (ETSI)*. It is similar to an *EN*, issued by *CEN* and *CENLEC*. An ETS is a voluntary standard, unless it is adopted as a *NET* or *CTR*. See also *I-ETS*.

European Telecommunications Standards Institute (ETSI): Set up on 31 March 1988 by *CEPT* to accelerate the formulation of standards and technical specifications in telecommunications, on behalf of *European Community (EC)* countries. It is an independent organisation, funded by its members who decide on its programme of work. However, the EC and *EFTA* may fund ETSI to produce specific standards of interest to the Community. There are many types of members of ETSI, such as telecommunications network operators, equipment manufacturers, users, private service providers, and research organisations. Figure E.19 shows

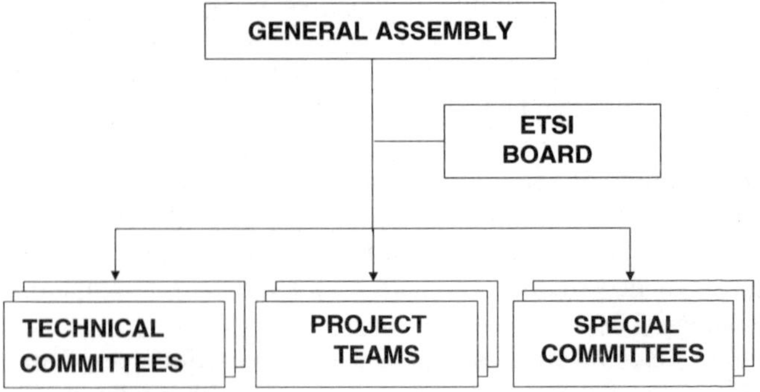

Figure E.19 ETSI structure

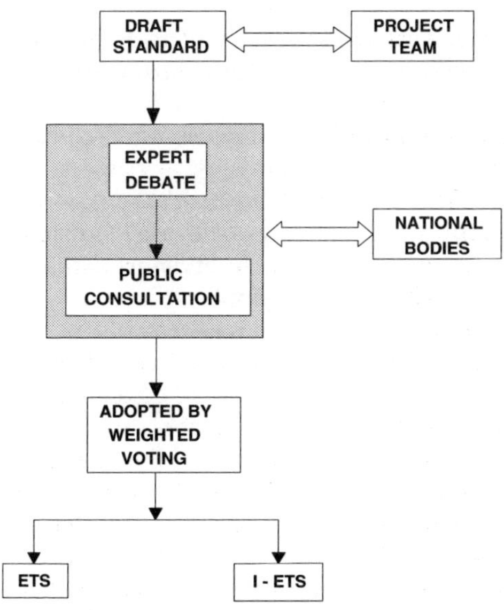

Figure E.20 ETSI standards making process

the structure of ETSI. The permanent Secretariat is based in Sophia, Antipolis, France, which is the headquarters of ETSI. The General Assembly is the governing body and determines its policy. All the technical work is carried out by the Technical Committees, Project Teams and Special Committees. Figure E.20 shows the standards making process within ETSI.

European Union (EU): See *European Community (EC).*

European Workshop for Open Systems (EWOS): Formed in December 1987 by several suppliers and users as the European forum for the development and promotion of *OSI functional standards*. It has a Secretariat in Brussels, under the auspices of *CEN/CENLEC*. The *European Commission* supports it and DG IX is a member of its steering committee. Membership of EWOS is primarily from the European *MAP* and *TOP* User Groups, CEN/CENLEC, national bodies, and *ECMA*. The EWOS technical programme is managed by a Technical Assembly, its standards work being done by Expert Groups. The EWOS works closely with the *AOW* and *OIW*, the three Workshops having set up a *Regional Workshop Co-ordinating Committee (RWS-CC)* to monitor their joint work programme and activities. Their outputs are primarily technical guides and functional standards, which are submitted to CEN/CENLEC for adoption as *ENVs* and to *ISO* for adoption as *International Standard Profiles (IPSs)*. Other outputs, which are not intended to be standards, are issued as Regional Workshop Technical Reports (RWS-TR) which are guides to the selection and implementation of standards. EWOS is also sometimes called the *European Workshop for OSI Standardisation*.

European Workshop for OSI Standardisation: See *European Workshop for Open Systems*.

EUTELSAT: *European Satellite Communications System*.

even parity check: A method of *error detection* using *parity check* in which the number of ones in a group of *binary digits* always adds up to an even number. See *odd parity check*.

EWOS: *European Workshop for Open Systems*.

exception condition: An abnormal condition which exists, the abnormal state having been defined in advance. In a communications system an exception condition could exist when the *network* cannot complete a request.

excess three code: A *binary coded decimal code* in which a decimal digit is represented by the binary equivalent of the digit plus three. This is shown in Table E.9.

exchange: Usually another name for a *Central Office (CO)*. Often known as a *telephone exchange*. Sometimes used to imply the *exchange area*.

exchange area: The geographical area served by an *exchange* in which there is a uniform *tariff* for all calls within the area.

Table E.9 Decimal, binary and excess three codes

Decimal	*Binary*	*Excess-three*
0	0000	0011
1	0001	0100
2	0010	0101
3	0011	0110
4	0100	0111
5	0101	1000
6	0110	1001
7	0111	1010
8	1000	1011
9	1001	1100

Exchange Carriers Standards Association (ECSA): A trade association of *wireline common carriers* in the USA. It was formed to provide a central public forum for the US telecommunications industry. Its *Standards Advisory Committee (SAC)* is responsible to *ANSI* for the *T1 Committee*. Now known as the *Alliance for Telecommunications Industry Solutions (ATIS)*.

exchange code: It is the *dialling code* assigned to an *exchange*. A *national numbering* system usually consists for three parts: the *area code*, the exchange code, and the *subscriber's local exchange* number.

exchange hierarchy: An interconnection of *exchanges* leading from the *local network* to international lines. See *class of exchange* and *class of office*.

exchange line: The *transmission line* coming from an *exchange* and connecting the *subscriber*, or a *PABX* or another exchange.

exhaustive transmission: Term used to describe the state in a *multiple access* system, using *token passing*, in which each *node* on the *Local Area Network (LAN)* is allowed to hold the *token* for a relatively long time, so that it, in effect, empties its *buffers*.

exit angle: Refers to the instance of a *electromagnetic wave*, such as light, passing from one surface to another, when the exit angle is the angle this ray of light makes with the normal to the surface at this point.

Expanded AMPS (EAMPS): An extension to the *Advanced Mobile Phone System (AMPS)* which uses an extra 10 MHz, so increasing the number of *channels* from 333 per service provider to 416.

expander: A device which restores *signals* to their original state, following *compression*.

expert system: A system which has been programmed to solve a certain class of problem (in which it is an 'expert') by use of *artificial intelligence*.

explicit reservation: A *multiple access* technique using *reservation ALOHA* in which one or more slots within a *frame* are divided into smaller slots called *reservation slots* (Figure E.21). Users contend for these slots, using a technique such as *ALOHA*, and if successful they place a reservation for a slot over a given number of frames. They can then transmit over these slots without risk of *contention* from other users. It is therefore required that each user keeps a record of the queue associated with each slot so that they know when to transmit their data. This mode of operation is called the reservation mode. When there are no remaining reservation slots in the queue (i.e. no stations waiting to transmit *data*) the whole frame is divided into smaller reservation slots, and it is now in the ALOHA mode.

exponential distribution: It is a continuous *probability distribution* with the form given by: $y = 1/\bar{x} \exp(-x/\bar{x})$, where $\bar{x}$ is the mean of the

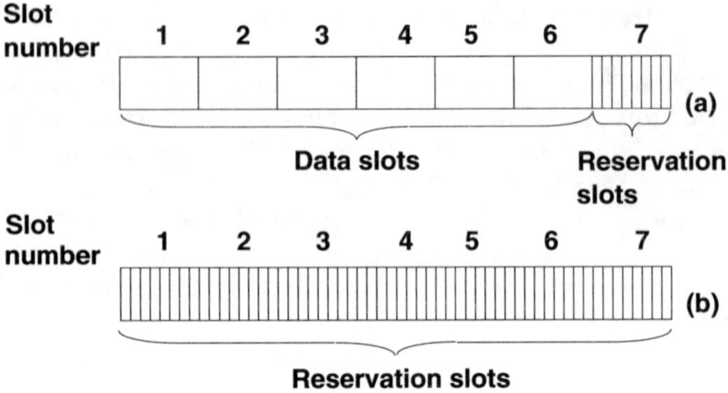

Figure E.21 Explicit reservation ALOHA system: (a) reservation mode; (b) ALOHA mode

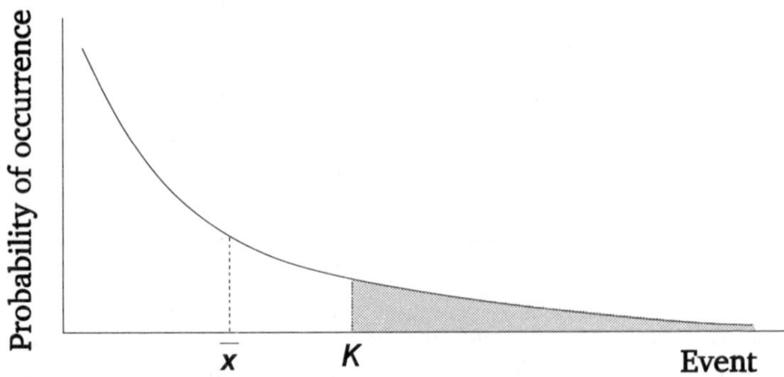

Figure E.22 The exponential curve

distribution. (Figure E.22.) For an exponential distribution 36.8% of the population is above the average value and 63.2% is below the average. Table E.10 shows the area under the exponential curve for different values of the ratio $K = x/\bar{x}$, this area being shown shaded in the figure. For example, suppose that the time between failures of a piece of equipment is found to vary exponentially. If results indicate that the mean time between failures is 1000 weeks, then what is the probability that the equipment will work for 700 weeks or more without a failure? Calculating K as 700/1000 = 0.7, then from the table the area beyond 0.7 is 0.497 which is the probability that the equipment will still be working after 700 weeks.

EXT: *End of Text.*

Table E.10 Area under the exponential curve

K	*0.00*	*0.02*	*0.04*	*0.06*	*0.08*
0.0	1.000	0.980	0.961	0.942	0.923
0.1	0.905	0.886	0.869	0.852	0.835
0.2	0.819	0.803	0.787	0.771	0.776
0.3	0.741	0.726	0.712	0.698	0.684
0.4	0.670	0.657	0.644	0.631	0.619
0.5	0.607	0.595	0.583	0.571	0.560
0.6	0.549	0.538	0.527	0.517	0.507
0.7	0.497	0.487	0.477	0.468	0.458
0.8	0.449	0.440	0.432	0.423	0.415
0.9	0.407	0.399	0.391	0.383	0.375

Extended Area Service (EAS): A service offered by *carriers* in which a *subscriber* can pay a fixed fee for obtaining coverage over a lager geographical area, without paying extra for *calls* made in this extended area.

Extended Binary Coded Decimal Interchange Code (EBCDIC): A *code*, used in IBM computers, which has a set of 256 *characters*, each character consisting of eight *bits*. It has been replaced by the *ASCII* code.

extended code: The addition of extra digits to a *code* so that its capabilities are increased, e.g. it can detect and correct a greater number of errors resulting from *transmission*

Extended Definition Television (EDTV): See *Enhanced Definition Television (EDTV)*

Extended Digital Subscriber Line (EDSL): A standard *Digital Subscriber Line* which has been extended beyond its specified limits by using *repeaters*.

extended numbering: Usually refers to the facility available in *ITU-T Recommendation* X.25 (see *X Series*) to allow an extension to the numbering of *frame* sequence. This permits larger *window sizes* to be use, for example in *satellite* systems which have larger *transmission* delays.

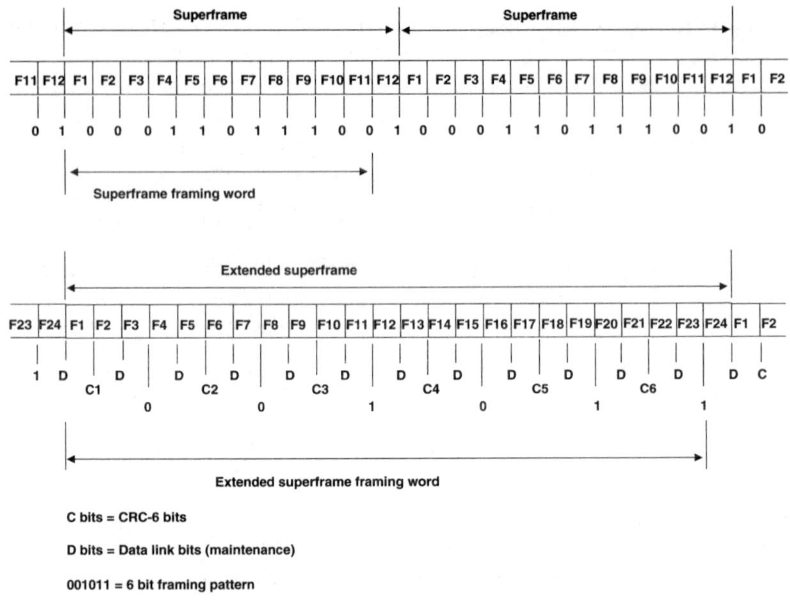

Figure E.23 Extended Superframe Format

Extended Superframe Format (ESF): A modification of the basic twelve frame *superframe* format used for *T1* transmission. ESF uses 24 *frames* in its definition of a superframe, as shown in Figure E.23, and it is replacing the older system. There are six bits in the *frame synchronisation* word, rather than 12. The C bits provide a six bit *Cyclic Redundancy Check (CRC)*, these being generated from the bits of the preceding frame. New *voice frequency signalling* capability is also available with ESF, by providing two additional bits for *Robbed Bit Signalling (RBS)*. These being known as the C and D bits and they complement the A and B bits of the superframe format.

Extended Total Access Communications System (ETACS): An extension to the *Total Access Communications System (TACS)* with additional *channel* allocation below the existing TACS channels.

Extensible Mark-up Language (XML): A proposed new standard for presenting information on the *World Wide Web (WWW)*. Its aim is to provide flexible control of content on WWW enabled devices. XML has the ability to define the type of information being marked and its location. It tells applications which parts of the contents to display, and where, and it works with *HTML* in displaying it.

extension: Usually refers to a *telephone* which is connected to a *switching centre*, such as a *PABX*, and has a unique *telephone number*. The term is also sometimes used to imply the opposite, i.e. an extension to an existing *line* in which two terminals have the same number.

extension bell: A bell, used on a *terminal* to provide the user with an audible indication of an *incoming call.*

extension circuit: A *circuit* which is used to provide a permanent connection from a *terminal* to another device, such as a *PABX*.

extension code: An increase on the basic capabilities of a *code* by the use of supplementary *characters.*

Exterior Gateway Protocol (EGP): The *protocol* which is used by *routers* to allow them to act as *gateways* between *Autonomous Systems* on the *Internet.* See also *gateway protocol* and *Border Gateway Protocol (BGP).*

external call: A *call* which goes beyond an internal switching system, such as a *PABX*, and uses an *external line.*

external line: Any *line* which connects a wide area, beyond a local area served by a system such as a *PABX*. Examples are *private circuits*, *trunk lines*, and *tie lines.*

external modem: A modem which is a stand-alone unit, i.e. not incorporated into another piece of equipment, such as a modem card which occupies a slot in a computer.

extinction ratio: A measure of the ability of a system to differentiate between two states. In an optical communication system it is the ratio of the optical energy in a logical 1 pulse to that in a logical 0 pulse; or the ratio of the energy received when a light source is on to when it is off.

Extranet: An interconnected group of *networks* which connect together a number of organisations which have an interest in communicating with each other, the networks being based on *protocols* and applications usually found within the *Internet.* Users which are not part of these organisations do not have access to this Extranet. Facilities also exist for ensuring that certain information can be kept secure and is not available to all members of the Extranet. See also *Intranet.*

Extremely High Frequency (EHF): The *electromagnetic spectrum* which lies in the *frequency range* from 30 GHz to 300 GHz. See also *Extremely Low Frequency (ELF).*

Extremely Low Frequency (ELF): The *electromagnetic spectrum* which lies in the *frequency range* from 30 Hz to 300 Hz. See also *Extremely High Frequency (EHF).*

extrinsic coupling loss: In *fibre optic transmission* systems, it is that part of the *coupling loss* which is not due to intrinsic factors. Examples are *angular misalignment loss* and *longitudinal displacement loss.*

eye: See *human eye.*

eye diagram: It is obtained by superimposing a sequence of received signals on each other, for example by recording the result on an oscillos-

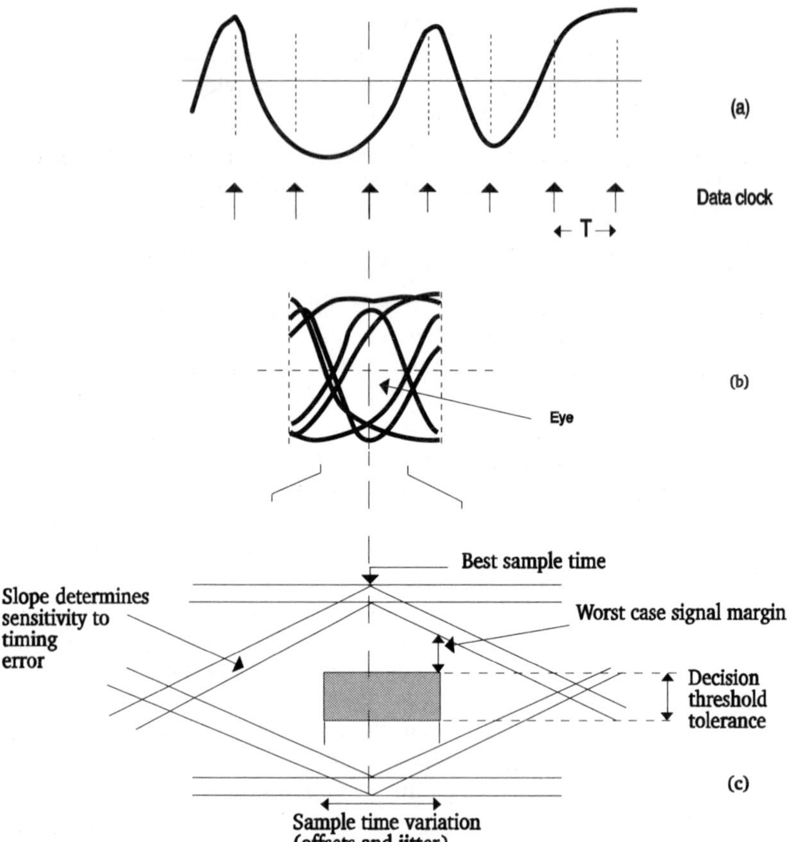

Figure E.24 Construction of a binary eye: (a) equalised binary signal; (b) eye diagram; (c) interpretation of the eye diagram

cope whilst triggering the oscilloscope timebase from the data clock, as shown in Figure E.24(a) and E.24(b). The eye diagram is known as such since the opening of the trace resembles the *human eye*. The diagram provides a large amount of information on the performance of the digital system, as shown in Figure E.24(c).

F

Fabry Perot laser: A *laser* in which two mirrors on the facets of its active layer are used for feedback of the light signals. See also *Distributed Feedback laser (DFB).*

facilities: A generic term used to describe *network* provided *hardware, software* or services.

Facilities Data Link (FDL): See *Embedded Operations Channel (EOC).*

Facilities Management (FM): The equivalent of *outsourcing* as used in the Information Technology (IT) industry, in which third parties are paid to manage an organisation's IT systems, such as computers.

facility request: The part of a *signal* which is concerned with defining and selecting the relevant *facilities* on a *network.*

facsimile (fax): Invented by Alexander Bain in 1843, facsimile is the technique for transmitting printed documents, including drawings and pictures, in electronic form over a *transmission medium*. The document is fed into a *scanner* which converts the black and white areas on the paper into *signals*. For *analogue signal* transmission. This is then modulated in a *modem* and sent over *PSTN* or *private circuit.* At the receiving end the process is reversed, a modem producing the signal which feeds into another *facsimile apparatus* which reproduces the original document. The most common type of facsimile is the *document facsimile system*, widely used within organisations. It is normally based on *ITU-T Recommendations*. Four Groups have been defined. Group 1 and Group 2 are for analogue transmission and Group 3 and Group 4 for digital transmission. Their characteristics are summarised in Table F.1. See also *photofax, weatherfax, pagefax, mobilefax, G3 facsimile* and *G4 facsimile*.

facsimile apparatus: The equipment used in *facsimile (fax) transmission.* Apart from the *document facsimile system* (see *facsimile*) there are many other types of equipment, as illustrated in Table F.2. All these use a separate unit for transmission and reception. In document transmission systems, used for business and commercial documents, a single unit does both transmission and reception.

facsimile bandwidth: The difference between the maximum and minimum *frequencies* used for *transmission* in a *facsimile* system.

facsimile baseband: Refers to the output from a *facsimile apparatus* immediately following the *scanning* process but before it has gone through any *modulation* onto a *carrier.*

facsimile gateway: A device, or *gateway*, which allows *facsimile* to be interworked with other forms of communications, such as *electronic*

Table F.1 Document facsimile types

	Public telephone (analogue)		*Public data (digital)*	
ITU-T Group/ Recommendation	G1/T.2	G2/T.3	G3/T.4/T.30	G4 Class 1/T.563
Transmission time	6 min (4 min option)	3 min	Around 30 sec (at 9.6kbit/s)	Around 5 sec (at 64kbit/s)
Data rate	Analogue		33.6–2.4kbit/s	64–2.4kbit/s
Modulation	FM	AM/VSB (vestige of upper sideband)	V.34, V.17, V.29, V.27ter	None (line driver and coder to suit network)
White	1300Hz (US 1500Hz)	Maximum carrier reversing phase		
Black	2100Hz (US 2400Hz)	26dB below white		
Synchronisation	±10 parts in 10^6	±5 parts in 10^6	Derived from data rate and sequence	
Index of cooperation	264 (176 option)	264	Not specified (264/528 option)	Not specified (549)
Scan line frequency	3Hz	6Hz	Not specified (around 400Hz)	Not specified (around 600Hz)
Horizontal resolution	Not applicable		1728 pels/215mm (8.04 pels/mm) (option 300 & 400pels/25.4mm)	200 pels/25.4mm (option 240/300/400)
Scan line length	215mm		215/255/303mm	219/260/308mm
Vertical resolution	3.85 lines/mm (2.57 opt)	3.85 lines/mm	3.85 lines/mm (option 7.7, 11.8 and 15.4 lines/mm)	Square with horizontal
Data compression encoding	None		One dimensional MH 2 dim MR, MMR JBIG & JPEG Uncompressed	2 dimensional MMR Uncompressed

Table F.2 Facsimile terminal types

Type and standard	*Page size (mm)*	*Nominal resolution (lines/mm)*	*Index of cooperation*	*Transmission time*
Photofax CCITT T.1	210 × 210 to 254 × 270	4 and 8 or 5 and 10	352 and 704	8 to 30 min
Weatherfax WMO	450 × 560	4 and 2	576 and 288	4 to 36 min
Pagefax proprietary	Up to 450 × 625	15 to 100	Not specified (1600 upwards)	1 to 3 min
Mobilefax proprietary	108 × 150	4	132	1 min
Military NATO Stanag 5000	215 × 297	4 and 8	Not specified (264/528)	20 sec to 20 min

mail and computer generated text. The gateway converts the electronic data into, for example, *ITU-T Recommendation* T.4 or T.30 for transmission as a facsimile signal.

facsimile switch: (1) A switching system, used within organisations, which allows *facsimile traffic* to be controlled and costed, including providing an archive of all facsimile transmissions. Figure F.1 shows an arrange-

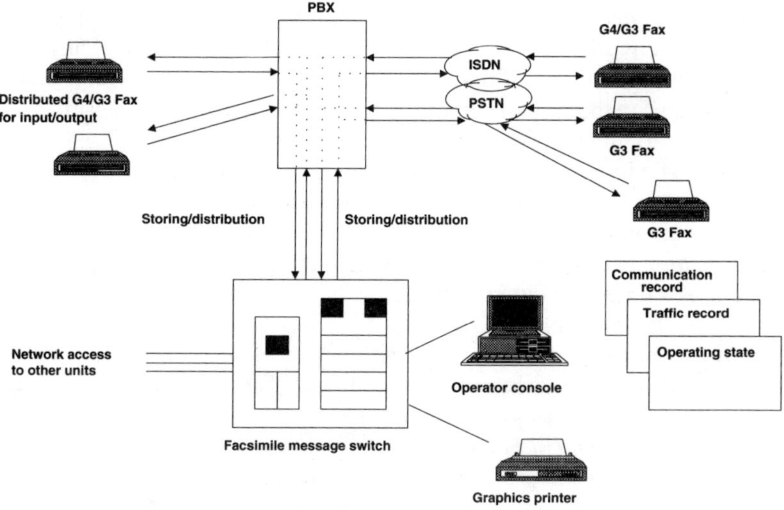

Figure F.1 G3 facsimile switch

ment for a *G3 facsimile* system in which all the facsimile traffic is routed through the central switch, associated with the organisation's *PABX*. **(2)** A switch which allows one line to be used for multiple devices, such as a *facsimile*, a *telephone* or a *modem*. The switch monitors the *incoming line* and switches in the appropriate device depending on the type of *traffic* sensed.

facsimile transceiver: The part of the *facsimile apparatus* which sends and receives the *facsimile signals*. It can either be *full duplex*, i.e. send and receive at the same time, or *half duplex*, i.e. send and receive at separate times.

facsimile transmission: See *facsimile*.

facsimile telegraphy: A form of *telegraphy* which uses *transmissions* based on *facsimile* to communicate text and images.

fade margin: A measure of the amount of *fading* which can occur in a *signal* whilst still maintaining an acceptable *Quality of Service (QoS)*. It is used to determine the unavailability of a system due to propagation effects. The probability (P) that fading will exceed a given margin (M) for a path length D (km) and at a frequency F (GHz) is given by $P = KQF^B D^C 10exp(-M/10)$, where D, Q, B and C are constants dependent on the propagation conditions.

fading: The loss of *signal* strength due to changes in propogation conditions, such as the characteristics of the *transmission medium*. It is mainly encountered in *microwave* and *radio transmission*. Causes of fading are losses in the transmission medium, obstructions in the transmission path, *multipath effects*, etc. See also *Rayleigh fading*.

fail safe: A circuit or system which is so designed that a failure of any part will not result in a catastrophic failure of the whole system, with damage being done to the system or harm to personnel. Generally the system may continue to function, although with reduced performance.

fail soft: The failure mechanism in which the system detects a failure and modifies its performance so as to prevent loss of critical *data* and malfunctioning of the equipment.

failure rate: The number of failures which occur in a system over a given period of time. A failure can be defined as being any malfunction of the system, including *errors* and faults.

fall time: The time needed for the *amplitude* of a *signal* to decrease over a given value, usually measured as the time to fall from 90% of the peak amplitude to 10%, as shown in Figure F.2.

FAP: *Frame Alignment Pattern*.

Faraday cage: In the protection of systems from *radiated emissions* the Faraday cage represents a perfect *shielding* method. It consists of a seamless box with no apertures, and made from material having zero resistance. Such a box does not exist in practice because all material has

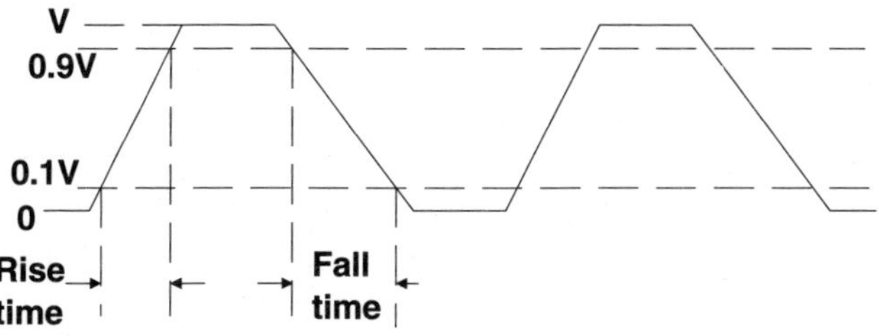

Figure F.2 Illustration of rise time and fall time

a finite resistance and it is not possible to build a structure without any discontinuities or apertures.

Far End Block Error (FEBE): A *signal* sent to the *transmitting terminal* to indicate that an error has occurred at the *receiving terminal.*

far-end crosstalk: *Crosstalk* which travels in the effected *channel* in the same direction as the *signal* in the effecting channel.

far-end echo: The *echo* caused by reflections of the *signal*, on a *transmission line*, which occurs when going from the *four-wire circuit* (the *trunk line* from the *exchange*) to the *two-wire circuit* (the *local line* connecting the *subscriber*). See also *near-end echo.*

far field sound: The sound which occurs from a source which is distant from the recording microphone. This sound is usually considered to be *noise*. See also *near field sound* and *noise cancelling microphone*.

FAS: *Frame Alignment Signal* or *Flexible Access System.*

fast Ethernet: A generic term used to describe *Local Area Network (LAN)* technologies which can carry *traffic* at *data rates* exceeding 10 Mbit/s, usually at 100 Mbit/s. Examples are *100Base-T* and *100VG-AnyLAN.*

Fast Fourier Transform (FFT): A mathematical technique which uses a fast *algorithm* for the efficient implementation of the *Discrete Fourier Transform (DFT)* in which the number of time samples of the input signal (N) are transformed into N frequency points and the required number of arithmetic operations is reduced to the order of $(N/2)\log_2(N)$.

Fast Frequency Shift Keying (FFSK): A modulation technique, used on *VHF* and *UHF* based radio systems, in which a *binary* 1 is represented by one cycle (e.g. at 1200 Hz, as shown in Figure F.3) and a binary 0 is represented by one and a half cycles (at 1800 Hz), with *Non-Return to Zero (NRZ)* coding.

Fast Packet Switching (FPS): Technique used to transfer *packets* of *data* at *data rates* which exceed that available from conventional *packet switching* (e.g. X.25; see *X Series.*) This is usually done by eliminating many of the *transmission* overheads. See also *Frame Relay (FR)* and *DQDB.*

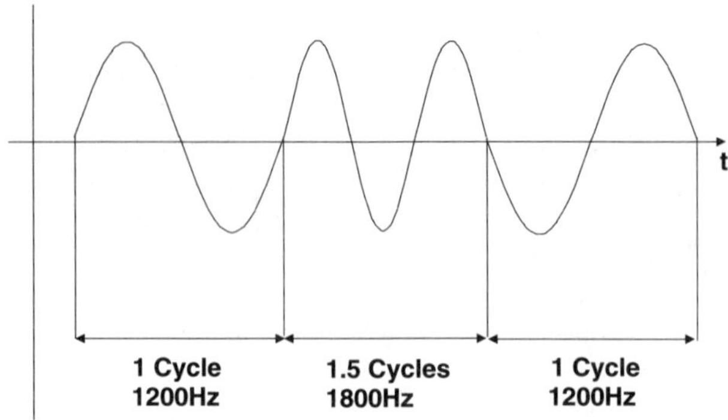

Figure F.3 Fast Frequency Shift Keying

fast select: A feature which allows user *data* to be included in a *packet* containing *call setup* and *call clearing* information. This feature is an essential requirement of the *ITU-T Recommendation* X.25 (see *X Series*).

fault: Generally refers to an error condition which causes the *Quality of Service (QoS)* being delivered to the *subscriber* to fall. This could be due to any number of causes, such as a permanent failure of a component or a temporary malfunction.

fault diagnosis: The process used in the detection of a *fault* and determining its cause.

fault management: The whole process of detection, diagnosis and correction of *faults* which occur on a *network*. It is one of the *network management functions* specified by the *ISO*.

fault masking: The situation which occurs when one *fault* hides another fault which occurs at the same time, and the second fault is only detected when the first one is rectified.

fault point: The locations on a *network* containing equipment which can exhibit *faults*. These locations therefore represent potential failure points on the network.

fault rate: A measure of the *reliability of service*. It is given by the number of *faults* which occur in the *network* over a given period of time.

fault report point: The location or locations on the *network* at which all *faults* which occur on the network are reported.

fault resilience: See *fault tolerance*.

fault tolerance: The ability of a component, system or *network* to continue to operate, within specified performance parameters, even if one or more *faults* occur. For example the failure of a *link* could result in *alternate path routeing*. Also called *fault resilience*.

FAW: *Frame Alignment Word*.

fax: See *facsimile*.

fax response: Generally refers to systems in which *facsimile* documents are automatically sent to callers. This is done, for example, by programming a *facsimile apparatus* such that when it receives a call from another facsimile machine it will send a pre-programmed document. Alternatively *voice processing* systems can be used to allow customers to call in and request information which is automatically faxed over to them by an associated facsimile apparatus.

FCC: *Federal Communications Commission.*

FCC Registration Number: A number attached to a *telephony* equipment which indicates that it is approved for connection to telephone lines. This has been specified in FCC docket 19528, part 68. The last three digits of this number specify the capabilities of the instrument. This is given in Table F.3.

FCS: *Frame Check Sequence* or *Forecasting Control System.*

FDDI: *Fibre Distributed Data Interface*

FDL: *Facilities Data Link.*

FDM: *Frequency Division Multiplexing.*

FDMA: *Frequency Division Multiple Access.*

FDMA/TDD: *Frequency Division Multiple Access/Time Division Duplex.*

FE: *Format Effectors.*

feature phone: A term applied to *telephones* which have additional built in features, such as *speed dialling*, etc.

FEBE: *Far End Block Error.*

FEC: *Feedforward Error Correction.*

FECN: *Forward Explicit Congestion Notification.*

Federal Communications Commission (FCC): Set up by the *Communications Act* of 1934 with the prime responsibility for regulating *traffic* carried by wire, radio, television, *satellite* or *cable*, both between US states and internationally to and from the US. The FCC is administered by five Commissioners appointed by the President and approved by the Senate. It is involved in a wide range of telecommunication standardisation and regulatory issues, although it does not produce standards of its own. It issues radio licences within the US to those complying with its rules, and also takes part in organisations involved in making telecommunication standards in the US, such as the *ANSI*, the *IEEE* and the *EIA*.

Federal Information Processing Standard (FIPS): Standards which have been developed by the US Federal government for information technology, and which are used by the Federal government for its own systems. In 1968 the *National Institute of Standards and Technology (NIST)* (formerly the *National Bureau of Standards (NBS)*) was given the responsibility for helping the Federal government make effective use of its vast base of computer and *Information Technology (IT)* equipment

Table F.3 FCC registration number

Code	*Capability*
Letters preceding last letter	
AL	Alarm dialler
AN	Answering machine
BR	Conference bridge
CI	Call forwarding
DI	Automatic dialler
DM	Modem
DT	Data terminal
MD	Computer modem
MU	Music on hold
MT	Multifunction telephone
RG	Extension ringer
SP	Speaker phone
TE	Telephone
WT	Cordless telephone
Last letter	
N	Device without dialling capability
R	Device with rotary (loop disconnect) dialling
T	Device with tone dialling

and for developing the Federal Information Processing Standards. Many FIPSs have been developed and published since then, the best known example is the *Data Encryption Standard (DES)* (FIPS 46) which has been adopted by ANSI as X3.92.

Federal Telecommunications System (FTS): The *private network* which is dedicated for use by the US Federal government and its civilian agencies, to call other Federal locations and to connect into the *PSTN*. It is also known as FTS 2000.

feedback error correction: A method for correcting *errors* in *transmission* in which only the receiver carries out the function of *error detection*. If errors are detected it informs the sender who would normally resend the *data*. The process is known as positive acknowledgement (or simply *acknowledgement ACK*) and *Negative Acknowledgement (NACK)*. See also *stop-and-wait ARQ*, *go-back-N ARQ* and *Feedforward Error Correction (FEC)*.

feeder: **(1)** A *line* carrying electrical power to an equipment. **(2)** A *transmission medium* which carries *signals* from a source to a distribution point. For example the radio signals which are fed from the radio source to an *antenna* for *transmission*.

feeder cable: A *cable* which acts as a *feeder*. Feeder cables are commonly used in *cable television* systems and carry the signals from the *head-end* to the main *amplifier* on the *trunk*.

Feedforward Error Correction (FEC): A method for correcting *errors* in *transmission* in which the receiver carries out *error detection* and *error correction* on the received *data*, without any reference to the sender. See also *feedback error correction*.

feed horn: The *feeder* used for *antennas*. The feed horn often forms part of the overall antenna performance and so careful design is required. Figure F.4 shows some of the constructions used with *reflector antennas*. The *polarisation* performance is important in feed horn designs, and the quality of the feed is often expressed by the level of the peak cross-polarisation.

feedthrough capacitor: A capacitor construction used for *RFI* shielding, where the inductance of the leads of a normal capacitor can reduce its effectiveness in this applications. In the feedthrough capacitor its outer body is screwed or soldered directly on to the metal material used for *EMC shielding*, as shown in Figure F.5(a). This allows the capacitor to provide effective operation well into the GHz *frequency range*. By separating the ceramic metallisation into two parts, and incorporating a ferrite bead, the through lead inductance is increased, allowing π-section filters to be obtained, as shown in Figure F.5(b).

female connector: A *connector* which has a cavity in which its contacts are located, the *male connector* being inserted into this cavity to engage the contacts. This mechanism is often used to ensure that the male and female connectors are fixed to each other and form a good electrical contact.

femtoseconds: 10^{-15} seconds.

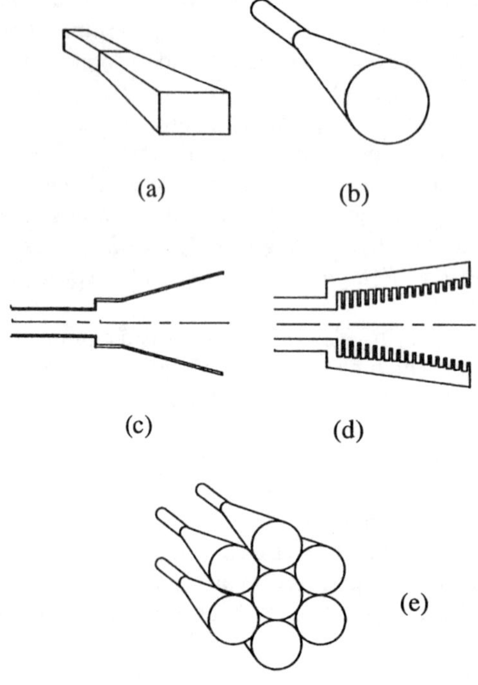

Figure F.4 Feed horns for reflector antennas: (a) rectangular or square; (b) small conical; (c) small conical with chokes; (d) conical corrugated; (e) antenna feed

FEP: *Front End Processor.*

FER: *Frame Erasure Rate.*

ferrul: A device which provides mechanical support and helps to align the stripped ends of *optical fibre cables.*

FF: *Form Feed.*

FFM: *Flat Fade Margin.*

FFSK: *Fast Frequency Shift Keying.*

FFT: *Fast Fourier Transform.*

FHSS: *Frequency Hopping Spread Spectrum.*

fibre: Thin filaments or wires, made from glass or plastic, which are used to carry *modulated* light *signals* for optical *transmission.* Transmission occurs by total internal reflection within the fibre. Also known as *fibre optic waveguide*. See also *singlemode fibre*, *multimode fibre optical fibre* and *fibre optic cable.*

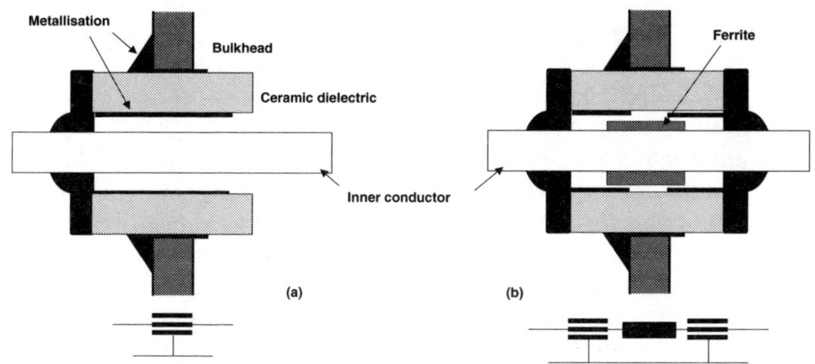

Figure F.5 Feedthrough capacitor: (a) simple capacitor; (b) π-section filter

fibre bandwidth: A measure of the maximum frequency which can be used for *transmission* over *optical fibres*. It is given as the frequency at which the *signal* level has fallen to below 3 *decibels* of its zero frequency value, i.e. it has halved.

FIA weighting: See *line noise weighting*.

Fibre Distributed Data Interface (FDDI): An *ANSI* and *ISO* defined *protocol* for a high speed *Local Area Network (LAN)* operating at 100 Mbit/s and using *token passing* for *multiple access* and *contention* control. The *network topology* consists of counter-rotating trunk rings with trees of cascaded *concentrators*. The *transmission* overheads are low, in the region of 10-20%, so the effective *throughput* of user *data* is about 80-90 Mbit/s. FDDI standards are defined by: Physical Medial Dependent (ANSI X3.166/ISO 9314-3); Physical layer Protocol (ANSI X.148/ISO 9314); Media Access Control (ANSI X.3.139/ISO 9314); Station Management (ANSI X.3.229). See also *Copper Distributed Data Interface (CDDI)*.

Fibre In The Loop (FITL): A generic term used to describe the use of *optical fibre* for part of the *network* connecting the *subscriber* to the *local exchange*. See, for example, *Fibre To The Home (FTTH)*, *Fibre To The Building (FTTB)*, *Fibre To The Curb (FTTC)* and *Hybrid Fibre Coax (HFC)*.

fibre loss: *Attenuation* of the *signal* travelling through an *optical fibre*. These losses can be of two types: intrinsic and extrinsic. Intrinsic losses are due to the fundamental properties of the material used to make the fibre, and includes *absorption* and *scattering*. Extrinsic losses are attributable to the actual fibre and its usage, and occur because of the practical problems of making and using fibres. Extrinsic losses include *bending loss*, losses at the permanent joins of fibres due to *splicing*, losses at *connectors*, and scattering at the core-cladding interface.

fibre optic: A generic term used to describe systems, subsystems and modules which are part of a *fibre optic communications* system.

fibre optic amplifier: See *optical amplifier.*

fibre optic attenuator: See *optical attenuator.*

fibre optic cable: See *optical fibre cable.*

Fibre Optic Connector (FOC): A *connector* which is used to connect two lengths of *fibre* together so that the optical *signal* can pass between them with low *fibre loss*. No *splicing* is used at the junction of the two fibres and the join can be readily made and broken. The join made using a *fibre optic splice* is permanent and cannot easily be broken and remade.

fibre optic communications: The use of light for communications. This can include *transmission* and *switching* of optical *signals* and these systems use *optical fibre* as the *transmission medium.*

Fibreoptic Link Around the Globe (FLAG): An underseas *fibre network* between Great Britain and Japan.

fibre optic splice: See *optical fibre splice.*

fibre optic splitter: A device used to divert some of the optical light power from one optical *transmission medium* to another. See also *fibre optic switch.*

fibre optic switch: An *optical switching* device, i.e. one in which all the light power travelling in one optical *transmission medium* is transferred to another. See also *fibre optic splitter.*

fibre optic waveguide: See *fibre.*

fibre pigtail: A short length of *optical fibre* which is normally permanently connected to an optical component. This pigtail can then be connected to an optical fibre *transmission line*, either by use of a *Fibre Optic Connector (FOC)* or a *fibre optic splice.*

Fibre To The Building (FTTB): A *Fibre In The Loop (FITL)* technology in which *optical fibre* forms the *network* from an *Optical Line Terminating unit (OLT)* in the *local exchange* to an *Optical Network Unit (ONU)* located on the *subscriber's* premises. Copper wiring, such as *twisted pair wire*, is then used to connect from the ONU to individual users within the building. The *data rate* over the optical fibre is clearly much higher than over the copper wire. The ONU performs the conversion between the electrical and optical *signals.*

Fibre To The Curb (FTTC): A *Fibre In The Loop (FITL)* technology, similar to *Fibre To The Building (FTTB)*, in which the *ONU* is moved closer towards the *local exchange*, being located on the curb close to *subscribers'* premises. *Optical fibre* is used to connect between the *local exchange* and the ONU and *twisted pair wire* connects the ONU to the subscribers. This is a cheaper system to FTTB since it spreads the cost of expensive optics and electronics over many subscribers.

Fibre To The Home (FTTH): A *Fibre In The Loop (FITL)* technology in which *optical fibre* is used in the *network* all the way from the *local exchange* to the *broadband* equipment, such as television or computer, on the *subscriber's* premises. Although providing the maximum amount of *bandwidth* to individual users, this system can be very expensive to implement compared to *FTTB* or *FTTC*.

field: **(1)** A collection of *bits*, usually occurring in a *frame* or a *record*, which convey *information*. **(2)** A force source which is a physical phenomena and can be varied. For example *electromagnetic radiation*. **(3)** A location where equipment is installed and working, e.g. not in the factory where it was built. **(4)** An open space, usually without any existing forms of communications.

field flyback: Part of the *flyback* used in the *scanning* system of a *Visual Display Unit (VDU)*. It is the period during which the *electron beam*, used for scanning, is turned off whilst the *raster* is moved to begin a new scan operation for the next line. Also called *line flyback*.

field frequency: The rate at which the complete *scanning* operation takes place during the creation of an image on a *Visual Display Unit (VDU)*. It is measured in *hertz*. For a television system with 525 lines (USA) it is equal to 15750 Hz and for a 625 line system (UK) it is equal to 15625 Hz. Also called *line frequency*.

field intensity: The *irradiance* of an *electromagnetic wave*, usually measured in watts per square metre. See *field strength*.

field programmable unit: A piece of equipment which can have some or all of its operating characteristics changed whilst installed in the *field*, i.e. it does not need to be returned to the factory to have this done.

Field Replaceable Unit (FRU): Refers to equipment or parts of an equipment which can be replaced when in the *field*, i.e. it does not need to be returned to the factory for this to be done. Usually it should be possible to carry out this replacement with basic tools and a lower level of skilled personnel.

field scan: The vertical movement used by the *raster* during *scanning*.

field strength: The magnitude of an electromagnetic, electric, magnetic or similar *field* at a given point. See also *field intensity*.

field wire: *Twisted pair wire* used as a temporary *transmission line*, usually for emergency or military use, such as for *telephony* or *telegraphy*, out in the *field*.

15 supergroup: See *hypergroup*.

figures shift: A special *character* in a *code*, which indicates that the subsequent characters in the *transmission* are to be treated as numbers or special characters (e.g. punctuation marks). Figures shift is operational until a *letters shift* occurs. Figures shift can also be implemented manually on a keyboard.

file: A collection of *records*. The file is usually structured to enable related records to be selected as a unit.

File Separator (FS): Special *character* used to differentiate boundaries of *files*.

file server: The *hardware* (usually a computer) and *software* which is used to store *files* centrally on a *Local Area Network (LAN)* so that they can be accessed by multiple users.

File Transfer, Access and Management (FTAM): An *Application Layer* standard within the *OSI Basic Reference Model*. It provides functions for accessing and modifying files on another system having an *open system architecture* and for transferring them between open systems.

File Transfer Protocol (FTP): The *Transmission Control Protocol/Internet Protocol (TCP/IP)* which allows files to be transferred between systems, such as between computers connected to the *Internet* or on a *Local Area Network (LAN)*.

fill characters: Extra *characters* which have been added to the data and do not form part of the main *message*. For example fill characters can be added in order to fill up a *buffer*, or to make the character count equal a certain value, or to prevent unauthorised entry in any space remaining.

filter: **(1)** A device which allows certain *frequencies* in the *signal* to go through but causes severe *attenuation* of other frequencies. See, for example, *high-pass filter*, *low-pass filter*, *band-pass filter*, and *band-stop filter*. **(2)** The term filter is also sometimes applied to any device which lets through certain signals whilst stopping others, the selection criteria being based on other considerations apart from frequency. For example, a *router* would examine certain *fields* in the message *frames* and let through some of the messages only.

finite state machine: A machine, or a mathematical model of a machine, which can only reach a finite number of states and transitions between these states. It is used in mathematical problem analysis.

FIPS: *Federal Information Processing Standard*.

firewall: *Software*, usually residing on a computer located on the *network*, which prevents unauthorised access in and out of the network. For example it can prevent users from outside an organisation from accessing resources on the organisation's network, or it can prevent users within the organisation from going outside their network to access expensive external services. The firewall can be programmed to be selective, e.g. prohibit access to all internal resources, to unauthorised users, or prohibit access to only some of these resources.

firmware: A programme (*software*) which is loaded into *hardware* (such as a memory chip) and cannot be changed by other software. It is frequently used in its existing form so it does not need to be changed. However, if required, it can be modified by other methods, e.g. changing the memory chip, or erasing its content electrically and reprogramming it.

first party clearing: A method of *call control* in which *call clearing* occurs when either the *called party* or the *calling party* terminates the call, e.g. goes *on-hook*. See also *calling party clearing*.

first party maintenance: The term used when the owner of the equipment carries out all the maintenance required on it.

FITL: *Fibre In The Loop*.

fixed assignment: A *non-contention multiple access* technique. Examples are *Frequency Division Multiple Access (FDMA)*, *Time Division Multiple Access (TDMA)*, Code Division Multiple Access (CDMA) and *Wavelength Division Multiple Access (WDMA)*.

fixed attenuator: An *attenuator* in which the amount of *attenuation* is fixed at manufacture and cannot be varied by the user.

fixed Earth station: An *Earth station*, used in a *satellite* communications system, which is fixed in one location, i.e. it is not portable and cannot be moved easily.

fixed equalisation: *Equalisation* technique used with some equipment, such as *modems*, in which the amount of equalisation is fixed, having been determined by a user via settings on the machine.

fixed loss loop: A classification system operated by the *FCC* which requires that the *signal* output from *modems*, which are registered with it, do not exceed 4 *dB*.

fixed microwave link: A *transmission link* which is set up using *microwave* radio, in which the microwave transmitters are fixed relative to each other.

fixed routeing: A method of *routeing* in which the *routeing tables* available in each *switching equipment* is fixed, usually at manufacture, and cannot be varied, for example to take account of *traffic* or *network* conditions.

Fixed Priority-Oriented Demand Assignment protocol (FPODA): A *Priority-Oriented Demand Assignment (PODA)* technique for *multiple access* in which fixed assignments are used to access the reservation slots.

Fixed Satellite Service (FSS): *Radio transmission* service, between *satellite Earth stations*, which operates in the 11.7 GHz to 12.2 GHz *frequency band*.

fixed tolerance-band compaction: A method of *data compression* in which *information* is only transmitted when the value of the *data* is either above or below a tolerance band. For example, in Figure F.6 the value of the data need only be transmitted during periods t_1 and t_2. When no transmissions are taking place the receiver assumes that the data is equal to some value within the tolerance band.

fixed virtual circuit: See *Permanent Virtual Circuit (PVC)*.

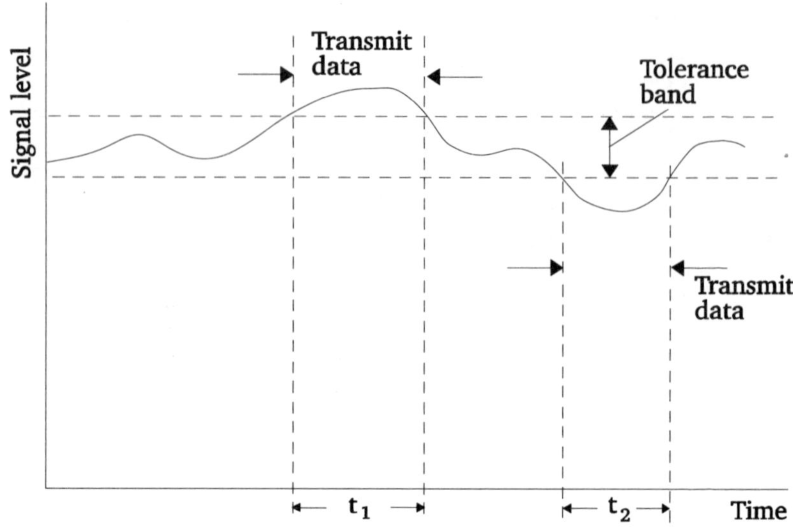

Figure F.6 Fixed tolerance-band compaction

fixed weight binary code: A *binary code* in which each numeral has a fixed number of logical 1s. For example these could be 0011, 0101, 0110, 1001, 1010, and 1100, i.e. a binary weight of two.

flag: A *signal* used to indicate the occurrence of an event or a condition. For example a fixed *bit pattern* is used in certain *protocols* to indicate the start and end of a *frame*.

FLAG: *Fibreoptic Link Around the Globe.*

flag sequence: The sequence of *bits* used as a *flag*. For example, the sequence 01111110 is used in many bit-oriented *protocols* to indicate the start and end of a *frame*.

flash: The term used for sending a short *signal* to obtain attention or additional services. This can be done, for example, by momentarily depressing and releasing the telephone switchhook or an associated manual switch.

flash converter: An *Analogue to Digital Converter (ADC)*, as shown in Figure F.7, in which one input to a chain of converters is fed from a fixed voltage derived from a resistor chain, with the other fed by the *analogue signal.* The number of comparators equals the number of levels which are to be coded in the output *digital signal.* In Figure F.7 a three bit output is assumed, so that eight levels are needed (000, 001, 010, 011, 100, 101, 110, 111). Flash converters are capable of very fast analogue to digital conversion rates.

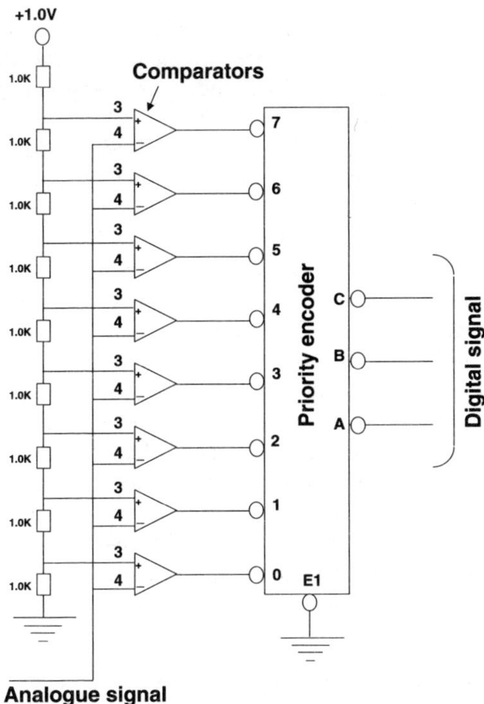

Figure F.7 3 bit flash converter

Flat Fade Margin (FFM): Term often used to describe the normal *fade margin* of a system. The FFM is often combined with the *Dispersive Fade Margin (DFM)* into a single term, called the *Composite Fade Margin (CFM)*.

flat rate service: A method of charging for telecommunications services in which a fixed *tariff* is charged (usually per month) irrespective of the amount of usage.

flat weighting: A *noise weighting* measurement technique in which measurements are made using a *frequency response* curve which has a flat amplitude-frequency characteristic over the *frequency range* being considered.

F layer: A layer of the *Earth's atmosphere*; see Figure E.4. During night this layer is relatively thin and is situated about 210 kilometres above the Earth's surface. During daylight ionisation due to the Sun's rays causes the F layer to increase in size and split into the F_1 and F_2 layers, at heights of 210 km and 300 km. Radio frequencies in the *High Frequency (HF)* band pass through the lower layers and are reflected from one of the F layers, so providing a means for very long distance radio communications. Frequencies in the *Very High Frequency (VHF)* range, i.e. above 30 MHz, generally pass through all the layers and escape into space.

Fletcher-Munsen curves: See *NPL curves*.

FLEX: *Protocol* developed by Motorola for *paging*. It can handle up to 2 million short *message addresses* and 5 million long addresses. *Interleaving* is used for both message and address *blocks* to protect against *burst errors*. Battery life is prolonged by use of very short synchronisation periods. The *code* has 128 *frames* per cycle, as shown in Figure F.8, and each cycle lasts 4 minutes. Each frame has a *synchronisation* block and eleven *information* blocks. The first synchronisation pattern contains information on the *bit rate* and type of *modulation*. This is followed by the frame information block, which provides the frame identity number and battery saving information. The second synchronisation pattern provides synchronisation to the new speed. Address, vector and message codewords follow the synchronisation blocks.

Flexible Access System (FAS): An *overlay network*, consisting mainly of *optical fibre*, introduced by BT in the UK as part of its public *network*. The prime objective was to enable *leased lines* to be controlled and so provide business users with a flexible option regarding mix of *circuits* used.

flexible routeing: The ability to vary the *routeing* of *calls* through the *network* depending on circumstances. This feature, provided by AT&T

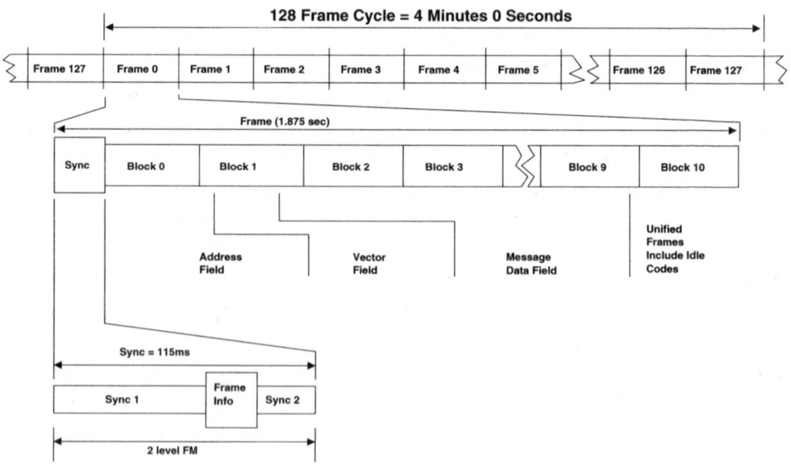

Figure F.8 FLEX code format

in the US, allows users, for example, to reroute calls to another number where a recorded message can provide additional information.

flicker: Variation in the intensity of an image on a *Visual Display Unit (VDU)* which gives the impression of an unsteady or flickering picture. This can be due to several causes, such as variations in the voltage, current, or *scanning line rate*.

flicker frequency: The *frequency* at which an image is presented to the *human eye*. If this frequency is too low then the eye will notice *flicker* in the image. As the frequency is increased the amount of flicker reduces until at about 50 Hz to 60 Hz this flicker will disappear and the eye will see a steady image.

floating: Usually refers to a *circuit* which has not been connected to *ground* and therefore does not have a fixed reference.

Floating point Operations Per Second (FLOPS): A measure of the processing power of a processor, such as a computer. It is the number of arithmetic operations it can carry out on floating point numbers, in a second.

flooding: The process of sending the same *packet* simultaneously through many different routes in a *Packet Switched Network (PSN)*, in order to ensure that it arrives at its destination.

flooding compound: A material which is used to fill voids in systems so as to prevent the ingress of contamination, such as moisture.

flood projection: A method of *scanning* used in *facsimile* equipment in which the document is fully illuminated and the *scanning spot* is defined by a masked area.

floor distributor: A more formal term for a *wiring closet.*

FLOPS: *Floating point Operations Per Second.*

flow control: The control of the *data* passing between two *terminals* to ensure that all the data is correctly received. For example, the *transmitting terminal* may be able to send data at a much higher *data rate* than the *receiving terminal* is able to process it. Flow control must then be used either to slow down the transmitting rate or to *buffer* the additional data which the receiving terminal cannot immediately process.

flow control information: *Information* which is interchanged between a *transmitting terminal* and a *receiving terminal* on the *network*, to enable effective *flow control* to take place.

fluorescence: The emission of *electromagnetic radiation* by certain materials when they absorb other forms of electromagnetic radiation. For example the light produced when a beam of electrons strikes a phosphor coated screen in a *Visual Display Unit (VDU).*

flutter: Rapid changes in the characteristics of a *signal*, such as its *amplitude*, *phase*, or *frequency*. This variation can occur due to various causes, such as changes in the *transmission medium*, and are often temporary.

flyback: Part of the *scanning* operation in which the *scanning beam* moves back to its start position so that the next scan can begin. See *field flyback* and *line flyback*.

flyback period: The time taken for *flyback*. During this period the *scanning beam* is turned off or blanked so the flyback period should be a short as possible.

flying spot scanner: A device in which a beam of light is used for *scanning* an object, such as a document. The reflected light from the object is detected by *photodetectors*, which generate *signals* that represent the object.

FM: *Facilities Management* or *Frequency Modulation.*

FM broadcast: Radio *broadcast* which uses *Frequency Modulation* of its transmitted *signals*.

F number: The number which relates the aperture of a optical lens in relation to the brightness of the object. The brighter the object the smaller the F number needs to be.

FOC: *Fibre Optic Connector.*

FOCC: Forward Control Channel.

Foiled Twisted Pair (FTP): *Twisted pair wire* which has a overall metallic coated plastic foil shield along its length. It exhibits better *EMC* performance than *Unshielded Twisted Pair (UTP)* cable. See also *Screened Foiled Twisted Pair (S- FTP)*.

follow on call: A *network* feature which allows a *call establishment* to be started on a new *call* before *call disestablishment* has been completed on the previous call. This feature could be used, for example, when several serial international calls are to be made to the same country.

footcandle: A measure of *illuminance*. One footcandle is equal to one *lumen* falling on an area of one square foot.

footprint: **(1)** The area of the Earth's surface over which an *antenna*, based on a *satellite*, can transmit *signals* with adequate power. **(2)** The area of the floor (or desk) occupied by a piece of equipment.

Forecasting Control System (FCS): Control system used within a *call centre* to forecast the requirement for staff and *trunks*, both to meet immediate and long term needs. It can also be used to prepare staff work schedules and for determining when these schedules are not adhered to. The FCS uses a variety of data, both current and historical, e.g. on the call *traffic*, call handling performance, etc.

foreign exchange service: A *network* provided service in which a users can be connected to a *Central Office (CO)* which is different from their *local exchange*. For example a *subscriber* in city A may be connected to a Central Office in city B (via a *private line*) rather than to his local exchange in city A. Now when callers dial his number from city B they will be making a local call, even though the call may be travelling several hundred miles to the subscriber in city A.

Format Effectors (FE): Special *characters* which are used in *data communications codes* to perform special control functions on devices such as printers. Examples are: *BS* for *Backspace*, *CR* for *Carriage Return*, *FF* for *Form Feed*, *LF* for *Line Feed*, *HT* for *Horizontal Tabulation, and VT* for *Vertical Tabulation.*

Form Feed (FF): A *control character* used to indicate to the system that a new form is to be displayed or printed. The most common application is to cause the printer, or the cursor of the *Visual Display Unit (VDU)*, to skip to the top of the next page.

forms traffic: Term normally used to describe the *traffic* which flows on a *network* in which a low level of traffic in one direction results in a much higher volume of traffic in the opposite direction. Example is the traffic used to query a *database* which results in a large amount of *data* flowing out of the database in answer to the query.

FORTRAN: It stands for FORmula TRANslator. It is a high level computer programming language, primarily used for mathematical operations.

fortuitous conductor: Any conductor which is available and provides an unplanned path for *signals*. Examples are water pipes, metal structural supports used in a building, etc.

fortuitous distortion: *Distortion* which occurs due to random causes and cannot be predicted.

forward busying: The *network* feature in which a *supervisory signal* is sent before any *address* signal. This seizes all the resources needed for the *call* in advance of the *call establishment*.

forward channel: The *channel* which carries *data* from the *calling terminal* to the *called terminal*. This usually consists mainly of user data and some control signals. It is also known as the main channel. See also *backward channel*.

Forward Control Channel (FOCC): The control channel used in *base station* transmitters for *trunked mobile radio*. See also *Reverse Control Channel (RECC)*.

forward direction: The direction in which *information* (as distinct from control and *supervisory signals*) flows. This is carried in the *forward channel*.

forward echo: An *echo* which travels in the same direction as the original *signal*. This echo can be caused by a defect in the *transmission medium* which causes a backward echo (known simply as an echo) which then meets a second defect and is reflected back as a forward echo.

Forward Error Correction (FEC): An *error correction* technique in which extra *bits*, which have been calculated according to an *algorithm*, are added to *transmission* by the *transmitting terminal*. These are used by the *receiving terminal* for both *error detection* and *error correction* without the need to ask for any *retransmissions*.

Forward Explicit Congestion Notification (FECN): Used in *frame relay* where a bit in the header of the *frame* is set to a logical 1 to indicate that the frame has met congestion on the *network*.

forward scatter: The *scattering* of an *electromagnetic wave* in which the component created by the scatter propagates in the same direction as the main incident wave.

forward signal: A *signal* sent in the *forward direction*, i.e. from the *transmitting terminal* to the *receiving terminal*.

FOT: *Frequency of Optimum Transmission*.

Fourier analysis: The mathematical analysis of *waveforms*, named after the French mathematician Emile Fourier who first proposed it. See, for example, *pulse train*, *triangular wave* and *sawtooth wave*.

four-wire circuit: A *transmission* system in which two *channels* are used to send *data* in one direction and two channels are used for data in the reverse direction (Figure E.9). Often these channels consist of separate

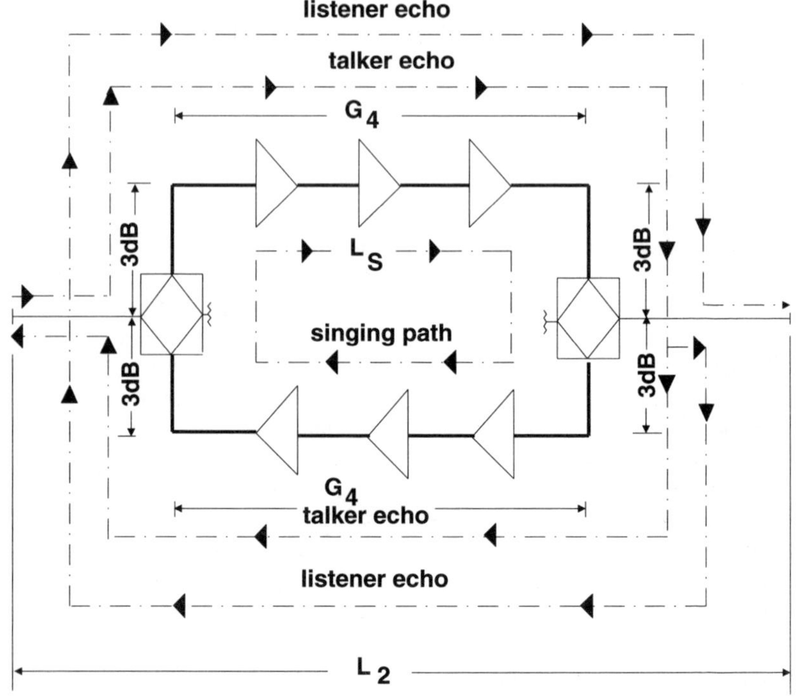

Figure F.9 Four-wire circuit

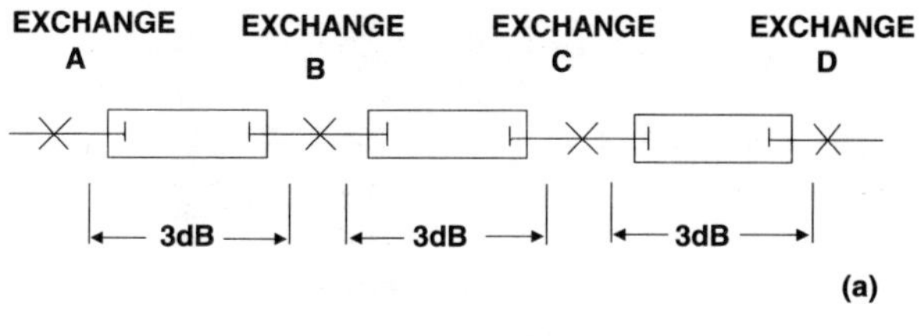

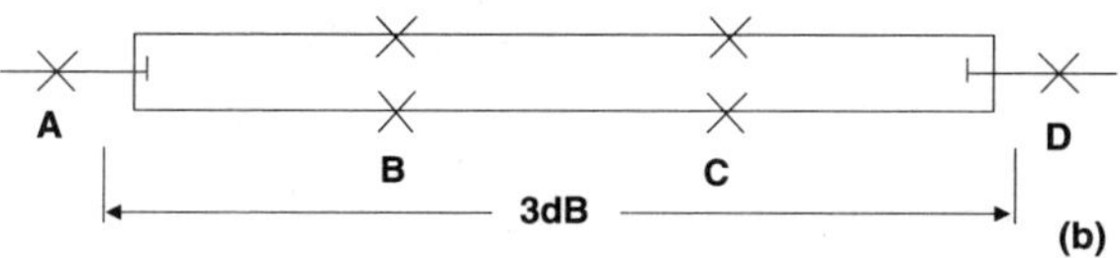

Figure F.10 Two and four-wire switching: (a) two-wire switching; (b) four-wire switching

transmission lines or wires. Four wire circuits are commonly used on *trunk lines* since these need to have *amplifiers* along the way, to compensate for *signal loss* in the long paths, and amplifiers tend to be unidirectional in operation. Figure F.9 shows a four-wire circuit arrangement, along with the *talker echo*, *listener echo* and *singing paths* through the system. A four-wire circuit usually connects to a *two-wire circuit* at each end via a *hybrid transformer*.

four-wire repeater: A *repeater* which is part of a *four-wire circuit* and has two *amplifiers*, one for each direction of *transmission*.

four-wire switching: Switching of *four-wire circuits*. If *two-wire switching* is used then the four-wire *trunk lines* must be converted to *two-wire circuits* at every exchange, as shown in Figure F.10(a). Because of the loss introduced at the four-wire/two-wire interface this would result in large overall *signal loss* along the *transmission path*. Using four-wire switching, as in Figure F.10(b), reduces this overall loss.

four-wire transmission: *Transmission over a four-wire circuit.*

fox message: Standard test message used to test keyboards since it uses all the letters of the alphabet. It reads: the quick brown fox jumped over the lazy sleeping dog. (Or: the quick brown fox jumped over the lazy dog's back.)

FPLMTS: *Future Public Land Mobile Telecommunication System.*

FPODA: *Fixed Priority-Oriented Demand Assignment protocol.*

FPS: *Fast Packet Switching.*

Fractional E1: A service offered by some *PTOs* in which a full *E1* (2 Mbit/s) *data rate* capability is provided to the *subscriber* but only a fraction of this (usually in 64 kbit/s slices) need be used, and only the *bandwidth* used is paid for.

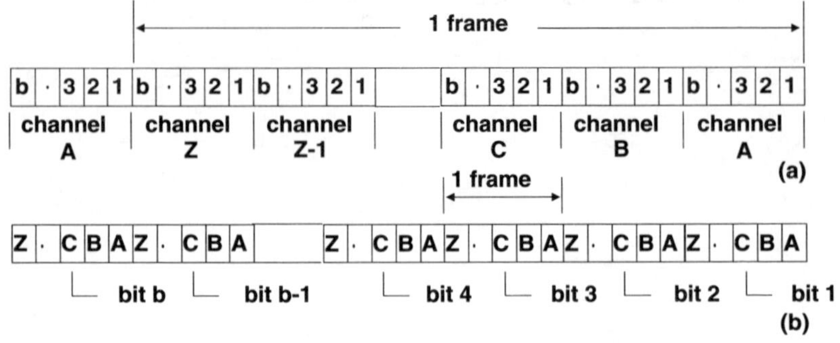

Figure F.11 TDM frame structure: (a) word interleaved; (b) bit interleaved

Fractional T1: A similar service to *Fractional E1* except that the *transmission medium* can carry a maximum of *T1* rates and a fraction of this can be used by the *subscriber*, up to the full amount.

FRAD: *Frame Relay Access Device.*

frame: A group of *bits* and *bytes* which are collected together in a recognised format for *transmission*. Some of the *data* carried in the frame represent user information and other parts are used for control and *signalling*. In a transmission system a frame is repeated many times to form the whole *message*. In *Time Division Multiplexing (TDM)* the frames can be formed in two ways, by word interleaving and by bit interleaving, as shown in Figure F.11. In word interleaving the interleaving is done on a *character* by character basis, whilst in bit interleaving it is performed on a bit by bit basis.

frame aligner: The circuitry within the *receiving terminal* which carries out the *frame alignment*. A simple bit by bit aligner is shown in Figure F.12. A test window, having the exact length as the *Frame Alignment Word (FAW)*, is used to detect the incoming FAW and to compare this with the value stored. If the two match then frame alignment has been achieved and the system counts a known number of *bits* and looks for the next FAW. If the test window does not match the stored FAW then the system continues searching the incoming signal until a match is found.

frame alignment: The technique used to ensure that there is *synchronisation* between the *frames* sent and received. It is used when *signals* go through *multiplexing*.

Frame Alignment Pattern (FAP): See *Frame Alignment Signal.*

Frame Alignment Signal (FAS): The *signal* used for *frame alignment*. This usually consists of a pattern of logical 1 and logical 0 *bits* and is also referred to as the *Frame Alignment Word (FAW)* or the *Frame Alignment Pattern (FAP)* or the *framing pattern*. The framing pattern is carried by

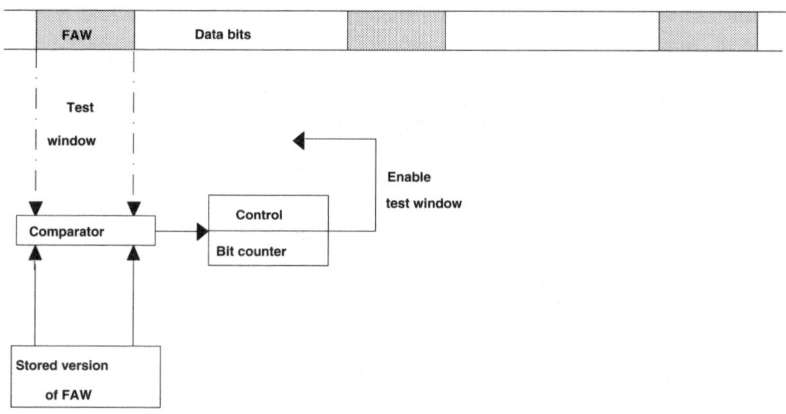

Figure F.12 Frame aligner

frames according to one of two systems: grouped framing patterns, in which a number of consecutive bits are used at the beginning of a frame; and distributed framing pattern, in which the pattern is spread over a frame on a bit by bit basis or over several frames at one bit per frame. For a *T1* circuit this consists of a pattern of six bits spread over several frames, in the sequence 101010. For an *E1* circuit it consists of seven bits in the sequence 0011011, in one frame, as shown in Figure F.13. The *ITU-T Recommendation* for frame alignment, for the 30 channel *plesiochronous* multiplexer hierarchy is given in Table F.4.

Frame Alignment Word (FAW): See *Frame Alignment Signal (FAS)*.

Frame Check Sequence (FCS): A sequence of logical *bits* transmitted in the *frame*, which is used by the *receiving terminal* to determine if an error has occurred during *transmission*. It does so by computing the *checksum*, using an agreed *algorithm* and then checking the value trans-

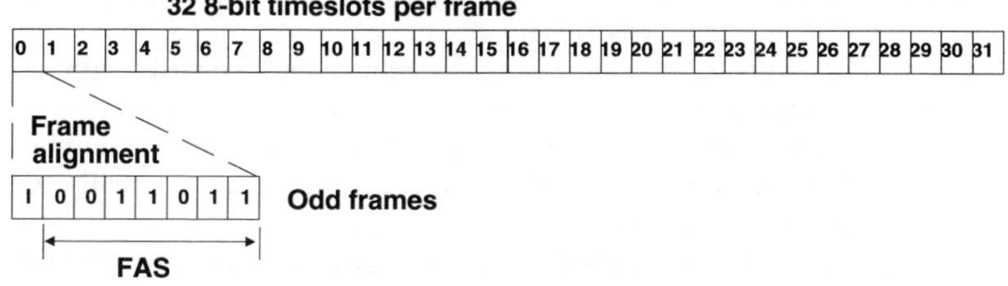

Figure F.13 Frame Alignment Signal in 30 channel PCM

Table F.4 ITU-T Recommendations for FAS for 30 channel PCM

Multiplexer level	*ITU-T Rec.*	*Frame length (bits)*	*Frame alignment word*
Primary (approx. 2Mbits/s)	G732	512	0011011 (7 bits)
Second (approx. 8Mbits/s)	G745	1056	11100110 (8 bits)
Third (approx. 34Mbits/s)	G751	1536	1111010000 (10 bits)
Fourth (approx. 140Mbits/s)	G751	2928	111110100000 (12 bits)

mitted to see if it is the same. If not an error is assumed. The FCS is normally transmitted at the end of the frame, before the closing flag. The number of bits in the FCS can vary depending on the *code* used. For *HDLC* it consists of 16 bits. See also *Cyclic Redundancy Check CRC).*

Frame Erasure Rate (FER): A measure of the *transmission* quality, in systems such as *GSM*. It is given by the number of bad *frames* received. See also *Bit Error Rate (BER)* and *Residual Bit Error Rate (RBER)*.

frame duration: The time occupied by one *frame*. The period between the start of one frame and the start of the subsequent frame.

frame format; The structure of a frame, i.e. its various components such as the *address field*, the control field, the *Frame Check Sequence (FCS)*, etc.

frame grabber: A device which examines *frames* as they pass and selects the ones whose *address* match its own. Used in *Video Display Units (VDU)* and other *video* applications, such as *teletext*, where frames are selected from a general *broadcast*.

frame level protocols: *Protocols* which are concerned primarily with the control of *frames*, such as its formation, routeing, and error checking.

frame rate: The speed of *transmission* of *frames*, measured as the number of frames passing a given point over a fixed period of time. Usually the frame rate is measured in frames per second.

Frame Relay (FR): A *Fast Packet Switching (FPS)* protocol which uses variable length frames for *data transmission*. Frame Relay can operate at much faster *data rates* than conventional *packet switching* systems. It uses a lightweight *protocol*, compared to X.25 (see *X Series*), which

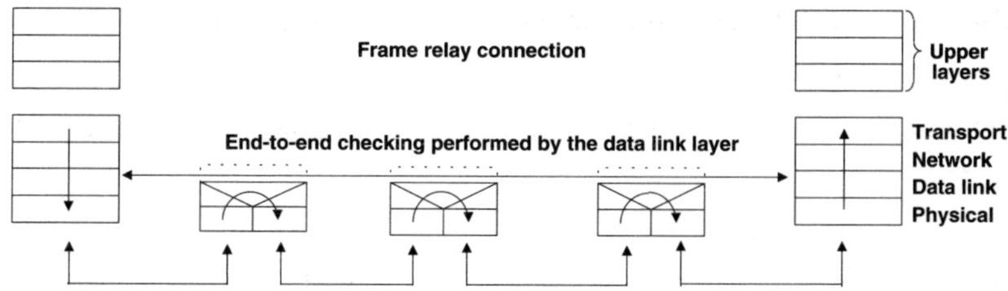

Figure F.14 Frame Relay model

means that it does not perform many of the basic control and error checking functions at each node. These are left to the end systems, so the *transmission path* must be of a higher quality (such as *optical fibre*) and relatively error free. Frame Relay only operates on the first two layers of the *OSI Basic Reference Model*, as shown in Figure F.14. The *routeing* path is part of the *Data Link Layer* and so the *network* addressing only works on a link-by-link basis. Nodes within the network contain *routeing tables* which show where frames should go, but no mechanism to check that they have safely arrived. Frame Relay is described in *ITU-T Recommendations* I.122 and I.441.

Frame Relay Access Device (FRAD): A device which performs *protocol conversion* by changing the input signal into a *Frame Relay* protocol so that it can be transmitted over a Frame Relay *network*.

Frame Relay Forum: Founded in 1991 with world-wide headquarters in Foster City, California. It is an association of some 150 vendors, users and carriers and its aim is to promote the implementation of Frame Relay in line with international standards. The Forum contains a Technical Committee, which takes existing standards and develops them into Implementation Agreements (IAs). These IAs are agreements between the Forum's members to apply the standards in a particular way, so as to enhance interoperability. Table F.5 gives some of these early IAs. The Technical Committee works through several subcommittees, each dedicated to a particular IA. The Forum also contains Marketing Committees in North America and Europe whose mandate is to spread the understanding and implementation of Frame Relay technology. They do this by producing newsletters and guides and by organising seminars and trade shows.

Frame relay User to Network Interface (FUNI): A *User-Network Interface (UNI)* which allows interworking between *ATM* and *Frame Relay (FR)* systems.

frame slip: The loss of *frame synchronisation* between the sent and received *frames*.

Table F.5 Some Frame Relay Implementation Agreements

Number	*Date*	Implementation Agreement Title
FRF.1.1	January 1996	User-to-Network (UNI)
FRF.2.1	July 1995	Frame Relay Network-to-Network (NNI)
FRF.3.1	June 1996	Multiprotocol Encapsulation (MEI)
FRF.4	n/a	Switched Virtual Circuit (SVC)
FRF.5	December 1994	Frame Relay/ATM Network Interworking
FRF.6	March 1994	Frame Relay Service Customer Network Management (MIB)
FRF.7	October 1994	Frame Relay PVC Multicast Service and Protocol Description
FRF.8	April 1995	Frame Relay/ATM PVC Service Interworking
FRF.9	January 1996	Data Compression Over Frame Relay

frame store: A memory unit which has the ability to store a *frame* of *data*. See also *page store*.

frame synchronisation: The alignment of a *timeslot* in the received frame with the same timeslot in the transmitted frame. Usually frame synchronisation is done by means of bits in the frame *header field* which also indicates the beginning of each frame.

frame synchronisation pattern: The sequence of logical 1 and logical 0 *bits* which are carried within a frame for *frame synchronisation*.

Framework Programme: Co-operative joint research programmes which are run between companies, universities and research establishments across Europe, and managed by the *Directorate General* DG XII working closely with DG XIII. The Fourth Framework Programme ran between 1984 and 1988 and had a total budget of ECU13.1 billion.

framing: The process of defining a *frame* so that *timeslots* within it can be identified. Framing is commonly used in *Time Division Multiplexing*

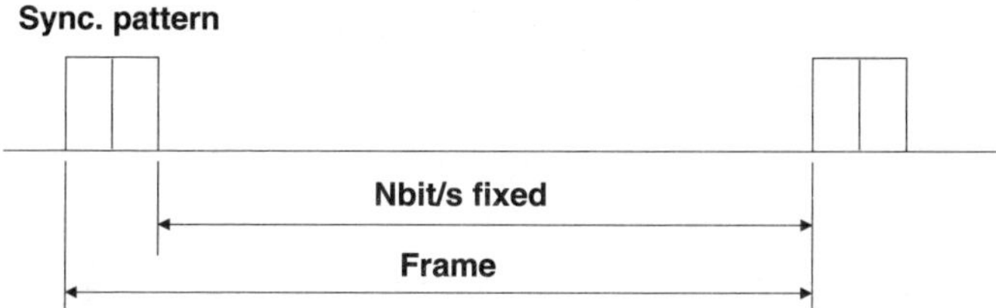

Figure F.15 Basic framing operation

(TDM) to ensure that *frame synchronisation* occurs. Framing is achieved by adding extra bits to the user data, usually at the start and end of a *frame*, to provide the reference, as shown in Figure F.15.

framing bits: The *bits* which are added to a *frame* to achieve *framing.*

framing pattern: See *Frame Alignment Signal (FAS).*

freephone service: *Network* provided *telephone* service in which the a special number is used, usually beginning with set letters such as 0800, which results in the called party accepting the *billing* costs. This is the same as the *800 service.* Although freephone services have been available in many countries for many years, international freephone started in June 1996 when the *ITU* published the standard for global 800 numbers. This can clearly only be used for numbers which terminate in a nation participating in the international freephone agreement. In the USA alone there are more than 100 million freephone calls placed every day.

free space: In its widest term this is a theoretical concept of an area which is completely devoid of any objects as well as any *electromagnetic radiation.* However in practical use it refers to an area which is devoid of solid materials, such as the space between two wires. See *free space communications.*

free space communications: *Transmission* which occurs through the atmosphere, i.e. does not use any solid *transmission medium.* Free space communications therefore covers all forms of radio, such as *cellular radio systems, microwave* and *satellite.* However it would normally mean that the *radio waves* were far away from the influence of the Earth's surface and any other reflecting or absorbing objects, including the transmitting and receiving *antennas.* Also known as *free space propagation.*

free space loss: The *attenuation* of the *signal* which occurs during *free space communications.* For a *half wave dipole antenna* free space loss in *decibels* is given by the *ITU* as $L_{fr} = 32.44 + 20\log f + 20\log d$, where f is the frequency in megahertz and d is the distance in kilometres. Figure F.16 give the free space loss for different frequencies and distances, based on this formula.

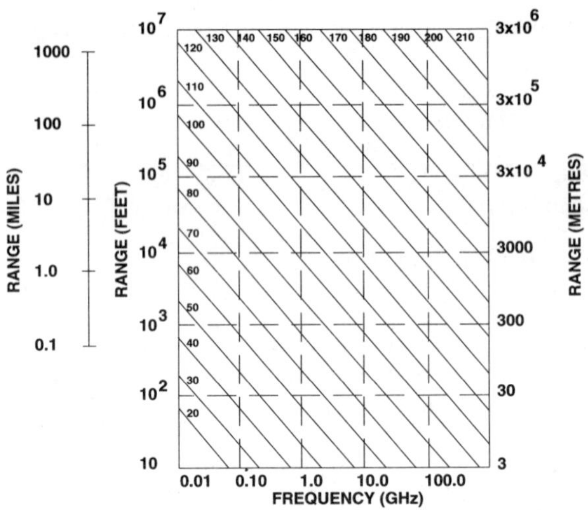

Figure F.16 Free space loss

free space propagation: See *free space communications*.

freeze frame: An image frame which has been selected from a *video* display and which is transmitted at a low speed so that the received *signals* do not give the impression of motion.

F region: The same as *F layer*.

frequency: It is the number of complete cycles of a *signal* or event which occur in a unit of time. Generally this time period is taken to be a second and the number of cycles which occur in a second are called *hertz*. Frequency is the reciprocal of *periodic time*.

frequency agile radar: A *radar* which has *frequency agility*.

frequency agility: The ability of a system to change its operating frequency, often automatically, to meet changing conditions. For example a *radar* could change its operating frequency to avoid a *jamming signal* or to reduce interference from another source.

frequency ageing: The gradual change in operating frequency of a device over time due to changes in the characteristics of its internal components.

frequency allocation plan: See *frequency assignment plan*.

frequency assignment plan: A plan showing the *frequencies* which have been allocated for use by different users or applications. For example frequencies have been allocated for use by emergency services, a *Wireless Local Area Network (WLAN)*, etc. and Table F.6 shows the frequency assignment plan, defined by the *ITU* for broadcasting satellite services in different regions of the world. Also known as *frequency allocation plan*.

Table F.6 Frequency assignment plan for broadcasting satellite service

	Region 1	*Region 2*	*Region 3*
Broadcasting downlink bands (GHz)	11.7–12.5	12.2–12.7	11.7–12.2
Feeder uplink bands (GHz)	14.5–14.8 and 17.3–18.1	17.3–17.8	14.5–14.8 and 17.3–18.1

frequency band: A continuous range of *frequencies* which lie between an upper and a lower limit. Frequency bands are often given names, for example, *Low Frequency (LF)* band (30 kHz to 300 kHz), *Medium Frequency (MF)* band (300 kHz to 3 MHz) and *High Frequency (HF)* band (3 MHz to 30 MHz).

frequency coordination: The agreed selection and use of *frequencies* for various applications and, in different geographical areas, so as to minimise interference between these frequencies.

frequency derived channel: A *channel* which has been obtained as a result of *Frequency Division Multiplexing (FDM)* another channel or *line*.

frequency converter: A device which changes the *frequency* of the *signal* from one value to another.

frequency distortion: The *distortion* which occurs in a *signal* due to different *frequencies* having different *attenuations* when propagating through a *transmission medium*.

frequency diversity: The use of a second *frequency channel* for carrying the *signal* in the event that the main frequency channel becomes unusable.

Frequency Division Multiple Access (FDMA): A *multiple access* technique in which the total *frequency band* is divided into segments and each user has exclusive use of one segment. This is shown pictorially in Figure F.17. *Guard bands* are used between each segment of the frequency band to prevent interference between users. There is clearly no *contention* in this system but it is not very efficient since if users are not using their allocated segments these lie idle and cannot be used by anyone else.

Frequency Division Multiple Access/Time Division Duplex (FDMA/TDD): A *Frequency Division Multiple Access (FDMA)* system in which *Time Division Duplex* operation is used to carry the *traffic* in both directions on one or more *channels*.

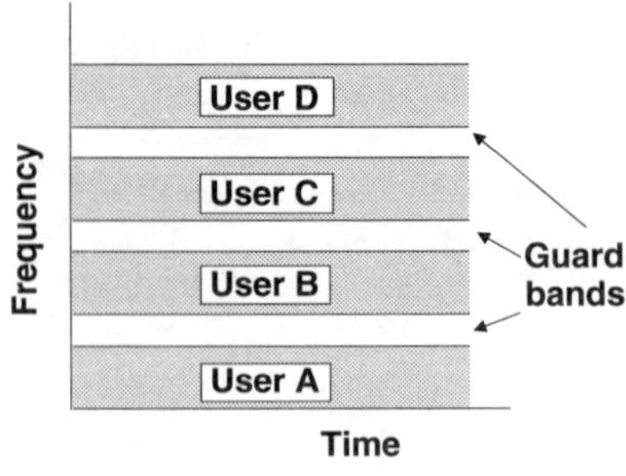

Figure F.17 User assignment within FDMA

Frequency Division Multiplexing (FDM): An earlier method of *multiplexing*, used with *analogue voice*, in which the total *frequency band* is divided into smaller groups, each carrying one *signal channel*. Figure F.18 illustrates the principle of FDM. Two channels, C_1 and C_2 having frequency bands of 0.3 to 3.4 kHz are mixed with separate *carrier signals* and only the lower *sidebands* are selected by the filters Fa1 and Fa2. The line *frequency spectrum* shows a *channel spacing* of $f_2 - f_1$. The process is reversed at the receiving end.

frequency drift: The slow, uncontrolled change in *frequency* caused both by component ageing (*frequency ageing*) and changes in environmental conditions.

frequency frogging: The process in which the frequency allocated to *carrier signals* are interchanged, usually to overcome some *transmission* defects, such as *crosstalk* or *singing*.

frequency guard band: A *guard band* formed by not using a small *frequency band* between two *channels* which are carrying *signals* so as to prevent interference between them.

frequency hopping: The technique used to continually change the *frequency* of the *carrier signal* during *transmission*. This is usually done

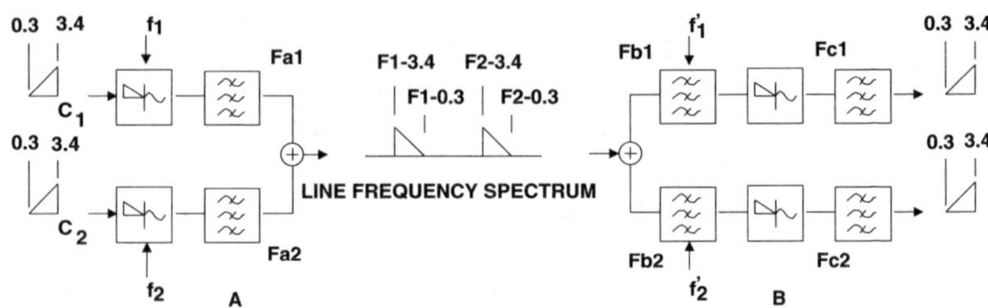

Figure F.18 Frequency Division Multiplexing

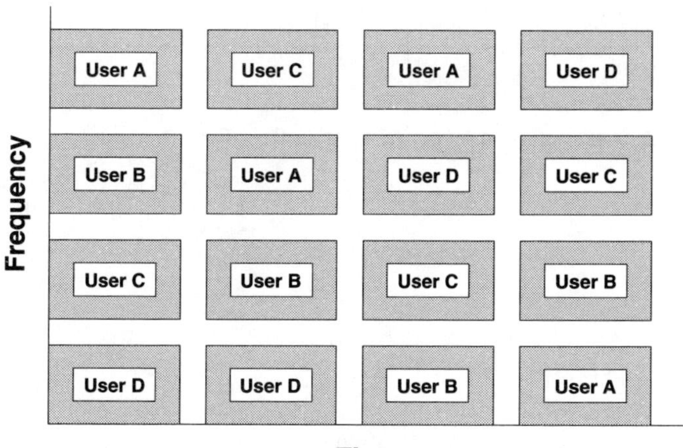

Figure F.19 User assignment using frequency hopping

according to a set *algorithm* which is known to the intended receiver, which hops in synchronism with the transmitter. *Spread spectrum* techniques are usually used for this. Figure F.19 shows a simplified representation of frequency hopping, and although all the users are shown to hop in step with each other this is not necessary in practice. Frequency hopping is generally used to prevent unauthorised monitoring of the signals and to minimise the effects of a *jamming signal.*

Frequency Hopping Spread Spectrum (FHSS): One of the techniques used for *spread spectrum* communications. It is based on the use of *frequency hopping* to spread the signal over the total *channel bandwidth.* For example a total bandwidth of W could be divided into N equal parts and different signals (less than N) could occupy one or more of these N parts, the actual frequency at which this part is located at any time being varied, as in Figure F.19.

Frequency Modulation (FM): A *modulation* technique in which the frequency of the *carrier signal* is caused to vary in proportion to the *modulating signal.* Figure F.20, for example, shows frequency modulation in which the modulating wave is a *digital signal.*

frequency offset: The amount by which the *frequency* of a *signal* varies in relation to the frequency of a reference signal, usually measured as a fraction of the reference frequency. If f_s is the frequency of the signal and f_r is the frequency of the reference, then the frequency offset of the signal is given by $(f_s - f_r)/f_r$.

Frequency of Optimum Transmission (FOT): The optimum frequency which can be used for *sky wave* and *ground wave* propagation systems,

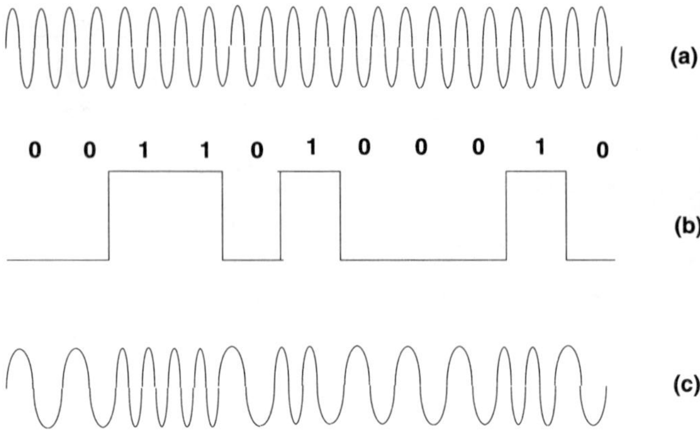

Figure F.20 Frequency Modulation: (a) carrier; (b) modulating waveform; (c) modulated waveform

which provides maximum *availability of service* and minimum *path loss*. Generally this is taken as 90% of the *Maximum Usable Frequency (MUF)*. Also known as the optimum transmission frequency.

frequency range: The difference between the lowest and the highest frequency in a *frequency band*. The terms frequency range and frequency band are often used interchangeably.

frequency response: (1) The behaviour of a *circuit* or system to different *frequencies*. For example an *amplifier* will normally have a frequency response in which it amplifies signals with different frequencies by different amounts. Similarly the *attenuation* of a *transmission line* will vary with frequency. **(2)** The difference in signal strength between frequencies in a *frequency band*.

frequency reuse: The ability to reuse the same frequency more than once in a system. Generally this is done in a *cellular radio system* by using different *cell repeat patterns* to avoid interference between adjacent *cells*. (See Figure C.10.)

frequency reuse ratio: In a system with *frequency reuse*, such as a *cellular radio system*, if there are N *cells* in the *cell repeat pattern*, D is the repeat distance between cells (as in Figure F.21) and R is the radius of a cell, then the frequency reuse ratio is given by $D/R = \sqrt{3N}$.

frequency scanning: The process of searching for a *frequency* over a *frequency band*. This may be done manually or automatically.

frequency sharing: Two users who are using the same frequency. Normally this sharing would involve some form of separation to prevent interference, for example by using the frequency at different times or in different locations.

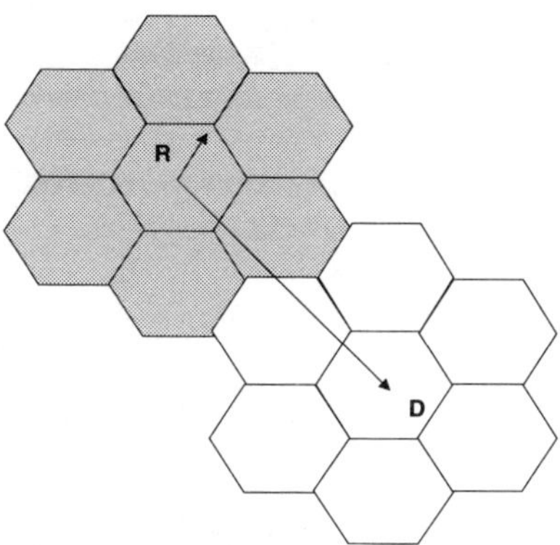

Figure F.21 Frequency reuse ratio

frequency shifting: (1) Changing the *frequency* of a *signal.* **(2)** Frequency modulation technique, used in *facsimile* systems, in which one frequency represents black and another represents white, grey scales being represented by in-between frequencies.

Frequency Shift Keying (FSK): A *modulation* technique in which the frequency of the *carrier* is shifted between set values by the *modulating signal.* FSK is usually used for transmission of *binary* signals and the frequency is shifted between two values for logical 1 and logical 0, as shown in Figure F.22. For large frequency deviations, compared to the digit rate, the spectrum is as shown in Figure F.23(a), being concentrated about the two frequencies f_1 and f_2. If the frequency deviation is reduced, as in Figure F.23(b), the *sidebands* about f_1 and f_2 merge.

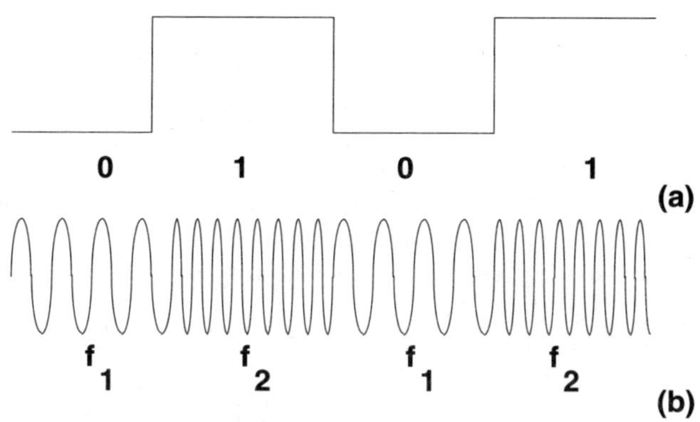

Figure F.22 Frequency Shift Keying: (a) baseband signal; (b) FSK signal

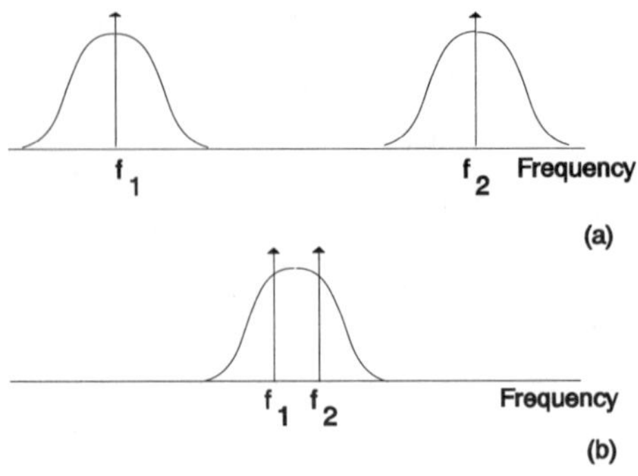

Figure F.23 Spectra for Frequency Shift Keying: (a) large deviation; (b) small deviation

frequency shift telegraphy: *Telegraphy* which uses *Frequency Modulation (FM)*, different frequencies being used to define the various states.

frequency spectrum: A portion of the *electromagnetic spectrum* which represents all the frequency components which are present in a *signal*. See *radio frequency spectrum*.

Frequency Subnet Number (FSN): Number allocated to individual *pagers* within *ERMES*.

frequency tolerance: **(1)** A measure of the accuracy of the *assigned frequency*, both in relation to the initial set up and subsequent changes, caused by factors such as *frequency ageing* and *frequency drift*. For a *frequency band* this is measured at the centre frequency. **(2)** The maximum change in *frequency* which a system can tolerate whilst still meeting its performance criteria. Frequency tolerance is measured in *hertz* or in parts per million.

frequency translation: **(1)** The shifting of a *frequency band* from one position in the *frequency spectrum* to another, no other changes occurring regarding the relationship of frequencies within the band. **(2)** The change in *frequency* of a *signal* as it travels down a *transmission medium*.

Fresnel equation: Equations which define the reflection and transmission performance when an *electromagnetic wave* is incident at the interface to two surfaces. For example, if a semi-infinite opaque screen is assumed to exist between a transmitter and receiver, as in Figure F.24, then Fresnel's integral states that the *attenuation* is related to free space, in amplitude and phase, by the complex function F, where F = $\frac{(1+j)}{2}\int_{\nu}^{\infty} e^{-j\frac{\pi}{2}(t^2+\nu^2)}\,dt$ and where $\nu = h\left(\frac{2(a+b)}{\lambda ab}\right)^{1/2}$.

Fresnel reflection: The reflection of light from the interface of two surfaces having different reflective indices. Fresnel reflection occurs, for

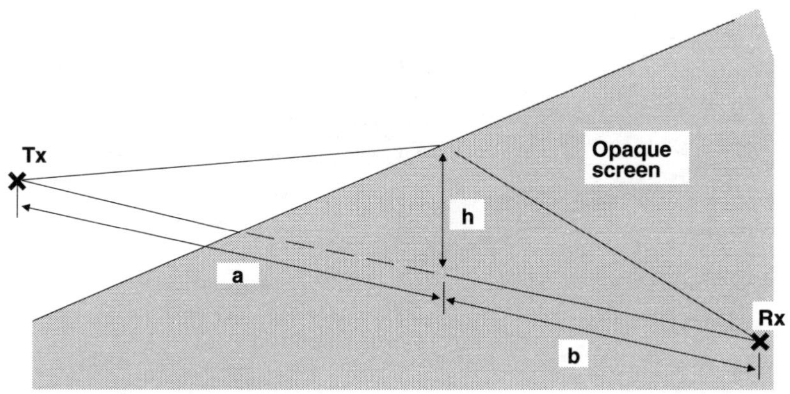

Figure F.24 Fresnel diffraction parameters

example, at a *optical fibre* splice or break point or when light enters or exists the fibre.

Fresnel zone: In radio communications between a transmitter and a receiver it is the surface of the ellipsoid of revolution with the transmitting and receiving *antennas* at the focal points in which the reflected wave has an indirect path half a miwavelength longer than the direct paths. This is illustrated in Figure F.25. There are several Fresnel zones, as shown in Figure F.26, for each of which the distance from the transmitter to a point on the surface is an additional half wavelength longer than the direct path.

Front End Processor (FEP): A unit which interfaces mainframe computers to communications systems and performs all the communications

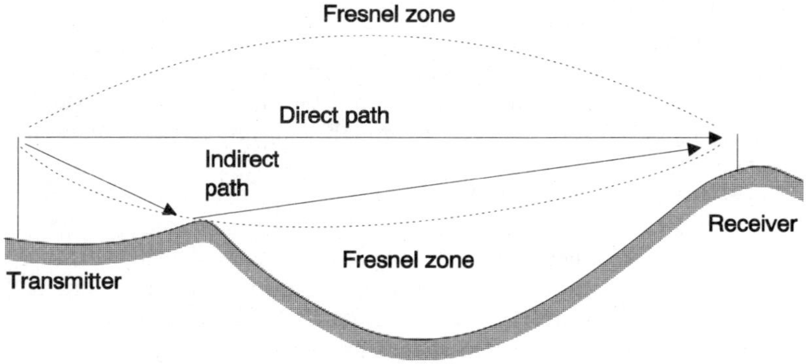

Figure F.25 Fresnel zone

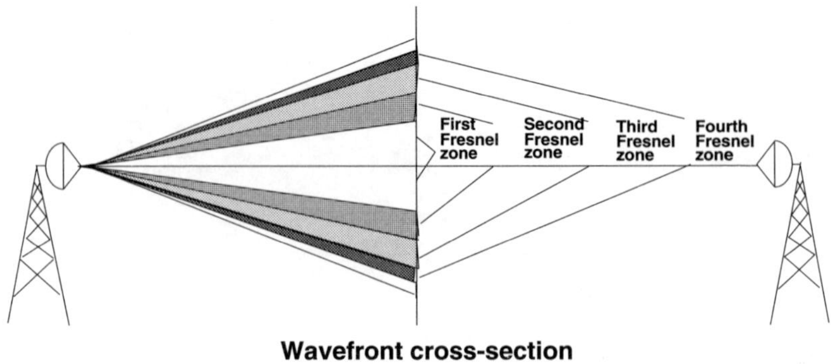

Figure F.26 Fresnel zones

related activities. This frees the computer from carrying out these tasks so that it can concentrate on processing. See Figure C.14.

FRU: *Field Replaceable Unit.*

FS: *File Separator.*

F Series: *ITU-T Recommendations* covering *telegraphy* and telematic service. This includes *telex, facsimile* and *videotex.*

FSK: *Frequency Shift Keying.*

FSN: *Frequency Subnet Number* or *Full Service Network.*

FSS: *Fixed Satellite Service.*

FTAM: *File Transfer, Access and Management.*

FTP: *File Transfer Protocol* or *Foiled Twisted Pair.*

FTS: *Federal Telecommunications System.*

FTTB: *Fibre To The Building.*

FTTC: *Fibre To The Curb.*

FTTH: *Fibre To The Home.*

full availability: Generally refers to a *switching* system in which there is a connection from all the switching groups from one stage to the next. This means that *calls* are connected through and *call blocking* does not connected.

Full Competition Directive: Article 90 of the *European Community (EC)* which was adopted on 29 February 1996. This mandated all member states to open up their public switched telephone markets by 1 January 1998. Some countries (Greece, Ireland, Portugal, Luxembourg and Spain) were allowed to opt for a later date to open up their markets to competitive voice network providers, and this has been referred to as a derogation.

full duplex: The simultaneous transmission of *signals* in both directions. See also *half duplex* and *simplex.*

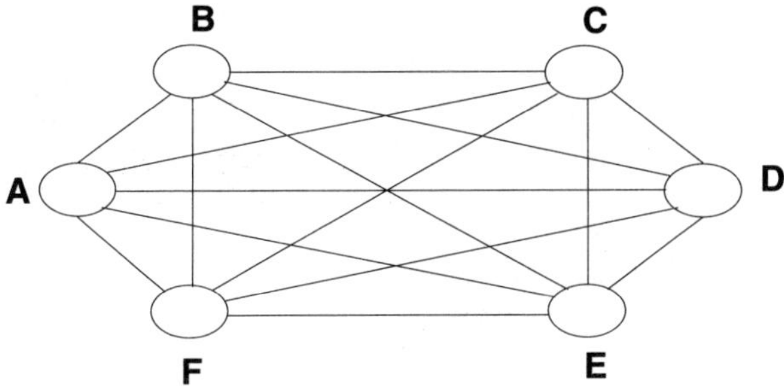

Figure F.27 Fully interconnected network

full echo suppressor: An *echo suppressor* which uses the *signals* travelling on one path, of the *four-wire circuit*, to control the suppression loss in the other path. See also *half echo suppressor*.

Full Service Network (FSN): Generic name for a *network*. providing *data*, *voice* and *video* services.

fully interconnected network: A *network* in which every *node* is directly connected to every other node, as in Figure F.27. If there are N nodes then the number of *circuits* needed to fully interconnect them is $1/(2N(N-1))$. This gives a cost which is approximately proportional to N^2, which can get very expensive if the number of nodes on the network is large. Other *network topologies*, such as the *star network* and the *partially interconnected network*, are then preferred. See also *network topology*.

fully provided route: A *transmission path* which can carry all the *traffic* required without having to depend on alternative routes when the traffic gets heavy.

functional profile: See *functional standard*.

functional specification: A document which defines the operating parameters of an equipment or system. The functional specification defines the characteristics, rather than the operating mode, i.e. what rather than how.

functional standard: Functional standards (also known as *functional profiles*, or *profiles*) specify a subset of the *base standards* for particular applications, often adding details which are missing from the base standards. (See Figure B.3.) It is these profiles which largely enable *interworking* between systems, for example those supplied by different manufacturers. Functional standards are usually developed by regional and national standards bodies, working with trade organisations and user groups. Agreed test specifications and test methods are also developed to verify interworking.

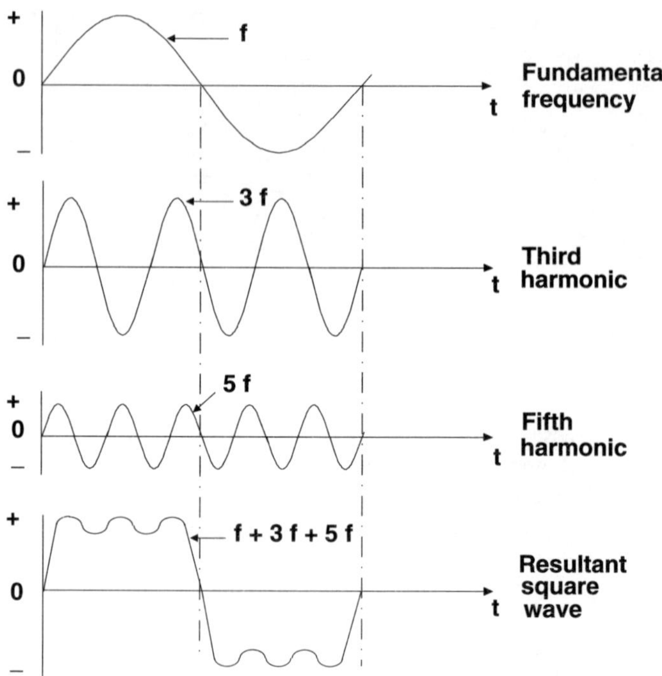

Figure F.28 Additions of harmonics to the fundamental frequency

functional test: A test carried out to verify that a device meets its *functional specification*. Functional tests are often specified as part of a *functional standard*.

fundamental frequency: The lowest *frequency* which exists in a *waveform*. A *sine wave* contains a single frequency, which is the fundamental, but any other waveform can be considered to be made up of a fundamental frequency plus *harmonic frequencies* which are a multiple of the fundamental. For example a waveform may consist of a fundamental frequency of f_0 and second and third harmonics at frequencies of $2f_0$ and $3f_0$ respectively. Figure F.28, for example, shows how these have been combined to give an approximate *square waveform*.

FUNI: *Frame relay User to Network Interface*.

fusable link: Usually refers to a length of wire which is designed to melt above a certain current and so disrupt the current. It therefore performs the same function as a *fuse*.

fuse: **(1)** A device which is designed to carry current with very low resistance provided it is below a specified value. Above this value the device will melt and disconnect the *circuit*, so protecting the system from excessive currents. It therefore acts as a protective device. There are many different types of fuses, such as High Rupturing Capacity (HRC) fuses which are designed to break a large amount of current. Fuses are also designed to have certain characteristics, such as soft failure, so that excessive voltage spikes are not generated by very sharp current interruption in inductive circuits. **(2)** To connect two items together, usually by melting them and then allowing them to cool whilst in contact with each other. This provides a good bond. See also *fusion splice*.

fusion splice: The join made between two optical *transmission media*, such as *optical fibre* made from glass or plastic, by forming a *fuse* between them. This is usually done by butting the two fibres together and melting the ends so that a join forms on cooling. Heat can be generated by various means, such as by an electrical arc or a flame. The technique is easy to apply and gives low *transmission loss*, below about 0.2 *decibels*. However the formation of the splice can reduce the tensile strength of the fibre at the join and it must be packaged to support it.

Future Public Land Mobile Telecommunication System (FPLMTS): *ITU-T* name for the third generation land mobile system, now renamed *IMT-2000*. See also *UMTS*.

G

gain: A measure of *signal* amplification. It is usually given by the ratio of the output *amplitude* of a signal to its input amplitude. This could be its current, voltage or power. The ratio can be measured as a fraction, percentage or in *decibels*. A ratio greater than unity, 100% or a positive value of decibels indicates a gain. Ratios less than unity, 100% or a negative value of decibels indicates a *loss*.

gain-bandwidth product: A measure of the performance of an *amplifier*, it is given by the product of the *gain* and the *bandwidth* that it can handle. The unit of gain-bandwidth is *hertz* if the gain is measured as a fraction and the bandwidth is in hertz.

gain hit: Errors caused in *data transmission* due to *signal* surges. Usually this occurs when the *gain* exceeds 3 *decibels* and lasts for more than four milliseconds. The maximum level of gain hits are specified in standards, the Bell standard calling for eight or less during a 15-minute measurement period.

gain saturation: A non-linear condition in which the *gain* of a device does not vary linearly with the input *signal*. This is illustrated in Figure G.1 where the output power increases linearly with input power, giving a constant value of gain, until gain saturation occurs when the output power flattens off. To avoid gain saturation the output power from an *amplifier* must be kept below a set value, which for the example shown is about 0 *dBm*.

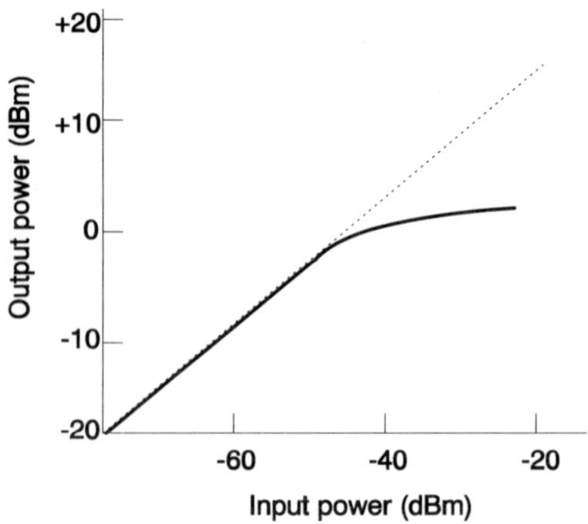

Figure G.1 Gain saturation

gamma corrected signals: The correction which is made in a *video* recording system, such as the camera of a television recorder, to compensate for differences between the source and the display tube characteristics. Compensation is applied to the three *primary colours*, red, green and blue.

GAP: *Generic Access Profile* or *Analysis and Forecasting Group*.

gap loss: The loss of power which occurs when two *optical fibres* are connected together, such as by a *fibre optic splice* or a *fibre optic connector*.

garble: Errors in the *message* which cause it to be unreadable or meaningless. This can be due to many causes, such as errors in *transmission* or reception of the signal, or incorrect *encryption*.

gas LASER: A *LASER* which uses a gas as its active material.

gateway: **(1)** A combination of *hardware* and *software* which performs a *protocol conversion* function, and so enables communications to occur between otherwise incompatible *networks*. Gateways operate in layers 4 to 7 of the *OSI Basic Reference Model*. Figure G.2 shows an example of the use of gateways to interconnect *Local Area Networks (LANs)* via an X.25 (see *X Series*) network, the gateways performing the protocol conversion between the various systems. **(2)** The entity which acts as an interface between two or more *electronic mail* systems, forwarding *messages* from one system to *addresses* in another system.

Gateway Protocol (GP): *Protocol*, used by *gateways* in *networks* operating on *TCP/IP* (e.g. the *Internet*) to interchange *information* such as on *routeing*. Also known as *Gateway-to-Gateway Protocol (GGP)*. See also *Border Gateway Protocol (BGP)* and *Exterior Gateway Protocol (EGP)*.

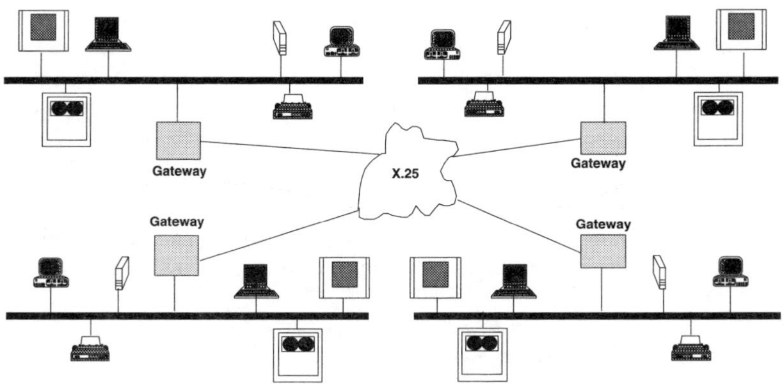

Figure G.2 Extended network using gateways

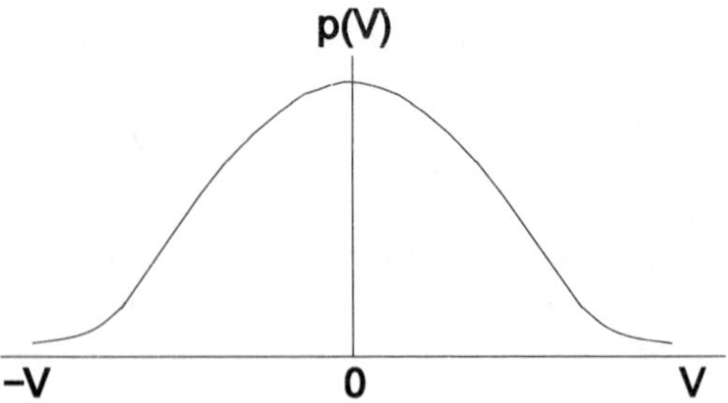

Figure G.3 Gaussian function

gateway switching centre: A *gateway* which is incorporated as part of a *switching centre* and interconnects different *networks*, for example networks in several foreign countries.

Gateway-to-Gateway Protocol (GGP): See *Gateway Protocol (GP).*

gating: The control of events such that they can occur only after certain other predefined events have occurred.

Gaussian beam: An *electromagnetic radiation*, such as light, whose power distribution resembles a *Gaussian distribution.*

Gaussian distribution: A statistical distribution function which has a dumbell shape, as in Figure G.3. See also *normal distribution.*

Gaussian Minimum Shift Keying (GMSK): A *Phase Shift Keying (PSK)* technique which uses a constant envelope continuous phase *modulating signal.*

Gaussian noise: *Noise* which occurs in electronic systems and is spread over a wide *frequency range*, occurring randomly in this range. It is often seen as *ambient noise*, *background noise* and hiss. See also *white noise.*

Gaussian tail area formula: It defines the probability (P_e) of making an error in the decision making process when transmitting in the presence of noise. If a is the voltage difference between the received signal and its decision threshold at the decision time and σ is the r.m.s. *noise voltage*, then the Gaussian tail area formula gives $P_e = \frac{1}{\sigma\sqrt{2\pi}} \int_a^\infty e^{-x^2/2\sigma^2}\, dx$.

G band: *Frequency range* between 4 GHz and 6 GHz, which is divided into ten equally spaced *frequency bands*, each occupying 200 MHz.

GBH: *Group Busy Hour.*

GDF: *Group Distribution Frame.*

GDN: *Government Data Network.*

GEA: Gigabit Ethernet Alliance.

GEN: *Global European Network.*

gender changer: A device which converts a *female connector* to a *male connector* and vice versa.

gender mender: A device which has the same type of connections (i.e. male/male or female/female) at both ends. This enables two cables with the same type of connectors (i.e. both *male connectors* or both *female connectors*) to be joined together.

general message: A *message* which has a wide distribution. For example a message sent to all ships in a given area, or to all aircraft in an area, etc.

general poll: A *polling signal* send to all *nodes* on a *network* to request *transmission* from those that are ready to do so.

General Packet Radio Service (GPRS): An *ETSI* standard for *GSM* which provides facilities for *Internet* access and *multimedia* transmission.

Generic Access Profile (GAP): *Common Air Interface (CAI) standard from ETSI* which is intended to ensure *interworking* between *DECT* equipment for *telephony* applications, provided by different manufacturers.

Generic Flow Control (GFC): The mechanism used for *flow control* in *ATM networks*. It is used to regulate the flow of user *information* towards the network at the *broadband User-Network Interface (UNI)* so as to overcome short term overload conditions. It is defined in *ITU-T Recommendation* I.150.

GEO: *Geostationary Earth Orbit.*

geometric mean: An averaging technique for numbers which follow a geometric progression or exponential law, such as rates of change. The geometric mean of N numbers, $X_1, X_2, ... X_N$ is given by $X_G = (X_1 \times X_2 \times X_3 \times .... \times X_N)^{1/N}$.

Geostationary Earth Orbit (GEO): See *Geostationary Satellite Orbit.*

geostationary orbit: The *orbit* followed by a *satellite*, rotating around a body which is itself rotating on its axis, in such as way that the path of the satellite remains fixed with respect to all points on the body being orbited. This is most commonly applied to a *Geostationary Earth Orbit (GEO)*. Also called *geosynchronous orbit.*

geostationary satellite: A *satellite* which is in a *Geostationary Satellite Orbit (GSO)*. In this position it has a line of sight coverage of a large part of the Earth. Three satellites, suitably located around the Earth, can cover most of its surface, as shown by Figure G.4, where the satellites are located at 30°W, 150°W and 90°E longitude.

Geostationary Satellite Orbit (GSO): The *orbit* followed by a *satellite* which is directly above the equator at a distance of about 35786 km from the ground, the period of the orbit being the same as the length of the *sidereal day*, about 23 hours, 56 minutes and 4 seconds. Moving in the same direction as the Earth's rotation causes the satellite to remain stationary as seen from points on the Earth's surface. Also referred to *Geostationary Earth Orbit (GEO)*.

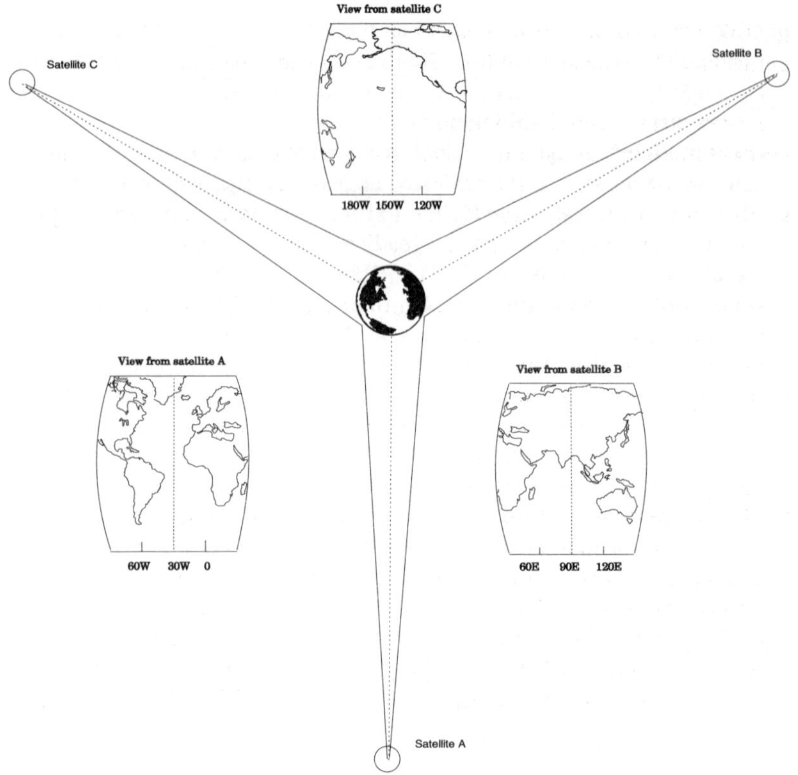

Figure G.4 Geostationary satellite coverage

geosynchronous orbit: See *geostationary orbit*.
GFC: *Generic Flow Control*.
G4 facsimile: Stands for Group 4 facsimile. See *Group x facsimile*.
GGP: *Gateway-to-Gateway Protocol (GGP)*
GHz: *gigahertz*.
giga: A thousand million, or a billion, or 10^9.
gigabit ethernet: An extension to the *ethernet* standard being developed by the *IEEE* 802.3 working group which enables transmission at 1000 Mbit/s.
Gigabit Ethernet Alliance (GEA): Alliance of equipment suppliers set up to develop standards and products for carying *ethernet traffic* at speeds of up to 1 Gbit/s.
gigabit LAN: Extension to *Local Area Network (LAN)* standards to enable them to operate at 1000 Mbit/s. Work is being done into extending *ethernet*, *100VG-AnyLAN* and *FDDI* standards.

gigabits: A thousand million *bits*.

gigaflops: A thousand million *Floating point Operations Per Second (FLOPS)*.

gigahertz: A thousand million hertz.

Gilbert model: Mathematical model used in *information theory* to characterise *error bursts* and *impulsive noise* in *transmission channels*.

Global European Network (GEN): Early joint venture between European *PTOs* to provide high speed *leased line* and switched services. See *METRAN*.

Global Mobile Personal Communications by Satellite (GMPCS): Generic term for *voice* and *data personal communications* using *satellites* in non-*geostationary orbit*.

Global Positioning System (GPS): A group of US Federal government sponsored *satellites* which circle the Earth and are used for military and civil applications, such as determining the location of objects and providing a *clock signal* for *timing* communication *networks*.

Global System for Mobile Communications (GSM): *Cellular radio system* developed as an initiative from *CEPT* for a mobile system which could be used throughout Europe. Although originally intended for operation over the 900 MHz *frequency band* it has now been adopted for the 1800 MHz (*PCN*) band in Europe and the 1900 MHz (*PCS*) band in North America. The basic architecture of GSM is similar to other cellular systems, as shown in Figure G.5. In addition GSM standards define

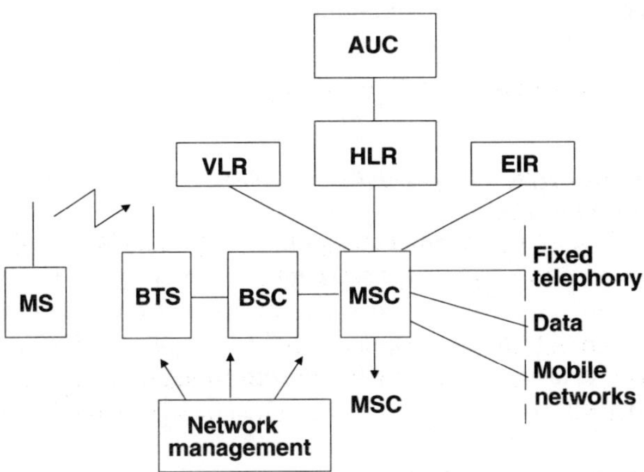

Figure G.5 GSM architecture

several interfaces, such as that between the *MSC* and the *BSC*, the *air interface* and the *signalling* interface, which is based on *ITU-T* No.7 signalling. The *BTS* and *BSC* carry out all the functions related to the radio *channel* management. The MSC, *VLR* and *HLR* are concerned with mobility management functions. The *authentication centre* works closely with the HLR and provides information for *authentication* of all *calls*. *Network management* monitors and controls the major elements of the GSM *network*.

global title: A numbering system, conforming to *ITU-T Recommendation* E.214, which is used to route *information* to appropriate *nodes* in a *cellular radio system*.

GMPCS: *Global Mobile Personal Communications by Satellite*.

GMSK: *Gaussian Minimum Shift Keying*.

GMT: *Greenwich Mean Time*.

GNMP: *Government Network Management Profile*.

go back N ARQ: An *Automatic Repeat Request (ARQ) algorithm*, used for *error correction*, in which a *Negative Acknowledgement (NAK)* causes retransmission of the errored word as well as the previous N–1 words. The value of N is usually chosen such that the time take to transmit the N words is less than the round trip delay from transmitter to receiver and back again. Therefore a *buffer* is not needed at the receiver.

Golay code: A *transmission* (23,12) triple *error correcting code*.

go path: The *transmission path* used to carry *data* outwards from a *terminal* or *node*. See also *return path*.

GoS: *Grade of Service*.

GOSIP: *Government Open System Interconnection Profile*.

Government Data Network (GDN): UK private *data network* for use by government departments.

Government Network Management Profile (GNMP): UK government procurement standard for *network management*, as part of *GOSIP*.

Government Open System Interconnection Profile (GOSIP): A subset of the *Open System Interconnect* standards, specified by the US government for its procurement. These were developed by vendors working with the *National Institute of Standards and Technology (NIST)*, and are specified in *Federal Information Processing Standard (FIPS)* Pub. 146.

GP: *Gateway Protocol*.

GPRS: *General Packet Radio Service*.

GPS: *Global Positioning System*.

graded index fibre: *Optical fibre* in which the refractive index decreases with distance from the core towards the cladding, this following an approximate parabolic law (see *graded index profile* and Figure G.6). The difference between the refractive index at the axial and the boundary of the core is about 1%. This makes the fibre self focusing and since the

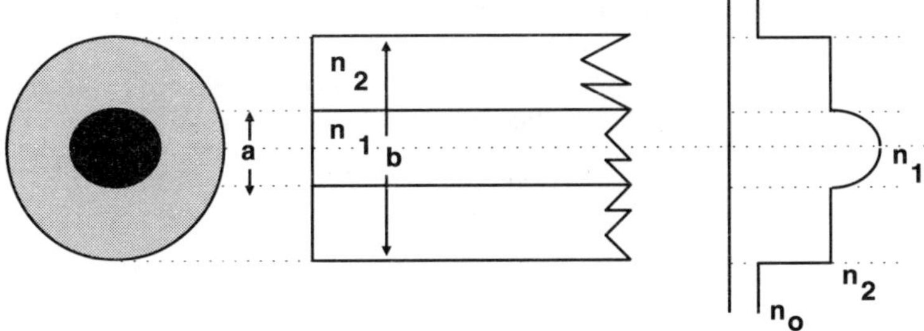

Figure G.6 Graded index profile

transit speed of light in the outer region is now faster than in the higher refractive index inner region there is equalisation of transit times and *dispersion* is reduced. See also *step index fibre*.

graded index profile: Any component, usually a *transmission medium* such as *optical fibred*, in which the *refractive index* varies over its cross section, as in Figure G.6. See, for example, *graded index fibre*. See also *step index profile*.

Grade of Service (GoS): **(1)** A general term applied to the quality of communications service provided. Examples are the quality of the *voice* in a *telephony* system, the *Bit Error Ratio (BER)* in a *data communications* system, etc. **(2)** In a *circuit switched network* the GoS defines the probability of a *call* being lost. Therefore a high GoS corresponds to poor service and vice versa. The GoS is calculated by *Erlangs loss formula* and is usually stated at the *busy hour*. A figure of 0.005, for example, would indicate that one call is lost in every 200, primarily due to the *switching equipment* being engaged on other calls.

grandfathered system: A system which was in existence before regulation was introduced and has therefore established rights and privileges which make it exempt from the new regulation. For example the *FCC* introduced new regulations for registration of equipment connected to the *PSTN*. However, it exempted those items which had been connected before 1 June 1978, provided they were not subsequently modified.

Graphical User Interface (GUI): The *software* and *hardware* system, part of a processing *terminal*, such as a computer, which allows users to manipulate items on the screen by pointing to icons, usually using a mouse, rather than by entering text via a keyboard.

graphic character: A visual representation of a *character*, excluding an alphanumeric or *control character*.

Graphic Interchange Format (GIF): A standard format for enabling graphical *data* to be encoded and displayed on different computer systems. First proposed by CompuServe.

Table G.1 Illustration of Gray code

Decimal	*Binary code*	*Gray code*
0	0000	0000
1	0001	0001
2	0010	0011
3	0011	0010
4	0100	0110
5	0101	0111
6	0110	0101
7	0111	0100
8	0100	1100
9	1001	1101
10	1010	1111
11	1011	1110
12	1100	1010
13	1101	1011
14	1110	1001
15	1111	1000

graphics terminal: A *terminal* which is able to handle and display both text and graphical images.

gravity switch: See *switch hook.*

Gray code: A *binary code*, in which there is only one *bit* change between successive *digits*. This is illustrated in Table G.1.

Greenwich Mean Time (GMT): The mean time measured at the 0° meridian which runs through Greenwich in England. It was formerly used as the standard reference time throughout the world but it has now been replaced by the *Coordinated Universal Time (CUT).*

Gregorian antenna: A *reflector antenna* which uses a sub-reflector to improve the overall performance. Figure G.7 shows its construction, the

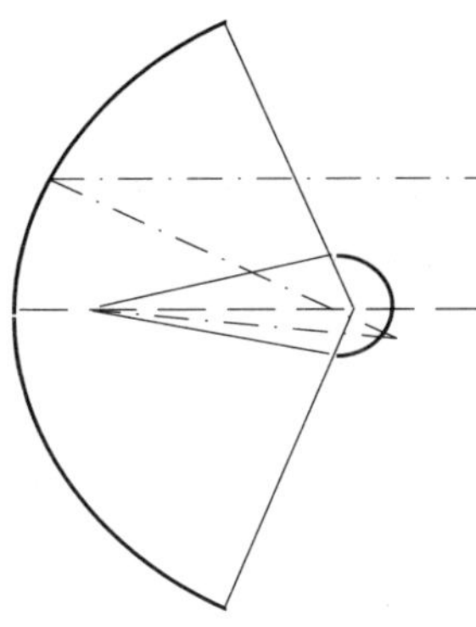

Figure G.7 Gregorian antenna

concave elliptical sub-reflector is placed on the outside of the parabola focus. See also *Cassegrain antenna.*

Gregorian calendar: The calendar in current use today, first introduced by Pope Gregory XIII in 1582.

grey scale: Shades of grey on an image, between pure white and pure black. These are usually implemented as step changes and different systems, such as *facsimile* can reproduce images in various numbers of grey scales, typically 16.

ground: **(1)** An electrically conductive material which is normally connected to the earth at some point. Also called a *ground conductor.* **(2)** Generally the term ground is used to imply zero resistance or zero voltage level and other voltages are measured with reference to this.

ground absorption: The loss of energy in *electromagnetic waves* due to *absorption* by the earth.

ground conductivity: Usually refers to the conductivity of the earth to electromagnetic waves. This varies considerable depending on the terrain and its measurement methods are given in *ITU-R Recommendations.*

ground conductor: See *ground.*

grounding: Connecting a *circuit* to *ground.* Also known as *earthing.*

ground loop: A fault condition which occurs when two points which are normally at *ground* are connected together and there is a potential between the points, which can cause a large amount of current to flow.

ground return: The use of the *ground* to provide a return path for currents. For example, in Figure G.8 one end of the battery B is connected to ground and so is one end of the resistor R, so when switch SW is closed

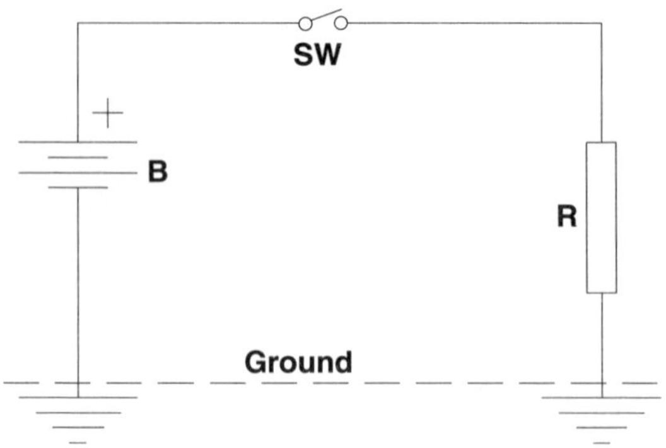

Figure G.8 Illustration of ground return

the current flows from the battery to the resistor and back to the battery through the ground path.

ground start: See *earth calling*.

ground station: See *earth station*.

ground wave: A *radio wave* which propagates along the surface of the Earth by reflection off its surface. It is generally used in the *Low Frequency (LF)* and *Medium Frequency* bands since *sky waves* at these frequencies are heavily absorbed in daylight by the *D Layer*. The *ITU-R* has published curves for ground wave field strength at different frequencies and surface conditions, as shown in Figure G.9. These curves

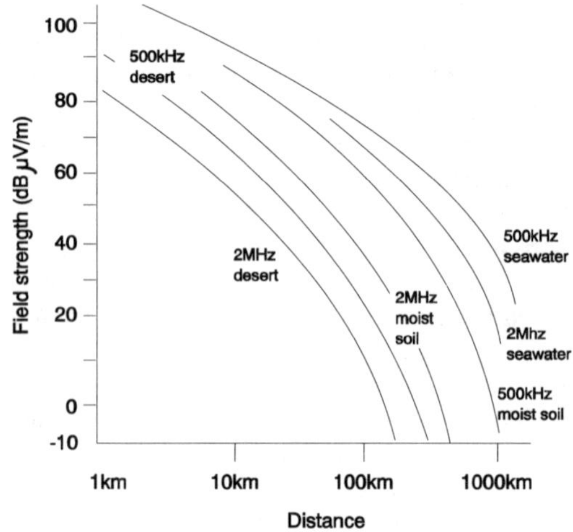

Figure G.9 Ground wave propagation: field strengths for different frequencies and surface conditions

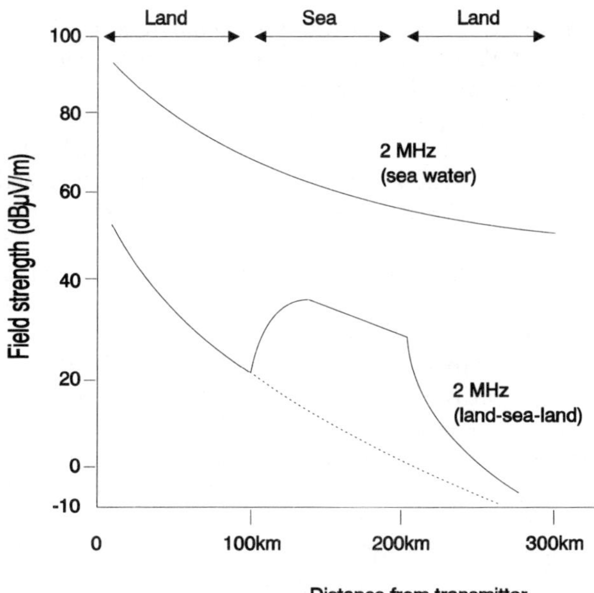

Figure G.10 Ground wave field strength for a hypothetical non-homogeneous path

assume a homogeneous surface. Discontinuities, such as a transition from land to sea, will cause variations in signal strength, as illustrated in Figure G.10. As seen the field strength increases over the sea.

group: **(1)** Generally refers to the 12 voice channels which make up a group in a *Frequency Division Multiplexing (FDM)* system. Also referred to as a *channel group*. See also *supergroup*, *master group* and *supermaster group*. See also *Group x facsimile*. **(2)** A collection of items, such as *characters*, making up a unit or users who interact in some way. See also *Group Separator (GS)* and *Closed User Group (CUG)*.

group abbreviated address calling: The use of *abbreviated address calling* for a *group address*, i.e. calling a group of users by *dialling* a simple number.

group address: An *address* which is common to several *stations* on a *network*.

Group Busy Hour (GBH): The *busy hour* applied to a specific *trunk group*.

group delay: A measure of the difference in *phase* shift when a *narrowband signal*, composed of a *group* of *frequencies*, passes through a *transmission medium*. See also *envelope delay distortion*.

Group Distribution Frame (GDF): *Distribution frame* used in *Frequency Division Multiplexing (FDM)* systems to terminate and interconnect *modulator* and *demodulator circuits.*

group index: For an optical wave propagating along a *transmission medium*, such as *optical fibre*, the group index (G_i) is given by the relationship $n - \lambda dn/d\lambda$, where λ is the *wavelength* of the source and n is the refractive index of the transmission medium, expressed as a function of wavelength. The ratio $dn/d\lambda$ therefore represents the rate of change of refractive index with wavelength. Group index is also equal to the ratio of the velocity of light in vacuum to that in the transmission medium (called the *group velocity*).

grouping: **(1)** The formation of *groups* in a *Frequency Division Multiplexing (FDM)* system. **(2)** The periodic error in the spacing of recorded lines which occurs in *facsimile* systems.

group selection; The process used, in an *automatic switching equipment*, for connecting an input *circuit* to one or more output circuits, in response to a control *signal.*

Group Selector (GS): The device, part of an *automatic switching equipment*, which carries out *group selection.*

Group Separator (GS): A *control character* used to separate a string of *characters* into logical *groups*, such as that used for communications *codes*.

Group Switching Centre (GSC): Terminology used in the UK for the *exchange* connecting the *local exchange* to the *trunk lines*. See Figure A.15. It is equivalent to the *primary trunk switching centre*, referred to by the *ITU-T*, and the US *Class 4* exchange. (See *class of office* and Figure C.18.)

group velocity: The velocity of propagation of optical energy along a *transmission medium* such as *optical fibre*. Group velocity (v_g) is given by the ratio of the velocity of light in vacuum (c) to the *group index* (G_i), i.e. $v_g = c/G_i$.

Group x facsimile: A series of *ITU-T Recommendations* for *facsimile transmission*, where x stands for 1, 2, 3 and 4. The original was for Group 1 facsimile, published as Recommendation T.2. It was intended for use on *analogue lines*. It used 1300 Hz for white and 230 Hz for black. No *data compression* was done and the transmission time for an A4 document (8.5 inches by 11 inches) was 6 minutes, although this time could be reduced if lower resolution was acceptable. Group 2 facsimile introduced *bandwidth compression* but was still designed for operation over analogue *telephone* lines and could transmit an A4 sheet in three minutes. It is specified in Recommendation T.3. Group 3 facsimile uses data compression and *redundancy checking*. It is specified in Recommendation T.4 and is designed for *digital* lines with the ability to transmit an A4 sheet in under one minute. Group 3 facsimile equipment have integrated *modems*. Group 4 facsimile is specified in Recommendation T.5 and T.6 and is also designed for operation over digital lines. It operates at 64 kbit/s can can use one of the *ISDN* channels. It can transmit an A4 sheet in six seconds.

GS: *Group Separator* or *Group Selector*.

GSC: *Group Switching Centre*.

G Series: A series of *ITU-T Recommendations*, a few of these being given in Table G.2.

GSM: *Global System for Mobile communications*.

Table G.2 ITU-T Recommendations, G Series
(Continued on next page)

Recommendation	*Description*
G.702	Digital hierarchy bit rates
G.703	Physical and logical characteristics of transmission over digital circuits
G.704	Physical and electrical characteristics of hierarchical digital interfaces
G.706	Frame alignment and cyclic redundancy check procedures
G.707	Synchronous digital hierarchy bit rates
G.708	Network mode interface for synchronous digital hierarchy
G.709	Synchronous multiplexing
G.721	32 kbit/s Adaptive Differential Pulse Code Modulation
G.726	Adaptive pulse code modulation. It supersedes G.721
G.728	Compression of 3 kHz audio to 16 kbit/s
G.747	Second order digital multiplexing
G.755	Digital multiplexing equipment at 139.264 Mbit/s and multiplexing three 44.736 Mbit/s sources

Table G.2 (Continued from previous page)

Recommendation	*Description*
G.802	Interworking between networks with different digital hierarchies and speech encoding laws
G.811	Timing on clocks used for plesiochronous operation of international networks
G.813	Error performance of international digital links forming part of an international ISDN
G.821	Performance of ISDN circuits
G.824	Jitter and wander in 1.544 Mbit/s digital circuits
G.921	Digital sections based on 2.048 digital hierarchy
G.960	ISDN basic rate access

GSM air interface: The interface, specified in *GSM* standards, for the physical link between the mobile and the *network*. It is summarised in Table G.3.

Table G.3 GSM air interface parameters

Frequency band mobile – base	890–915MHz
Frequency band base – mobile	935–960MHz
124 radio carriers spaced by 200kHz	
TDMA structure with 8 timeslots per radio carrier	
Gaussian Minimum Shift Keying (GMSK) Modulation with BT = 0.3	
Slow frequency hopping at 217 hops per second	
Block and convolutional channel coding with interleaving	
Downlink and uplink power control	
Discontinuous transmission and reception	

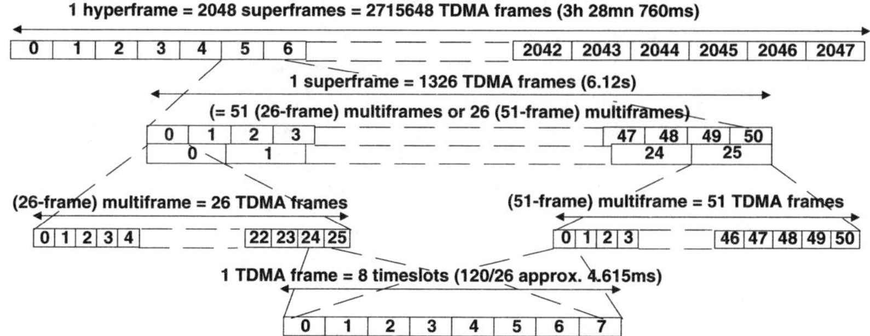

Figure G.11 GSM framing

GSM control channels: See *GSM logical channels.*

GSM framing: The *frame* structure used for *GSM*, as shown in Figure G.11. Each *TDMA* frame is combined into a multiframe of 26 or 51 *timeslots* and this is further combined into superframes and hyperframes.

GSM logical channel: The *channels* which are carried within the *GSM* physical channels (see *GSM air interface*). These are illustrated in Table G.4. There are two principle logical channels: *GSM traffic channels* and

Table G.4 GSM logical channels

Traffic Channels (TCH)		*Control Channels (CCH)*			
Speech	Data	Broadcast CCH (BCCH)	Common CCH (CCCH)	Standalone Dedicated CCH (SDCCH)	Associated CCH (ACCH)
Full rate TCH/F	TCH/F9.6 TCH/F4.8 TCH/F2.4	Frequency Correction (FCCH) Synchronisation (SCH)	Paging Channel (PCH) Random Access (RACH)		Fast (FACCH) Slow (SACCH)
Half rate TCH/H	TCH/H4.8 TCH/H2.4		Access Grant (AGCH)		

GSM control channels. Traffic channels can be *voice* full rate (22.8 kbit/s) or half rate (11.4 kbit/s); or *data* at various *transmission rates*, 2.4 kbit/s, 4.8 kbit/s and 9.6 kbit/s. The *control channels* consist of four basic types: the Broadcast Control Channel (BCCH), the Common Control Channel (CCCH), the Standalone Dedicated Control Channel (SDCCH), and the Associated Control Channel (ACCH). Further subdivisions are specified in *ETSI* standards.

GSM signalling: The *signalling* system used within *GSM*. This includes the signalling between the mobile *station*, the *base station*, the *Base Station Controller(BSC)*, the *Mobile Switching Centre (MSC)*, the *Home Location Register (HLR)* and the *Visitor Location Register (VLR). See Figure G.5.*

GSM traffic channels: See *GSM logical channels.*

GSO: *Geostationary Satellite Orbit.*

G3 facsimile: Stands for Group 3 facsimile. See *group x facsimile.*

guard band: An unused *frequency band* which is used to separate *channels* carrying *information* to prevent interference between them, such as *crosstalk.* Guard bands are used in systems such as *Frequency Division Multiplexing (FDM)* and *Time Division Multiple Access (TDMA)*, as shown in Figure F. 17 and Figure G.12. See also *guard time.*

guard time: In a *Time Division Multiplexing (TDM)* system it is the period between *blocks* of *information* when no *transmission* occurs.

GUI: *Graphical User Interface.*

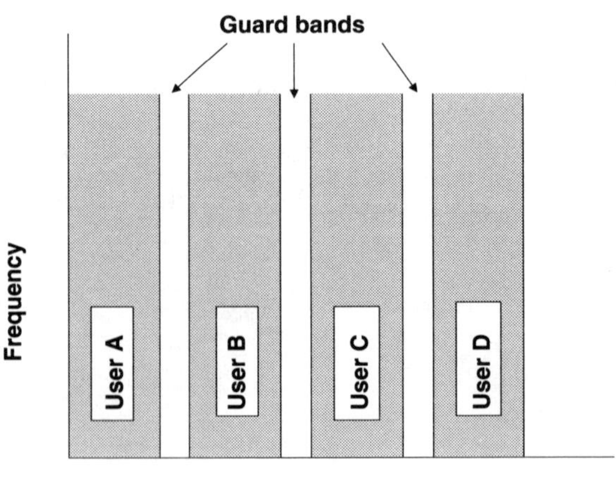

Figure G.12 Guard bands within TDMA user assignments

H

Haas effect: The effect in which the human brain assumes that a sound is coming from the direction in which it first hears it, even though subsequent sound levels also come from a different direction and may be slightly louder. This is used, for example, in live performances, where any loudspeakers, used to relay sounds from a stage, are timed to reproduce the sound after a delayed interval. The amount of delay needed varies with the difference in sound level and Table H.1 gives the approximate values.

Hadamard code: A linear *block code* in which the *code words* are the rows of a n × n matrix (called the Hadamard matrix) n being even and each row differing from the other rows by n/2 positions.

half bridge: See *remote bridge*.

half duplex: A method of *transmission* in which the *line* is able to carry *data* in both directions, but only in one direction at a time.

half duplex circuit: A *circuit* which operates in *half duplex* mode.

half echo suppressor: An *echo suppressor* in which the *speech signal* on one *line* controls the *echo* from the other line. See also *full echo suppressor.*

half power point: The point at which the power level is half that at some other point, usually the point of maximum power.

half reflection: *Reflections* which occur at the junction of two surfaces in which half the incident power is transmitted and the other half is reflected.

halftone: **(1)** In a printed document it is the variation in density formed by different sizes and shapes of the evenly spaced dots on the document. **(2)** In *facsimile transmission* it is the density of the reproduced document compared to that of the original.

half wave dipole antenna: See *half wavelength dipole.*

half wavelength dipole: An *antenna* which has a length equal to one half the *signal* operating *wavelength.* It is centre fed and has a current distribution which is approximately sinusoidal as in Figure H.1. The

Table H.1 Time delay and level difference for Haas effect

Time delay (ms)	5	10	20	30	40
Level difference (dB)	7	10	10	9	7

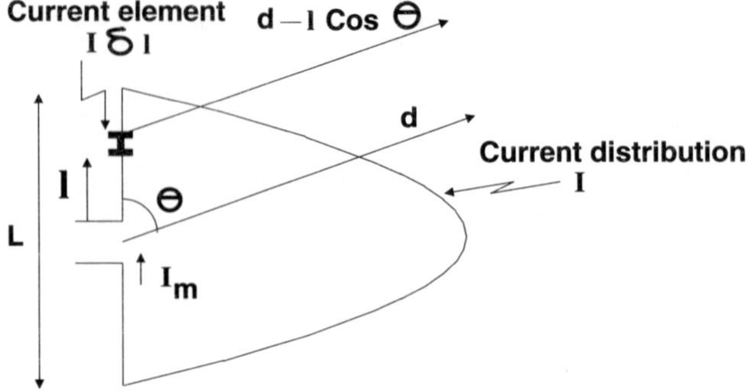

Figure H.1 Have wavelength dipole

electrical field (E) and magnetic field (H) components are given by E = ηH and H = $[I_m/2\pi d][\cos(\pi\cos\theta/2)/\sin\theta]$.

Hamming code: An *error correction* and *error detection code* named after its inventor, R.W. Hamming of Bell Laboratories. It is a distance 3 (see *Hamming distance*) *linear block code* so it can be used for single error correction or dual error detection. For a *binary* Hamming code the *code word* length is equal to $n = 2^r - 1$, where r is the number of *parity bits*, and the *message* length is therefore equal to n – r.

Hamming distance: It is the number of *bit* positions which are different between two *code words*. For example if the two words are 011011 and 110011 then the Hamming distance is 2, i.e. two bit positions differ. It indicates the number of bits which need to change, due to *transmission* errors, before the two words cannot be differentiated. Hamming distance is also known as *signal distance*.

Hamming weight: The number of non-zero symbols in the *code word*. For a *binary* word this is equal to the number of binary 1 positions. Therefore the word 011011 has a Hamming weight of 4.

handoff: In a *cellular radio system* it is the process of transferring the radio link, to the mobile, from one radio transmitter, situated in a *cell*, to a transmitter in another cell, as the mobile moves from one cell to the next.

handset: Strictly, that part of a *telephone* which the user lifts and places close to the face for listening and talking into. However the term is generally applied to the whole telephone instrument. See also *headset*.

handsfree answerback: The facility which, when it is activated, automatically answers a *call* after a number of rings and switches the receiver into *handsfree operation* so that the user can listen and speak without having to lift the *handset*. Used in *cellular radio systems*, operated in vehicles, and in intercom systems.

handsfree dialling: *Dialling* using a *handsfree telephone*.

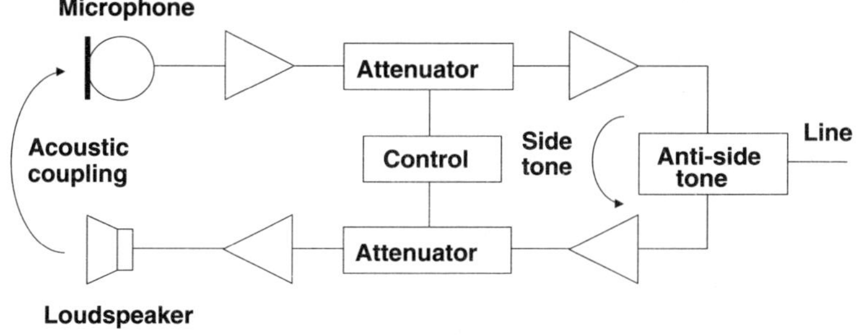

Figure H.2 Handsfree telephone

handsfree operation: Calling using a *handsfree telephone*.

handsfree telephone: A *telephone* which can be operated without having to lift the *handset*. Generally this still involves manually *dialling* the number but the handset does not need to be lifted for dialling, talking, or listening. A loudspeaker and microphone are built into the instrument and careful design is needed, as shown in Figure H.2, to prevent acoustic coupling between the two. Also known as *loudspeaking telephone*.

handshake protocol: The *protocol*, used during *call setup*, which identifies the two communicating *terminals* and sets up the conditions for the exchange of *information* which follows. Operating parameters, such as the *code* to be used, the *data rate*, etc. are exchanged and agreed upon during the handshake stage. See also *call establishment*.

hang up: The process of terminating a *call*, usually by hanging the *handset* back on to the base of the *telephone* instrument, so that the *switch hook* is depressed.

hang-up signal: The *signal* sent to the *exchange* when the user *hangs up*. This informs the exchange that the *call* has been completed and it should be cleared.

HA1 receiver weighting: An older method of *noise weighting* used in noise measurement across HA1 receivers of a 302 or similar type of instrument.

hard copy: Information which has been produced in a permanent form, such as on paper. For example a computer print out, *facsimile* pages, etc.

hard copy printer: A device which can produce a *hard copy*.

hard decision decoding: Decoding in which the input *signal* to the *decoder* is manipulated so as to produce a *digit* stream. The decoder then performs *error detection* and *error correction* using this stream. See also *soft decision decoding*.

hardware: Any equipment which is physical and whose operation can be seen. See also *software* and *firmware*.

hardware flow control: The *flow control* facility, available in *packet switching networks*, in which, after a *receiving terminal* has sent a *clear to send signal*, it can drop the command at any time if *traffic* becomes too heavy. Also known as *out-of-band flow control*. See also *software flow control*.

hardwire: A fixed connection between pieces of equipment. The operation of these items relative to each other cannot change, as is possible when they not hardwired but are, for example, under software control.

harmonica: A device used to convert a twenty-five pair cable into individual two, three or four pair cables.

Harmonically Related Carrier (HRC): A method for reducing the effects of *transmission* defects, such as noise and non-linearity, commonly used in applications such as *cable television* systems. Each *carrier* used is an exact multiple of the spacing between them so that carrier intermodulation results in outputs which fall on other carrier frequencies, such as the 2nd and 3rd order rather than within the *frequency band* of the *video signal* being transmitted. See also *Incrementally Related Carrier (IRC)*.

harmonic distortion: *Distortion* of a output *signal* which results in it having frequencies which were not present in the input. Harmonic distortion is normally caused by non-linearities in the system. If V_1, V_2, etc. are the r.m.s. values of the harmonic voltages present in the output and V_t is the total r.m.s output voltage, then the harmonic distortion (HD) is given by : $HD = (V^2{}_1 + V^2{}_2 + ... + V^2{}_n)^{1/2} / V_t$

harmonic frequency: A *signal* whose *frequency* is an integer multiple of the *fundamental frequency*. For example, the third harmonic has three times the fundamental frequency, etc.

harmonic mean: The mean or average of a series of numbers. If there are n numbers equal to x_1, x_2, x_3, ..., x_n, then their harmonic mean (M_h) is given by $M_h = \dfrac{n}{\sum_{r=1}^{n} \frac{1}{x_r}}$.

This averaging method is used with dealing with rates or speeds or prices. As a rule when considering items such as A per B, if the figures are for equal As then the harmonic mean is used, but if they are for equal Bs then the arithmetic mean is used.

harmonic ringing: The use of *harmonic frequencies* for *ringing*. This is used in *party lines* since, by tuning individual ringers to a different frequency, only the called *telephone* will ring.

harness: An assembly of wires and cables which are bundled together and carry *signals* or power to and from equipment.

hash total: The output from an *algorithm* applied to a set of *data fields* primarily to carry out *error detection* or *error correction*.

Hata's formula: An empirical formula for predicting the *path loss* in radio communication, such as *cellular radio systems*. The path loss L_p, in *decibels* is given by $L_p = 69.55 + 26.16 \log(f_c) - 13.82 \log(h_b) - a(h_m) + (44.9 - 6.55 \log(h_b)) \log R$, where f_c is the *carrier frequency* in MHz, h_b is the height of the *base station antenna*, h_m is the mobile antenna height, R is the radial distance in kilometres, and $a(h_m)$ is the mobile antenna height correction factor.

Hayes AT protocol: One of the first *protocols*, introduced by Hayes Microcomputer Products Inc. in 1981, for *modems* which allows automatic *dialling*. It is designed for *asynchronous transmission*. *ITU-T Recommendation* V.25 bis covers both asynchronous and *synchronous transmission*.

H channel: One of the three main *channel* types defined by the *ITU-T Recommendation* I.412 for *ISDN*. It is defined at various *bit rates* as shown in Table H.2.

HDB3: *High Density Bipolar 3-zero*.

HDLC: *High Level Data Link Control*.

HD-MAC: A *High Definition Television (HDTV)* system proposed by the European Eureka EU-95 project. It was intended to introduce a system suitable for *Direct Broadcast Satellite (DBS)* systems. This standard has now largely been overtaken by other HDTV systems based on all-digital broadcasting.

HDSL: *High bit rate Digital Subscriber Line*.

HDTV: *High Definition Television*.

head-end: **(1)** In a *Cable Television (CATV)* system it is the location where all the material for services on the *network* are generated. The head-end is linked to remote aerial sites and production studios, and collects this material before multiplexing it and launching it onto the network to users. **(2)** In a *Local Area Network (LAN)* or *Wide Area Network (WAN)* the head-end is the device which provides centralised functions and

Table H.2 H channel

Level	*Gross bit rate (Mbit/s)*	*Effective bit rate (Mbit/s)*
H0	0.384	0.384
H11	1.544	1.536
H12	2.048	1.920
H21	34.368	32.768
H22	139.264	135.168

allows *nodes* on the network to send and receive *messages* on a single cable.

header: *Control field*, added to the start of a *message* and containing *information* regarding the message, for example, the *address* of the *sending terminal* and the *receiving terminal*, the *routeing* information, *synchronisation*, etc.

Header Error Control (HEC): A *field* consisting of eight *bits* which is part of the *ATM* cell structure, using *Cyclic Redundancy Check (CRC)* for control of errors in the *header*. It is also used for cell delineation, i.e. identifying the start of each cell within a *bit stream*. See Figure A.21.

header information: See *header*.

head-on collision: See *collision*.

headphone: See *headset*.

headset: Part of a *telephone* system, a combination of earpiece and microphone which is worn over the head and allows *handsfree operation*. Also known as a *headphone*.

hearing exposure: The maximum level of sound which an average person can be exposed to without causing hearing defects. This level will vary with exposure time. Below 90 *dBA* the exposure can be continuous but above this value the maximum duration is approximately as in Table H.3.

hearing threshold: The level of sound which a person with average hearing is just able to detect. Below this threshold the sound will not be noticed.

heavy duty connector: A *connector* which is designed to operate in adverse conditions, such as outside a distribution box, or to carry large amounts of current.

HEC: *Header Error Control*.

helical antenna: An *antenna* whose body is in the form of an helix. The circumference of the helix, relative to the *wavelength* of the transmitted *signal*, determines the radiation direction. When it is one wavelength the maximum radiation is along the axis of the helix and when smaller than one wavelength it is normal to this axis.

helical ray: The light which propagates along the axis of an *optical fibre*, having a *graded index profile*, by winding around the axis of the fibre as it travels its length.

HEMP: *High altitude Electromagnetic Pulse*.

Table H.3 Permissible hearing exposure times above 90dBA

dBA	90	93	96	99	102	105
Hours	8	4	2	1	0.5	0.25

hertz (Hz): The SI unit for *frequency*, named after the scientist Heinrich Hertz. In a *periodic signal* it is equal to the frequency of the *waveform* having a *periodic time* of one second.

Hertzian wave: *Electromagnetic wave* which can be propagated without use of a *waveguide*, and which has a frequency below 3000 GHz.

heterodyne: To produce new *frequencies* by mixing two or more *signals* in a non-linear manner. The new frequencies produced are equal to the sum and difference of the mixed frequencies.

heterodyne repeater: A *repeater* which converts the received *signal* into an intermediate *frequency* before further processing.

heterogeneous network: A *network* which is made up of dissimilar components, such as those using different *protocols*. An heterogeneous network normally contains equipment supplied by different manufacturers and which do not use international standards.

heterojunction: The junction between two semiconductor layers having differences in their doping levels, atomic composition and conductivity. These junctions are points of discontinuity and provide means for controlling the level and direction of radiation.

heuristic: The technique of arriving at a solution by trial and error, and using empirical methods which cannot be proved by theory. Heuristic is opposite to *algorithm*.

heuristic routeing: *Routeing* within a *network* which uses the data received from earlier *transmissions*, such as the *delay time* being experienced on certain *links*, to determine the route for subsequent transmissions.

hexadecimal: A numbering system which uses 16 states, the numbers 0 to 9 and then the letters A to F. See Table H.4.

HF: *High Frequency*.

HFC: *Hybrid Fibre Coax*.

HFDF: *High Frequency Distribution Frame* or *High Frequency Direction Finding*.

hi-fi: *high fidelity*.

hierarchical network: See *network hierarchy*.

High altitude Electromagnetic Pulse (HEMP): An *Electromagnetic Pulse (EP)* which is produced at a high altitude, usually greater than 120

Table H.4 Hexadecimal notation

Decimal	0	1	2	3	4	5	6	7	8	9	10	11	12	13	14	15
Hexadecimal	0	1	2	3	4	5	6	7	8	9	A	B	C	D	E	F

km above the Earth's surface, i.e. clear of the lower levels of the *Earth's atmosphere*.

High bit rate Digital Subscriber Line (HDSL): The first *Digital Subscriber Line (DSL)* system, standardised by the *ANSI* T1E1.4 committee in 1992. As with all DSL systems it is intended to provide *broadband transmission* over the conventional copper *local loop*. HDSL uses the 2B1Q *line code* and supports unrepeatered corporate *E1* (2 Mbit/s) and *T1* (1.5 Mbit/s) links for distances up to 4 km. Unlike other DSL techniques HDSL uses two copper pairs, rather than one pair, and also does not carry *POTS*. It is used mainly to provide E1 or T1 *leased lines* in areas having a high number of business customers.

High Definition Television (HDTV): The transmission and reception of television *signals* which have a much higher resolution than conventional television (which is 625 lines in the UK and 525 lines in North America). Generally HDTV systems provide about twice the horizontal and vertical emitted resolution, so having about four times the number of *pixels* compared to conventional television. HDTV systems aim to provide the same picture quality as that obtained from projected 35 mm film.

High Density Bipolar 3-zero (HDB3): A *line coding* scheme, similar to *B8ZS* used in the US, which is used in Europe with *E1* systems. One of the problems with *Alternate Mark Inversion (AMI)* is that it does not overcome the problem of *synchronisation* which arises if several successive *binary* zero *bits* are being transmitted. Most receiver systems depend on obtaining periodic transitions in the input signal for synchronisation. This is overcome by HDB3 which replaces groups of four binary zeros by groups of four ternary symbols of which the last is non-zero and is transmitted with the same polarity as the last non-zero symbol, i.e. a violation of the alternating law of the AMI code. This avoids the appearance of more than three consecutive zeros, as shown in Figure H.3.

high fidelity: A system which is able to process and reproduce sound which is a good representation of the original sound.

High Frequency (HF): Part of the *radio frequency spectrum* covering the *frequency band* from 3 MHz to 30 MHz.

High Frequency Direction Finding (HFDF): Direction finding systems which use transmitters operating in the *High Frequency (HF)* range.

High Frequency Distribution Frame (HFDF): A *distribution frame* used with *High Frequency (HF) signals*, such as *supergroup* systems.

high frequency ground: A *ground* used for *High Frequency (HF) signals*.

high gain antenna: An *antenna* whose power *gain* has been increased, primarily by confining its emissions into a smaller area.

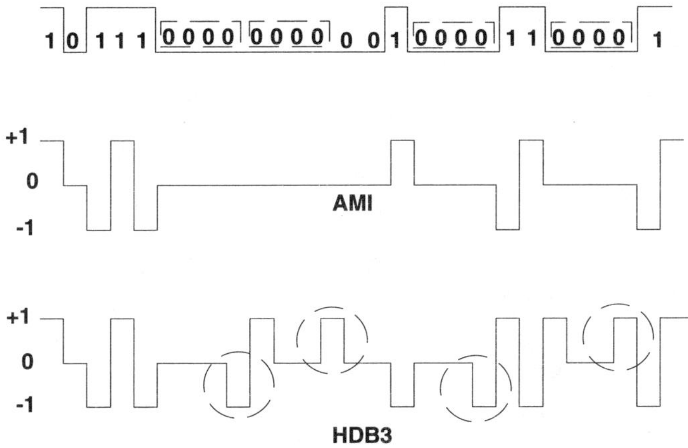

Figure H.3 AMI and HDB3 codes

High level Data Link Control (HDLC): *Bit* oriented *protocols*, developed by the *ISO* and operating at the *Data Link Layer* of the *OSI Basic Reference Model*. It ensures that *data* can be transmitted over a *link* with *flow control* and *error control*. Figure H.4 shows the structure of a basic HDLC *frame*. The *flag* identifier is *hexadecimal* 7E. The *CRC* allows error checking of the content of all frames.

High Level Language (HLL): Computer programming language which is written in the same form irrespective of the computer system on which it is to run. This is then interpreted by another system, called a compiler, which is different for different makes of computers, into a form which individual computers can understand.

high level protocol: A *protocol* which contains more than just the basic functions, such as those needed for the *transmission* of *data*.

highlighting: To emphasise parts of text or image on a display, such as a *Visual Display Unit (VDU)*. Highlighting can be done in several ways, such as changing colour, thickening lines, *reverse video*, flashing areas, etc.

high order multiplexing: The process of applying *multiplexing* in tandem so that the output from each stage contains all the previous inputs,

Flag	Address	Control	128 bytes of user data	CRC	CRC	Flag

Figure H.4 HDLC information frame

Table H.5 High order multiplexing

Multiplex level	*Europe (kbit/s)*	*USA (kbit/s)*	*Japan (kbit/s)*
0	64	64	64
1	2048	1544	1544
2	8448	6312	7876
3	34368	44736	32064
4	139264	274176	97728
5	565148		397200

building up a hierarchy of multiplexed *channels*. Three principal *PDH* systems are currently available, in Europe, USA and Japan, as shown in Table H.5, which is based on *ITU-T Recommendation* G.702 (see *G Series*). Each system has a base speed of 64 kbit/s from which it multiplexes upwards to the higher order systems.

high pass filter: A *filter* which allows *frequencies* above a specified value to pass through with very little attenuation, but severely attenuates all frequencies below this value. See also *band pass filter* and *low pass filter*.

High Performance Parallel Interface (HIPPI): *ANSI* standard X3T9.3 for a *bus* connecting computers to high speed peripherals. It is widely used to interconnect *nodes* on a *network* and can handle *data rates* of 800 Mbit/s over 32 twisted pair copper wire.

high resolution: A display system, such as a *Visual Display Unit (VDU)*, which is able to show the details of an image on its screen.

high resolution display: See *high resolution*.

high resolution graphics: Standards used to display graphics which enable them to be shown with *high resolution*.

High Speed Circuit Switched Data Service: A new *data* service proposed for *GSM* which offers *data rates* of up to 64 kbit/s by combining several *timeslots* together for a *call*.

high-usage group: A *trunk* group which is designed to carry a large volume of *data*, and is therefore intended to be the first choice for any *traffic*. It may be backed up by auxiliary routes for overflow traffic at peak times.

highway: A *transmission path* or *bus* which carries several *channels* of *information* with some form of separation between them, such as by allocating different *timeslots* to each channel.

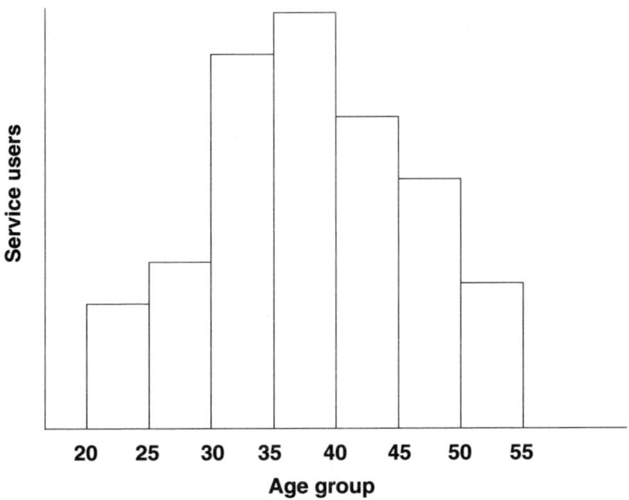

Figure H.5 A histogram

HIPPI: *High Performance Parallel Interface.*

histogram: A method for representing the distribution of frequency of occurrence of an event. For example, Figure H.5 shows the number of *subscribers* to a service provided by a *PTO*, these being broken down into age groups. There are clearly very few users below 20 years and above 55 years, the most popular ages being between 35 to 40. The area of the rectangles in a histogram represent the frequencies in the different groups. Histograms are commonly used in display systems indicating the colour density in an image.

hit: (1) A short duration disturbance which causes errors to occur in a communications system. Also known as a *line hit*. **(2)** A successful occurrence during data processing, such as the retrieval of information from a *database*.

HLL: *High Level Language.*

HLR: *Home Location Register.*

hogging: The situation which arises when a *node* on a *network* tries to continue to transmit *data* to the exclusion of other *nodes*. For example, in a *token passing* network a user may hold on to the *token* so that no other node can transmit data.

hold: The feature, usually available with a *PABX*, which allows a user to temporarily leave a *telephone* connection without disconnecting it. This user can return to the *call* when required.

holding time: The length of time that a *telephone circuit* is in use for each *call*. It covers the *call establishment* and *call disestablishment* times.

home carrier: The *carrier* with whom the *subscriber* is registered. The term is usually used in *cellular radio systems* to indicate the carrier who bills the *subscriber*.

home directory: The directory, located on a file server, to which a user is connected when he/she first logs on to the *network*.

Home Location Register (HLR): A master *database* used with mobile applications, such as *cellular radio systems*, which contains details of all *subscribers* to that service, such as their number, services covered by their subscription, their most recently known location, etc. This database is accessed when is call is made. See also *Visited Location Register (VLR)*.

home page: A location on the *World Wide Web (WWW)* which contains information about a topic, organisation or site.

HomeRF: A proposed standard for Wireless Local Area Networks, for use in the home or small office environment, developed by a consortium of companies. It has a range of about 50 metres, with a bit rate of about 2 Mbit/s and uses the *ISM frequency band* at 2.4 GHz.

homes passed: Term used in *Cable Television (CATV)* to indicate the number of homes which could be provided with CATV services, since the main cable has been laid past their homes and they only need the final cable drop to be fully connected.

homogeneous network: A *network* which has *nodes* which operate in the same way, e.g. use the same *protocol*. Usually this would contain equipment supplied by a single vendor or which conforms to international standards.

hook switch: See *switch hook*.

hoop-up wire: Wire used for low current and low voltage connections and is usually located within an enclosed cabinet.

hop: **(1)** The passage of a *message* from one *node* on a *network* to another node. **(2)** The movement of a *radio wave* from the Earth to the *ionosphere* and back again, or from the Earth to a *satellite* and back again, or between two satellites in *geostationary orbit*.

hop count: **(1)** A count of the number of *nodes* a *message* goes through. Hop count is normally used to ensure that the message does not spend too much time in the *network* as this would increase the *traffic* on the network. A counter is normally set to a value when the message is launched into the network and it is decreased by one each time the message passes through a node. When the count reaches zero the message is either cleared or discarded. **(2)** The number of times a *radio wave* is reflected from the *ionosphere* back to Earth.

Horizontal Radiation Pattern (HRP): The *radiation pattern* from an *antenna* measured along the horizontal plane. Figure H.6 shows the

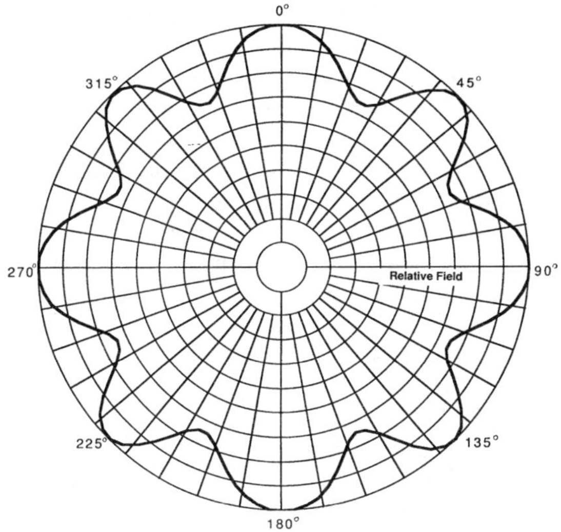

Figure H.6 Horizontal radiation pattern of a UHF broadcast antenna

Horizontal Radiation Pattern for a typical *UHF broadcast* antenna. See also *Vertical Radiation Pattern (VRP)*.

horizontal redundancy check: See *longitudinal redundancy check*.

horizontal resolution: The number of picture elements which occur, per unit distance. For example during a horizontal *scan* in a *facsimile* system, or along a horizontal line on a display screen.

Horizontal Tabulation (HT): Also commonly known as TAB. It is a *Format Effector (EF)* which causes the printer print position, or the cursor in a *Visual Display Unit (VDU)*, to move forward to the next predetermined position.

horn antenna: An *antenna* which has the shape of a circular or rectangular tube, the size of the tube increasing towards the open end through which the *radio waves* are launched.

horn feed: See *feed horn*.

host: See *host processor*.

host interface: The interface between a *host* and the *network* to which it is connected.

host node: The *node* which forms the *host* on the *network*.

host processor: A processor which is located on a *network* and provides services to other *nodes* connected to the network, such as access to *databases*, programming languages, etc. Often shortened to host. See also *server*.

host table: A *file* which contains the names and *addresses* of *hosts* which are resident on a *network*.

hot line: A *transmission line* which connects two *terminals* such that when one goes *off-hook* it automatically connects to the other terminal, without the need for any *dialling*.

hot standby: Equipment which is powered on and is waiting, ready to take over operation if the system it is shadowing were to fail for any reason.

house cable: A cable which is located within a building, or a group of buildings, and is used to interconnect communications equipment within the building and to outside lines.

housekeeping information: *Information* which is generated and used by a system to enable it to perform efficiently and which does not form any part of a user application.

HRC: *Harmonically Related Carrier.*

HRP: *Horizontal Radiation Pattern.*

HRX: *Hypothetical Reference Connection.*

H Series: *ITU-T Recommendations* which cover the use of *transmission channels* for applications other than *telephony*. A few of these are given in Table H.6.

HT: *Horizontal Tabulation.*

Table H.6 ITU-T Recommendation, H Series
(Continued on next page)

Number	*Description*
Characteristics of transmission channels used for other than telephone purposes	
H.11	Circuits in the switched telephone network
H.12	Telephone type leased circuits
H.14	Group links for wide spectrum signals
H.15	Supergroup links for wide spectrum signals
Use of telephone type circuits for voice frequency telegraphy	
H.21	Composition and terminology of international systems
H.22	Transmission requirements of international links
H.23	Basic characteristics of telegraph equipment used in international systems

Table H.6 (Continued from previous page)

Number	*Description*
Telephone circuits or cables used telegraph transmission	
H.32	Simultaneous transmission of telephony and telegraphy on telephony circuits
H.34	Division of frequency band between telegraphy and other services
Telephone types circuits used for facsimile telegraphy	
H.41	Phototelegraph transmission on telephone type circuits
H.43	Document facsimile transmission on leased telephone type circuits
Characteristics of data signals	
H.52	Transmission of wide spectrum signals on wideband group links
H.53	Transmission of wide spectrum signals on wideband supergroup links
Characteristics of visual telephone systems	
H.100	Visual telephone systems
H.120	Codecs for videoconferencing using primary digital group transmission
H.140	Multipoint international videoconferencing system
Audio visual services: transmission, multiplexing and synchronisation	
H.221	Frame structure for a 64 to 1920 kbit/s channel
H.223	Multiplex protocol for low bit rate multimedia communications
Audio visual services: systems aspects	
H.230	Frame-synchronous control and indication signals

Table H.6 (Continued from previous page)

Number	*Description*
H.231	Multipoint control units using digital channels up to 1920 kbit/s
H.233	Confidentiality system
H.234	Encryption key management and authentication systems
Audio visual services: Communication procedures	
H.242	Communications between audio-visual terminals using digital channels up to 2 Mbit/s
H.243	Communications between three or more audio-visual terminals using digital channels up to 2 Mbit/s
H.245	Control protocol for multimedia communications
Audio-visual services: coding of moving video	
H.261	Video codec for audio-visual services at n × 64 kbit/s
H.263	Video coding for low bit rate communications
Audio-visual services: system and terminal equipment	
H.320	Narrowband visual telephone systems and terminals
H.321	Adaptation of H.320 visual telephone terminals to the B- ISDN environment
H.322	Visual telephone systems and terminals for Local Area Networks
H.323	Videoconferencing over packet switched Local Area Networks
H.331	Broadcasting type audio-visual multipoint systems and terminals

HTML: *HyperText Markup Language*.
HTTP: *HyperText Transfer Protocol.*
hub: (1) The centre of a *network* to which other branches connect. Also known as a *star network*. **(2)** A device which connects other branch *nodes*

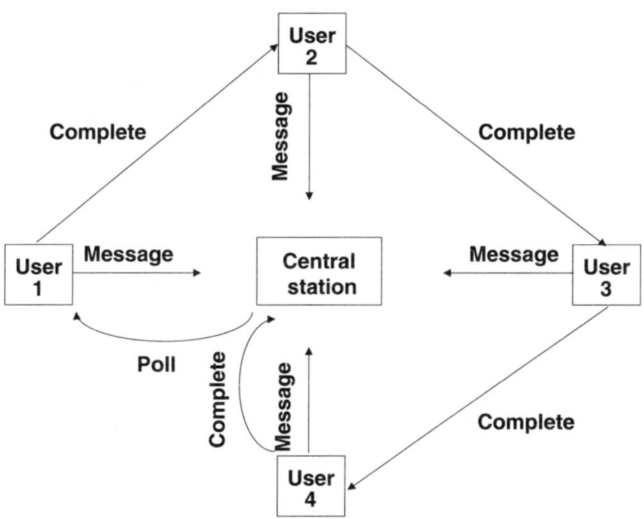

Figure H.7 Illustration of hub polling

on a *Local Area Network (LAN).* Several hubs can be interconnected within the overall network.

hub card: A card which can be inserted into a computer and enables it to perform the functions of a *hub*.

hub polling: A *polling* technique used for *multiple access*, as shown in Figure H.7. The central station commences the *poll cycle* by sending a poll to the first user. If this user has any *data* to transmit it puts it onto the common line for the addressed user to retrieve. When all data has been completed (or if there was no data in the first place) a 'complete' *signal* is sent to the next user in line (which need not be the next adjacent user). The 'complete' signal is treated as a poll by the recipient, and so on. Finally the last user in the chain returns the 'complete' signal to the central station, so ending the poll cycle.

hue: One of the characteristics used to define a colour. Hue is the dominant electromagnetic *wavelength* which is present in the colour.

Huffman coding: A *data compression* technique in which frequently occurring sequences are assigned short codes and rare occurrence are assigned longer codes. Modified Huffman coding is commonly used in Group 3 facsimile (see *Group x facsimile*). Also referred to as *Variable Length Coding (VLC)* or *entropy coding*.

human ear: A section through the human ear is shown in Figure H.8. The outer ear is called the pinna. The ear drum is the tympanum, which is attached to a chain of bones, the ossicles. These ensure that the maximum amount of sound as possible enters the cochlea.

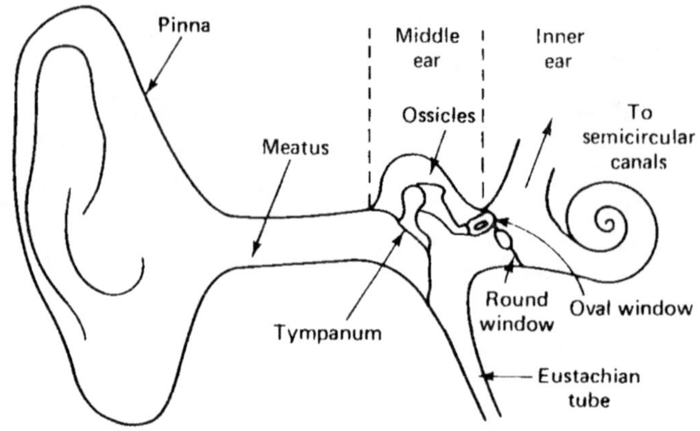

Figure H.8 Section through the human ear

human eye: Figure H.9 shows a section through the eye. The greatest change in *refractive index* occurs at the air-cornea interface. The air has a refractive index of 1.000 whilst the cornea has an index of 1.376 and the subsequent components have a value between 1.336 and 1.406.

100Base-T: A *fast Ethernet* technology. It is an upgrade to conventional *Ethernet* or *10Base-T* and can run at 100 Mbit/s. The same *packet*

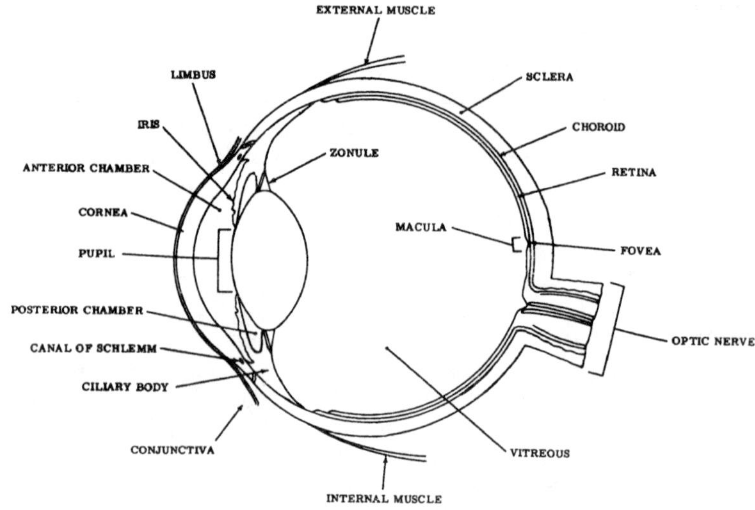

Figure H.9 Section through the human eye

structure is used, i.e. *CSMA/CD*. It operates on *Unshielded Twisted Pair (UTP) cable, Shielded Twisted Pair (STP)* cable and *optical fibre*. 100Base-TX uses 2 pairs of UTP or STP; 100Base-T4 uses UTP with 4 pairs of wires; and 100Base-FX uses *multimode fibre*.

hundred call seconds: See *Cent Call Second (CCS)*.

100VG-AnyLAN: A *fast Ethernet* technology, based on *IEEE* standard 802.12, and operating at 100 Mbit/s. It uses a *demand assignment multiple access* system and can achieve higher performance than *CSMA/CD*. It can work with 2 or 4 pair *Unshielded Twisted Pair (UTP)* wire.

hunt group: A group of *lines* which are accessed during *hunting*.

hunting: The process, used by *switching equipment*, to find a free *line* for a *call*. The selector will move through the *hunt group*, trying different lines until a free line is found.

Huygen's principle: Huygen's principle states that every point on an *electromagnetic wave* acts as a source of a secondary wavelet. This provides a means for analysis of several phenomena.

hybrid: **(1)** A device made from a number of different technologies. **(2)** A *hybrid transformer* or *hybrid coil*.

hybrid coil: See *hybrid transformer*.

hybrid connector: A *connector* which interfaces to both *optical fibre* and electrical *cables*.

Hybrid Fibre Coax (HFC): A *Fibre In The Loop (FITL)* technology in which *optical fibres* are used to distribute *broadband signals* and *POTS* to several smaller *nodes* serving a few hundred customers and from there to the individual *subscribers* over *coaxial cable*. This arrangement is commonly used in *Cable Television (CATV)* systems.

hybrid network: **(1)** A *network* in which some *lines* carry *analogue signals* and some carry *digital signals*. **(2)** The use of both *private lines* and public lines as part of the overall network, as shown in Figure H.10. The

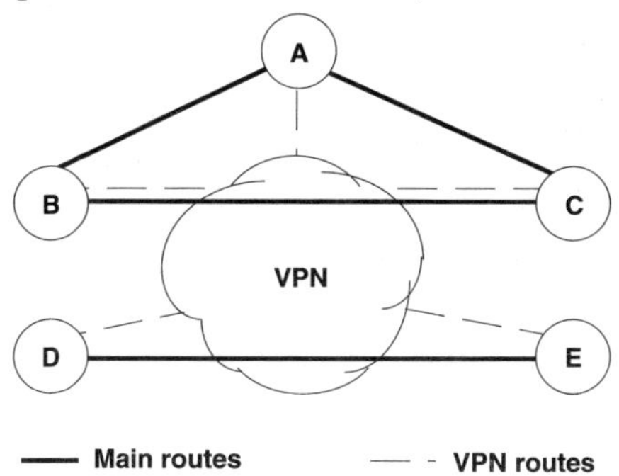

Figure H.10 A hybrid network of private and public links

public lines are part of the *PSTN* and form a *Virtual Private Network (VPN)*. **(3)** A mix of *transmission medium* used in the network, such as *coaxial cable* and *fibre optic cable*, etc. See also *Hybrid Fibre Coax (HFC)*.

hybrid router bridge: See *brouter*.

hybrid spread spectrum: *Spread spectrum* system which uses a combination of *Direct Sequence Spread Spectrum (DSSS)* and *Frequency Hopping Spread Spectrum (FHSS)*.

hybrid transformer: A transformer which is used to connect a *four wire circuit* (such as the *trunk lines*) to a *two wire circuit* (such as that connecting the *subscriber*). The hybrid transformer provides impedance matching and prevents *echoes* at the two wire/four wire interface. Commonly called a *hybrid coil*.

hypergroup: The 15 *supergroup* assembly formed in *Frequency Division Multiplexing (FDM)*, as shown in Figure H.11. These 900 *channel* hypergroups are combined into larger groups up to 60 MHz.

hypermedia: Documents and *files* which contain links to other *multimedia data*, which can be accessed at random by pointing to markers in the original document or file. The new data addressed may be in the existing or another document or file. Also sometimes referred to as *hypertext*.

hypertext: The system used to create *hypermedia*. It is also used to navigate between hypermedia information in a logical manner.

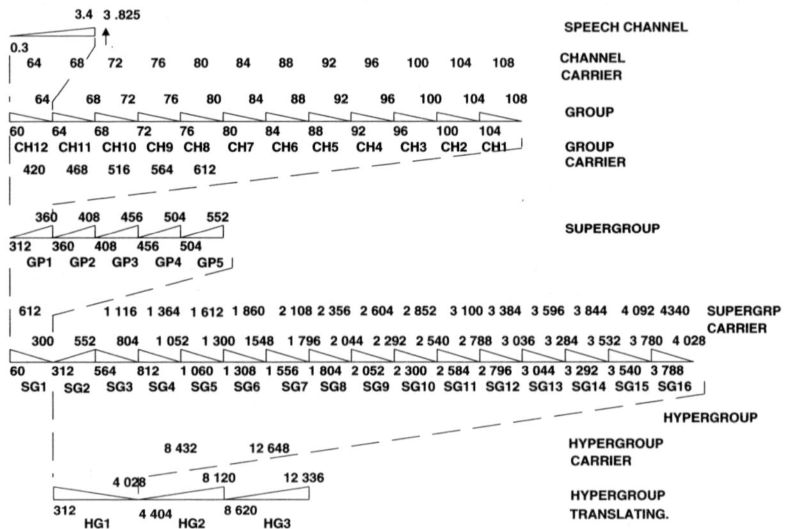

Figure H.11 Hypergroup

HyperText Markup Language (HTML): The most commonly used *programming language* for creating *World Wide Web (WWW)* documents so that they can be downloaded and interpreted by a *web browser*.

HyperText Transfer Protocol (HTTP:) The *protocol* used to transfer documents, created in *HyperText Markup Language (HTML)*, between local and remote systems.

hypothetical reference circuit: A theoretical *circuit*, specified by the *ITU-T*, which enables design and operational issues to be studied. These circuits cover several areas, such as *telephony*, television, *telegraphy*, *data networks*, etc.

Hypothetical Reference Connection (HRC): A theoretical circuit, specified in *ITU-T Recommendation* G.821, for *error performance* measurements. It is shown in Figure H.12 where the low grade sections would typically be the copper *local loop* to the *subscriber*, the medium grade sections would be the connection from, for example, the *local exchange* to a *trunk exchange*, and the high grade section would be the long distance lines consisting, for example, of *optical fibre*. Each of these sections are allocated a portion of the total errors.

Hz: *Hertz*.

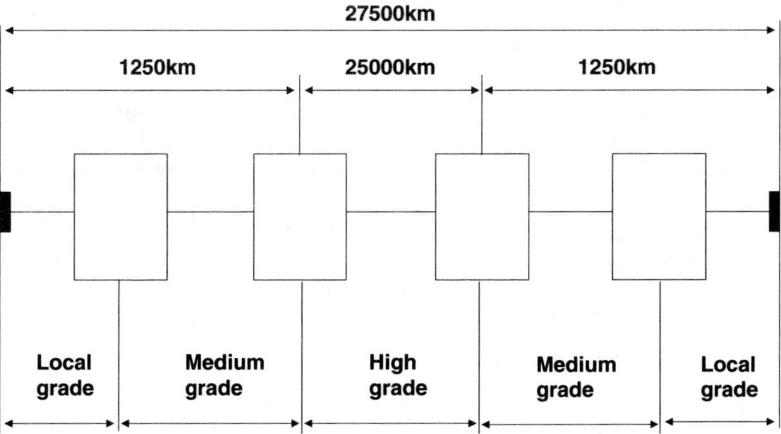

Figure H.12 Hypothetical Reference Connection

I

IA: *International Alphabet.*
IAB: *Internet Architecture Board* or *Internet Activities Board.*
IA5: *International Alphabet No. 5.*
IANA: *Internet Assigned Numbers Authority.*
IA2: *International Alphabet No. 2.*
I band: The *frequency band* between 8 GHz and 10 GHz.
IBC: *Integrated Broadband Communications.*
ICA: *International Communications Association.*
ICAO: *International Civil Aviation Organisation.*
ICI: *Incoming Call Identification.*
ICMP: *Internet Control Message Protocol.*
icon: A picture or pictorial representation of an object, task or concept. This can usually be selected from a computer screen to start a programme.
IDA: *Integrated Digital Access.*
IDC: *Insulation Displacement Connector.*
IDCT: *Inverse Discrete Cosine Transform.*
IDDD: *International Direct Distance Dialling.*
IDE: *Integrated Digital Exchange.*
identification: The process which positively identifies an entity, such as a *call*, a *user* or a resource on a *network.*
Identified Outward Dialling (IDOD): A feature, usually provided by a *switching* system such as a *PABX*, which allows all outgoing toll *calls* to be recorded for *billing* purposes.
identifier: *Characters* used within the *message* for *identification* purposes.
identity message: The *message* which allows the *calling party* or the transmitting device to be identified.
IDF: *Intermediate Distribution Frame.*
IDFT: *Inverse Discrete Fourier Transform.*
idle byte: See *idle cell.*
idle cell: The *cells* which contain no *information* and are used in *transmission* systems, such as *ATM*, to ensure that a certain *data rate* is maintained when there are insufficient user data to transmit. (See Figure V.4.) These idle cells are discarded at the receiving end. Also called *idle byte* and the process of idle cell insertion is often referred to as *idle channel loading.*
idle channel: A *transmission channel* which is not carrying any *information*, although it is ready for use and may be carrying *idle characters.*
idle channel loading: See *idle cell.*
idle channel noise: The *noise* present in an *idle channel.*

idle character: *Control characters* which are transmitted on *idle channels*, when there is no useful *information* to be sent. It indicates that there is no useful information present and fulfils control functions, such as filling *timeslots* and maintaining timing.

idle condition: See *idle state*.

idle line termination: An electrical network which is applied to a *line* when it is in an *idle state*. This is normally done automatically by switches and maintains the impedance of the line at a predefined value.

idle state: The condition which exists in an *idle channel* or line i.e. it is not currently in use but is ready and waiting for a *transmission*.

idle time: The time over which a facility has been in an *idle state*. Often specified as a percentage over a certain period.

IDN: *Integrated Digital Network*.

IDOD: *Identified Outward Dialling*.

IDTV: *Improved Definition Television*.

IEC: *International Electrotechnical Commission* or *Interexchange Carrier*.

IEEE: *Institute of Electrical and Electronic Engineers*.

IEEE 802: The *IEEE* committee which developed the 802 series of standards for *Local Area Networks (LAN)*. See *IEEE standards*.

IEEE standards: Standards developed by the *Institute of Electrical and Electronic Engineers*. A few of these are given in Table I.1.

IESG: *Internet Engineering Steering Group*.

IETF: *Internet Engineering Task Force*.

I-ETS: *Interim European Telecommunications Standard*.

IF: *Intermediate Frequency*.

IFL: *International Frequency List* or *International Facilities Licence*.

IFRB: *International Frequency Registration Board*.

IG: *Interframe Gap*.

IGC: *Intergovernment Conference*.

IHL: *Internet Header Length*.

ILF: *Injection Laser Diode*.

ILF: *Infra Low Frequency*.

illegal character: A *character* which does not conform to the *algorithm* required of the *code* being used.

illuminance: **(1)** In an optical system it is the density of *luminous power* falling on a surface. It is often expressed in *lumens* per square centimetre. **(2)** In a radio system, such as *radar*, it is the amount of energy falling on the target, measured in watts per square metre.

IM: *Intermodulation*.

image antenna: An hypothetical *antenna*, used for mathematical analysis, in which an image of the antenna is assumed to exist relative to the earth, i.e. the antenna is situated as far below the earth as it is located above it.

Table I.1 IEEE standards

IEEE Standard	*Description*
488	Parallel interface bus used to interconnect equipment. Also known as a GPIB
802.1	LAN network management at the hardware level. Includes bridging standards
802.2	A data link layer standard used with IEEE 802.3, 802.4 and 802.5 standards
802.3	Local Area Network standard for access at the physical layer using CSMA/CD
802.4	Local Area Network standard for access at the physical layer using token passing. Used with bus topologies
802.5	Local Area Network standard for access at the physical layer using token passing. Used with ring topologies
802.6	Standard for Metropolitan Area Networks. Introduces Distributed Queue Dual Bus (DQDB)
802.7	Standard for broadband networks
802.8	Report on installation and use of fibre optic cables
802.9	Standard for integrated voice and data on a LAN
802.10	Standard for LAN/WAN security
802.11	Committee dealing with Wireless Local Area Network (WLAN) standards
801.12	Standard for 100VG-AnyLAN

image compression: The reduction in the amount of *data* needed to represent a picture or drawing following *imaging*. This means that less *capacity* is needed for storage and *transmission*.

image intensifier: Equipment which increases the *luminance* of objects which have low *intensity*.

image plane: The geometrical plane in which an image is considered to lie.

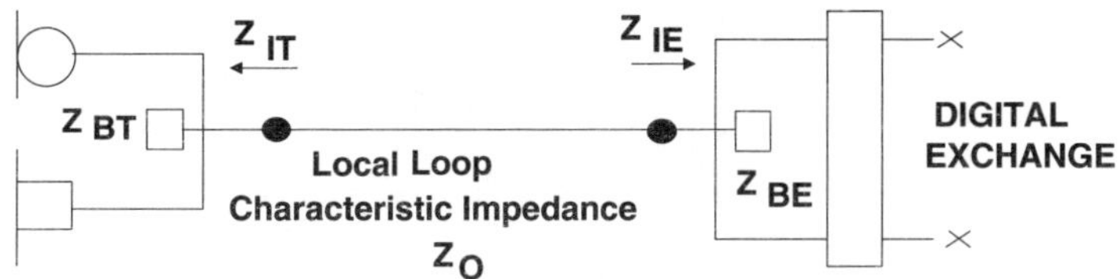

Figure I.1 Local impedance strategy

imaging: The representation of a object, such as a picture or a drawing, on a medium, such as film, a *Cathode Ray Tube (CTR)*, etc. Generally this is in the form of *data* which can be subsequently processed, e.g. stored and transmitted.

IMD: *Intermodulation distortion.*

IMEI: *International Mobile Equipment Identity.*

IMM: *Ineractive Multimedia.*

IMP: *Interface Message Processor.*

impedance matching transformer: A transformer which acts as additional impedance in order to archive a given aim, such as balancing a *transmission line* or preventing reflections from it. Most transformers use the effect of mutual inductance between two or more windings to achieve this.

impedance strategy: The strategy used in a *transmission plan* to obtain optimum performance. Impedances associated with telephones, local lines and digital exchanges (see Figure I.1) all affect many of the transmission parameters, such as *echo*, *sidetone* and *transmission loss*. Optimum transmission conditions are reached when all the impedances (Z_{BT}, Z_{IT}, Z_O, Z_{IE}, Z_{BE}) are equal or as close to each other as possible.

implicit reservation ALOHA: A *Reservation ALOHA (R-ALOHA)* system for *multiple access* in which the transmission slots are grouped into *frames*, the number of slots exceeding the number of *users*. Each user is nominated as being the owner of a slot position within a frame, as in Figure I.2 (Frame 0). Each user has priority for transmission on its slot. If there are more slots than users then all users contend for the spare slots, using *ALOHA* or *slotted ALOHA*. Whilst an owner is transmitting on its allocated slot no other user can interfere. However, if a slot falls idle for a frame, then in the next frame other users can contend for its use. Once a slot has been so seized only the owner can get it back by transmitting on it. A *contention* occurs and in the next frame only the owner is allowed to transmit. Therefore in Figure I.2 Frame 1 owners A, C E and H transmit. In the following frame (Frame 2) A and E successfully seize another slot each. A collision has also occurred on slot 4. In Frame 3 slot 4 is vacant, which means that the owner was not involved in the earlier

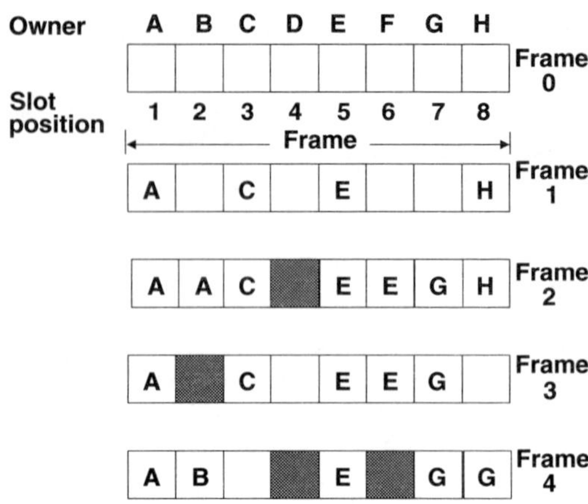

Figure I.2 Implicit reservation ALOHA

collision so each station waits to make sure. Also a collision has now occurred in slot 2, which is due to the owner wanting its slot back. In Frame 4 owner B commences its transmission in its slot. Several users have contented for slot 4 so collision has again occurred. In addition the owner has forced a collision in slot 6 to get it back.

Improved Definition Television (IDTV): Improvements in the *transmission* or reception of standard television systems which enhance its performance but still operate within the overall confines of the conventional systems, such as *NTSC*, *PAL* and *SECAM*, and do not make existing *television receivers* or other equipment obsolete.

Improved Mobile Telephone System (IMTS): An older mobile radio system, used in the USA, which was replaced by the *Advanced Mobile Phone System (AMPS)*.

impulse: Generally refers to a surge of energy (electrical, magnetic, or electromagnetic) which occurs at random and lasts for a short duration.

impulse hits: Errors or *hits* in transmitted *data* cause by *impulse noise*. Bell standards allow no more than 15 impulse hits in a 15 minute interval.

impulse noise: *Noise* which has the characteristics of an *impulse* i.e. high amplitude and short duration, occurring at random and lasting for a short duration. Generally impulse noise is considered to be the noise which is 12 db or more above the r.m.s. value and lasts for less than 12 ms. Impulse noise causes annoyance on *voice circuits* but it can result in serious errors in *data transmission*.

impulse noise counter: A test instrument used to measure *impulse noise*. It enables the noise threshold to be adjusted in steps and measures the number of times this threshold is exceeded in a set time interval.

impulsive interference: Interference due to *impulse noise*, which can result in annoyance or errors.

IMS: *Interactive Multimedia Service.*

IMSI: *International Mobile Subscriber Identity.*

IMTS: *Improved Mobile Telephone System.*

IMT-2000: *International Mobile Telecommunications 2000.*

IMUX: *Inverse Multiplexer.*

IN: *Intelligent Network.*

INAP: *Intelligent Network Application Part.*

in-band signalling: *Signalling* which is carried within the same *bandwidth* of the *call* being controlled. See also *Channel Associated Signalling (CAS)* and *bit robbing*.

in-call handover: In a *cellular radio system* the *signal* strength from the mobile to its *base station* is continually monitored and if this falls below a pre-set value the *network* will ask neighbouring base stations to monitor the signal from the mobile. If their signal strength is greater than that at the current base station the network requests the mobile to switch frequencies to that of the new base station and this base station now controls the call. This process of in-call handover between base stations is illustrated in Figure I.3.

incoming call: A *call* which is being received by a *terminal*. Also refers to a call received by an *exchange* and which will be switched through to a terminal.

incoming call handling: The *call handling* process used to deal with *incoming calls*.

Incoming Call Identification (ICI): A feature, usually provided by *switching* systems such as a *PABX*, which allows an attendant to visually determine the type of *trunk* or service group used by a *call*.

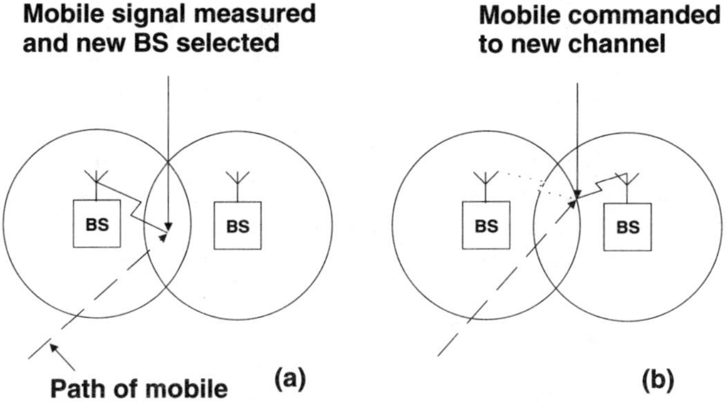

Figure I.3 In-call handover

incoming call rate: The frequency at which *calls* are being received by a *node* on a *network*, measured over a period of time.

incoming lines: The *transmission lines* which bring *messages* into a *node* on the network or to a *switching centre*. The calls from the switching centre are then sent out to the appropriate node over *outgoing lines*.

incoming trunks: *Trunk circuits* which connect to an *exchange* and carry *signals* into that exchange. See also *outgoing trunks*.

Incrementally Related Carrier (IRC): *Transmission* system in which each *channel* in the spectrum is placed on Incrementally Related Carriers, i.e. equal spacing between channels. This ensures that the *intermodulation (IM)* outputs fall exactly on third order *carrier frequencies* rather than within the actual *signal* band, where they would cause interference. See also *Harmonically Related Carrier (HRC)*.

independent clock: A *clock* which is not *synchronised* with any other *timing* source. This clock would normally be derived from an accurate *frequency* generator and would act as the *master clock* for the *node* or a part of the *network*.

Independent Sideband transmission (ISB): A *Double Sideband transmission (DSB)* method in which each *sideband* carries different *information*. The *carrier* may be transmitted or suppressed.

Independent Telephone Company (ITC): A *PTO* who is not connected with a *Bell Operating Company (BOC)* in the USA. There are close to 1500 ITCs in the USA.

index matching material: Material which has the same *refractive index* as the device it is being applied to. For example, the material could have the same refractive index as the *core* of an *optical fibre* and this would ensure that *Fresnel reflections* from the end of the fibre are reduced.

index of cooperation: See *cooperation index*.

index of refraction: See *refractive index*.

index profile: The variation of the *refractive index* of a *transmission medium*, such as *optical fibre*, along its cross section. This can vary in several different ways; see, for example, *step index fibre* and *graded index fibre*.

indirect access: The access which a *PTO* gains to its customers by going through another PTO's *local access network*. This is often done because many new PTOs have not installed their own local network and usually *interconnect charges* will have to be paid to the incumbent operators for this privilege.

indirect routeing: The *routeing* of *messages* between two *nodes* on a *network* in which the most direct route between the two nodes is not taken, for example if that route is congested or faulty.

indirect signalling: Terminology used to describe *signalling* systems which do not vary line conditions directly (such as by voltage or current) to carry the signalling information. Examples are *Channel Associated Signalling (CAS) and Common Channel Signalling*. See also *direct signalling*.

individual call sign: A *call sign* which is linked to a single transmitting *station* and can identify that station.

individual presentation control: One of the service classifications of *B-ISDN* in which, for *distribution services*, the *user* has control of the presentation. For example, *information* from a central source could be *broadcast* to many recipients and it would have cyclic repetitive sequences so that users can control the start and sequence in which to view the material.

individual reception: The reception of signals by individuals. For example, in a *broadcast satellite service antennas* mounted on individual homes can receive the *signals*.

inductance: The phenomena in which a conducting material provides impedance to the flow of *Alternating Current (AC)* through it. This is due to the magnetic field generated by the current which induces an electromotive force in the conductor.

Industrial Organisation (IO): Organisations which are recognised by the *ITU* and who can attend its meetings. There are about fifty of these recognised, mainly trade organisations and user groups, such as the *International Telecommunications User Group (INTUG)*.

Industrial, Scientific and Medical (ISM): Refers to the dozen *frequency bands*, in the ranges from 902 MHz to 928 MHz, 2.4 GHz to 2.5 GHz, and 5.8 GHz to 5.9 GHz, which have been identified by the *ITU*, in collaboration with the *International Electrotechnical Commission (IEC)*, for use in industrial, scientific and medical applications. These frequencies have also now been allocated for other radio use, such as by the FCC for unlicensed *spread spectrum* communications, but interference from ISM equipment must be tolerated in these bands.

inference engine: The part of an *Artificial Intelligence (AI)* system which carries out the reasoning process.

inFLEXion: A *two-way pager* system developed by Motorola Inc., a trial system being demonstrated in Dallas in 1997. Digitally compressed speech is transmitted to the pager at a data rate of 112000 bits per second. It is able to receive up to 32000 *characters* and transmit 5000 characters.

information: Meaning and knowledge assigned by a *user* to *data* which is received. Different *users* can sometimes obtain different information from the same data, based on their knowledge and past history.

information bearer channel: **(1)** A communications *channel* which has the required *bandwidth* to carry all the *information* needed to meet a *user's* requirements, such as the *data*, control information, *synchronisation* information, etc. **(2)** A basic *transmission channel* provided by a

PTO which has the *bandwidth* needed to carry the user's data but no other enhanced services.

information bit: A *data bit* which carries *user information* but no overheads, such as control, synchronisation, etc.

information channel: The *channel* used to transfer *information* between *users*.

information content: The part of a *message* which contains *user information* rather than *signals* required by the *network* for control, *synchronisation*, etc.

information feedback system: A *transmission* system in which the *receiving terminal* returns some of the *data blocks* to the *transmitting terminal*, so that it can confirm that this was correctly received, performing an *error detection* function.

information provider: Generally used to describe commercial organisations who provide *information* over the telecommunications *network*. See also *value added service provider*.

information rate: See *data transfer rate*.

information retrieval: Processes and systems used to obtain *information* which has been stored in *data processing systems*, usually as *databases*.

information security: Processes and systems which are in place to prevent unauthorised or accidental access to confidential *information*.

information separators: Special *characters* used to separate the different *fields* which form the *information* being processed.

information service: Commercial service provided (for a fee) on a variety of topics, such as travel information, weather information, etc. See also *value added services*.

information superhighway: A term sometimes used to describe the *Internet*.

Information Technology (IT): A wide ranging generic term covering all aspects of computing, *data processing systems*, including *hardware*, *software* and *firmware*.

Information Technology Association of New Zealand (ITANZ): An organisation representing manufacturers and suppliers of *Information Technology (IT)* equipment and services in New Zealand. It seeks to influence decision makers, such as government and PTT, on behalf of its members. To this aim it sets up task forces to deal with particular issues.

Information Technology Steering Committee (ITSTC): A joint work group set up by *ETSI*, *CEN* and *CENELEC* for sharing standards tasks. The ITSTC administers all work on *Information Technology (IT)* and CEPT is also represented on it. It uses Expert Groups, formed as required, to develop standards. Included are ITAEGS for standards, ITAECM for manufacturing, ITAEGT for telecommunications, and ITAEGC for certification.

information theory: The branch of science dealing with all aspects of *information*, such as its creation and *transmission*.

information transfer: The movement of *information* between *nodes* on a *network*.

information transfer transaction: The process of transferring *information* between *nodes* on a *network*. It may be considered to consist of three phases: the set up phase, including agreement on the *data* to be transferred; the actual *information transfer* phase; and the disengagement phase.

Infra Low Frequency (ILF): *Frequencies* in the *frequency band* from 300 Hz to 3 kHz.

infrared (IR): The part of the *electromagnetic spectrum* below the visible red region. It can be considered to be made of three separate regions: the near infrared region, from 0.75 μm to 3 μm, the middle infrared region from 3 μm to 30 μm and the far infrared region from 30 μm to 100 μm.

infrared transmission: *Transmission* which uses *infrared (IR)* light as the *carrier signal*.

initialise: To set up the system into a known state, usually at the start of an operation.

Injection Laser Diode (ILD): See *LASER*.

INMARSAT: *International Maritime Satellite organisation*.

in-octet signalling: Term sometimes applied to *signalling* systems which use *bit robbing*.

in-plant communication: A communications system which is located in a building or a local area and is completely made from *private lines* i.e. it does not use any lines or equipment owned by a *PTO*.

in-plant equipment: Equipment which is used for *in-plant communications*.

input channel: A *channel* used to feed *data* into a device, system or *network*.

input data rate: The *data rate* at which a *network* or device can accept *transmissions*.

input/output: Moving *data* into and out of a device, system or *network*. Usually refers to the communications between a central processor and its peripherals, such as *terminals* and printers.

input unit: The device, such as a *terminal* or *keyboard*, used to feed *data* into a device, system or *network*.

inquiry/response signals: The *signal* which interrogates a system (inquiry), such as a computer *database*, and the results which it obtains (response).

insertion gain: The *gain* in a transmitted *signal* which occurs due to the insertion of a device, such as an *amplifier*, into the *transmission medium*. It is measured as the ratio of the power at a point with the device present to that when it is not present, expressed in *decibels*. This would give a positive result. See also *insertion loss*.

insertion loss: (1) The *loss* in a transmitted *signal* which occurs due to the insertion of a device, such as an *attenuator*, into the *transmission medium*. It is measured as the ratio of the power at a point with the device present to that when it is not present, expressed in *decibels*. This would give a negative result. See also *insertion gain*. **(2)** In a *optical fibre transmission* system the insertion loss is the total *overall loss* in the system and is due to many causes, such as *absorption*, *dispersion*, *microbends*, *scattering*, *splicing*, etc.

inside plant: Equipment which is located in a building, such as in a *Central Office (CO)*, and is not located outside, where it could be exposed to the elements, such as on telephone poles or buried underground.

inside wiring: Wiring which is located inside a customer's premises, for example going from one floor to the next.

in-slot signalling: *Signalling* which occurs within the *timeslot* with which it is associated. See, for example, *Channel Associated Signalling (CAS)* and *bit robbing*.

instantaneous traffic level: The number of *calls* which are ongoing at any instance of time.

Institute of Electrical and Electronic Engineers (IEEE): An international organisation with over a third of a million members in about 150 countries. It has headquarters in New York and is structured into ten regions: Regions 1 to 6 in the US; Region 7 in Canada; Region 8 in Europe; Region 9 in South America; and Region 10 in Asia Pacific. If was formed in 1963 by a merger between the American Institute of Electrical Engineers (formed in 1884) and the Institute of Radio Engineers. The aim of the IEEE is to advance the theory and practice of engineering within the electronic and electrical disciplines, which includes *Information Technology (IT)* and *telecommunications*. It is structured into specialised divisions: Circuits and Systems; Industrial Applications; Communications Technology; Electromagnetic and Radiation; Computer; Engineering and Human Environment; Energy and Power Engineering; Signals and Applications; Systems and Control. It is active in standards making, with close to a thousand standards published, the best known being that produced by the 802 committee which defines standards for data communications, primarily in *Local Area Networks (LAN)*. See *IEEE standards*.

Insulation Displacement Connector (IDC): A connector with jaws which punch down into the *cable* insulation, so making contact with the internal metal core without requiring the outer insulation to be first stripped.

Integrated Broadband Communications (IBC): *RACE* research programme for the introduction of *broadband* communications systems

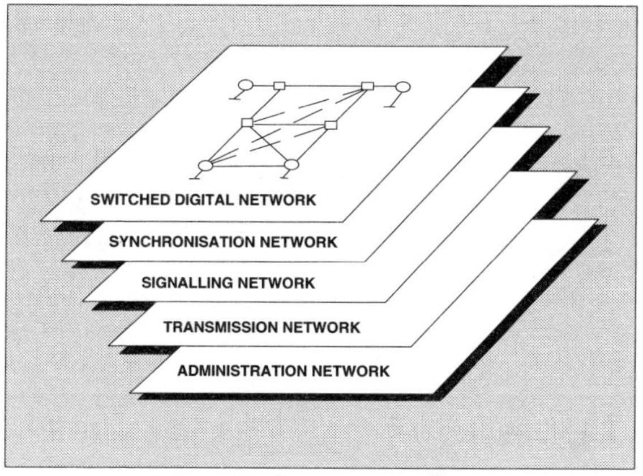

INTEGRATED DIGITAL NETWORK

Figure I.4 Components of an IDN

within European countries. It ran from 1988 to 1995, covering two phases.

Integrated Digital Access (IDA): An early *ISDN* service operated by BT in the UK. The pilot system was started in June 1985 based on System X switches in the City of London. It used *ITU-T Signalling System No. 7.* This was subsequently extended to other *exchanges* in major centres around the UK.

Integrated Digital Exchange (IDE): See *digital exchange.*

Integrated Digital Network (IDN): A *network* in which all the major components operate on *digital signals*, such as *switching*, *transmission*, *signalling*, etc., as shown in Figure I.4.

integrated optical circuit: See *Optical Integrated Circuit (OIC).*

integrated modem: A *modem* which is an integral part of another device, such as a card which plugs into a *terminal* or a computer, and is contained within the same case as these devices.

Integrated Services Digital Network (ISDN): A series of *ITU-T Recommendations* (see *I Series*) covering the *transmission* of *voice* and *data* over the *local loop*. Two types of access are specified, *Basic Rate Access (BRA)* at 144 kbit/s, consisting of two *B channels* and a *D channel*, and *Primary Rate Access (PRA)* at 2.048 Mbit/s (1.544 Mbit/s in the USA) consisting of 30 B channels and a D channel (23 B channels and one D channel in the USA). Figure I.5 shows an ISDN arrangement.

Integrated Services Local Area Network (ISLAN): A *Local Area Network (LAN)* which can carry an integrated service, such as *voice*, *data* and *video*.

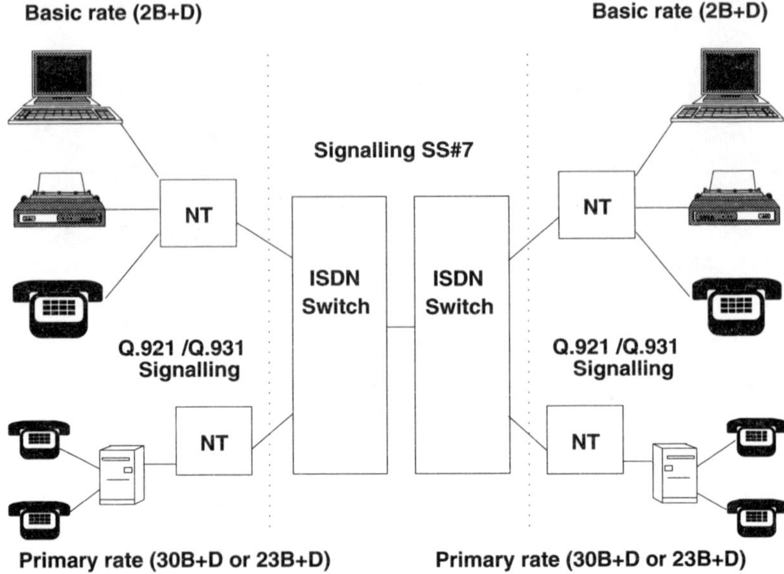

Figure I.5 The Integrated Services Digital Network

Integrated Services Private Branch Exchange (ISPBX): A *PBX* whose operation is based on the *Integrated Services Digital Network (ISDN).*

Integrated Voice-Data Terminal (IVDT): A *terminal* which can handle both *voice frequencies* and *data signals.*

integration: The process of combining different pieces of *hardware* and *software* to form one unified system.

intelligent controller: A device which controls a group, or cluster, of devices, providing a control interface between them and the *network.* See also *cluster controller.*

Intelligent Network (IN): Three concepts define an Intelligent Network: separation of service and call processing, so that services can be set up quickly and changed quickly to meet customer requirements; use of standard, reusable software modules and development tools, which allow services to be set up quickly; use of modularisation and standards based interfaces, so that IN modules can be bought from different vendors and still work with each other. Conventional *networks* typically use a separate *signalling* system to allow switches within the network to communicate with each other. In an IN separate computers, working through a separate *data* network, are used to control all the operations in the service network. The large amount of software associated with the computer based intelligence means that more complex tasks can be carried out. Examples are the *freephone service*, ability to direct *calls* to different offices depending on time of day or location of the caller, etc. Several key elements go into the making of an IN: the *Service Switching Point (SSP)*, the *Service Control Point (SCP)*, the *Intelligent Peripheral (IP)*, the *Service Management Point (SMP)*, and the *Service Creation Environment (SCE)*.

Intelligent Network Application Part (INAP): Standard for *Intelligent Networks (IN)*, developed by the *ITU-T* and *ETSI*. This defines modularisation and standard based interfaces, so that IN modules can be bought from different vendors and still be made to work with each other. It performs a similar function to the *Advanced Intelligent Network (AIN)* standard from *Belcore*.

Intelligent Peripheral (IP): One of the key components of an *Intelligent Network (IN)*, the others being the *SSP*, the *SCP*, the *SMP*, and the *SCE*. The Intelligent Peripheral acts as the interface between the user of the IN service and the IN system itself. It will obtain *data* from the *user*, providing instructions as needed, e.g. 'enter PIN number'.

intelligent terminal: A *terminal* with a level of processing power so that it can carry out some of the *data* processing either before or after *transmission*.

Intelligent Time Division Multiplexing (ITDM): *Multiplexing* in which *timeslots* are allocated to *users* according to their requirements rather than in a fixed sequence, as is done in conventional *Time Division Multiplexing (TDM)*. See also *statistical multiplexing*.

intelligible crosstalk: *Crosstalk* in which intelligible *signals* are coupled from one *line* to another.

INTELSAT: *International Telecommunication Satellite organisation*.

itemised bill: *Billing* in which the *subscriber* receives a breakdown of all the *calls* which have been made. Usually this would provide the *telephone number* of all *trunk calls* but all *local calls* would be collected together under a single entry.

intensity: The strength of an *electromagnetic radiation*, measured as the square of the electrical *field strength*. Intensity is proportional to *irradiance*.

intensity modulation: A *modulation* technique in which a *light source* forms the *carrier signal* and its power output is varied by the *modulating signal*.

interactive mode: A method of communications with a system in which the *user* communications consists of a series of interrogations to which responses are received. The user sends a *signal* and then waits for a response before sending another signal. The content of a query is determined by the content of an earlier response. Also called a *real-time system*.

Interactive Multimedia (IMM): *Multimedia* applications in which the *user* can interact with the system, such as for *Video On Demand*, *video* games, and on the *Internet*.

Interactive Multimedia Service (IMS): *Interactive Multimedia (IMM)* service provided to a *subscriber* by a *PTO*.

interactive service: One of the *service classifications* proposed by the *ITU-T*, the other being *distribution service*. Interactive services use *interactive mode* for communications between the *user* and the system. It is further subdivided into *conversational service*, *messaging service* and *retrieval service*.

Interactive Voice Response (IVR): An *interactive mode* in which communications between the system and a caller takes place by the system playing a recorded *message* and requesting that the caller press one of several keys on a *MF* telephone in reply. Depending on the key pressed the system then plays another message, so the user and the system interact by messages and keypad until the caller has obtained the desired information.

Interactive Voice Response Unit (IVRU): The unit which carries out the *Interactive Voice Response (IVR)* communications.

interbuilding cable: *Cable* which runs between buildings located on the same campus. This can be buried or overhead.

interbuilding cable entrance: The point at which the *interbuilding cable* enters the building.

intercept: To acquire control of a transmitted *signal* and then to either terminate it, delay it, redirect it to another location or number, or to learn its contents and then send it on to its original destination, so that the interception goes undetected by the recipient.

intercepting trunk: The *trunk* to which a *call* is connected when an *intercept* has occurred because the original call could not be completed due to several reasons, such as the *line* being called is faulty or not available.

interchange circuit: A *circuit* which allows communications to occur between two, usually disparate, systems. Examples are communications between *optical fibre* and copper based systems, and between a *Data Terminal Equipment (DTE)* and a *modem*.

interchange signal: The *signal* passed along the *interchange circuit*, between systems having different functions.

interchange specification: The specification of the *interchange circuit* connecting two systems. This specification allows communications to occur between the systems, for example between a *Date Terminal Equipment (DTE)* and a *modem*. See also *interface specification*.

intercom: A system which allows *voice* communications between *users* situated in the same location, such as a building, aircraft or ship. Usually interconnection is made by *dialling*, or automatically when one of the receiver is lifted or a simple *signal* is sent, such as by pushing a button.

interconnect charges: The charge which one *PTO* makes to another PTO for using its communications facility to carry *traffic* belonging to the second PTO's customers. Usually this would be between a long-distance *carrier* and a local carrier (see *access charges*) or between national and international carriers.

interconnect company: Term used in the USA to describe an organisation which sells, installs and maintains equipment for connection to telephone *lines.*

interconnection: Connecting together compatible systems, such as those belonging to competing *carriers*, in order to carry a *signal.*

Interconnection Directive (ID): See *European Interconnection Directive.*

Interexchange Carrier (IEC or IXC): US term for any telephone operator licensed to carry *traffic* between *LATAs*, interstate or intrastate.

interexchange signals: *Signals* which are passed between *exchanges* during *call establishment* and *call disestablishment.*

interface bus: In an equipment the *bus* which transports *signals* between different internal modules, such as the memory unit, the processing unit, the input/output devices, etc.

interface specification: The specification which, if followed, will allow two systems to interact with each other in the required manner. This can cover electrical, mechanical and functional definitions. See also *interchange specification.*

interface compatibility: The state which exists when a device meets the *interface specification* of other equipment to which it is connecting.

Interface Message Processor (IMP): A processor used to connect *nodes* to a *network*. This is commonly used in a *Packet Switched Network (PSN)* to enable routeing of *packets* between incompatible devices.

interface processor: A processing device, such as a computer, whose prime function is to provide the interface needed for connecting components to a *network*, for example by carrying out *protocol conversion.*

interference: The interaction of two *signals* such that the quality of one or both of them is impaired.

interference coupling: The mode used for *interference* to occur between systems. Generally this is by *conducted interference* or *radiated interference* as illustrated in the example of Figure I.6.

interference filter: *Filter* used to prevent *interference*, usually by blocking the *signals* at the interfering *frequency* from being conducted to the equipment being protected by the filter.

interference limit: **(1)** The maximum amount of *interference* which an equipment can stand without malfunctioning. **(2)** The maximum amount

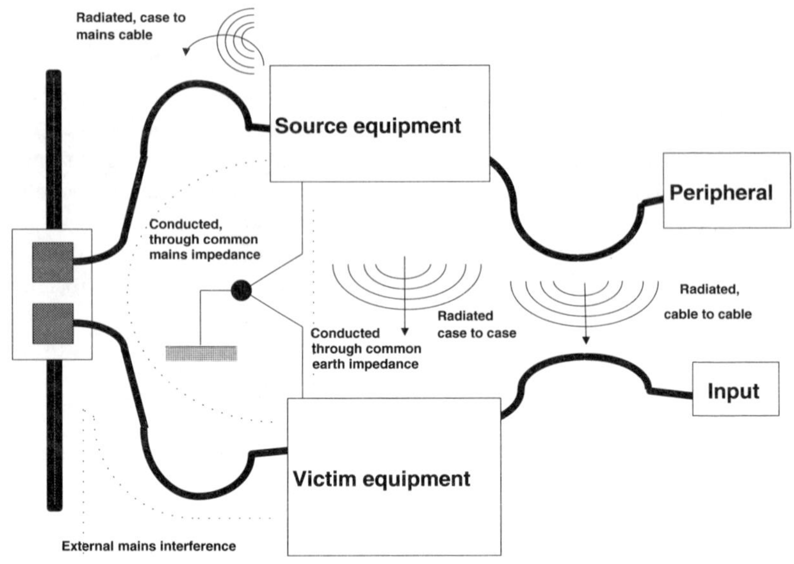

Figure I.6 Interference coupling

of *interference* which is permitted by a *regulatory body* in certain *frequency bands* and on certain classes of *users* or equipment.

interference pattern: The *signal* pattern produced by *interference*. Usually seen as regularly spaced lines or curves on a display.

interferometer: A device used to measure changes in distance, *refractive index* or *wavelength* by comparing the relative *phase difference* between two or more beams through changes in an *interference pattern*. Examples are the Fabry Perot interferometer and the *Michelson interferometer*.

Interframe Gap (IG): The time period which occurs between successive *frames* or *packets* in a *transmission* sequence. This will vary depending on the *protocol* used.

Intergovernment Conference (IGC): Conference, stipulated in the *Maastricht Treaty*, which met in 1996 to consider revisions to the Treaty, including enlargement of the *European Community (EC)* to cover up to 28 nations.

Interim European Telecommunication Standard (I-ETS): A standard issued by *ETSI* when it is only likely to be required for an interim period, for example, because the standard is a provisional solution and a more advanced standard is to be prepared, or if a standard is immature and needs a trial period. See also *European Telecommunication Standard (ETS)*.

interlaced scanning: The process of *scanning* used in television and a *Video Display Unit (VDU)* in which the *scanning beam* first traces the odd numbered lines on the screen and then goes back and traces the even numbered lines. This reduces the *bandwidth* required for *transmission* since the two scans form different *fields*. Since the eye cannot effectively differentiate different lines of a single frame this only introduces a small amount of interline *flicker*.

interlacing: See *interlaced scanning*.

inter-LATA: A *call* which originates in one *Local Access and Transport Area (LATA)* and terminates in another LATA. Under the original rules of *divesture* a *Regional Bell Operating Company (RBOC)* was not allowed to provide inter-LATA service, this being provided by the *Interexchange Carrier (IEC or IXC)*.

inter-layer interface: The *interface specification* between layers of a structured *network architecture*, such as the *Open System Interconnect (OSI)* reference model.

interleaved code: A *transmission code* in which two successive code symbols are separated by a number of *bits* (the *Hamming distance*). This makes it less susceptible to *burst errors* and enables *error correcting codes* to operate more effectively. For example, if the separation between code symbols is ten bits then a burst error of ten bits in length would only effect one bit in each *code word* which can be detected and corrected by most error correcting systems. See also *Hamming code*.

interleaving: The process of alternate *transmission* of parts of separate *messages*, these being separated by *Time Division Multiplexing (TDM)*. Interleaving can involve *bits*, *characters*, *bytes*, *blocks*, etc.

interlock code: A *code*, sent as part of a *call request packet* during *call setup*, to indicate that the *calling terminal* is part of a *Closed User Group (CUG)*. It prevents other *calls* from interrupting a CUG session.

Intermediate Distribution Frame (IDF): A *distribution frame* used as an extension to a *Main Distribution Frame (MDF)*, to interconnect equipment situated in areas remote from the MDF.

intermediate equipment: Equipment, usually located between a *DTE* and a *DCE*, which performs functions specified in *interface specifications*.

intermediate exchange: See *transit exchange*.

Intermediate Frequency (IF): The *frequency* to which a *signal* is moved as a temporary measure during *signal processing*, such as the *transmission* or reception phase.

intermediate node: A *node* which is connected to two other nodes, i.e. it is not an end node in the *network*, and is capable of *routeing traffic* between these nodes.

intermediate system: **(1)** Any system, such as an *intermediate node*, which routes *data* between other systems. **(2)** Sometimes used to describe a *relay system*.

intermodal dispersion: Also known as *modal dispersion*. See also *intramodal dispersion*.

intermodulation (IM): *Modulation* of the *frequencies* in an *electromagnetic wave* so as to produce frequencies which are a sum and difference of the original *signals* as well as the harmonics above the *fundamental frequencies*. It occurs in *Frequency Division Multiplexing (FDM)* systems.

intermodulation distortion: Non-linear *distortion* in which the *amplitude* of a *signal* at one *frequency* is affected by the amplitude of a signal at another frequency, the combined signal consisting of frequencies which are the sum and difference of the two original frequencies. (See Figure I.7.)

intermodulation noise: *Noise* in a system resulting from non-linear effects (*intermodulation (IM)*) which generates spurious signals having frequencies which are the sum and difference of the original frequencies in the *transmission path*.

intermodulation product: The sum and difference frequencies produced as a result of *Intermodulation (IM)*.

internal call: A *call* which is placed to numbers controlled by a *PABX* and not one which passes through a *Central Office (CO)*. See also *external call*.

internal clock: A *clock*, used for *synchronisation* and *timing*, which is located within the equipment being controlled.

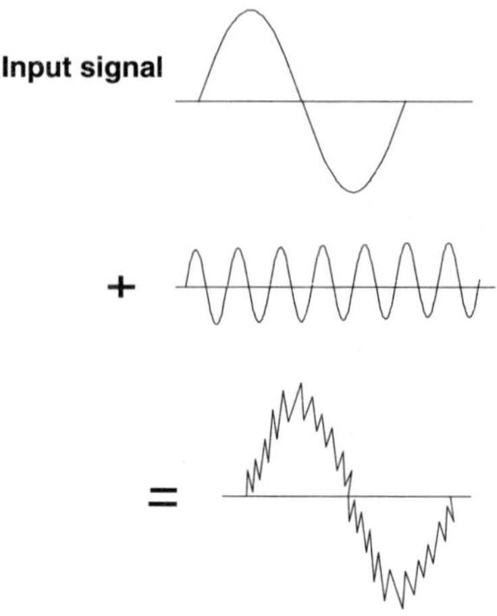

Figure I.7 Intermodulation distortion

internal reflection: In a *transmission medium*, such as *optical fibre*, the *reflection* of the *signal* which occur internally, usually between the *core* and the *cladding*, as the signal travels along the length of the medium. See also *total internal reflection*.

International A: The frequency given to the musical note A and adopted as a standard. It corresponds to 440.00 Hz.

international access code: The *dialling code* which comes before the *country code* of a *telephone number*, when making an *international call*. These codes vary depending on the country from which the *call* originates and is used to obtain access to an outgoing *international exchange*. Also known as *international dialling prefix*.

International Alphabet No. 5 (IA5): A *transmission code*, first agreed internationally by the *ITU* in 1932, and intended for *data communications* using *telegraphy*. It uses five *bits* and, by use of the shift *character*, permits 64 characters to be generated. It has the facility for automatic retransmission to correct for errors. Variants are permitted to the IA5, such as the *ASCII*.

International Alphabet No. 2 (IA2): An international *code* for *telex transmission* developed by the *ITU*, and shown in Figure I.8. It grew out of the *Murray code*. See also *Morse code*.

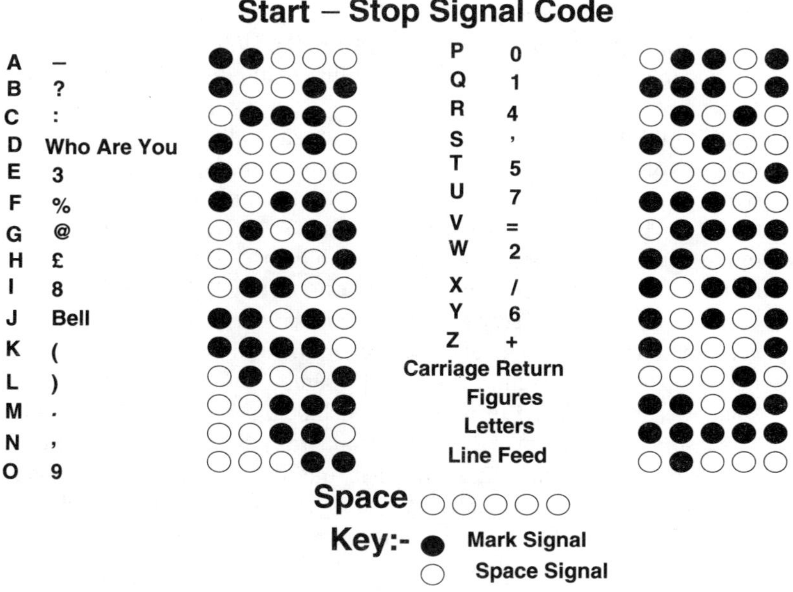

Figure I.8 International Alphabet No. 2

international call: A *call* which is made between one country and another.

international call sign: A *call sign* allocated by the *ITU* for international identification of *radio transmitters*.

international circuit: A *circuit* which carries an *international call*, interconnecting *exchanges* located in different countries.

International Civil Aviation Organisation (ICAO): The organisation responsible for coordination of civilian aviation activities, such as the allocation of *frequency bands* for national civil aviation purposes, and the use of *satellite* navigation systems, such as the *Global Positioning System (GPS)*.

International Communications Association (ICA): Formed in 1948, it is a member of the *International Telecommunications User Group (INTUG)*. It is the largest telecommunications users' group in the USA, representing the big telecommunications users from government, industry and educational establishments. The ICA sets standards, holds exhibitions and seminars for its members, and provides them with other relevant information.

international congestion signal: A *congested signal* sent on an *international circuit* to indicate the failure of *call setup*.

international dialling prefix: See *international access code*.

international directory assistance: *Directory assistance* in connection with *international numbers*.

International Direct Distance Dialling (IDDD): *Direct Distance Dialling* applied to *international calls*, i.e. the ability to directly dial *international numbers*.

international distress frequency: The frequency allocated for the transmission of a *distress call*. The international radio telegraph distress frequency has a nominal frequency of 500 kHz, the international radio telegraph lifecraft nominal frequency is 8364 kHz, and the nominal international radio telephone distress frequency is 2182 kHz.

International Electrotechnical Commission (IEC): Founded in 1906 with the prime aim of carrying out standardisation work for domestic and commercial electrical and electronic equipment. It has close to 50 member nations, each country being represented by a committee. Draft standards are produced from over 80 technical centres and their related subcommittees, with over 3000 IEC standards published. These are then adopted by member countries for their internal use. In the USA this is done by *NIST* and in the UK by *BSI*. There are three main technical groups within the IEC, dealing with a number of Technical Committees. Group A covers general subjects and industrial electronics. Group D covers electronics, components and application of *Information Technology (IT)*. Group C covers safety, measurements and consumer goods. See also *ACET*, *ACOS* and *ACEC*.

international exchange: An *exchange* connecting two *international circuits* and carrying *international calls.*

International Facilities Licence (IFL): A licence granted to a *PTO*, by the *Department of Trade and Industry (DTI)* in the UK, allowing it to provide international services from and to the UK.

international frequency allotment: *Frequency* plans, issued by the *ITU* for certain *frequency bands*, which extend the principle of *international radio channelling* by allotting *channels* to national administrations for use by stations within their control.

international frequency assignment: *Frequency* plans, issued by the *ITU*, which specify details of assignments in certain *frequency bands*. This includes location of transmitting stations, etc.

International Frequency List (IFL): List of international assigned *frequencies*, condensed from the *Master International Frequency Register (MIFR)*, published twice a year by the *ITU* in microfiche form. List of newly notified national frequency assignments, together with an update of the IFL, is sent out weekly to national administrators by the *International Telecommunications Union/Radiocommunications Bureau (ITU/RB)*.

International Frequency Registration Board (IFRB): An organisation of the *ITU* which determines policy on international frequencies and maintains the *Master International Frequency Register (MIFR)*.

international gateway: An *Central Office (CO)* switch which is used to transfer *traffic* between a national *network* and an international network.

international gateway operator: A *PTO* who has been licensed to operate an *international gateway*. In Japan examples are Kokusai Denshin Denwa (KDD), International Telecom Japan (IJT) and International Digital Communications (IDC).

international leased line: A *leased line* which geographically spans more than one country.

International Maritime Satellite organisation (INMARSAT): An international organisation, with members from 55 countries, which operates a series of *satellites* for communications with mobile stations. The first satellite communications system was MARISAT, operated by the Communications Satellite Cooperation, which provided communications to merchant ships at sea. INMARSAT took over this system in 1982 and in 1987 it extended this service to aircraft and mobile land stations.

International Mobile Equipment Identity (IMEI): Identification of mobile equipment, used within *GSM*, to prevent fraudulent use. It is used in conjunction with the *Equipment Identity Register*.

International Mobile Subscriber Identity (IMSI): Part of the *GSM signalling* architecture, associated with the *Mobility Management (MM)* sublayer. It is used in the identification of *subscribers*.

International Mobile Telecommunications 2000 (IMT-2000): Proposed new *ITU-T* standard for mobile *telecommunications* which is planned to come into service in the year 2000. The activity started as the *Future Public Land Mobile Telecommunications System (FPLMTS)* which changed its name in 1996 to IMT-2000. Work on the standard is being done within two working groups, ITU-R TG 8/1 and ITU-T SG 11/3. The first group is concerned with the development of the radio interface and the second group with the terrestrial *network* architecture.

international Morse code: See *Morse code.*

international number: The *telephone number* of a *subscriber* which excludes the *international access code* but includes the *country code.* Examples of international numbers, defined by the *ITU-T* are given in Table I.2. The zone numbers refer to the *world numbering zones.*

international numbering plan: See *world numbering plan.*

International Organisation for Standardisation (ISO): See *International Standards Organisation (ISO).*

international prefix: See *international dialling prefix.*

international radio channelling: *Frequency* plans, issued by the *ITU* for certain *frequency bands*, which specify the *carrier frequencies* which

Table I.2 Examples of international numbers

Zone	*Country*	*Country code*	*No. of digits in national number*	*Total no. of digits*
1	USA	1	10	11
1	Canada	1	10	11
2	Egypt	20	8	10
2	Liberia	231	6	9
3	France	33	8	10
3	Portugal	351	7 or 8	10 or 11
4	UK	44	9 or 10	11 or 12
4	Switzerland	41	8	10
5	Brazil	55	9	11
5	Ecuador	593	7	10

should be used and the type of *traffic* which they should carry, including some other parameters.

International Record Carrier (IRC): Term used in the USA to describe a *common carrier* who operates *international gateways* for *traffic* between the USA and other countries.

International Record Carrier selection: The ability for the *subscriber* to select the IRC to be used.

International Satellite Organisation (ISO): Organisation providing international communications via *satellite*. Examples are *INTELSAT*, *INMARSAT* and *EUTELSAT*.

International Simple Resale (ISR): The operation of hiring international telecommunication *capacity* in bulk from a *PTT* and then selling it on for a profit to organisations and other *subscribers*.

International Simple Reseller (ISR): A organisation which carries out *International Simple Resale (ISR)*.

International Standard: A standard which has been agreed for international use. International Standard, however, usually refers to an *ISO standard*, i.e. a standard published by the *International Standards Organisation (ISO)*.

International Standardised Profile (ISP): *Profile* or *functional standard* issued by the *Joint Technical Committee 1 (JTC1)* of the *ISO* and *IEC*. These profiles can also be developed by other bodies, such as the regional workshops on *OSI* (the *OIW*, the *EWOS* and the *AOWS*) and then submitted to JTC1 for ratification and publication.

International Standards Organisation (ISO): Founded on 23 February 1947 following agreement between standards organisations from 25 nations meeting in London in October 1946. It has now grown to standards bodies from close to 100 countries. The first ISO standard was published in 1951. ISO is a non-treaty organisation, based in Geneva, and a body of the United Nations. Its technical work is done by about 300 technical bodies and 2000 working groups. It has published over 10000 *International Standards* and drafts. Membership of ISO is primarily made up of national standards making bodies, such as ANSI (USA), BSI (UK) and DIN (Germany). All standards developed by ISO are published as International Standards. It is the responsibility of the individual national standards organisations to promote and distribute these standards within their own countries. The organisation of ISO consists of: a General Assembly which approves all standards; a Council which administers the ISO; a Technical Board, which appoints technical committees and monitors their work; Technical Advisory Groups (TAGs) which are established by the Technical Board and advise it on specific matters; Technical Committees (TCs), Sub Committees (SCs) and Working Groups (WGs) which carry out the standards work; an Executive

Board; and a Central Secretariat. The telecommunications equivalent of ISO is the *ITU-T* and the two organisations work closely together in areas of common interest.

International Subscriber Dialling (ISD): A development of *Subscriber Trunk Dialling (STD)* which allows a *subscriber* facilities for *dialling* an *international number* without operator intervention. Also called *international trunk dialling.*

international switching centre: *Switching centre* which carries *international calls* over *international circuits.*

International Table of Frequency Allocations (ITFA): Administered by the *ITU-R* it records the allocation of all the *frequency bands* between 9 kHz and 275 GHz. In each region a band may be allocated exclusively to a service or shared between services. When shared the status of each can be defined as either a *primary service*, a *permitted service* or a *secondary service.*

International Telecommunication Satellite organisation (INTELSAT): An international organisation, founded in 1964 and owned by over a hundred governments, which operates a *network* of about two dozen communications *satellites.* It is managed by *Comsat.*

International Telecommunications Union (ITU): Founded in 1865 as the Union Telegraphique, with the prime aim of developing standards in telecommunications. In 1947 it became a specialised agency for telecommunications, of the United Nations, under the UN Charter Articles 57 to 63, when it was also renamed the International Telecommunications Union. The ITU's headquarters are located in Geneva at the Place des Nations. The ITU has three prime aims: (i) To encourage interconnectivity of telecommunication equipment and services by promoting and establishing technical standards. (ii) To promote the best use of scarce telecommunications resources. This is especially important in the use of the *radio frequency spectrum*, which the ITU controls via the *Master International Frequency Register (MIFR).* (iii) To encourage the growth of telecommunications in less developed countries. The ITU is an inter-government organisation and every state which is a member of the United Nations Organisation may also become a member of the ITU. Only Administrations, i.e. government departments concerned with telecommunications, can be ITU members. The UK is represented by the *Department of Trade and Industry (DTI)*, most countries are represented by their *PTT* which is state owned, and the USA is represented by a complex mix of government, agencies and suppliers. In addition other organisation groupings are recognised, who can attend meetings but cannot generally vote: *Recognised Private Operating Agency (RPOA)*, *Scientific and Industrial Organisation (SIO)*, and *Industrial Organisation (IO).* The ITU is funded by voluntary contributions from its member

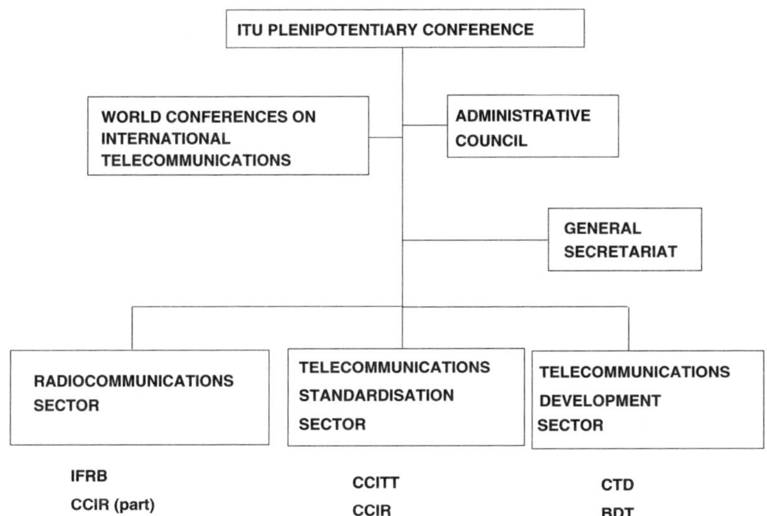

Figure I.9 The ITU structure

countries. Figure I.9 shows the organisation of the ITU. This new structure became effective from 1 July 1994. The standards activities are concentrated in the *International Telecommunications Union Telecommunications Standardisation Sector (ITU-T)*, and combines the standards making activities of the former *CCIR* and the *CCITT*. The new *International Telecommunications Union Radiocommunications Sector (ITU-R)* combines the rest of the former CCIR activities with those of the *IFRB*. The *International Telecommunications Union Development Sector* combined the activities of the Telecommunications Development Bureau (BDT) and the Centre for Telecommunications Development (CTD), which existed in the earlier organisation structure.

International Telecommunications Union Development Sector: The ITU Development Sector incorporates the *CTD* and the *BDT* which existed in its previous structure. It provides telecommunications training and technical assistance, and carries out studies on economic, financial or regulatory policy to help developing countries.

International Telecommunications Union Radiocommunications Sector (ITU-R): The aim of the ITU-R is to ensure the efficient and equitable use of the *radio frequency spectrum* and of *satellite orbits*. It does this by: (i) Allocating frequency and maintaining the *Master International Frequency Register (MIFR)*. (ii) Publishing *Recommendations* on operational procedures and technical characteristics for radiocommunications services and systems. Table I.3 shows the main series of

Table I.3 ITU-R Recommendations

Series	*Recommendation*
BO Series	Broadcasting satellite service (sound and television)
BR Series	Sound and television recording
BS Series	Broadcasting service (sound)
BT Series	Broadcasting service (television)
F Series	Fixed service
IS Series	Inter-service sharing and compatibility
M Series	Mobile, radiodetermination, amateur and related satellite services
P Series	Radiowave propagation
RA Series	Radioastronomy
S Series	Fixed satellite service
SA Series	Space applications and meteorology
SF Series	Frequency sharing between the fixed satellite service and the fixed service
SM Series	Spectrum management
SNG Series	Satellite news gathering
TF Series	Time signals and frequency standards emissions
V Series	Vocabulary and related subjects

Recommendations. The ITU works through Study Groups in preparing these Recommendations, some of these being shown in Table I.4. (iii) Providing information and tools to help countries to manage their radio frequency spectrum. (iv) Holding *World Radiocommunications Conferences (WRC)* to agree the use of the radio frequency spectrum. Figure I.10 shows the organisation of the ITU-R. See also *Regional Radiocommunications Conference (RRC)*, *Radiocommunications Advisory Group (RAG)* and *Radiocommunications Assembly (RA)*.

Table I.4 ITU-R Study Groups

Study Group	*Activity*
SG 1	Spectrum management
SG 3	Radiowave propagation
SG 4	Fixed-satellite service
SG 7	Science services
SG 8	Mobile, radiodetermination, amateur and related satellite services
SG 9	Fixed service
SG 10	Broadcasting service (sound)
SG 11	Broadcasting service (television)

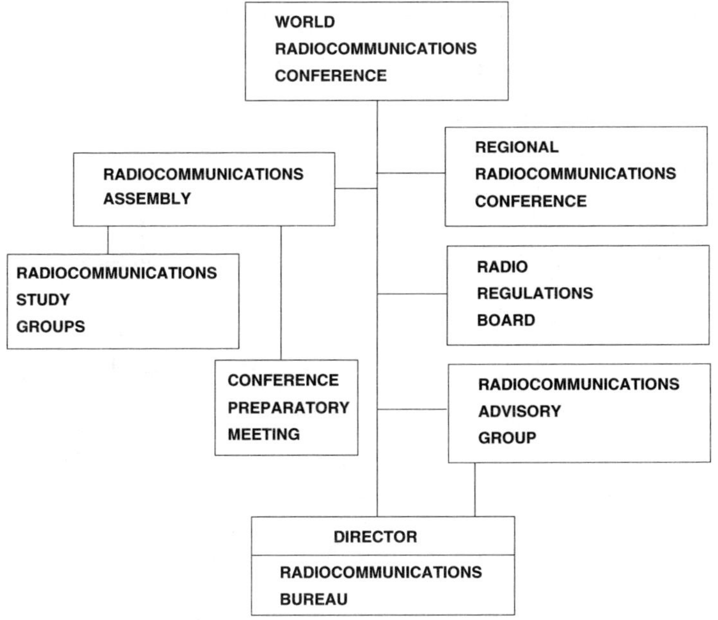

Figure I.10 ITU-R organisation

International Telecommunications Union Telecommunications Standardisation Sector (ITU-T): The CCITT was set up in 1925 as an agent of the *ITU* to carry out studies into the technical problems of *telephony* and it was combined with parts of the *CCIR* to form the *ITU-T* when the ITU organisation was changed in 1994. Article 13 in the constitution of the ITU defines the role of the ITU-T as being to study technical, operating and *tariff* questions and to issue *Recommendations* on these with a view to standardising *telecommunications* on a worldwide basis. All the standardisation work of the IUT-T is carried out within Study Groups, some of these being shown in Table I.5. Recommendations published are in a set of series, some of these being given in Table I.6.

Table I.5 ITU-T Study Groups
(Continued on next page)

Study Group	*Activity*
I	Service definition
II	Network operation
III	Tariff and accounting principles
Regional Tariff Groups of Study Group III	
GR TAF	Tariffs (Africa)
GR TAL	Tariffs (Latin America)
GR TAS	Tariffs (Asia and Oceania)
GR TEUREM	Tariffs (Europe and the Mediterranean Basin)
IV	Network maintenance
V	Protection against electromagnetic effects
VI	Outside plant
VII	Data communications networks
VIII	Terminals for telematics services
IX	Television and sound transmission
X	Languages for telecommunications applications
XI	Switching and signalling

Table I.5 (Continued from previous page)

Study Group	*Activity*
XII	Transmission performance of networks and terminals
XIII	General network aspects
XIV	Modems and transmission techniques for data
XV	Transmission systems and equipment

Table I.6 ITU-T Recommendations
(Continued on next page)

Series	*Recommendations*
Series A	Organisation of the work of the ITU-T
Series B	Means of expression (definitions, symbols, classifications)
Series C	General telecommunication statistics
Series D	General tariff principles
Series E	Overall network operation, telephone service, service operation and human factors
Series F	Telecommunication services other than telephone
Series G	Transmission systems and media, digital systems and networks
Series H	Line transmission of non-telephone signals
Series I	Integrated Services Digital Networks (ISDN)
Series J	Transmission of sound programme and television signals
Series K	Protection against interference
Series L	Construction, installation and protection of cable and other elements of outside plant

Table I.6 (Continued from previous page)

Series	*Recommendations*
Series M	Maintenance: transmission systems, telephone circuits, telegraphy, facsimile
Series N	Maintenance: international sound programme and television transmission circuits
Series O	Specifications of measuring equipment
Series P	Telephone transmission quality, telephone installations, local line networks
Series Q	Switching and signalling
Series R	Telegraph transmission
Series S	Telegraph services terminal equipment
Series T	Terminal characteristics and higher layer protocols for telematic
Series U	Telegraph switching
Series V	Data communication over the telephone network
Series X	Data networks and open system communication
Series Z	Programming languages

International Telecommunications User Group (INTUG): Formed in 1974 to represent the national telecommunications user organisations from about 20 countries, including the USA, UK, Australia and Japan. It is active in promoting the interests of its members, and lobbying associations such as the *ITU*, *CEPT*, and *PTTs*. Any person or group can join, except PTTs and manufacturers, who need to be represented as individual members.

international telegraph alphabet: See *International Alphabet No. 2* and *International Alphabet No. 5*.

international trunk dialling: Same as *International Subscriber Dialling (ISD)*.

International Virtual Private Circuit (IVPC): A *Virtual Private Circuit* which operates across two or more countries.

International X.25 Infrastructure (IXI): A *Packet Switched Network (PSN)* provided by PTT Telecom of The Netherlands. It uses *leased lines* from other *PTOs* in order to provide a high speed international backbone interconnecting national research *networks*.

Internet: A large collection of interconnected *networks* and *gateways* throughout the world which operate as one large network and use the *Internet Protocol (IP)*. The Internet grew out of the *Advanced Research Projects Agency Network (ARPANET)* and the *National Science Foundation Network (NSFNet)*. See also *Intranet* and *Extranet*.

Internet Activities Board (IAB): Initial body set up to administer the *Internet*, primarily associated with the National Science Foundation and the US Department of Defence. See *Internet Architecture Board (IAB)*.

Internet addresses: The *address* used to differentiate *users* on the *Internet*. The Internet is designed to use an addressing system based on 32 *bits*. There are three main types or classes of addresses on the Internet. Class A addresses divide the 32 bits into groups with the most significant 8 bits defining the Class A network address and the remainder 24 bits being used for addresses within each of these networks. Since there are very few Class A addresses they are only given to very large organisations. Class B addresses use the most significant 16 bits for specifying the network and the remainder 16 bits for addresses within the networks. Class C addresses use 24 bits for specifying the networks and 8 bits for individual addresses within these. A fourth class, Class D, was subsequently added to the above three classes. This allows *multicasting* and introduces the concept of a *subnet address*.

Internet Architecture Board (IAB): The body responsible for setting the overall direction and strategy for the *Internet*. See also *Internet Society*.

Internet Assigned Numbers Authority (IANA): The body responsible for allocating overall *Internet* numbers, which is the primary means for *addressing* on the Internet. This allocation is done on a hierarchical basis, and national and regional authorities, such as the *Internic* in the US and *Reseaux IP Europeens (RIPE)* in Europe, allocate address space in their areas.

Internet Control Message Protocol (ICMP): Part of the *Internet Protocol (IP)* which is used to relay information on errors and control messages. The ICMP *frame* has a simple *header* structure, as shown in Figure I.11.

00-01	02-03	04-05	06-07	08-09	10-11	12-13	14-15	16-17	18-19	20-21	22-23	24-25	26-27	28-29	30-31
Type				Code				Header Checksum							

Figure I.11 The Internet Control Message Protocol

Table I.7 ICMP message types

Decimal	*Protocol*
0	Echo Reply
3	Destination Unreachable
4	Source Quench
5	Redirect
8	Echo
9	Router Advertisement
10	Router Solicitation
11	Time Exceeded
12	Parameter Problem
13	Timestamp
14	Timestamp Reply
15	Information Request
16	Information Reply

The first eight bits are used to define the message types, as in Table I.7. The next eight bits are used to further refine the information contained in the type field. This is then followed by a sixteen bit *checksum*.

Internet datagram: The *datagram*, or basic unit of information transfer, used on the *Internet*.

Internet domains: The use of words are preferred by *users* to numbers, in *addresses*. These name spaces are divided on the *Internet* into domains. These are primarily by type of user (as in Table I.8) and by geography, a few of these being shown in Table I.9. The top level domains are administered by organisations within the countries concerned. In the UK this is Nominet.

Internet Draft: See *Internet Standard*.

Internet Engineering Steering Group (IESG): The body which approves all technical standards produced by the *Internet Engineering Task Force (IETF)*. It is composed of directors from the various technical areas of the IETF.

Table I.8 Internet domains by type of user

Domain	*Description*
com	Commercial organisation
edu	Educational establishment
gov	US Government organisation
mil	The US military
net	Network organisation
org	Non-commercial
int	International (multinational) organisation

Table I.9 Internet domain by geography
(Continued on next page)

Domain	*Country*
ar	Argentina
at	Austria
au	Australia
be	Belgium
br	Brazil
ca	Canada
ch	Switzerland
cl	Chile
cn	China
de	Germany
ec	Ecuador
ee	Estonia

Table I.9 (Continued from previous page)

Domain	*Country*
eg	Egypt
es	Spain
fr	France
gb	Great Britain
hk	Hong Kong
ie	Ireland
in	India
it	Italy
jp	Japan
kw	Kuwait
lu	Luxembourg
mx	Mexico
nl	Netherlands
nz	New Zealand
pl	Poland
se	Sweden
sg	Singapore
th	Thailand
tw	Taiwan
uk	United Kingdom
us	United States
yu	Yugoslavia
za	South Africa

00-01	02-03	04-05	06-07	08-09	10-11	12-13	14-15	16-17	18-19	20-21	22-23	24-25	26-27	28-29	30-31
Version		IHL		Types of Service				Total Length							
Identification								Flags	Fragment Offset						
Time to Live				Protocol				Header Checksum							
Source Address, +															
Destination Address, +															
Options												Padding			

Figure I.12 Internet Protocol header record

Internet Engineering Task Force (IETF): Part of the *Internet Society* the IETF primarily concentrates on the development of *Internet Standards.* It is composed of a number of technical groups, each having its own director structure. All standards produced by the IETF are approved by the *Internet Engineering Steering Group (IESG).* See also *Internet Architecture Board (IAB).*

Internet Header Length (IHL): A four bit *field* in the *Internet Protocol (IP),* which specifies the length of the *datagram* in 32 bit words. This is a minimum of 5, where no options are present, and a maximum of 15, i.e. *binary* 1111.

Internet network class: Refers to the class which forms the *Internet address.*

Internet Protocol (IP): *Protocol* specified in RFC 791, the header record being show in Figure I.12. The first four bits specify the version number of the protocol. The next four bits are the *Internet Header Length (IHL).* The Type of Service field is used to determine the precedence and the type of service, as in Figure I.13. The precedence field (Table I.10) is used to determine the order in which the *packets* are dealt with, for example if there is queuing due to congestion. Type of service deter-

0-2	3	4	5	6	7
Precedence	D	T	R	C	–

Figure I.13 Type of service in the IP

Table I.10 IP precedence settings

Setting			Precedence Use
0	0	0	Routine
0	0	1	Priority
0	1	0	Immediate
0	1	1	Flash
1	0	0	Flash Override
1	0	1	Critic-ECP
1	1	0	Inter-network Control
1	1	1	Network Control

mines the route to be taken. D is the most direct route, T is the route with the highest throughput, R is the most reliable route and C is the lowest cost route. The protocol field is used, on reaching the destination, to specify the higher level application to which the *packet* should be directed. Some of these have been reserved for special applications, a few being shown in Table I.11.

Internet Service Provider (ISP): Organisations who provide access to the *Internet*, and other Internet services, for commercial gain. Usually they manage the local servers which connect into the Internet, and are granted a batch of *Internet addresses* which they allocate to their clients.

Internet Society (IS): A not-for-profit organisation consisting of interested companies and individuals, who has, since 1992, administered the *Internet*. Detailed work on the Internet is done in two primary groups, the *Internet Architecture Board (IAB)* and the *Internet Engineering Task Force (IETF)*.

Internet Standard: The standards process can be started by a *Draft Standard* being submitted to the *Internet Engineering Task Force (IETF)* by any interested party. This is reviewed for 60 days and if there is sufficient interest in the subject then formal work begins on the Internet Standard. This is then sent out as an Internet Draft via a *Request For Comment (RFQ)* and following input and revisions the standard is submitted for approval by the *Internet Engineering Steering Group (IESG)*.

Table I.11 Commonly used protocol values

Decimal	*Protocol*	*Common acronym*
1	Internet Control Message	ICMP
3	Gateway-to-Gateway	GGP
4	IP in IP (encapsulated IP)	IP
5	Stream	ST
6	Transmission Control Protocol	TCP
8	Exterior Gateway Protocol	EGP
17	User Datagram	UDP
29	ISO X.200 Transport Class 4	ISO-TP4
89	Open Shortest Path First	OSPF

Internet Watch Foundation (IWF): A group, set up by members of the *Internet* service industry in the UK, to monitor the content of information provided on the Internet. If a complaint is received it is investigated and if it fails to meet the guidelines laid down the member hosting the content must take steps to remove it and to remove all service from the offender.

Internetworking: The process of connecting several *networks* together so that they can operate as one network, and *messages* can be passed between them, even though each network may be a separate entity with different *addresses*, operating *protocols*, etc.

internetwork router: A *router* which carries out *internetworking*. It only passes messages to the *network* which is being addressed and operates at the *Network Layer* of the *OSI Basic Reference Model*.

Internic: See *Internet Assigned Numbers Authority (IANA)*.

internodal: A activity, such as a *call*, which occurs between two *nodes* on a *network*. See also *intranodal*.

interoffice call: A *call* which passes through two or more *Central Offices (COs)*. This is often used to set the cost of the call and for *billing*. See also *intraoffice call*.

interoffice trunk: A direct *trunk* line which connects two *Central Offices (COs)*.

interoperability testing: The process of testing a system to ensure that the various elements, usually provided by different manufacturers, work with each other within their performance limits.

inter-register signalling: A *signalling* system used in a *PABX*. It uses twelve *frequencies* in the range from 540 Hz to 1980 Hz and operates on *two wire circuits* or *four wire circuits*.

interrogation: The process in which a *signal* is sent to determine the state of a device, or to determine some other *information*, or to carry out a specific task. An interrogation signal usually requires a reply.

interrupt: To temporarily suspend the operation of a system so that some other task can be carried out. After this has been completed operation can resume from the point at which it was suspended.

interrupt driven: A system which operates on the basis of receiving a series of *interrupts* from a *user*, e.g. in an *interactive mode*.

interrupted isochronous transmission: See *burst isochronous transmission.*

Interrupt Request (IRQ): The *signal*, within a processing systems such as a computer, which causes an *interrupt* to occur.

Intersymbol Interference (ISI): *Interference* which occurs during *transmission* when a transmitted pulse interferes with other received pulses. This can occur, for example, in *optical fibre* transmission where the pulse can spread in width as it travels the length of the fibre so that it interferes with other pulses and these cannot then be differentiated. The pulse *waveform* is usually shaped to minimise this interference. According to the Nyquist the minimum transmission *bandwidth* which meets this condition is one which passes all frequencies up to ½T and stops all others, as shown by the dashed lines in Figure I.14. In practice this is achieved by using an *equaliser* (Figure I.15) which gives an overall transmission characteristics of a raised cosine, as shown by the solid line of Figure I.14.

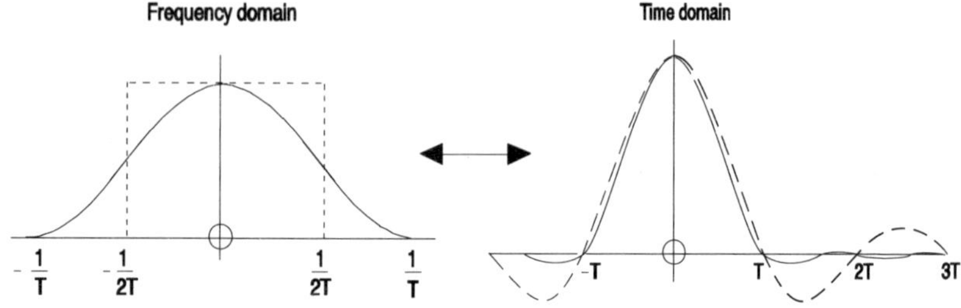

Figure I.14 Waveform shaping for zero Intersymbol Interference

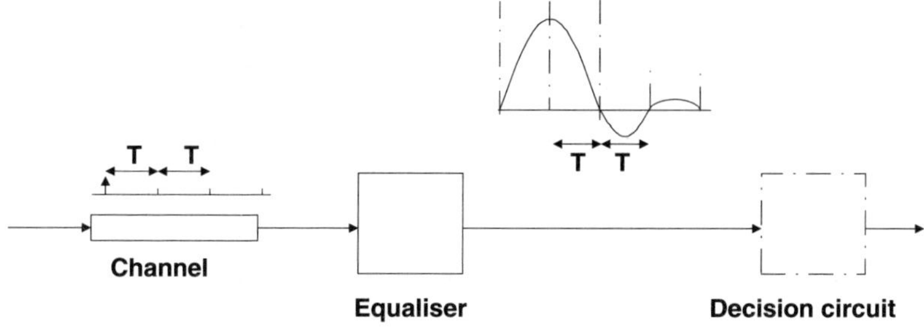

Figure I.15 Using an equaliser for waveform shaping for zero ISI

intertoll trunk: The *trunk* connecting two *toll offices*, these usually being located in different *telephone exchanges*.

interval: The time between two events, such as a pulse interval, *frame* interval, etc.

interworking: The process in which two systems or devices can work with each other, e.g. a *terminal* in one *network* interworking with a terminal in another network. See also *interoperability testing*.

intra-LATA: A *call* which originates and terminates within the same *Local Access and Transport Area (LATA)*.

intramodal dispersion: One of the two elements of *dispersion* which occurs in an *optical fibre*, the other being *intermodal dispersion*. Intramodal dispersion is caused by the finite width of the optical source and its interaction with the properties of the *waveguide* material (known as *material dispersion*) and the structure of the waveguide itself (known as *waveguide dispersion*).

intramodal distortion: *Distortion* resulting from *intramodal dispersion*.

Intranet: A series of interconnected *networks* which use similar technologies and *protocols* as the *Internet*, being connected to the Internet but intended for use within an organisation only. Usually the Intranet would be protected from unauthorised access by use of *firewalls*. See also *Extranet*.

intranodal: Events and communications which take place within a single *node*.

intranode address: The *address* used for *users* situated within the same *node*.

intranode routeing: *Routeing* of *data* between *users* situated on the same *node*.

intraoffice trunk: A *trunk* used to make connections within the same *Central Office (CO)*.

intrinsic coupling loss: The loss which occurs in an *optical fibre* communications system due to mismatch between fibre parameters when two fibres are joined together. This could be due to differences in dimensions, *refractive index*, etc. Also called intrinsic joint loss.

intrinsic noise: The *noise* which is inherent in a system, such as a *transmission line*, and has not been caused by other external factors.

intrusion-resistant cable: A *cable* which has been protected such that unauthorised access to the *data* which it is carrying can be prevented.

INTUG: *International Telecommunications User Group*.

inventory management: A part of *accounting management*, which is one of the five groups of *network management* functions classified by *ISO*. Inventory management is the function which keeps track of the individual elements being managed on the *network*, their characteristics, asset values and ownership, and contractual information.

Inverse Discrete Cosine Transform (IDCT): An *algorithm* used in *digital video coding*, defined in *ITU-T Recommendation* H.261 (see *H Series*). See also *Discrete Cosine Transform (DCT)*.

Inverse Discrete Fourier Transform (IDFT): A mathematical technique used in the analysis of digital signals. See also *Discrete Fourier Transform (DFT)*.

Inverse Multiplexer (IMUX): Equipment which can split several *multiplexed channels* into independent outputs.

inverse square law: The strength or *intensity* of an *electromagnetic radiation* at any point varies inversely as the square of its distance from the source of the radiation.

Inward Wide Area Telephone Service (IN-WATS): See *WATS*.

IN-WATS: *Inward Wide Area Telephone Service*.

IO: *International Organisation*.

ionosphere: A layer of the *Earth's atmosphere* extending from about 50 km to 500 km above the surface of the Earth. It contains ions and free electrons, generated by several causes, such as the ionisation effects of solar radiation and cosmic particles. The layer is continuously changing, to reflect atmospheric conditions, and it has a marked influence on radio communications. At very low *frequencies* the *radio waves* are trapped by this layer and travel around the Earth as if confined in a *waveguide*. At higher frequencies the signals are reflected off the ionosphere as *sky waves*, and at very high frequencies they pass through the layer.

ionospheric absorption: The *absorption* of *electromagnetic waves* which occurs within the *ionosphere*.

ionospheric reflection: The bending of *radio waves* back towards the Earth as they travel through the *ionosphere*. This occurs due to *reflection* and *refraction* as the waves pass through the ionised layer.

ionospheric scatter: The *scattering* of *radio waves* as they pass through the *ionosphere*, this scattering forming an important element of *radio propagation* within the layer.

ionospheric sounding: Obtaining information about an *ionosphere* layer by directing *radio waves* at it and measuring the characteristics of the received signals, such as the time delay before it is received, its strength, etc. This can provided valuable information which will affect *radio propagation*, such as the height of the layer, its ionisation, etc.

IP: *Intelligent Peripheral* or *Internet Protocol.*

IP address: The same as *Internet address.*

IR: *infrared.*

IRC: *International Record Carrier* or *Incrementally Related Carrier.*

IRQ: *Interrupt Request.*

irradiance: A measure of the amount of *electromagnetic radiation* falling on, or passing through, a surface from a source. It is usually measured in watts per square centimetre.

IS: *Internet Society.*

ISB: *Independent Sideband transmission.*

ISD: *International Subscriber Dialling.*

ISDN: *Integrated Services Digital Network.*

ISDN address: The *address* used within an *ISDN network*, which incorporates the *ISDN number* with additional ISDN sub-addresses, as in Figure I.16.

ISDN interfaces: The interface points defined for an *ISDN network.* See *User-Network Interface (UNI)*, *S interface*, *T interface* and *R interface.* Also known as ISDN reference points. See also *Basic Rate Interface (BRI)* and *Primary Rate Interface (PRI).*

ISDN number: The *address* of a *subscriber* on an *ISDN network.* Defined in *ITU-T Recommendation* I.331 it consists of three parts: a *country code*, a *national Destination Code (NDC)* and the *subscriber's* number. See Figure I.16.

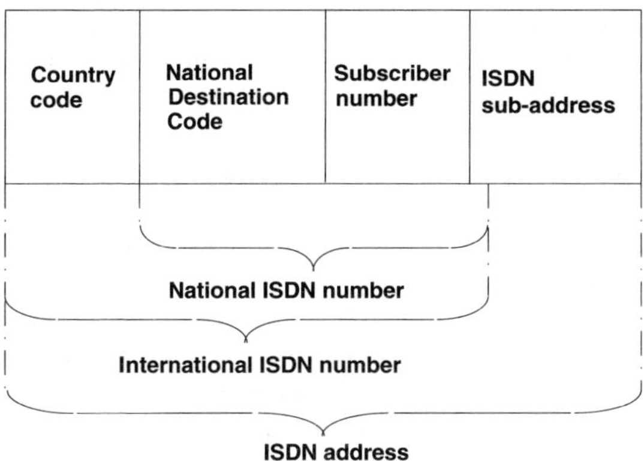

Figure I.16 ISDN address structure

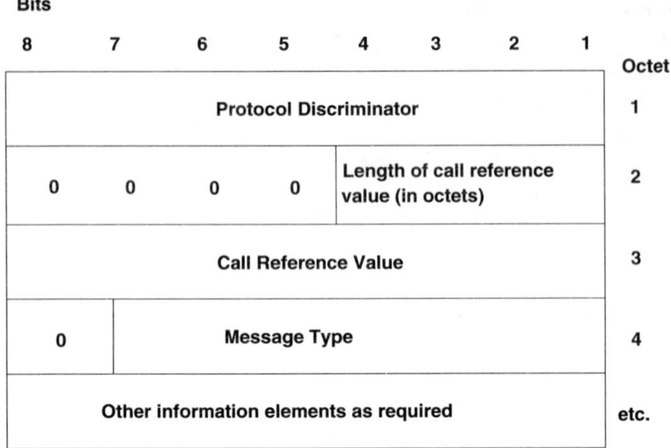

Figure I.17 ISDN signalling message structure

ISDN signalling: The *signalling* used within *ISDN networks*. Figure I.17 shows the signalling message structure. The protocol discriminator gives the *D channel* the ability to provide additional *protocols*, if required. The call reference value identifies the call with which the message is associated. The message type indicates the type of message being sent, such as *call establishment*, *call disestablishment*, etc.

ISDN terminal: A *terminal* which operates on an *ISDN line* and can, usually, provide access to both *voice* and *data*. These terminals can operate off the *U interface* or the *T interface*.

I Series: The *ITU-T Recommendations* published for *ISDN*. There are many of these, which have been structured into six main parts, as shown in Figure I.18, covering the I.100, I.200, I.300, I.400, I.500 and I.600 series. Also include are I.700 for *ATM* aspects.

ISI: *Intersymbol Interference*.

ISLAN: *Integrated Services Local Area Network*.

ISM: Industrial, Scientific and Medical.

ISO: *International Satellite Organisation* or *International Standards Organisation*.

isochronous signal: A *signal* in which the time between any two significant instances is equal to one or more multiples of the unit interval.

isochronous transmission: *Transmission* in which *timing* is transmitted on the same *channel* as the *data*, so that the timing information is derived from the data and a separate timing or *clock* lead is needed. Isochronous transmission is essentially *start-stop transmission*, but the ends of the link are both clocked.

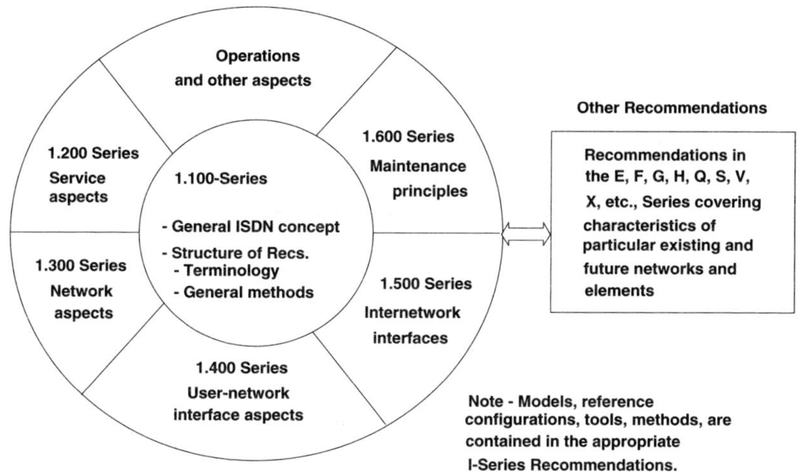

Figure I.18 ITU-T Recommendations, I Series

ISO reference model: See *OSI Basic Reference Model.*

ISO standard: Standards published by the *International Standards Organisation (ISO)*, a few of these being shown in Table I.12.

isotropic antenna: An *antenna* which radiates and receives equal signal strengths in all directions. It is a theoretical antenna, and cannot be achieved in practice, but it is used as a reference against which the performance of other antennas are measured for directional properties.

ISP: *Internet Service Provider* or *International Standardised Profile.*

ISPBX: *Integrated Services Private Branch Exchange.*

ISR: *International Simple Resale* or *International Simple Reseller.*

IT: *Information Technology*

ITANZ: *Information Technology Association of New Zealand.*

ITC: *Independent Telephone Company.*

ITDM: *Intelligent Time Division Multiplexing.*

ITFA: *International Table of Frequency Allocations.*

ITSTC: *Information Technology Steering Committee.*

ITU: *International Telecommunications Union.*

ITU Development Sector: *International Telecommunications Union Development Sector.*

ITU-R: *International Telecommunications Union Radiocommunications Sector*

ITU-R Recommendations: Standards published by the *International Telecommunications Union Radiocommunications Sector (ITU-R).* See Table I.3.

Table I.12 ISO standards

Standard	*Description*
ISO 646	Character set for information interchange
ISO 1155	Longitudinal Redundancy Check (LRC)
ISO 1745	Character oriented protocol for operating procedures
ISO 2110	DTE/DCE 25 pin connector, including pin assignments
ISO 2593	Pin allocation for connectors used with high speed DTE
ISO 3309	HDLC frame structure
ISO 4335	HDLC procedure elements
ISO 4902	HDLC unbalanced classes of procedures
ISO 4903	DTE/DCE 15 pin connector, including pin assignments
ISO 6159	HDLC unbalanced procedures
ISO 6256	HDLC balanced procedures
ISO 6936	Conversion between ITU-T ITA2 and ISO 646
ISO 8072	OSI transport service definition
ISO 8073	OSI transport protocol
ISO 9660	CD-ROM directory structure and contents

ITU-T: *International Telecommunications Union Telecommunications Standardisation Sector.*

ITU-T Recommendations: Standards published by the *International Telecommunications Union Telecommunications Standardisation Sector (ITU-T).* See Table I.6

IVPC: *International Virtual Private Circuit.*

IVR: *Interactive Voice Response.*

IVRU: *Interactive Voice Response Unit.*

IWF: *Internet Watch Foundation.*

IXC: *Interexchange Carrier.*

IXI: *International X.25 Infrastructure.*

J

jabber: The situation which exists, usually in a *Local Area Network*, when a *transmitting station* continues *transmissions* beyond the time allowed by the *protocol* being used.

jabber control: The ability of a *station* to interrupt a *transmission* which is going on beyond its allocated time.

jabber frame: A *transmission frame* whose length exceeds the value allowed by the *protocol* being used.

jack: The socket into which a plug fits to make an electrical connections. Used, for example, in telephone wiring.

jacket: The coating used over a *transmission medium* to protect its interior from handling damage or from the environment.

jammer: A device used to generate a *jamming signal.*

jamming signal: A *signal* which is used to cause *interference* with another signal so that it cannot be received error free and so cannot be understood. Usually applies to signals transmitted as *radio waves* and the jamming signal prevents reception in a given *frequency band.* The jamming signal is usually of greater power than the signal being jammed.

jam signal: The *signal*, usually used in a *Local Area Network (LAN)*, which is sent to inform *nodes* connected to the *network* that they must not transmit, for example because a *collision* has occurred.

JANET: The Joint Academic Network which forms the communications backbone linking UK educational and research establishments in the UK.

Japanese Industrial Standard (JIS): See *Japanese Industrial Standards Committee.*

Japanese Industrial Standards Committee (JISC): Formed in 1949 to advise the Japanese government on industrial standards. It contains over 1000 Technical Committees, drawn from a wide range of Japanese academic and industrial organisations, and from the government and users. Its standards cover all aspects of industrial products, from building materials to electrical goods. It is responsible for marking goods which conform to its standards with the JIS (Japanese Industrial Standard) mark.

Java: A popular computer programming language developed for applications on the *World Wide Web (WWW).* It allows interactive objects to be added to *Internet* sites.

Java Virtual Machine (JVM): A computer or processor *operating system* which allows it to run *software* written in the *Java* language and provides the universal portability of the Java *code.*

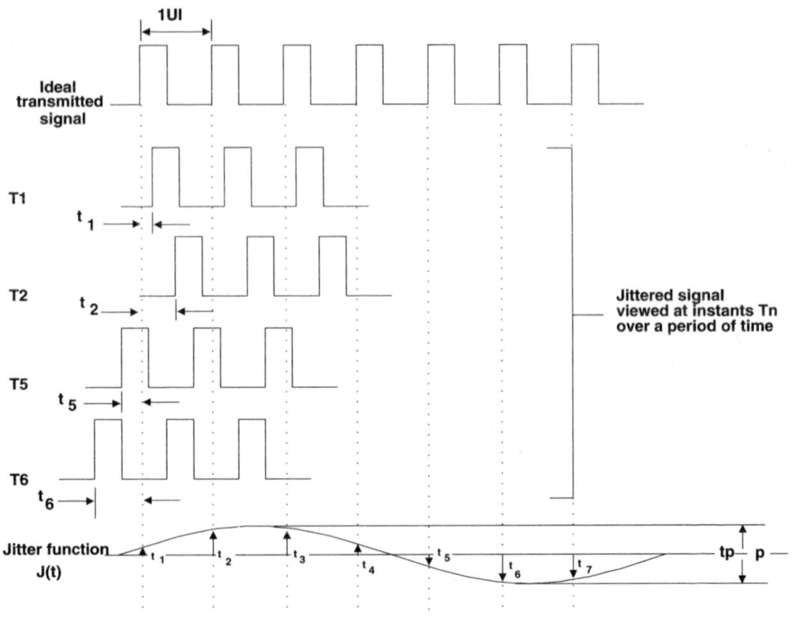

Figure J.1 Jitter modulation on a digital signal

J band: The *frequency band* in the range from 10 GHz to 20 GHz.

JESI: *Joint European Standards Institute.*

JIS: *Japanese Industrial Standard.*

JISC: *Japanese Industrial Standards Committee.*

jitter: Small and random variations in time or phase of a transmitted *signal* which can result in errors or loss of *synchronisation*. Figure J.1, for example, shows the effect of jitter on a transmitted *pulse train*. This random variation can result in errors in the sampled waveform at various instances in time. There are several causes of jitter, the most common being the variation in the *clock* signal, as shown in Figure J.2. The unwanted component of the clock causes unwanted *amplitude modulation* and *phase modulation* of the output signal.

jitter accumulation: The build up or accumulation of *jitter* in a *signal* as it moves through sequential components in a *network*, such as *regenerators*.

jitter limits: The amount of *jitter* permitted in *transmission* systems, usually specified by published standards, such as in *ITU-T Recommendations* G.823 and G.824. Figure J.3 shows typical jitter limits from these Recommendations.

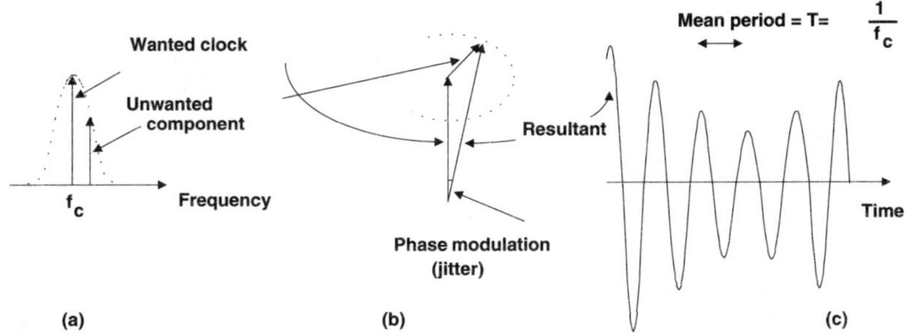

Figure J.2 Jitter due to interference: (a) spectrum from clock selecting filter; (b) phasor representation; (c) resulting waveform

Job Transfer and Manipulation (JTM): Communications *protocols* used to perform tasks in an *open system interconnect environment*. It enables work to be submitted on one open system and to be run on another system.

Johnson noise: See *thermal noise*.

Joint European Standards Institute (JESI): The joint organisation formed by *CEN* and *CENELEC* to carry out activities in *Information Technology (IT)*.

Joint Photographic Experts Group (JPEG): The standards group which produced the most commonly used standard for coding of continuous tone still images, which has been endorsed by the *ISO* and the *IEC*. Figure J.4 shows the basic arrangement of a JPEG encoder. See also *MPEG*.

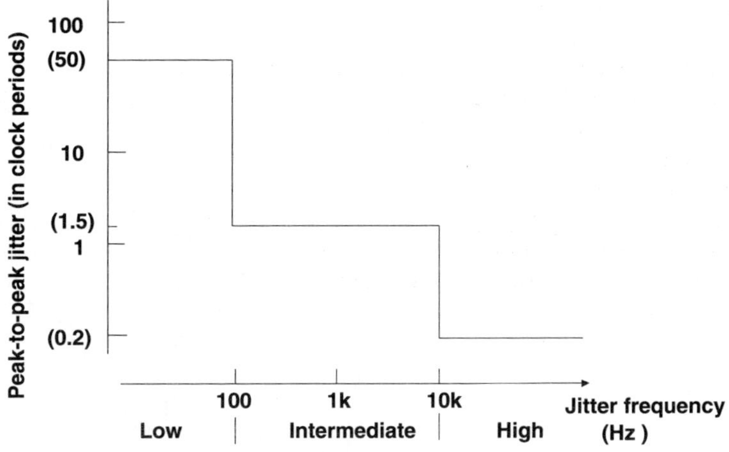

Figure J.3 Typical jitter limits for digital transmission

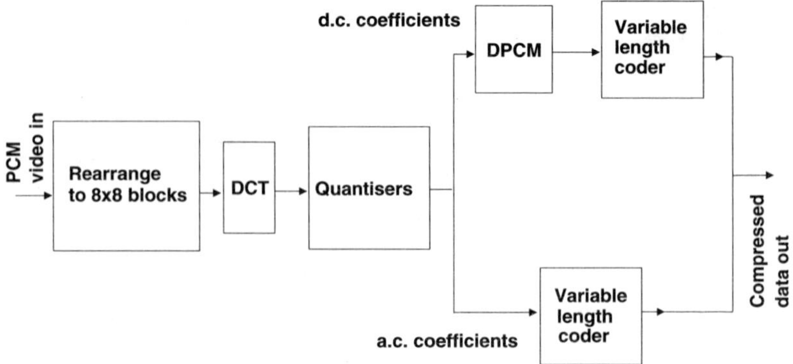

Figure J.4 JPEG encoder

Joint Presidents' Group (JPG): Group set up by *ETSI* to share mandated standardisation work between itself, *CEN* and *CENELEC* in the *Information Technology Steering Committee (ITSTC).*

Joint Tactical Information Distribution System (JTIDS): Secure information distribution system for use by the US military, which has capabilities to minimise the effects of *jamming.*

Joint Technical Committee on Information Technology (JTC1): Joint Technical Committee set up by the *International Standards Organisation (ISO)* and the *International Electrotechnical Commission (IEC)* to develop generic *Information Technology (IT)* standards. This committee incorporates ISO TC 97 ad IEC TC 83 and Table J.1 shows some if its

Table J.1 ISO/IEC JTC1 subcommittees
(Continued on next page)

Sub-committee	*Activity*
SC 1	Vocabulary
SC 2	Coded character sets
SC 6	Telecommunications and information exchange between systems
SC 7	Software engineering
SC 11	Flexible magnetic media for digital data interchange

Table J.1 (Continued from previous page)

Sub-committee	*Activity*
SC 14	Data element principles
SC 15	Volume and file structure
SC 17	Identification cards and related devices
SC 18	Document processing and related communication
SC 21	Open systems interconnection, data management and open distributed processing
SC 22	Programming languages, their environments and system software interfaces
SC 23	Optical disk cartridges for information interchange
SC 24	Computer graphics and image processing
SC 25	Interconnection of information technology equipment
SC 26	Microprocessor systems
SC 27	IT security techniques
SC 28	Office equipment
SC 29	Coding of audio, picture, multimedia and hypermedia information
SC 30	Open electronic data interchange

subcommittees. JTC1 is responsible for producing *International Standardised Profiles (IPS).*

Joint Technical Programming Committee (JTPC): Committee set up by the *International Standards Organisation (ISO)* and the *International Electrotechnical Commission (IEC)* which has the task of ensuring that overlaps of work between the two organisations are avoided.

JPEG: *Joint Photographic Experts Group.*

JPG: *Joint Presidents' Group.*

J Series: *ITU-T Recommendations* on the transmission of sound and television *signals*. A few of these are given in Table J.2.

JTC1: *Joint Technical Committee on Information Technology.*

Table J.2 J Series Recommendations
(Continued on next page)

Recommendation	*Description*
General: sound transmission	
J.11	Hypothetical reference circuit
J.13	Definitions for international circuits
J.16	Measurement of weighted noise
J.17	Pre-emphasis on sound circuits
J.19	Test signal for measurement of interference
Performance: sound transmission circuits	
J.21	15 kHz sound circuits
J.23	7 kHz sound circuits
J.24	Signal modulation by interference from power supplies
J.26	Test signals on international circuits
J.27	Signals for alignment of international circuits
Characteristics: sound equipment and lines	
J.31	15 kHz sound circuits
J.33	6.4 kHz sound circuits
J.34	7 kHz sound circuits
Characteristics: equipment for analogue coding of sound signals	
J.41	High quality transmission on 384 kbit/s channels
J.42	Medium quality transmission on 384 kbit/s channels
J.43	High quality transmission on 320 kbit/s channels
J.44	Medium quality transmission on 320 kbit/s channels

Table J.2 (Continued from previous page)

Recommendation	*Description*
Digital transmission of sound	
J.51	General principles
J.52	High quality transmission on one to three, 64 kbit/s channels (mono signals)
J.56	High quality analogue transmission at 320 kbit/s over mixed analogue and digital channels
J.57	Studio quality transmission over H1 channels
Characteristics: television transmission	
J.61	Circuits for international connections
J.63	Test signals in field blanking interval of monochrome and colour signals
J.67	Test and measurement for circuits carrying MAC packet signals or HD-MAC
Television transmission over metallic lines and radio relay links	
J.73	12 MHz system for simultaneous television and telephony transmission
J.75	Interconnection of systems using coaxial pairs and radio relay links
J.77	Television signals over 18 MHz and 60 MHz
Digital transmission of television signals	
J.83	Multiprogramme system for cable distribution
J.84	Multiprogramme system for SMATV networks
J.85	General principles for transmission over long distances
J.86	Mixed analogue and digital transmission over long distances

Table J.2 (Continued from previous page)

Recommendation	*Description*
Television transmission: specific recommendations	
J.91	Privacy in long international circuits
Transmission of video, sound and data	
J.100	Tolerancing between sound and video transmission
J.101	Measurement and test for teletext signals

JTIDC: *Joint Tactical Information Distribution System.*

JTM: *Job Transfer and Manipulation.*

JTPC: *Joint Technical Programming Committee.*

judder: The effect seen in *facsimile transmission* when unequal *scanning* results in overlap of the reproduced picture elements.

jumper wire: A wire used to connect two points together, such as on a *distribution frame*. A jumper wire is also used as a temporary connection, such as for test purposes.

junction call: A call made between *local exchanges*, using *junction circuits.*

junction circuit: A *circuit* connecting two *local exchanges.* Also used to describe the *link* between a local exchange and a *trunk exchange*. In the USA refers to an *interoffice trunk.*

junction network: The *network* connecting a group of *local exchanges* and linking these to the main trunk centre.

justification: The process of shifting *bits* in a *frame* or shift register so that they occupy a specific position, usually to align with bits in another frame or register.

justifying bit: The additional *bits* inserted into a *frame* or shift register during the *justification* process and not forming part of the *information* being carried. See also *bit stuffing*.

JVM: *Java Virtual Machine.*

K

Ka band: That part of the *electromagnetic spectrum* covering the *frequency band* from about 20 GHz to 30 GHz. It is used for *satellite* communications.

K band; The portion of the *electromagnetic spectrum* in the *frequency range* from about 10 GHz to 12 GHz, and used for *satellite* communications.

KDC: *Key Distribution Centre.*

Kelvin: The unit of absolute temperature. Absolute zero is at -273^{0} Celsius.

Kendal notation: Used in the definition of queues. Generally queues are categorised by the notation A/B/m/S/P, where these symbols are defined as in Table K.1. The most frequently used interarrival and service distributions are given in Table K.2. Therefore, for example, D/G/1 indicates a queue with deterministic arrivals, general or arbitrary service time, and one server.

kermit: A widely used asynchronous file transfer protocol which was originally developed by Columbia University in New York City.

Kerr cell: A device whose *refractive index* can change in proportion to the square of the applied field strength. It is used to *modulate* a light source.

kernel: The part of the *operating system* of a processor which carries out the basic system level functions, such as switching between tasks.

key: **(1)** The physical switch on a *keyboard* or *telephone keypad* which is used to enter *data* into the system. **(2)** A *character* or group of characters used to identify a data record, such as in a *database*. **(3)** Symbols or special characters used for data *encryption*. See *data encryption key*.

Table K.1 Kendall notations in common use

Symbol	*Item*
A	Interarrival distribution
B	Service distribution
m	Number of servers
S	Storage capacity
P	Customer population

Table K.2 Distribution symbols used with Kendall notations

Item	*Symbol*
Deterministic	D
Erlangian with k degrees of freedom	E_k
General	G
Hyperexponential with j degrees of freedom	H_j
Markovian	M

keyboard: A device used to manually input data into equipment, such as a *data terminal* or *Visual Display Unit (VDU)*. It would normally consist of a standard number of keys which follow a recognised physical layout, such as *QWERTY*.

Keyboard Send and Receive (KSR): Equipment which is used to enter *data* into the system by means of a *keyboard* and to output it by means of a printer.

Key Distribution Centre (KDC): The centre where *keys* used for *encryption* are stored and from which it is distributed, in electronic form, to users. The centre would also incorporate a *key generator*.

key generator: A device which is used to create *data encryption keys* in electronic form, based on a special *algorithm*.

keying: The generation of a *signal* by *modulation* of a *carrier*, usually by *phase modulation* or *frequency modulation*. For example, *Phase Shift Keying (PSK)* and *Frequency Shift Keying (FSK)*.

key management: The overall management of *keys* used for *encryption*. For example their generation, storage, distribution and eventual destruction.

keypad: **(1)** The *keys* on a *DTMF telephone* which are used for *dialling* and sometimes for communications with a computer based system. **(2)** The group of keys which are grouped together on one end of a *keyboard* and used for special purposes, such as numerical entry, cursor movements, etc.

key phone: A *telephone* which incorporates a *switching* function, obtained by pressing *keys*.

key pulsing: Sending *telephone* calling *signals* using the *keys* on a *keypad*, such as for *Dual Tone Multifrequency signalling*.

Key Service Unit (KSU): The heart of the *key telephone system* which terminates the incoming lines from the *exchange* and connects to the individual *telephones*.

key set: A *user terminal* which has an extended *keypad* which can be used for multifunctions or multilines. Examples are *call pickup*, *auto-dialling*, *intercom*, etc. Also called *Key Telephone Set (KTS)*.

key system: See *Key Telephone System (KTS)*.

Key Telephone Set (KTS): See *key set*.

Key Telephone System (KTS): A system in which the *telephones* have *keypads* which allow direct access to a variety of services and would normally include direct access to the *Public Switched Telephone System (PSTN)*. Also used with reference to a *PABX* with limited facilities, such as the number of *trunk lines* or *telephone* extensions which can be interfaced.

Key Telephone Unit (KTU): Building blocks which go to from part of the *Key Telephone System (KTS)*. Examples are line cards, *key set*, *EIA* interface, etc.

keyword: **(1)** A word used to search for *information*, such as retrieving a file or record from a *database*. **(2)** A significant word, such as one which defines the contents of a *database* or which is reserved for special use in a *programming language*.

kHz: *KiloHertz*.

kilobit: One thousand or 1024 *bits* depending on whether *transmission* rate (kilobits per second) or storage capacity is being referred to, respective-ly.

kilobyte: One thousand or 1024 *bytes*, depending on whether *transmission* rate (kilobytes per second) or storage capacity is being referred to, respectively.

kiloHertz: One thousand *hertz*.

kilosegment: A measure of *data traffic* used for *billing* purposes in *Packet Switched Networks (PSN)*. It equals 64000 *characters*.

kiloStream: A 64 kbit/s and below, point to point *leased line digital* service, first provided by BT in the UK in January 1983. It is similar to the *Dataphone Digital Service (DDS)* from Bell in the USA. (Figure K.1.) The *Network Terminating Unit (NTU)* encodes *subscriber data* for *transmission* to the *local exchange* where it is *multiplexed* into an *E1 signal*. Data rates can vary from 2.4 kbit/s to 64 kbit/s. Subscriber data is framed, as shown in Figure K.2, known as envelope encoding, and this provides the *signalling* and control information needed by the *network* for maintenance.

knowbots: Computer programme developed for carrying out automatic searches on distributed *databases*, such as on the *Internet*.

KSR: *Keyboard Send and Receive*.

KSU: *Key Service Unit*.

KTS: *Key Telephone System* or *Key Telephone Set*.

KTU: *Key Telephone Unit*.

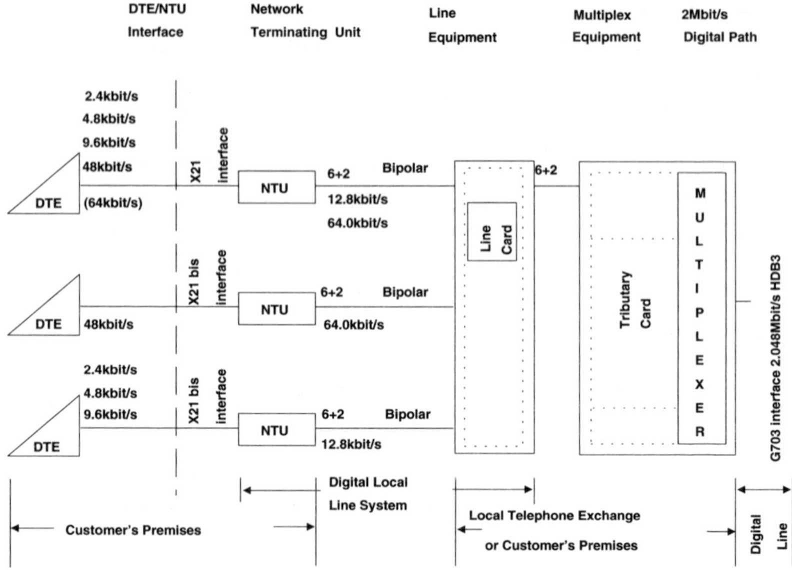

Figure K.1 KiloStream structure

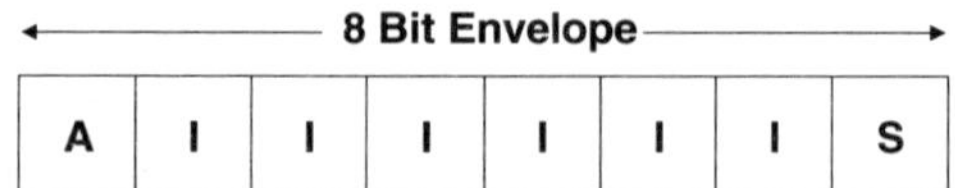

A = Alignment bit which alternates between 1 and 0 in successive envelopes to indicate the start and stop of each 8-bit envelope

S = Status bit which is set or reset by the control circuit and checked by the indicator circuit

I = Information bits

Figure K.2 KiloStream 2-bit envelope encoding

Ku band: The part of the *electromagnetic spectrum* which lies in the *frequency band* from about 12 GHz to 14 GHz. The Ku band is used for many communications *satellites*.

L

label: Symbols which are used to identify an item, such as a *message* or a *file*. It can form part of a *message* or be attached to it and contains *information* regarding the message, such as its nature, identification, etc.
labelling algorithm: A *algorithm* in which labels are used on individual *nodes*, the labels being updated as required. The technique is commonly used for problems such as shortest path routeing.
LADT: *Local Area Data Transport.*
LAMA: *Local Automatic Message Accounting.*
lambert: A measure of *luminance* and equal to $10^4/\pi$ *candela* per square meter. See also *lumen.*
Lambert's cosine law: The law give a measure of *luminous intensity* from a source falling on a surface. If I_v is the intensity at an angle θ from the normal to the surface and I_{nv} is the intensity along the normal, then Lambert's cosine law states that $I_v = I_{nv} \cos \theta$.
Lambert's law: See *Lambert's cosine law.*
lamp: Usually refers to an optical source, such as an electrical bulb or *Light Emitting Diode (LED)*, which is mounted close to a piece of equipment and provides a visual indication of the state of the equipment, e.g. error state, open or closed, etc.
LAN: *Local Area Network.*
landline: *Transmission medium* which travels over the ground (including suspended over it or buried under it) or inland waters. It includes all forms, such as *twisted pair wire*, *coaxial cable*, *fibre optics* and *microwave.* It does not include underseas transmission systems, *satellite* systems, or radio based systems, such as *cellular radio systems.*
Land Mobile Satellite Service (LMSS): A *satellite* service in which mobile *Earth stations* are used.
Land Mobile Service (LMS): A radio service between a *base station* and mobile land stations or directly between the mobile land stations.
Land Mobile Station (LMS): A mobile radio station on land which forms part of the *Land Mobile Service (LMS)* or the *Land Mobile Satellite Service (LMSS).*
land station: A radio station which is only used when it is stationary and may form part of a mobile radio system.
LANE: *Local Area Network Emulation.*
LAP: *Link Access Procedure.*
LAPB: *Link Access Procedure Balanced* or *Link Access Procedure B channel.*
LAPD: *Link Access Procedure D channel.*

LAPM: *Link Access Procedure Modem.*

Large Scale Integration (LSI): Semiconductor device with a large number of devices on a single dice. Generally this is assumed to be up to 1000 circuits. The density is greater than *Medium Scale Integration (MSI)* but less than *Very Large Scale Integration (VLSI).*

LASER: LASER stands for *Light Amplification by Stimulated Emission of Radiation*, although the acronym is more commonly used and is written in lower case, i.e. as laser. A laser is a device which can produce a single *frequency*, coherent beam of narrow light of high intensity. It does this by stimulating molecules to higher energy levels. Figure L.1 shows one type of LASER structure. Optical feedback is required and this is done by incorporating reflective surfaces onto the semiconductor material.

laser beam: The beam of light produced by a *LASER*. It is *monochromatic radiation* and has very low divergence and high *irradiance*.

laser chirp: The shift in the *laser frequency* between individual pulses, caused by defects.

laser diode: A semiconductor device which emits a *laser beam* when it is excited with current above a fixed value.

lasing: The process which can take place in certain materials, such as semiconductors, gases and liquids, which cause them to emit a *laser beam* when excited. Lasing occurs in a *LASER*.

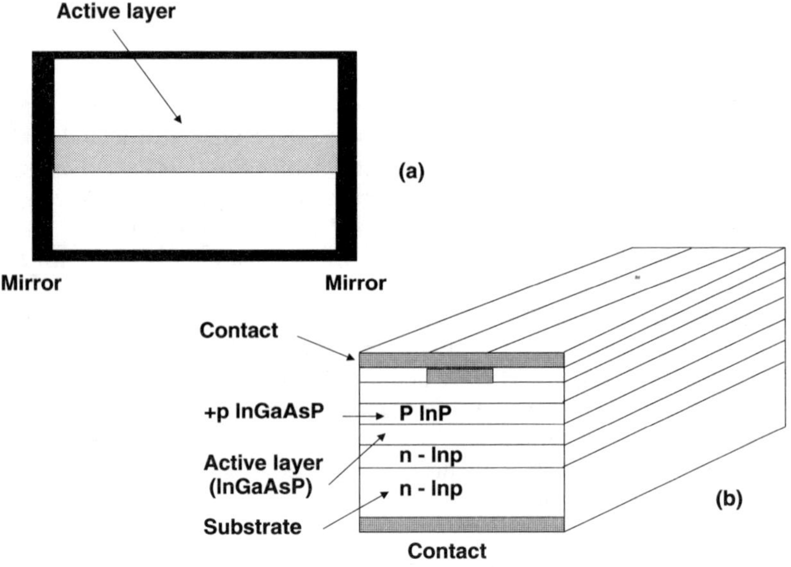

Figure L.1 LASER: (a) schematic; (b) long wavelength LASER stucture

LASS: *Local Area Signalling Service.*

last exchange: During a *call* it usually refers to the *Central Office (CO)* or *exchange* which is located closest to the *called terminal.*

last mile: A colloquial term used to refer to the *access loop*, i.e. the line connecting the *subscriber* to the *local exchange.*

last number redial: A feature which allows the *user* to redial the last number which had been dialled, usually by depressing a *key* on the *telephone keypad.* Also referred to as last number dialled.

LATA: *Local Access and Transport Area.*

latency: **(1)** The *delay time* between the request by a *station* for *network* access, and this access being granted. **(2)** The delay, measured in *bits*, for the *signal* to propagate through the *transmission medium.*

lateral communications: Communications which occurs between entities which are at the same operational level, such as within the same *layer* of the *OSI Basic Reference Model.* See *layered architecture.*

lateral offset loss: The loss which can occur at the junction of two *optical fibres* which have not been aligned accurately.

launch angle: **(1)** The angle at which a beam of light is emitted from a light source, such as a *Light Emitting Diode (LED)*, measured with reference to the normal at the surface. **(2)** The angle at which light enters an *optical fibre*, measured with reference to the axis of the fibre.

launch loss: The loss of optical power which occurs when light from a source is coupled into a *transmission medium*, such as *optical fibre*, due to defects, such as *lateral offset loss* and *Fresnel reflection.*

layer: **(1)** Layers in the *Earth's atmosphere*, which have a significant effect on radio communications. See Figure E.4. **(2)** The layer within a *layered architecture*, such as the *OSI Basic Reference Model*, in which each entity within a layer communicates by means of *protocols* whilst those in different layers communicate via interfaces. Each layer provides services to the layer above and received services from the layer below, as in Figure L.2.

layered architecture: A communications system architecture in which the total system is built up of *layers*, each layer having a clearly defined function and building on the layers below. The *OSI Basic Reference Model* is the most popular layered architecture.

layered protocol; A *protocol* which is designed to receive services from a lower *layer* and deliver services to a higher layer, as in the *OSI Basis Reference Model.*

LBA: *Least Busy Alternative.*

L band: The part of the *electromagnetic spectrum* with a *frequency band* approximately in the region of 1 GHz to 1.6 GHz.

LBO: *Line Build Out.*

LBRV: *Low Bit Rate Voice.*

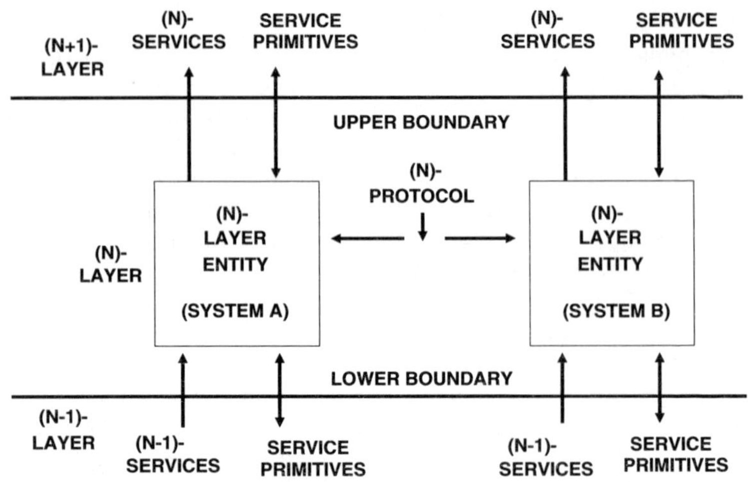

Figure L.2 Concept of a layer

LCD: *Liquid Crystal Display.*
LCI: *Logical Channel Identifier.*
LCN: *Logical Channel Number.*
LCR: *Least Cost Routeing.*
LDAP: *Lightweight Directory Access Protocol.*
LDM: *Limited Distance Modem* or *Linear Delta Modulation.*
LDN: *Listed Directory Number.*
LD signalling: *Loop Disconnect signalling.*
leaky cable: A *coaxial cable* with slots precisely cut or milled into its outer conductor. This enables it to radiate *radio waves* when it is carrying a *signal.* Leaky cables provide a means for radio communications within a building where conventional *antennas* would not be as efficient due to obstructions. Figure L.3, for example, shows the use of a leaky cable, or *radiating cable*, used with a *wireless PABX* to communicate with mobile *handsets*.
leased circuit: See *leased line*.
leased line: A *transmission line* which has been leased from a *PTT* and is for the exclusive use of an organisation. It would normally connect several of the organisation's sites together, carrying *traffic* between these sites. Also known as a *private line*.
Least Busy Alternative (LBA): Routeing strategy in which the routeing for a *call* is first attempted on the most direct route. If this is not available then other available routes are searched and the least busy alternative route is chosen.

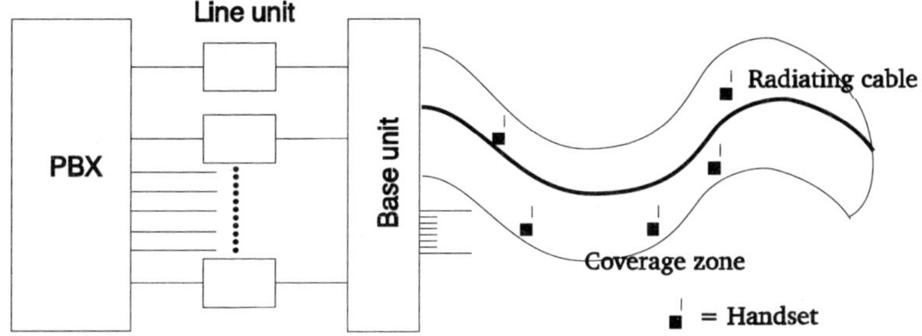

Figure L.3 Wireless PABX using radiating (leaky) cable

Least Cost Routeing (LCR): A *routeing algorithm* which sends *calls* via the *route* which will incur the least cost. Therefore it needs to have information available, such as alternate routes, the cost charged by the *PTT* depending on distance and time of day, etc. It can include the use of different long distance carriers for parts of the route, depending on their costs. This feature is usually available in a *PABX*.

Least Significant Bit (LSB): In a *binary number* or *character* the *bit* position corresponding to the lowest value, i.e. 2^0.

LEC: *Local Exchange Carrier.*

LED: *Light Emitting Diode.*

Lempel-Ziv algorithm: Part of the V.42 bis *data compression protocol*, developed by the *ITU-T*. This builds dictionaries of *code words* for *characters* which recur, the code words using less space than the characters. These are then used when the characters occur.

LEO: *Low Earth Orbit.*

LES: *Local Area Network Emulation Server.*

letters shift: A special *character* in a communications *code* which causes all following characters to be treated as alphabetic characters. This remains in force until a *figures shift* occurs.

level: **(1)** The absolute amount of a parameter, such as the *signal* level or *noise* level. **(2)** The position within an *hierarchical network* or system. Often also referring to the *layer* within the hierarchy.

LF: *Low Frequency* or *Line Feed.*

LHCN: *Long-Haul Communications Network.*

liberalisation: Usually refers to the process adopted by governments in which other telecommunications operators are granted licences to offer services in competition with the incumbent *PTT*.

life cycle: All the stages through which a system or equipment goes through during its entire useful life.

life cycle pricing: The pricing policy adopted by a supplier during the life of a service or produce. An example is shown in Figure L.4. At the

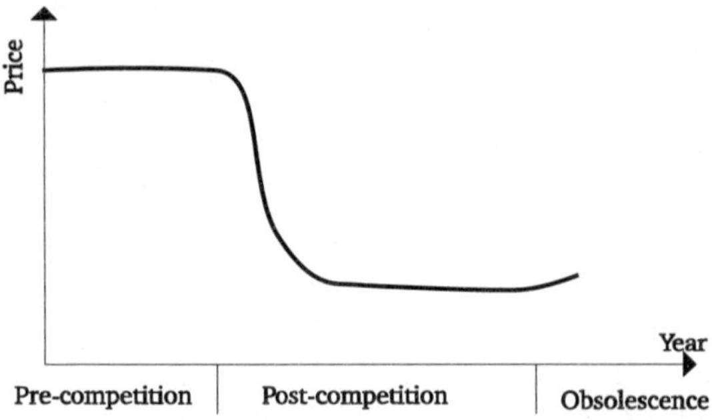

Figure L.4 Price variations during a product's life

pre-competition stage there are very few suppliers and the price can be kept high. As competition builds up, at the post-competition stage, the price is forced down until it is low enough so that new suppliers do not enter the market. Eventually the product or service will reach obsolescence and the number of supplier left will be few so that the price may actually rise.

light acceptance cone: See *acceptance cone*.

Light Amplification by Simulated Emission of Radiation: See *LASER*.

Light Emitting Diode (LED): A p-n junction semiconductor device which emits light when current is passed through the junction in the forward direction. It is often used as an indicator in the front of equipment.

light frequency: The *frequency* of the *electromagnetic radiation* which forms light. It is equal to the *frequency range* from 1 THz to 100 THz.

light guide: *Waveguide*, such as *optical fibre*, used for light *transmission*. Also called a *light pipe*.

light pipe: See *light guide*.

light velocity: The velocity of light, which in vacuum is equal to 299792.5 km/s.

light wave: *Electromagnetic radiation* which covers the *optical spectrum*, i.e. a *wavelength* between about 0.3 μm and 3 μm.

lightwave communications: A communications system in which light is used to carry the *information*. It would include all the elements of the system, such as the light source (e.g. *LASER*), the *transmission medium* (e.g. *optical fibre*), and the light detector.

Lightweight Directory Access Protocol (LDAP): A simplified *Directory Access Protocol (DAP)* used in applications to access *directory services* such as X.500.

Lightweight Presentation Protocol (LPP): The *Presentation Layer* used in *CMOT*, defined in RFC 1085, which provides a means for carrying *OSI* applications over the *TCP/IP* environment.

Limited Distance Modem (LDM): A device used to connect *DTEs* together by conditioning and boosting a *digital signal* so that it can be transmitted over relatively short lines, which are further than that obtained by conventional signals, such as RS-232. The device is more of a *line driver* than a *modem* since no *modulation* or *demodulation* occurs.

limiter: A device which automatically operates on a *signal* and prevents certain of its characteristics, such as *amplitude*, from exceeding a preset value.

limiter hardness: A measure of the extent to which a *limiter* prevents changes in the output *signal* as the input signal varies. A hard limiter would allow very little variations whilst a soft one would give a less sharp limiting value.

limit test: Tests carried out on a component or system which covers the limits of its rated characteristics, usually to determine any malfunctions or weaknesses which may exist.

line: The *transmission medium* connecting two or more points, usually the *local exchange* to the *subscriber terminal.*

line access point: The physical point at which a *terminal* can be connected to a *line*.

linear array: An *antenna* with a number of equispaced elements, usually of identical dimensions and all with the same feed.

linear block code: A class of *code* used for *error control coding*, of which the *even parity check* code is an example. See also *block code*.

Linear Delta Modulation (LDM): *Delta modulation* technique in which a series of linear segments of constant slope provide the input time function.

linear polarisation: *Polarisation* of an *electromagnetic wave* in which the electrical or magnetic field *vector* lies wholly in a single plane.

Linear Predictive Coding (LPC): A *voice compression* technique in which a *block* of *speech* is processed to extract pitch period and format information and then these are transmitted. Good quality can be achieved with data rates as low as 4.8 kbit/s.

Line Build Out (LBO): The *attenuation* which is introduced into the last span of a *T1* circuit, primarily to prevent *crosstalk* effects with other systems in the same *cable* enclosure.

line circuit: See *line*.

line code: A *code* which provides the equivalence between digits generated in a *terminal* and the symbols which are transmitted over a *channel*. The line codes are chosen to suit the characteristics of the *transmission channel* and Figure L.5 shows the relationship between line coding and channel coding functions. The main purpose of a line code is to provide reliable transmission, both in terms of *timing* and *error control*. There are

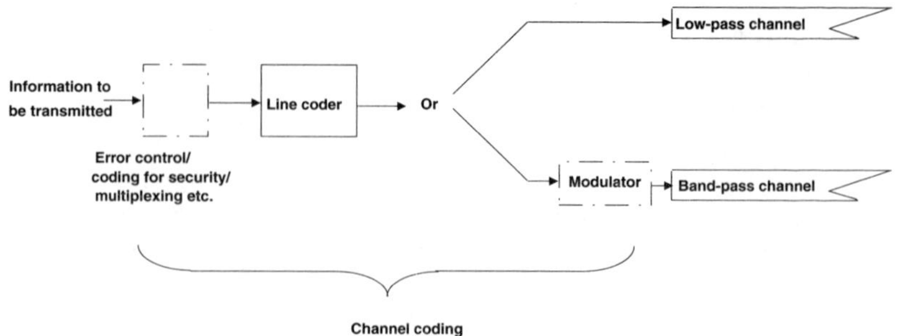

Figure L.5 Relationship between line code and other channel coding functions

many types of line codes, such as *Manchester code* and *Alternate Mark Inversion (AMI)*.

line code violation: A measure of *transmission errors* in which the rules for the *line code* are broken.

line conditioning: Techniques used for improving the performance of *transmission lines*, such as by controlling *distortion*, *attenuation* and *delay time* at different *frequencies*. This enables *data* at higher *bandwidth* to be transmitted. *Carriers* normally charge a fee for line conditioning.

line control: The methods used, such as *signalling*, for different *terminals* connected to a *transmission line* to communicate with each other.

line driver: A device which enhances the *signal* and so enables transmissions beyond the limit of a system. For example a line driver can be used to enable RS-232 transmission to occur for distances greater than the 50 feet specified by the standard.

Line Feed (LF): A *Format Effector (FE)* which causes the print head on a printer, or the *cursor* on a *Visual Display Unit (VDU)*, to move down to the start of the next line.

line finder: The first element of a *switching equipment* which identifies a *calling terminal* and connects it to the switching system for the commencement of *call setup*.

line flyback: See *field flyback*.

line frequency: See *field frequency*.

line hit: See *hit*.

line identification signal: A *signal* which is sent to the *calling terminal* or the *called terminal* to allow identification of the *called terminal* or the *calling terminal*, respectively.

line impedance: The impedance presented to a *signal* at the input terminals of a *line*. This is usually determined by the characteristics of the *transmission line* and the *frequency* or *bandwidth* of the signal.

Line Impedance Stabilising Network (LISN): An artificial mains *network*, used for *conducted emission* measurements below 30 MHz, to define the impedance of the mains supply.

line isolator: A device which is used to interface a *node* to the *network*, which has the capability of protecting the network from extraordinary conditions which may arise in the node, such as voltage spikes.

line loading: **(1)** Addition of external components, such as inductors, to a *transmission line* in order to improve its characteristics, such as for long distance audio circuits. **(2)** The amount of *traffic* being carried on a transmission line, measured as a percent of its total *capacity*.

line noise weighting: *Noise* on the *line* has a different effect on *telephone users* depending on the *frequency* involved. Experimental line noise weighting curves exist which capture this and standard line weighting *networks* have been produced which also take into account factors such as the effect of attenuation caused by the *telephone handset*.

Line Of Sight (LOS): Refers to systems using the *electromagnetic spectrum* for communications in which the *transmission path* between the *calling terminal* and the *called terminal* is not obstructed. Examples are *microwave*, *infrared*, etc.

line period: See *scanning period*.

line powered: Equipment which derives its power supply down the *transmission line*. Generally refers to equipment located on the *subscriber's* premises which is powered from the *local exchange* down the *access line*.

line protocol: The rules or *protocol* used to control the *data communications* between two or more *nodes* connected to a *transmission line*.

line rate: The speed or rate of *data transmission* along a *communications channel*. Also called *line speed*.

line restoral: The facility, available in some equipment such as a *modem*, in which the modem switches the *transmissions* to a back-up line if the main *transmission line* fails, and then continually tests the failed line, restoring the *traffic* back to it once the line has recovered.

line route map: An overall *network* plan which shows the types of *transmission medium* used along different *paths*, including the location of network *nodes*, *exchanges*, etc.

line scan: The horizontal movement of the *raster* during the *scanning* operation used in a *Visual Display Unit (VDU)*.

line seizure: The operation in which a *terminal* appropriates use of a *transmission line* and so prevents other users from using it.

line side: The side of the equipment which is connected out towards the *transmission path*, rather than inwards to other equipment such as *patch panels* or switches.

line speed: See *line rate*.

line splitter: A device which allows several *terminals* to access facilities on a common *transmission line.*

line status indicator: An device, usually located on a *telephone*, which indicates the status of the lines connected to it, e.g. whether they are *busy*, *on-hook*, etc.

line switching: Term sometimes used to describe *circuit switching.*

line turnaround time: The *delay time* between a *terminal* on one side of a *transmission line* sending *information* to a terminal on the other side, and the roles being reversed, i.e. information being sent in the reverse direction. This is usually measured on a *half-duplex circuit.*

link: The complete *transmission channel* between two *nodes*, which includes the *transmission medium*, the *protocols* used, and all other associated equipment.

Link Access Procedure (LAP): The *data link* level *protocol* specified within the *ITU-T Recommendation* X.25 (see *X Series*) for a *Packet Switched Network (PSN).*

Link Access Procedure Balanced (LAPB): A *Link Access Procedure (LAP)* which is based on the *HDLC protocol.*

Link Access Procedure D channel (LAPD): A *frame* oriented *protocol* designed for end-to-end *signalling* for *circuit switching* over the *ISDN.* It is an extension of *HDLC* and operates in the *data link* level.

Link Access Procedure Modems (LAPM): *Protocol* operating at the *data link* level which has been published by the *ITI-T* for *error correction* and *data compression* within *modems.*

link budget: The maximum *signal* loss which can be tolerated between the signal source and receiver, for the receiver to correctly interpret the signal. A greater loss would result in *errors* in the received *data.* The link budget is usually specified in *decibels.*

link by link signalling: *Signalling* in which *signals* are processed in each intermediate point when passing from one end to the other of the system.

link control: The overall control procedures used to ensure that *transmission* over a *link* occurs efficiently and without *errors.* See also *Data Link Control (DLC).*

link header: The *bits* added to the start of a *packet* by the *Data Link Control (DCL)* for *error control.* See *link trailer.*

link layer: Refers to the *Data Link Layer* of the *OSI Basic Reference Model.* Also called *link level.*

link level: See *link layer.*

link trailer: The *bits* added to the end of a *packet* by the *Data Link Control (DCL)* for *error control.* See *link header.*

Liquid Crystal Display (LCD): A display which is formed of a thin layer of cells filled with liquid crystal fluid. When a current is passed through a cell it causes the crystals to rotate and turn the cell opaque so that it

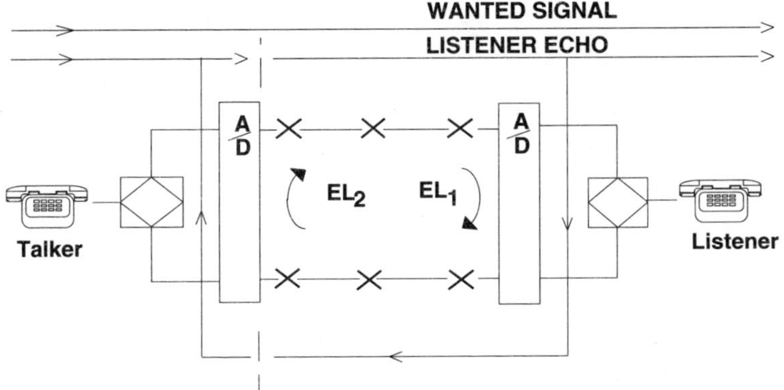

Figure L.6 Signal to listener echo path

reflects light from the front, or obstructs light which is shining from behind.

LISN: *Line Impedance Stabilising Network.*

Listed Directory Number (LDN): The main *telephone number* of a *subscriber* which is normally listed in a telephone directory.

listener echo: The *echo* caused by reflections of the signal in a *four wire circuit*, as shown in Figure L.6. This causes the listener to hear the wanted signal along with a delayed and attenuated version of the original signal. Listener echo has a more detrimental effect on *data transmission* in the speech band than on *voice* transmission. The signal to listener echo ratio is given approximately by the sum of the two echo losses, i.e. $EL_1 + EL_2$, in *decibels.*

Listener Sidetone (LSTR): Sidetone effects which are related to the *Sidetone Masking Rating (STMR)* and is given by the relationship LSTR = STMR + D, in *decibels*, where D is a constant which takes into account the sensitivity of the *telephone* to room noise. It is determined by the acoustic of the *handset* and the type of microphone used, and is specified in *ITU-T Recommendation* G.111.

Little's law: A law used in *queuing theory.* If customers arrive at a rate of λ, as in Figure L.7, and remain an average time of E[D], and if the average number of customers in the system is E[N], then Little's law states that $\lambda E[D] = E[N]$.

LLC: *Logical Link Control.*

LMDS: *Local Multipoint Distribution Service.*

LMI: *Local Management Interface.*

LMS: *Land Mobile Service* or *Land Mobile Station.*

LMSS: Land Mobile Satellite Service.

LNP: *Local Number Portability.*

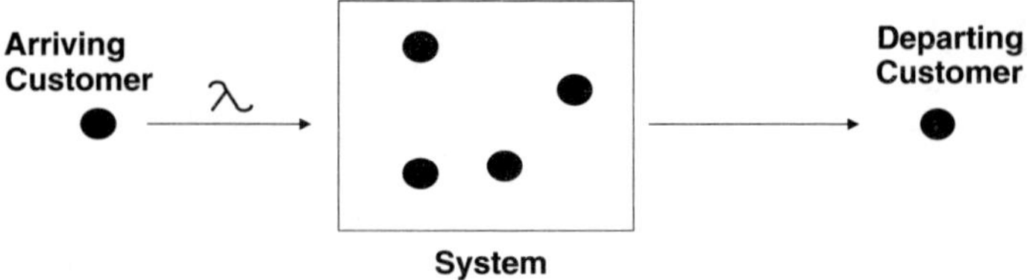

Figure L.7 Little's law

load: (1) A device which consumes power and is attached to a *circuit* or *line*. **(2)** The act of connecting a power consuming device to a circuit or line. **(3)** To enter *data* into a processor or *database*. **(4)** To increase the *traffic* on a circuit or line.

load balancing: Distributing the total *traffic* across different *transmission lines* so that none of them are overloaded and the overall transmission performance is improved.

loaded line: *Transmission line* which has been fitted with *loading coils* in order to improve its long distance *voice frequency* characteristics. This reduces the *amplitude distortion* and give a relatively flat *attenuation-frequency* curve, as in Figure L.8. This also limits the line *bandwidth*.

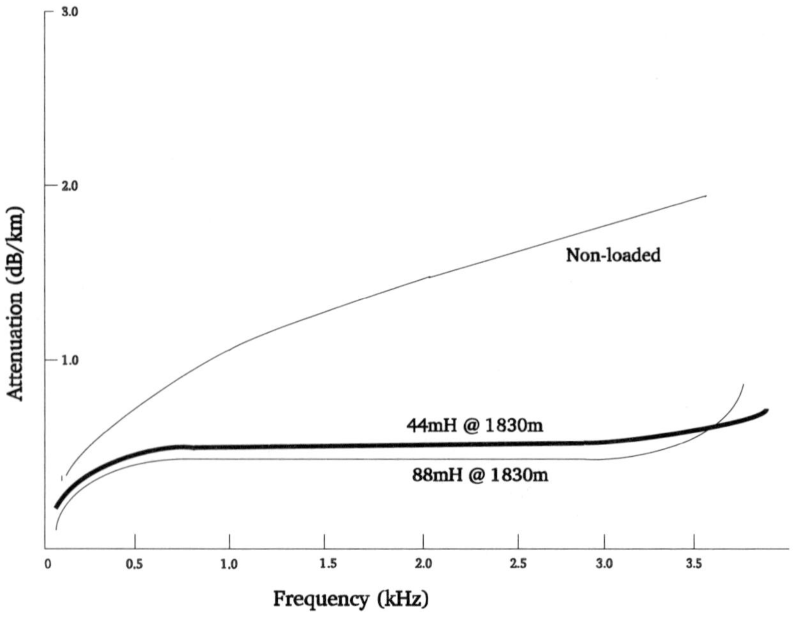

Figure L.8 Effected on loading on a line

loading: **(1)** Connecting *loading coils* to a *line* or *circuit*. **(2)** Increasing the *traffic* on a *transmission line*.

loading coil: Coils or inductors which are connected to a pair of *transmission lines* in order to improve their *voice* transmission characteristics by making *attenuation* as constant as possible across the *voice frequency* band.

load sharing: **(1)** Dividing the processing task between two or more computers. **(2)** Dividing the *traffic* between two or more *transmission paths*.

lobe: **(1)** A part of the *radiation pattern* from an *antenna*. **(2)** The cable which is used to connect a device to an access unit in a *Local Area Network (LAN)*.

Local Access and Transport Area (LATA): The geographical area which is the responsibility of a *Regional Bell Operating Company (RBOC)*. The LATA was defined following *divesture* of AT&T by the *Modification of Final Judgement (MFJ)*. When *telephone circuits* have their start and finish points within a LATA they are the sole responsibility of the local telephone company concerned. When they cross a LATA's boundary, i.e. go inter-LATA, they are the responsibility of an *Interexchange Carrier (IEC or IXC)*.

Local Area Data Transport (LADT): The system for sending and receiving digital data over the existing *local loop* between the *subscriber* premises and the *local exchange*. Introduced by AT&T in the USA.

Local Area Network (LAN): A *network* which covers a relatively small geographical area, such as a building, group of buildings or campus, and is owned and operated, usually, by the user, although in some cases its management may be subject to *outsourcing*. It may connect to the *PSTN* via *gateways* but is does not form part of the PSTN and is therefore not subject to the same level of regulation.

Local Area Network Emulation (LANE): The *ATM Forum* standard for use of *ATM* in a *LAN* environment. It is the method by which existing systems, such as *Ethernet* and *token ring*, which use the *Media Access Control (MAC) protocol*, can employ the facilities offered by *ATM*.

Local Area Network Emulation Server (LES): One of the two *server* entities used within *LANE*. The LAN Emulation Server provides *address* resolution whilst the Broadcast Unknown Server (BUS) handles *network broadcasts*.

Local Area Signalling Service (LASS): *Central Office (CL)* services provided in the USA within a *LATA* for voice and data. This includes features such as *Calling Line Identification (CLI)*, *call forwarding* and automatic *callback*.

Local Automatic Message Accounting (LAMA): The system, available in a *PABX* or *Central Office (CO)* switch, which allows information on all *trunk calls* and *long-distance calls* to be automatically recorded for *billing* purposes.

local bridge: See *bridge*.

local call: A *call* between *terminals* connected to the same *local exchange*. This means that *local lines* are used and no *trunk lines*.

local call rate: The *tariff* charged for *local calls*, which in some regions of the USA are free, a flat charge being made for this facility.

local Central Office: See *local exchange*.

local circuit: The *circuit* connecting the *subscriber to the local exchange*.

local clock: A *clock* which is close to the item being supplied, e.g. the clock in a *Central Office (CO)* or *node*.

local composite loopback: A *loopback test* in which the output of the local *multiplexer* is connected back into the input.

local dataset: A device which conditions the *signal* from an RS-232 interface so that it can be sent over a metallic *transmission medium* without *interference* between adjacent pairs of wire.

local exchange: The *exchange* or *Central Office (CO)* to which the *subscribers* are connected using *local lines*. The local exchange may be connected to other local exchanges and to *trunk exchanges*.

Local Exchange Carrier (LEC): Term used in the USA to describe *common carriers* who provide services within a *local service area* or *LATA*.

local exchange loop: See *access line*.

local line: See *access line*.

local loop: See *access line*.

Local Management Interface (LMI): A *protocol* within *Frame Relay (FR)* which enables the interchange of *information* between the *network* and a *subscriber* regarding the *link* and *Permanent Virtual Circuits (PVC)*.

local mode: Operation of a device when it is not connected to other devices or to the *network*. Useful *information* is still being obtained. Also called *standalone mode*.

Local Multipoint Distribution Service (LMDS): A *wireless local loop* system which has been developed to provide a wider range of *broadband* services than *Multichannel Multipoint Distribution Service (MMDS)*, such as home shopping, remote working, *videoconferencing*, etc. It has a greater *bandwidth* capability than MMDS but more stringent *Line Of Sight (LOS)* requirements and a smaller coverage pattern. Like MMDS it does not have a significant upstream bandwidth or point-to-point capabilities.

local network: Usually refers to the *access network* connecting the *subscriber* equipment, such as *telephone* or *PABX*, to the *local exchange*. In *ITU-T* terminology the local network includes the above as well as the *junction network*, i.e. the network interconnecting a group of local exchanges and connecting these to the *trunk* centre.

local number: The *telephone number* which identifies a *subscriber* on the same *local exchange* as another subscriber.

Local Number Portability (LNP): Term used in the USA to describe *Number Portability (NP)*.

local office: See *local Central Office*.

local service area: A geographical area in which *calls* are charged at the *local call rate*.

local signal: *Signals* which pass between a *subscriber* and the *local exchange*. These signals are usually used for *call setup*.

local switching centre: Same as a *local exchange*.

Local System Environment (LSE): Term used to describe the parts of a system operating within the *open network architecture*. It includes all the basic processing, information and communications components.

local telephone network: Same as *local network*.

location register: The register within a *Mobile Switching Centre (MSC)* which keeps track of all mobiles within its area. When a mobile is moving between areas it monitors the *signalling channel* in its *cell* and will retune to the signal from another cell if it is stronger than in its present cell, as shown in Figure L.9. The mobile then registers its new location in the location register. Also called *mobile location register*.

lock-in frequency: The *frequency* at which a closed loop system is able to start to track an input *signal*.

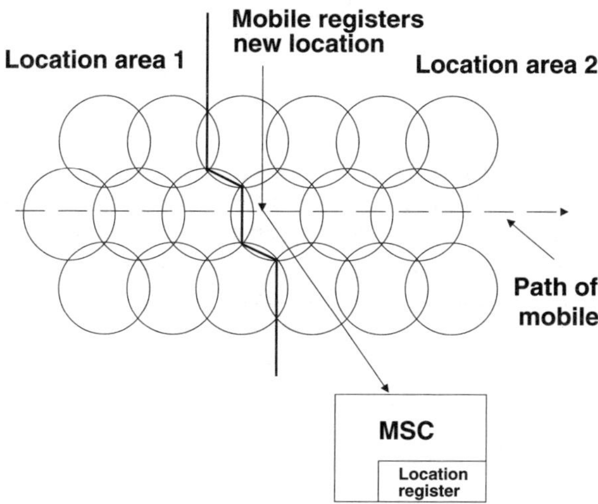

Figure L.9 Location registration

locking: The process of ensuring that two users cannot simultaneously access and change the content of a shared facility, such as a *database*, so that they unintentionally overwrite each other's changes.

lock-in range: The *frequency range* over which a closed loop system is able to start to track an input *signal*.

lock on: The process by which an *antenna* is able to track an object or a *signal*. Examples are *radar* systems which track objects and *Earth stations* which track *satellite* signals.

lock out: The problem in some voice operated *telephone* systems, such as a speakerphone, in which noise on the line or from one end prevents one or both of the *users* from getting through.

lock up: A fault state in which the system unexpectedly stops functioning completely. This could be due to a malfunction within the system or an external condition, such as an overload on its resources.

logical channel: Concept used in a *Packet Switched Network (PSN)* for several virtual connections being established simultaneously over the same physical *line*. See also *Virtual Channel (VC)*.

Logical Channel Identifier (LCI): In a *packet* based communications system, such as a *Packet Switched Network (PSN)* or *Frame Relay (FR)* it is the *field* within the packet which indicates the *Virtual Circuit (VC)* to be used by the packet.

Logical Channel Number (LCN): A unique number assigned to *virtual channels* in a *multiplexing* or *packet switching* communications system, so that the *calls* can be identified and traced.

logical connection: A *path* between two *terminals* which uses a *Virtual Circuit (VC)*.

logical data link: A *data link* between two *terminals* which uses a *Virtual Circuit (VC)*.

logical interface: The rules or *protocols* which defines the way that two devices operate when communicating with each other.

Logical Link Control (LLC): A *data link* level *protocol*, developed by the *IEEE* 802 committee. It is based on *HDLC* and forms the basis of all the IEEE standards for a *Local Area Network (LAN)*. See also *electrical interface* and *mechanical interface*.

logical ring: A *network* which is considered to be a *ring network* although it may have a different *network topology*. For example Figure L.10 shows a *bus topology* using *token passing* and it may be considered to be a logical ring network, the *tokens* moving between *nodes* as shown by the dotted lines.

log-in: Same as *log-on*.

log-off: The process of closing an operating session and disconnecting from a *terminal* or *network*.

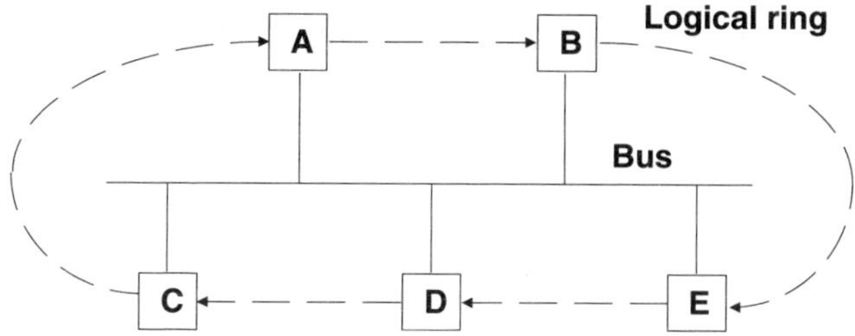

Figure L.10 A logical ring

log-on: The process of commencing a session, usually following a series of defined steps, and connecting to a *terminal* or *network*.

log-out: The same as *log-off.*

long circuit: A *circuit* which covers a long length and usually requires *echo suppressors.*

long-distance call: A *call* between two *terminals* which are served by different *local exchanges*. The call is therefore outside a *local service area.*

long-distance carrier: A *common carrier* which provides services between *LATAs*. This can also include *international calls*. See also *Interexchange Carrier.*

long-distance line: A *line* which is used to carry a *long-distance call.* Usually this would include one or more *trunk lines.*

long-haul: Usually refers to a *long-distance line*.

Long-Haul Communications Network (LHCN): A communications *network* which covers a long distance, usually nationally or internationally, using *trunks* and passing between several *Central Offices (COs).*

longitudinal balance: A measure of the equality of the electrical signal between a pair of wires in a *telephone circuit*, measured with respect to *ground.*

longitudinal displacement loss: The optical power loss which occurs at the junction of two *optical fibres* which are perfectly aligned along their axis but in which there is a gap between their endfaces.

longitudinal jitter: A defect in a *facsimile transmission* caused by an irregular *scanning* speed. Also called *longitudinal judder.*

longitudinal judder: See *longitudinal jitter.*

longitudinal parity check: A *parity check* performed on a stream of *binary digits*, which are received in serial.

Longitudinal Redundancy Check (LRC): A method of *error detection* in *data transmission* in which the data is formed into *blocks* and a *Block Check Character (BCC)* is computed and added on the basis of an *odd parity check* or an *even parity check.* This is then computed at the receiving end and used to check the integrity of the transmitted data. See also *Cyclic Redundancy Check (CRC)* and *Transverse Redundancy Check (TRC)*.

longitudinal signal: A *signal*, usually unwanted, which is impressed across both wires of a *two wire transmission* line. It may be eliminated by balancing the lines. See also *longitudinal balance*.

long-term store: A device which is used to store *data* for a relatively long time. It would normally consist of passive medium which does not need a power supply to maintain its *information*.

long wave: An *electromagnetic wave* which has a *wavelength* which is greater than 1 μm. Also used to describe the *Low Frequency (LF) band*.

loop: **(1)** The *twisted pair wire* connecting a *subscriber terminal* to the *local exchange*. Also called the *local loop* or *access loop*. **(2)** An electrical circuit which is closed on itself. **(3)** A computer programme which repeats a set of instructions depending on a set of conditions.

loop antenna: An *antenna* whose construction consists of one or more *loops* of wire. Both ends of the loop are connected to the receiving equipment.

loopback test: A test in which the transmitted *signal* is returned to the transmitter, usually after it has passed through all or a part of the communications *network*. This received signal is compared with that which was transmitted so that diagnostic tests can be carried out on the system.

loop calling: A simple *signalling* system, as shown in Figure L.11. In the idle condition the *Central Office (CO)* provides a *Direct Current (DC)*

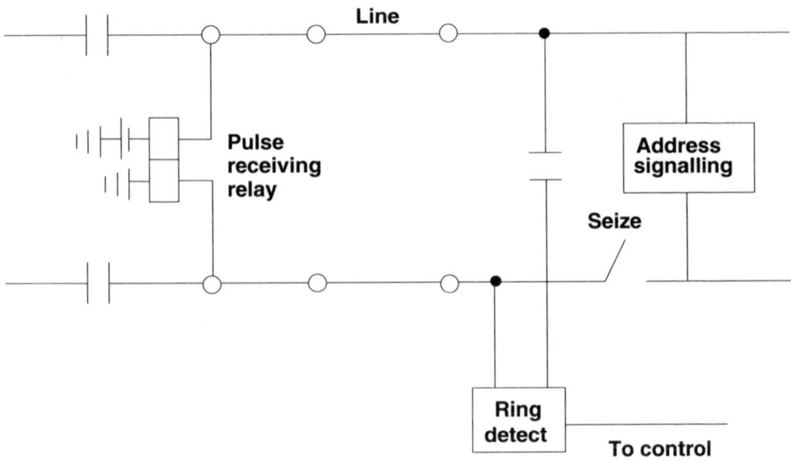

Figure L.11 Loop calling

voltage across the two lines. To initiate a call the *terminal*, for example a *telephone* or *PABX*, seizes the *exchange* circuit by closing a contact and looping the line. The exchange replies by providing a *dial tone* superimposed on the normal DC feed. The sequence of actions for *incoming calls* and *outgoing calls* is given in Table L.1.

loop calling PABX: A *PABX* using *loop calling*.

loop calling signalling: See *loop calling*.

loop checking: A technique in which the *data* received by the *receiving terminal* is returned to the *transmitting terminal* so that it can be compared with the original *transmission*. This enables a check to be made on the accuracy of the transmission.

loop current: The current which flows in the *access line* when a device connected to it goes *off-hook*. This enables the equipment in the *local exchange* to identify that a *call* is being made.

loop dialling: Same as *Loop Disconnect (LD) signalling*.

Loop Disconnect (LD) signalling: See *dial pulse signalling*.

loop network: A *network* in which all the *nodes* are connected serially and the *network topology* forms a *loop*. Also called a *ring network*.

loop start: Method of *signalling* in which the *off-hook* condition results in a switch closing, which closes a *loop* and allows *Direct Current (DC)* current to flow.

loop test: Same as *loopback test*.

loop timing: A *synchronisation* method in which the *clock* is derived from the *signal* coming into the equipment.

LOS: *Line Of Sight* or *Loss Of Signal*.

loss: **(1)** The reduction in the *amplitude* of a *signal*, usually due to *attenuation* in the *transmission path*. The opposite of *gain*. **(2)** The dissipation of power without any useful work having being done. **(3)** The *calls* which do not go through, usually due to meeting a *congested system*. See *lost call*.

loss budget: The amount of *loss* in the transmitted *signal* which can be tolerated within the various elements of a *network*, so that the signal will eventually be received within the tolerance of the receiving equipment.

loss-frequency distortion: The *distortion* caused by the the difference in *loss* which occurs in the *transmission medium* at different *frequencies*.

loss-frequency response: The variation of *loss* within a *circuit* at different *frequencies*.

loss-of-frame alignment: An error in a *Pulse Code Modulation (PCM)* system in which the *receiving terminal* is unable to determine the start of a *frame* in the received *signal* and cannot therefore align itself to this.

Loss Of Signal (LOS): An error condition on a *T1* circuit where the *DS-1* signal is absent for longer than a defined time.

Table L.1 DEL and loop calling PABX
(Continued on next page)

Signalling state	*Exchange to telephone set or PABX*	*Telephone set or PABX to exchange*
Idle	Feed to the line – apply –48V to the *b* wire and earth potential to the *a* wire	Apply disconnection to the line (no loop)
Outgoing call (subscriber to exchange)		
Seize the line	Apply normal feed to the line	Apply loop to the line
Proceed to dial	Apply normal feed to the line and a dial tone	Apply loop
Dial digits	Apply normal feed. Dial tone removed when first digit detected	Apply loop disconcertion pulses representing the dial digits
Remote subscriber answer	Apply reverse feed	Apply loop
Conversation phase		
	Reverse feed	Apply loop
Incoming Call		
Remote subscriber calling	Apply ringing – battery potential on *a* wire and 75V AC on *b* wire	No loop
Subscriber answers	Ring current removed – normal feed	Apply loop
Clear down (clear forward)		
Subscriber clears down	Reverse feed	Remove loop
Exchange confirms	Remove feed for a short time	No loop
Idle	Normal feed	No loop

Table L.1 (Continued from previous page)

Signalling state	*Exchange to telephone set or PABX*	*Telephone set or PABX to exchange*
Clear back (from conversation phase)		
Remote subscriber clears down first	Remove feed for a short time	Loop
Calling subscriber clears	Normal feed	Remove loop

lossy medium: A *transmission medium* in which a significant *loss* occurs in the *signal* as it passes through it.

lost call: A *call* which cannot be completed, for example due to *congestion* or a failure of the system. A call which cannot be completed due to the *called terminal* being *busy* is not considered to be a lost call.

lost calls cleared: A strategy used in *call handling* within *switching systems* in which *lost calls* are either discarded by the *network* or voluntarily leave. These calls may return at a later time. The terminology is used in *teletraffic theory*.

lost calls delayed: A strategy used in *call handling* within *switching systems* in which *lost calls* are delayed or held within the system and are queued until they are served. The terminology is used in *teletraffic theory*.

lost calls held: A strategy used in *call handling* within *switching systems* in which *lost calls* are held for a time equal to the average *holding time*. The terminology is used in *teletraffic theory*.

Loudness Rating (LR): A standardised method for expressing and measuring the *transmission loss* which occurs in a *speech path*. It provides a single value relating to the loudness with which a listner perceives speech emitted by a talker. See *Overall Loudness Rating (OLR)*.

loudspeaker: A device which converts *analogue* electrical *signals* into *sound waves* which can usually be heard at a long distance from the device.

loudspeaking telephone: See *handsfree telephone*.

Low Bit Rate Voice (LBRV): A *speech compression* technique which allows *voice transmission* at under 64 kbit/s.

low delay circuit: A *circuit* in which *signals* passing through it experience a relatively small *delay time*.

Low Earth Orbit (LEO): An *orbit* a *satellite* uses which is close to the Earth's surface, at a distance between 700 km and 1500 km in altitude. Because of the shorter distance which radio signals need to travel less power is needed when communicating with satellites in Low Earth Orbit, compared to those, for example, in *Geostationary Orbit (GEO)*, so smaller portable equipment can be used.

Low Earth Orbit Satellite (LEOS): A *satellite* which is in *Low Earth Orbit (LEO)*.

lower sideband: As a result of *modulation* of a *signal* the lower sideband has a *frequency* which is lower than that of the *carrier* and the *upper sideband*.

Lowest Usable Frequency (LUF): Generally refers to the lowest *frequency* which can be effectively used in a *radio transmission* system. The LUF is determined by the *path loss* and by the *noise* level at the receiver site. See also *Maximum Usable Frequency (MUF)* and *Frequency of Optimum Transmission (FOT)*.

Low Frequency (LF): The *electromagnetic spectrum* in the *frequency range* from 30 kHz to 300 kHz. Also referred to as *long wave*.

low frequency cutoff: The lowest *frequency* which can be effectively transmitted through a *waveguide*.

low level control: Control functions which are carried out in the lower two layers, i.e. (*Physical Layer* and *Data Link Layer*), of the *OSI Basic Reference Model*. Examples are *bit stuffing*, *Frame Check Sequence (FCS)* calculations, etc.

low level language: *Programming language* which can be directly executed by a computer, although it does not contain words which are meaningful to a human. Usually consists of the *object code* from a *compiler*. See also *machine code*, *machine language* and *High Level Language (HLL)*.

low level protocol: *Protocol* used in the lower layers of the *OSI Basic Reference Model*, such as the *Physical Layer* and the *Data Link Layer*.

low pass filter: A *filter* which allows all *frequencies* below a specified value to pass through with negligible *attenuation*, but severely attenuates all frequencies above this value. See also *high pass filter* and *band pass filter*.

Low Probability of Intercept (LPI): Refers to a *signal* which is *coded* such that it cannot be easily *intercepted*, e.g. *deciphered* or effected by a *jam signal*. Example is a *Direct Sequence Spread Spectrum (DSSS)* signal.

low resolution: A *Visual Display Unit (VDU)* which has poor *resolution* and cannot be used to display complex graphical images. See also *high resolution.*

low-speed modem: A device which operates at less than 1.2 kbit/s and is used to connect a *Data Terminal Equipment (DTE)* to the *network.*

low-speed Morse: *Morse code* which is operated at less than its rated speed, e.g. below 18 words per minute.

LPC: *Linear Predictive Coding.*

LPI: Low Probability of Intercept.

LPP: *Lightweight Presentation Protocol.*

LR: *Loudness Rating.*

LRC: *Longitudinal Redundancy Check.*

LSB: *Least Significant Bit.*

LSE: *Local System Environment.*

LSI: *Large Scale Integration.*

LSTR: *Listener Sidetone.*

LUF: *Lowest Usable Frequency.*

lumen (lm): The SI unit for *luminous power.* It is equal to $1/4\pi$ of the power emitted by a source having a *candlepower* of one *candela.* The symbol for a lumen is lm.

lumen-second: A measure of energy, it is equal to one *lumen* flowing for one second.

luminance: A measure of the brightness of a light source. It is the *luminous intensity* per unit area, leaving, passing through or arriving at a surface in a given direction. The surface area is the projected area as seen from the specified direction. It is expressed in *candelas (cd)* per square centimetre (cd/cm^2).

luminous existence: The total *luminous flux* divided by the surface area of the source. It is expressed as *lumens* per square centimetre (lm/cm^2).

luminous flux: The time rate of flow of luminous energy emitted from a light source, measured in all directions from the source. It is also known as *luminous power* and is expressed in *lumens.*

luminous intensity: The *luminous flux* per unit solid angle, travelling in a given direction. It is expressed in *lumens* per *steradian* (lm/sr).

luminuous power: See *luminous flux.*

lux: The SI unit for *illuminance.* It is equal to one *lumen* passing through one square metre of surface which is normal to the direction of propagation.

M

Maastricht treaty: Treaty which formed the *European Union (EU)* in November 1993. It created the three-pillar structure with a new Common Foreign and Security Policy (CFSP). It codified the co-operation in the field of Justice and Home Affairs (JHA) and it expanded the scope of the *EEC* to include provisions for Economic and Monetary Union, with a single currency from the end of this century.

MAC: *Medium Access Control* or *Multiplexed Analogue Component.*

machine code: *Code* or instructions which are written in a form or language which a processing unit, such as a computer or communications controller, can directly interpret and execute, without the need for a *compiler*. Also known as *machine language*. See also *High Level Language (HLL)*.

machine independent: A programme, operation or procedure which does not depend on the *hardware* characteristics for successful operation. Example is *HyperText Markup Language (HTML)* which allows the *Internet* to be accessed from a variety of computing and communications platforms.

machine language: See *machine code*.

mach number: A measure of speed relative to that of sound through the same medium. If v_1 is the speed of the body in a medium and v_2 is the speed of sound in the same medium, then the body has a mach number of $M = v_1/v_2$.

MACP: *Motion Adaptive Colour Plus.*

macrobend loss: The loss of *signal* (such as light) which occurs in a *transmission medium* (e.g. *optical fibre*) due to bends in it which form a curve with a radius less than a few tens of millimetres.

magneto board: Used in early forms of *manual exchanges* in which the caller gains the attention of the operator by using a hand operated generator, to send an *Alternating Current (AC) signal* which is detected at the exchange.

magnetooptic: The effect which results from a magnetic field changing the *refractive index* of a *transmission medium* to which it is applied.

magnetostriction: The effect caused by a magnetic field on some material in altering their physical dimensions.

magnetostriction delay line: A *delay line* which uses the *magnetostriction* effect. The change in dimension caused by a magnetic field pulse causes a mechanical shock wave to travel the length of the material, which is detected at the other end after a delay.

mail box: A storage location, usually on a *mail server*, which holds *messages* until requested by the addressee.

mail server: A *server*, which is used to control an *electronic mail* system.

Main Distribution Frame (MDF): A *distribution frame* which is used as the interface between the external *PTT* cable and the PTT *switching* equipment. It provides a cross-connect facility, a termination point for cables, a test and isolation point, and has overvoltage protection components. An MDF may carry copper or *fibre optic* cable. See also *Building Distribution Frame (BDF).*

main feeder: A *cable* which originates from a point on the street and terminates in a *subscriber's* premises. It belongs to the *common carrier.*

mainframe: A large, high performance, computer whose processing power is shared by several other computers and *terminals.*

main lobe: The *lobe*, from the *radiation pattern* of an *antenna*, which contains maximum *irradiance* compared to other lobes.

mains filter: A *filter* which is connected to the main *Alternating Current (AC)* supply, usually to prevent *conducted interference* from going down the mains lines.

Main Switching Centre (MSC): Part of the *analogue network* used in the UK (see Figure A.15), it formed the highest *switching centre*. Together with the *DSC* it was called a *trunk transit exchange* and the *network* interconnecting them the trunk transit network.

Maintenance Control Circuit (MCC): A *Voice Grade Circuit (VGC)* which is used by personnel for the maintenance of a *telecommunications network.*

majority logic decodable code: A form of *cyclic code* used for *error correction* in *data transmission.*

male connector: One half of a *connector* which has protruding elements which fit into the cavity in a *female connector*, so making a good electrical contact.

MAMI: *Modified Alternate Mark Inversion.*

MAN: *Metropolitan Area Network*

Managed European Transmission Network (METRAN): A *CEPT* initiative for the introduction of a high speed *transmission backbone* across Europe.

Managed Data Network Services (MDNS): An earlier proposal by *CEPT* for *one stop shopping*, with supporting tools, which has now been discontinued.

managed object: A concept used in *OSI network management*. The managed object forms an abstract representation of a resource looked at from the perspective of management. It may, for example, be a physical item, such as a *multiplexer*, or it may be logical, such as a connection. Managed objects contain *attributes*, each attribute having a value. Figure

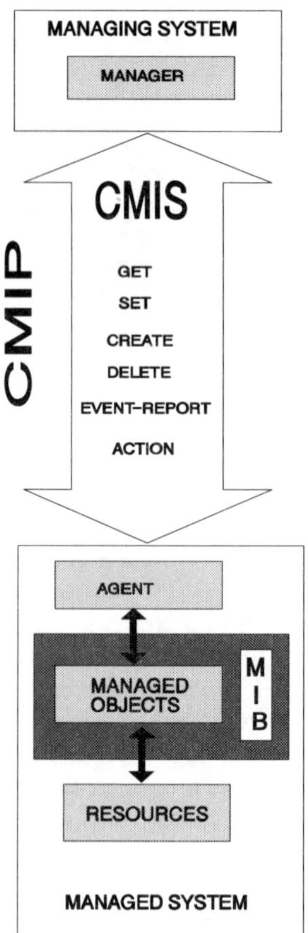

Figure M.1 Illustration of a manager and agent process

M.1 illustrates the position of managed objects within a management system and their relationship with resources and agents (see *agent process*).

managed object class: The collection, or set, of *managed objects* which share identical properties is referred to as a *managed object class*. Every occurrence of a managed object, which conforms to the same class, is called a *managed object instance*. Managed object classes are important since, if a management system knows the class of a managed object, then it knows unambiguously the *management services* which it can expect to be provided by the object.

managed object instance: See *managed object class*.

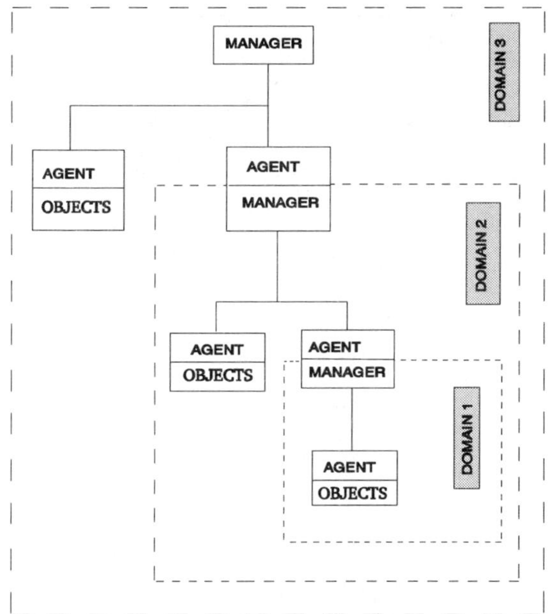

Figure M.2 Manager and agent within management domains

management: Generally used to refer to *network management*. See also *accounting management*, *security management*, *performance management*, *configuration management* and *fault management*.

management domain: The *OSI network management* system uses the concept of a *manager process* and an *agent process*. A complex system would have several managers and several agents. A manager can be a managing process for its own agents and an agent for another managing process. Managers and agents can therefore form a hierarchy of management domains, as shown in Figure M.2, each domain defining the scope of its manager.

management hierarchy: Usually refers to the different levels within *network management* systems, where the higher levels obtain a more global view of the managed *network* than lower levels. Examples are element level management, network level management, and service level management.

Management Information Base (MIB): The set of *managed objects*, made visible by the *agent process* to the *manager process*, is called the Management Information Base, as illustrated in Figure M.1. Because the *data* relating to managed resources are often stored in a physical *database*, this database is often spoken of as the MIB. The MIB provides a

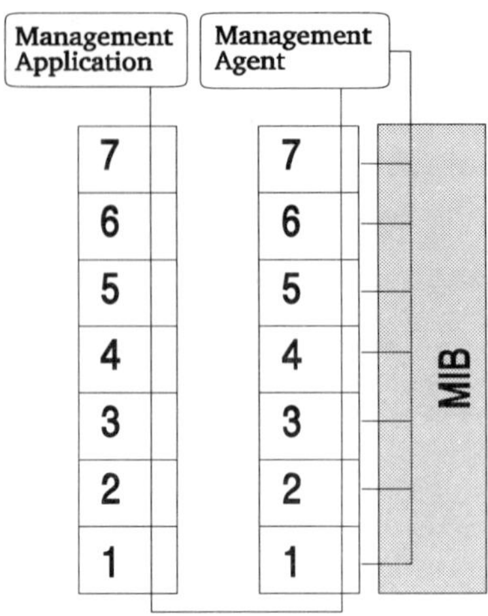

Figure M.3 A full stack MIB

method for relating managed objects to each other. It is by manipulation of the virtual object held within the MIB that the real database held within the *node* can be examined or changed. The MIB contains a virtual image of how a *network* is behaving at any time. It should not be confused with the *configuration* databases, which are located at *network management centres* and which keep information on how the node should be behaving. The MIB must span all *seven layers* of the *OSI Basic Reference Model*, as shown in Figure M.3, since it must be able to access and change the management information in all layers of the stack. A problem can arise when controlling devices which do not have a full seven layer implementation, such as *modems*, *bridges* and *routers*. Two solutions are possible in these instances: use of a thin stack or of a *proxy agent*. Figure M.4 shows a device which uses a thin stack. In this arrangement the device which does not have the full seven layers only performs the minimum functionality needed to support network management.

Management Information System (MIS): The collection of systems, such as inventory control, order processing, *billing*, etc., including the *Information Technology (IT)* systems on which they run, which are used by managers in the management of their business.

Management Information Tree (MIT): A concept used within *OSI network management* systems. Within the *Management Information Base (MIB)* the *managed objects* are organised in a containment hierarchy or tree structure, as in Figure M.5, where the subordinate *nodes* on the tree are contained within superior objects. This tree is called a Management

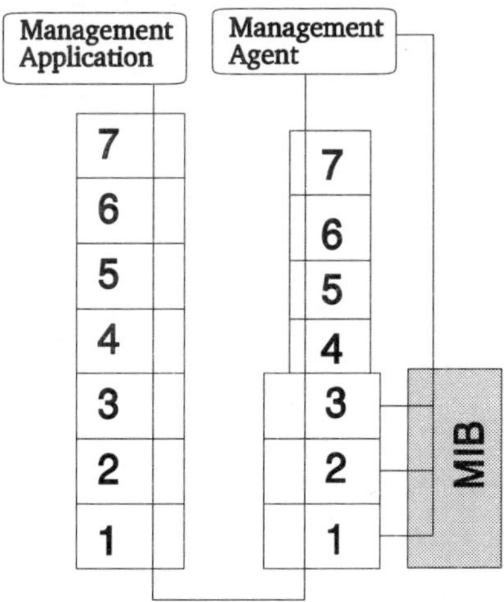

Figure M.4 A thin stack MIB

Information Tree. Operations on managed objects within the MIB need a knowledge of this containment hierarchy.

management traffic; *Traffic*, flowing in a *network*, which is primarily used for the management of tasks or functions on the network. It does not contain any *information* being generated by the *users* on the network.

manager process: A concept used in *OSI network management*. The manager process is responsible for management of the network, working through the *agent process*. (See Figure A.4.) The manager process interacts with remote agents using the *Common Management Information Service (CMIS)*, which is transmitted by the *Common Management Information Protocol (CMIP)*. (See Figure M.1.)

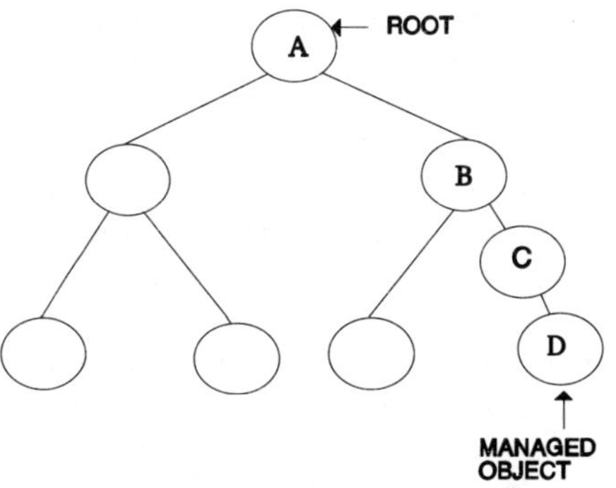

Figure M.5 Containment relationship of objects within a MIB

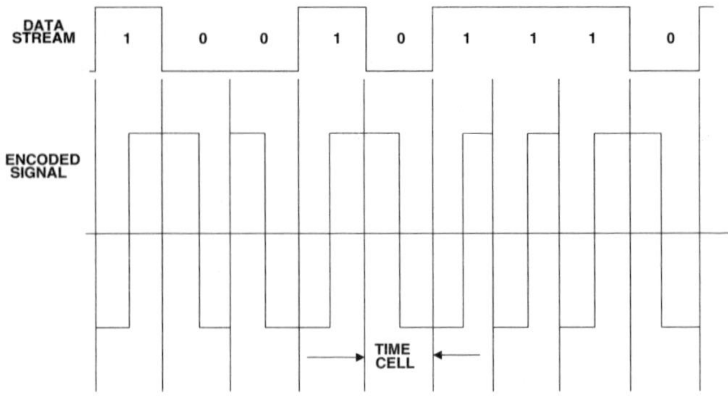

Figure M.6 Manchester encoding

Manchester code: A *bit by bit code* (*line code*), which is also called a WAL1 code or a diphase code. *Information* is encoded into a two *bit* symbol, for *transmission*, as in Figure M.6. A negative to positive transition in the middle of a bit period indicates a *binary* 1, whilst a positive to negative transition indicates a binary 0. Because of these transitions the *clock* can be derived from the transmitted *signal*, i.e. it is self clocking or *self timing*. This is a simple line code which is used in applications such as *Ethernet Local Area Networks (LAN)*.

Manchester encoding: The application of *Manchester code*.

Man-Machine Interface (MMO): The interface between an operator and the system, which could be a computer, an application running on a processor, etc. It usually refers to the part of a *network* which carries out this interface, for example interpreting commands, providing a *Graphical User Interface (GUI)*, etc.

manual answering: An operating mode in which human intervention is needed for successful *call setup*. For example a called *modem* may require an operator to lift its receiver, which would indicate to the calling modem that it could start its *data transmission*. See also *manual calling*.

manual calling: An operating mode in which human intervention is needed for calling another *terminal*. For example an operator may be required to lift the receiver and dial the *address* of the called terminal before *call setup* can be completed. See also *manual answering*.

manual exchange: An *exchange* in which the connection between an *incoming line* and an *outgoing line* is completed by a human operator. For example the operator may be required to link two lines together using a piece of cord which plugs into jacks on a switchboard.

manual telegraphy: *Telegraphy* in which the elements of the *code* are keyed individually by a human operator. These are transmitted as they are keyed and are not stored for subsequent *transmission*.

Manufacturing Automation Protocol (MAP): A *protocol* used for communications between automation equipment used in a manufacturing environment. Initially developed by General Motors it is very similar to *IEEE* 802.4. See also *Technical and Office Protocol (TOP).*

many-to-many call: A call made between several *calling terminals* and *called terminals* so that there is more than one simultaneous *user* at each end.

MAP: *Manufacturing Automation Protocol* or *Mobile Automation Part.*

mapping: The logical association of one or more parameters in different *networks.*

marine broadcast station: A *transmitting station* which makes broadcasts, at scheduled intervals, of items of interest to shipping, e.g. time, weather conditions, etc.

marine telephone: A *telephone* system which operates on a *frequency range* assigned for marine use and can contact other marine telephones. Access to land-based telephones is normally via an operator.

MARISAT: A *satellite* communications system, set up in 1976 by the Communications Satellite Corporation, to provide *voice* and *data* services to shipping. This function was taken over by *INMARSAT* in 1982.

Maritime Mobile Satellite Service (MMSS): A *satellite* communications systems which uses *Earth stations* located on board ships. The communications system may include other sea based systems, such as survival craft and radio beacons.

Maritime Radionavigation Service (MRNS): A radio *broadcast* service which is used for safe navigation of shipping.

mark: **(1)** One of two states used in a *binary* communications system, the other state being *space.* Usually represented by a binary 1. **(2)** In *telegraphy* it represents the closed state, when current is flowing in the *circuit*, the other condition being an open state. **(3)** In a *transmission* sequence a mark is used to indicate the last *bit* of a *character*, the first bit being a space.

marker: **(1)** A control device, part of a *crossbar switch* which determines the *path* a *call* should take through the *exchange*, by operating the appropriate *crosspoints.* **(2)** A symbol on a *Visual Display Unit (VDU)* which has a distinctive shape and is used to indicate a position on the screen.

mark-hold: The *transmission* of a continuous *mark* condition to indicate that there is no *traffic* on the *line.*

mark inversion: See *Alternate Mark Inversion (AMI).*

Markov process: Mathematical model used in *queuing theory* in which a future event (such as the route through an *exchange* or arrival of *calls*) is determined by the previous events. A discrete time Markov process is called a Markov chain.

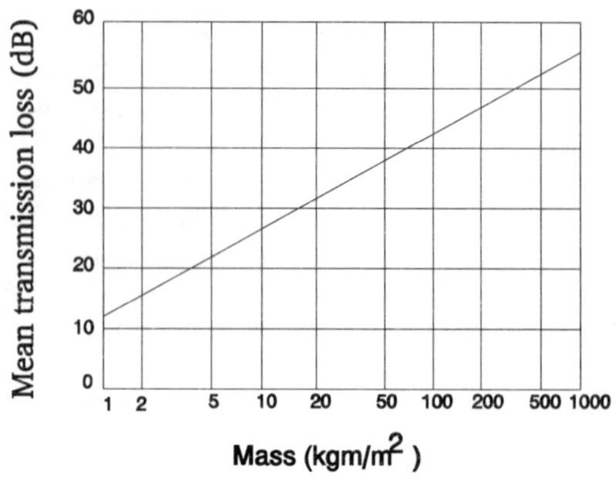

Figure M.7 The Mass Law

Markov chain: See *Markov process*.

mark-space: A *signalling* method which uses a *mark* and a *space* for the prime elements. See, for example, *double current signalling*.

MASER: *Microwave Amplification by Simulated Emission of Radiation*.

mask: A pattern which is applied to a *signal* in order to let through portions of the signal only, depending on the pattern arrangement. The mask usually consists of a pattern of *binary* 0 and 1 *bits*.

Mass Law: The physical law which gives the relationship between the loss of transmitted sound (in *decibels*) through a solid object and the mass of the object (in kilograms per metre square). Figure M.7 shows this relationship and Table M.1 gives typical mass values for some common materials.

mass storage: Equipment which can store large amounts of *data*.

Table M.1 Typical mass per unit area

Material	*Mass in kgm/m^2 surface area*
6mm glass	15
100mm breeze block	150
12mm plasterboard with studs	60
115mm brickwork	160
150mm concrete	300

master clock: The main *clock*, or *timing* device, in the *network* which is used for *synchronisation* of all other clocks and *nodes* connected to the network.

mastergroup: The grouping of *voice channels* used in *Frequency Division Multiplexing* systems. In Europe five *supergroups* are combined into a mastergroup, spanning the *frequency range* from 812 kHz to 2044 kHz. The North American Bell system combined ten supergroups into a master group, giving a capability of carrying 600 voice channels.

Master International Frequency Register (MIFR): A register of all *frequency* assignments made worldwide. This register is maintained by the *International Frequency Registration Board (IFRB)*, an organ of the *International Telecommunications Union (ITU)*. The MIFR is published twice a year in microfiche form by the ITU as the *International Frequency List (IDL)*. Updates to this list are circulated weekly to administrations.

master-slave: An operating mode between two *stations* on a *network* where the *timing* and control for one station (the *slave station*) is derived from the other station (the *master station*).

master station: **(1)** The *station* on a *network* which has been given the responsibility, by the *control station*, for the data transfer and control of *slave stations*. Often this is done by *polling*. **(2)** In a *token ring* network the station which is responsible for recovering from an *error* condition.

Mast Antenna Television (MATV): A *Cable Television (CATV)* system which uses a local *antenna* for the provision of television programmes to a hotel or apartment block. Also know as Master Antenna TV. See also *SMATV*.

material dispersion: The *dispersion* of an *electromagnetic wave* (such as light) as it travels through a *transmission medium* (such as *optical fibre*) due to the variations in the *refractive index* of the medium for the different *frequencies* contained in the wave.

MATV: *Mast Antenna Television*.

maximum acceptance angle: The maximum angle between the incident ray and the longitudinal axis of the *transmission medium* for the ray to be transmitted by *total internal reflection* within the medium. See *acceptance cone*.

maximum calling area: The maximum geographical area which an *access line* can be used for making *calls*. This area is determined by the characteristics of the line.

maximum justification rate: See *maximum stuffing rate*.

maximum stuffing rate: The maximum rate at which *bit stuffing* can be applied to a system. Also known as *maximum justification rate*.

Maximum Usable Frequency (MUF): The highest *frequency* which can be effectively used in radio based *transmission* systems which rely on the reflections off the *ionosphere* for communications. Since ionospheric conditions vary the MUF is taken as the median frequency which is applicable to 50% of the days in a month. See also *Lowest Usable Frequency (LUF)* and *Frequency of Optimum Transmission (FOT).*

Maxwell's equations: Fundamental equations which are used to calculate the behaviour of *electromagnetic waves* in a *transmission medium*, such as the transmission of light in *optical fibre* or of *radio waves* in the atmosphere.

M band: The *frequency band* from 60 GHz to 100 GHz.

mB1C: A *line code* which is an example of a *block code* of the *bit* insertion type. A single bit is added to a *block* of m *information* bits so as to provide long term polarity balance, i.e. eliminate *DC* components. In practice m has the value from about 7 to 23.

mBnB: A *line code* which is an example of a *block code*. In this system m *binary* source *bits* are mapped into n binary bits for *transmission*. n and m can take several values but for redundancy n should be greater than m. Often n = m+1.

MBONE: *Multicast Backbone.*

MCC: *Maintenance Control Circuit.*

MCM: *Multi-Carrier Modulation.*

MCN: *Micro-Cellular Network.*

MCU: *Multipoint Control Unit.*

MD: *Mediation Device.*

MDF: *Main Distribution Frame.*

MDNS: *Managed Data Network Services.*

MDSL: *MPEG Syntatic Description Language* or *Medium bit rate Digital Subscriber Line.*

mean busy hour: The sixty minute period during a day in which the *traffic volume* is greater than in any other sixty minute period.

mean deviation: It is a measure of the *deviation* of a number from the average number in a series. It is found by taking the mean of the differences between each individual number in the series and the *arithmetic mean*, or *median*, of the series. Negative signs are ignored. See also *standard deviation.*

mean down time: A measure of the reliability of a system it is equal to the average time that a system is unusable due to a system fault. See also *Mean Time To Repair (MTTR).*

mean holding time: The average time for which calls occupy a system so that it is *busy* and not available to other *users*.

Mean Time Between Failures (MTBF): The average time between consecutive failures of a piece of equipment. It is a measure of the reliability of the system.

Mean Time Between Outages (MTBO): The mean time between equipment failures which are of such severity that the *telecommunications* service cannot be maintained and it results in an *outage*.

Mean Time To Failure (MTTF): The average time between the equipment being placed into service for the first time and its first failure. It is a measure of the infant mortality rate of a system.

Mean Time To Repair (MTTR): The average time needed to correct a failed piece of equipment.

Mean Time to Service Restoration (MTSR): The average time needed to restore a service following a system *outage*.

mechanical interface: The physical joint between two devices, such as the *male* and *female* join between two halves of a *connector*. A good mechanical interface is usually a prerequisite for a good *electrical interface*.

mechanically despun antenna: An *antenna* which is mounted on a platform whose position is continually changed such that the antenna is pointing in a fixed direction irrespective of the movements of the equipment on which it is located. For example an antenna on a *satellite* is moved such that it is always pointing to the Earth irrespective of the movements of the satellite.

mechanical splice: The method used to join two *optical fibres* together in which their ends are pressed together and glued. The adhesive used is transparent and has a *refractive index* which matches that of the fibres. See also *fusion splicing*.

media converter: Device which converter *signals* from one type of *transmission medium* to another. For example it interface 10Base 2 to *10Base T*, *twisted pair wire* to *fibre optic cable*, and *singlemode fibre* to *multimode fibre*. Media converters operate primarily in *Physical Layer* of the *OSI Basic Reference Model*, taking signals from one transmission medium and sending it to the other.

median: An averaging technique in which the median or 'middle one' of a series of numbers is found by placing all the figures in order and choosing the one in the middle or, if there are an even number of items, the mean of the two central numbers. It is a useful technique for finding the average of items which cannot be expressed in figures, e.g. shades of colour.

Mediation Device (MD): A device, used within the *Telecommunications Management Network (TMN)*, which performs a *Mediation Function*.

Mediation Function (MF): The function, used within the *Telecommunications Management Network (TMN)*, for adapting, filtering and condensing information which passes from the *Network Element Functions (NEF)* and sometimes from the *Q Adapter Function (QAF)*, to meet the requirements of the *Operations Systems Function (OSF)*. See *reference point* and Figure R.7.

medium: See *transmission medium.*

Medium Access Control (MAC): See *Medium Access Control layer (MAC)*

Medium Access Control (MAC) Layer: Mainly part of the *Data Link Layer* of the *OSI Basic Reference Model*, it uses the services of the *Physical Layer* to support *topology* dependent functions. For example, in *DECT* (Figure D.2), the MAC Layer selects the radio *channels* and then establishes and releases the communication *link*. It also carries out *multiplexing* and *demultiplexing* of all *information* into burst *packets*.

Medium bit rate Digital Subscriber Line (MDSL): A *Digital Subscriber Line (DSL)* technology which provides a symmetrical 1 Mbit/s to 3 Mbit/s *data transmission rate* over the *local loop*. First proposed by AT&T Paradyne in 1996 and called the *Symmetrical Digital Subscriber Line (SDSL)*.

Medium Earth Orbit (MEO): A *satellite orbit* which is higher than a *Low Earth Orbit (LEO)* but lower than a *Geostationary Earth Orbit (GEO)*. These satellites are often in inclined (at about 50^{0}) elliptical or circular orbits with short orbital periods, of around 2 hours, and have an altitude of about 8000 kilometres at *apogee*. Continuous global coverage can be provided with about 12 such satellites in circular orbit and with fewer satellites in elliptical orbit.

Medium Frequency (MF): Part of the *electromagnetic spectrum* in the *frequency range* from 300 kHz to 3 MHz.

Medium Interface Connector (MIC): (1) The *connector* which is located at the interface point of the *bus* interface unit and a *terminal* in a *Local Area Network (LAN)* or a *Wide Area Network (WAN)*. This point is also known as the *bus interface point*. **(2)** The connector which interfaces a *fibre optic cable* to another fibre cable or to a *Fibre Distributed Data Interface (FDDI) node*.

medium interface point: See *medium interface connector*.

Medium Scale Integration (MSI): A semiconductor device which has about 100 circuits integrated onto a single silicon dice.

meet me conference: A *conference call* in which participants join the conference by dialling a perdetermined code.

megabit: A unit of size, equal to one million *binary digits* or *bits*.

megabyte: A unit of *data* size or storage size, equal to 1048576 *bytes*.

megahertz: A unit of *frequency* equal to one million *hertz*.

megaphone: A hand held device which is used to magnify *voice*. It is usually shaped like a cone, the voice being entered at the narrower end. Megaphones may be unpowered or powered, in which case they act as a sound *amplifier*.

MegaStream: High speed *transmission* system offered by BT in the UK which uses the *Plesiochronous Digital Hierarchy (PDH)* at 2.048 Mbit/s speeds. It uses the *E1 frame* structure.

Member of the European Parliament (MEP): The *European Parliament (EP)* has five hundred and seventy six MEPs who are directly elected by the citizens of Member States of the *European Community (EC)*. They belong to European political groupings and sit accordingly in Parliament.

memoryless channel: A concept used in *Information Theory*. A *channel* is said to be memoryless when the output from the channel is directly related to the input at that time, i.e. it is not connected with any event which happened in the past. See also *Markov process*.

memory unit: A device used for the storage of *signals*, *data*, etc., for processing and later usage.

menu: An organised selection of options presented to the *user*, either in a displayed form or as spoken text using a *voice processing* system. The user can then select items from this menu, using a variety of methods, such as a *keyboard*, a *mouse*, or a spoken word.

MEO: *Medium Earth Orbit.*

MEP: *Member of the European Parliament.*

mesh network: A *network* of interconnected *nodes* in which each node is interlinked by at least two *paths*. See *network topology*.

mesochronous signals; Two signals whose significant instances occur at the same average rate, i.e. they coincide.

message: Any sequence of *characters* which are used to convey *information* or *data*. The message may be in *plain text* or it could have gone through *encryption*. Generally the message will be in an agreed format and will consist of the *address*, the main body of the text, and a *trailer* to specify the end of the message. Special characters may be included for control functions, such as *error detection*, *synchronisation*, etc.

message buffering; Placing *messages* in a temporary storage area from which they can be sent at a later time, for example to suit the characteristics of the *transmission channel*, or to allow higher priority messages to be sent first.

message centre: A central location where *messages* can be received and then forwarded to the appropriate *terminals*.

message feedback system: Same as *information feedback system*, the *messages* being sent back to the originator for comparison, those which are in error being retransmitted.

message format: The rules or *protocol* used for the construction of a *message*, e.g. *header information*.

Message Handling System (MHS): A set of standards for use in *electronic mail*, where the *message* may consist of *data*, *text*, *speech* or graphics.

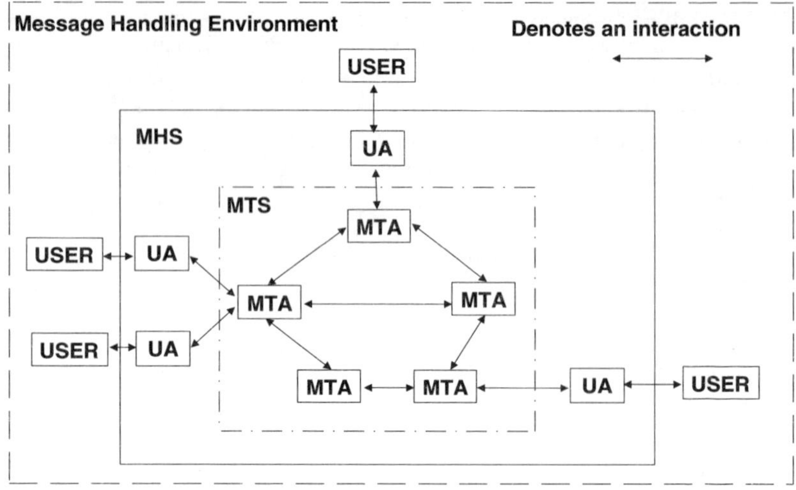

Figure M.8 Functional view of a MHS model

MHS is defined in *ITU-T Recommendation* X.400 (see *X Series*), the *ISO* equivalent being the *Message Oriented Text Interchange Standard (MOTIS)*. Figure M.8 gives a functional view of the MHS model. The key elements of this are the *Message Transfer Agent (MTA)* and the *User Agent (UA)*.

message header: The *header* which forms part of the *message*.

message lifetime: Message lifetime ensures that a *message* does not spend too long a time within a *network*. In this technique when a message enters a network a lifetime is set up for it and when this expires without the *sending terminal* receiving an *Acknowledgement (ACK)* this terminal resends the message via another route. See also *hop count*.

Message Oriented Text Interchange Standard (MOTIS): The *ISO* equivalent of the *ITU-T Message Handling System (MHS)*.

message processing time: The total time measured from the instant the *sending terminal* completes the *message* to when it is *displayed on the receiving terminal* for the *user* to read.

message routeing: The process of *routeing* a *message*, which includes the selection of the *path* or *channel*.

message service: Part of the *ITU-T* classification for *services*, such as in the *ISDN* standard defined in *ITU-T Recommendation* I.211 (see *I Series*). Message services are a subset of *interactive services*, the other subsets being *conversational services* and *retrieval services*. Message services are only concerned with the 'send' part of an action, the 'retrieval' being carried out by a separate service element, the retrieval service.

Examples of message services are logging of *voice* onto a *voice mail* system, sending *electronic mail*, and sending *facsimile* messages. For this service accuracy is important since a human operator is not available to compensate for lost *data*. *Delay time* is less important unless the system is waiting for an *Acknowledgement (ACK)*. Also known as *messaging service*.

message sink: The final *user* for which the *message* is intended. Also called *receiving terminal*.

message source: The originator of a *message*. Also called *sending terminal*.

Message Store (MS): Introduced in the 1988 issue of the *ITU-T Recommendation* X.400 (see *X Series*), a Message Store allows a *Message Transfer Agent (MTA)* to store a message it cannot deliver immediately to a *User Agent (UA)*. This could occur, for example, if the User Agent is on a *Personal Computer (PC)* which has been temporarily switched off.

message switched exchange: An *exchange* which forms part of a *message switched network*.

message switched network: A public *network* in which the individual switches take responsibility for handling the *message*, routeing it to its final destination based on the *address information* contained within its *header*. It is primarily used for *text* messages, such as in *telex*. See also *circuit switched network* and *Packet Switched Network (PSN)*.

message switching: A system in which *data* is transferred between the *message source* and *message sink* by a process of storage of the complete message within the *network* and then forwarding it to one or more destinations as they become free. Message switching can be a relatively slow and so it is primarily used for administrative types of messages. See also *circuit switching* and *packet switching*.

Message Telephone Service (MTS): USA term for a long distance *telephone* service.

message transfer: The transfer of a *message* from the *message source* to the *message sink* once the *transmission path* has been set up. See also *data transfer*.

Message Transfer Agent (MTA): Part of the *ITU-T Recommendation* X.400 for *electronic mail* systems. As illustrated in Figure M.8 the MTA and the *User Agent (UA)* are the *software* and *hardware* processes which enable *users* to send and receive *messages*, by moving these between locations.

Message Transfer Part (MTP): Part of the *ITU-T Recommendation* for *Signalling System No. 7 (SS7)*, it incorporates the lower *layer* functions.

Message Unit (MU): The unit used for *billing* of *local calls* in the USA. It is dependent on several parameters, such as duration of the call, time of day and the day of the week, and distance.

messaging service: See *message service.*

metallic circuit: A *circuit* which is completely made from metal (i.e. no *optical fibre* or atmosphere) and in which the *ground* or earth is not involved.

metallic voltage: A voltage applied between two metallic conductors and not between a metallic conductor and *ground.*

meteor burst communications: A communications system which uses the principle of *scattering* of *radio waves* from the ionised trails of meteors. Meteors that have usable trails have a diameter between 0.2 mm and 2 mm. The trails last for about half a second to several seconds and occur at a height of about 120 km. Bursts of high-speed *data* are used, lasting for about a second and with a duty cycle of about 5%. A *frequency range* of 30 MHz to 100 MHz will give communications over a distance of 200 km to 2000 km. Meteor burst systems are difficult to intercept, due to the limited angle of scattering, so they are used for military applications. Also known as *meteor scatter communications.*

meteor scatter communications: Same as *meteor burst communications.*

metering pulse: Pulses generated by an *exchange* at regular intervals which are used to determine the cost of a *call.* Metering pulses can also be sent by an exchange to a *PABX*, for example in a hotel or a multi-tenant environment where the PABX owner charges *users* for their calls.

METRAN: *Managed European Transmission Network.*

metric wave: An *electromagnetic wave* having a *frequency range* between 30 MHz to 300 MHz i.e. a *wavelength* between 2 metre and 10 metres.

Metropolitan Area Network (MAN): A communications *network* which covers a geographical area in-between that of a *Local Area Network (LAN)* and a *Wide Area Network (WAN).* Usually this is taken to include a metropolitan area, such as a large city and its suburbs. It may be used to interconnect two or more LANs and it operates at speeds faster than a WAN. *IEEE* standard 802.6 specifies the *Distributed Queue Dual Bus (DQDB)* as the *protocol* for a MAN.

MF: *Mediation Function, Medium Frequency* or *Multifrequency signalling.*

MFJ: *Modification of Final Judgement.*

MHS: *Message Handling System.*

MHz: *megahertz.*

MIB: *Management Information Base.*

MIC: *Medium Interface Connector.*

Michelson interferometer: An *interferometer* which uses two mirrors (or other reflecting surfaces), as shown in Figure M.9. As one mirror moves

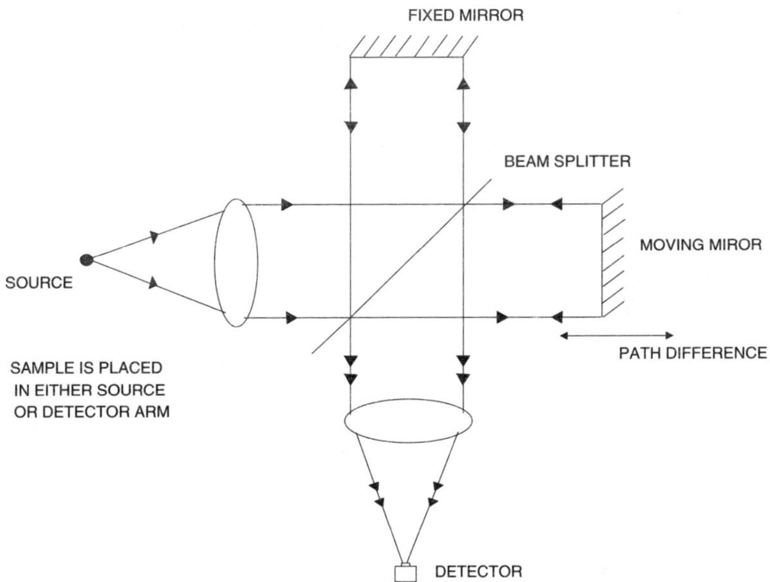

Figure M.9 Michelson interferometer

the detector records the *interference pattern* produced by the superposition of the two beams which are split and recombined by the beam splitter. The Michelson interferometer is used to measure the displacement of two fibres relative to each other where the ends of the fibres take the place of the two mirrors in Figure M.9.

microbend: Usually refers to very small bends in a *fibre optic cable* which are too small to be seen with the naked eye. These bends may be caused during the manufacturing, packaging, shipping or installation processes and result in a small amount of *signal loss* during *transmission.*

microbend loss: The *loss* of transmitted light in a *fibre optic cable* caused by imperfections or *microbends* in the fibre.

microcell: Very small *cells* used for *cellular radio systems*, the cell size often being as small as 0.5 km in diameter. Use of microcells increases the *capacity* available in the system.

Micro-Cellular Network (MCN): A *cellular radio system* which uses *microcells*, usually to provide a *Personal Communications Network (PCN)* type of service.

microcode: A group of computer instructions which are fixed in memory and perform detailed processing operations. See also *firmware.*

Microcom Networking Protocol (MNP): A de facto standard *data compression* and *error correction protocol*, developed by Microcom Corporation, which is widely used for *modems*. It offers *file* transfer and interactive applications at several levels or classes, each class adding further features.

microcomputer: Term sometimes used to refer to a *Personal Computer (PC)* or to a *microprocessor*.

micron: A measure of small dimensions, equal to on millionth of a meter. Often used to indicate the size of items such as a *fibre optic core*, a silicon dice, etc.

microphone: A *transducer* which converts *sound waves* into *analogue* electrical signals and vice versa. The sound is normally in the *audio frequency band*.

microprocessor: Dense electronic circuitry (*VLSI*) built onto a single silicon chip which has the processing power of a small computer, and consists of a Central Processor Unit (CPU), memory, input-output circuitry, etc.

microsecond: A unit of time, equal to one millionth of a second.

microwave: *Electromagnetic waves* in the *frequency band* from about 1 GHz to 100 GHz. Microwaves are widely used in *Line of Sight (LoS)* radio communications systems.

Microwave Amplification by Simulated Emission of Radiation (MASER): A device which generates *electromagnetic radiation* in the *microwave frequency range*. See also *LASER*.

microwave circulator: A device used in *microwave transmission* for coupling an *antenna* to a receiver and to a transmitter.

microwave loss formula: A formula used to predict the *loss* which occurs in a *microwave transmission* system, assuming a transmitter which radiates uniformly in all directions and a receiver which has no directivity. If f is the *frequency* of transmission in *gigahertz* and L is the length of the *transmission path* in miles, then the loss in *decibels* is given by $\text{Loss} = 96.6 + 20 \log_{10} f + 20 \log_{10} L$.

Microwave Pulse Generator (MPG): A device which generates electrical pulses at *microwave frequencies*.

microwave relay station: A *relay station* which is used to transmit *microwave signals* over long distances. The *antennas* are mounted on high towers, to avoid obstacles and the curvature of the earth, since *Line of Sight (LOS)* communications needs to occur.

mid-fibre meet: The property which allows equipment supplied by different vendors to interwork at the *optical fibre* level, i.e. to be able to freely transmit *signals* from one equipment to another. Mid-fibre meet is possible with *Synchronous Digital Hierarchy (SDH)* and *SONET* equipment which use *fibre optic cables* for *transmission*. Also referred to as *mid-span meet*, where also other *transmission medium* can be involved, such as *coaxial cable*.

mid-span meet: Same as *mid-fibre meet.*

MIFR: *Master International Frequency Register.*

MIME: *Multimedia Internet Mail Extension.*

minicomputer: A computer which is intermediate in size and processing power between a *microcomputer* and a *mainframe* computer.

Mini-Manufacturing Automation Protocol (MINI-MAP): A cut down version of the *Manufacturing Automation Protocol (MAP)* used for low cost applications. It consists of only the *Physical Layer*, *Data Link Layer* and the *Application Layer.*

MINI-MAP: *Mini-Manufacturing Automation Protocol.*

minimise: The procedure which causes the normal *traffic* passing through a *network* to be considerably reduced in order to give priority to additional traffic which is expected as a result of an emergency.

minimum bend radius: The maximum amount that a *transmission medium*, such as a *fibre optic cable*, can be bent, measured in terms of the mimimum radius at the bend, before it is permanently damaged. See also *critical radius.*

minimum signal level: The minimum value of a *signal* which can be received correctly by an equipment.

Mini Slotted Alternating Priorities (MSAP): A *multiple access protocol* which uses *reservations* to avoid *contention* and *collisions.* It is very similar in operation to the *Broadcast Recognition (BRAP)* protocol, as illustrated in Figure B.14.

Ministry of Posts and Telecommunications (MPT): The *regulatory body* within Japan who is responsible for the country's post, *telecommunications* and *broadcasting* services.

MIS: *Management Information System.*

mismatch loss: One of the elements of *transmission loss*, it is introduced at analogue interconnection points and is determined by the impedance at various points of a connection.

MIT: *Management Information Tree.*

MLHG: *Multiline Hunt Group.*

MLMA: *Multi-Level Multi-Access protocol.*

MLP: *Multilink Procedure.*

MLPP: *Multilevel Precedence and Preemption.*

MM: *Mobility Management.*

MMDS: Multichannel Multipoint Distribution Service.

MMI: *Man-Machine Interface.*

MMSS: *Maritime Mobile Satellite Service.*

mnemonic code: Computer instructions which are written in a form which is easy for a human to understand and recall. For example PRT to mean print, SAV for save, etc. These instructions need to be converted by the computer into a form (i.e. *machine code*) which it can understand.

MNP: *Microcom Networking Protocol* or *Mobile Number Portability.*

Mobile Application Part (MAP): The *signalling* interface defined in the *GSM* architecture. It is based on *ITU-T Signalling System No. 7 (SS7)* and it allows the mobile *user* to roam between *networks.*

mobile communications system: Systems which allow communications to take place between mobile *terminals* and between fixed and mobile terminals. Examples are *cellular radio systems*, *Personal Mobile Radio (PMR)*, *pagers*, etc.

mobile data: A generic term used to describe various systems which are used to transmit *data* rather than *voice* to and from mobile *terminals.* It can be based on several technologies, such as *cellular radio systems*, *Private Mobile Radio (PMR)*, etc.

mobile Earth station: An *Earth station* which is part of the *Mobile Satellite Service (MSS)* and can be used either whilst in motion or from different fixed locations.

mobilefax: *Facsimile transmission* which uses a *radio channel* so that documents can be transmitted and received whilst the facsimile unit is mobile. *Cellular radio systems* are usually used for this.

mobile location register: See *location register.*

mobile mounting kit: Usually refers to the equipment which is used to connect a cellular telephone in a vehicle. The telephone uses the same power supply and *antenna* as the car radio so that reception is improved. These kits also allow the *terminal* to operate as a *handsfree telephone*, for safety.

Mobile Number Portabililty (MNP): *Number Portability (NP)* applied to *mobile phones.*

mobile phone: Generic term used to describe any *telephone* which uses *radio waves* for communications. For example, telephones used in a *cellular radio system*, in *Private Mobile Radio (PMR)*, in *Cordless Telephony (CT)*, etc.

Mobile Satellite Service (MSS): The use of *satellites* to provide mobile services, such as *voice*, *data*, position location, etc.

Mobile Station ISDN number (MSISDN): One of the elements of a *numbering plan* for a *cellular radio system.* It is the *telephone number* of a *subscriber* and conforms to the conventional number format as specified in *ITU-T Recommendation* E.164. This can be mapped onto the *International Mobile Subscriber Identity (IMSI)* by the *Home Location Register (HLR).*

Mobile Switching Centre (MSC): *Exchange* used in a *cellular radio system*, which is equipped to connect to several *base stations*, as shown in Figure C.12. The MSC connects to other MSCs, and to the *Public Switched Telephone Network (PSTN)*, by fixed *lines*, and so provides the interconnection between the *wireline system* and the *wireless system.*

Mobile Telephone Switching Office (MTSO): Term used within *AMPS* to describe a *Mobile Switching Centre (MSC).*

Mobility Management (MM): Part of *layer* three of the *GSM signalling* model, it contains elements needed to support *user* mobility. This includes functions such as *authentication,* location tracking, registration, and management of the *International Mobile Subscriber Identity (IMSI).*

modal dispersion: The *dispersion* of an *electronic wave* (such as light) as it travels along a *transmission medium* (such as *optical fibre*) caused by the different paths taken through the medium by the various *modes.* This results in a rounding of the edges of a pulse waveform.

modal distribution: For a single *wavelength* in a *transmission medium* it is the number of *modes* supported and their propagation time differences. For multiple wavelengths, as in a *Wavelength Division Multiplexing (WDM)* system, it is the separation in the modes which are supported by the medium.

modal loss; For a *mode* of an *electromagnetic wave* propagating in a *transmission medium* it is the *loss* which occurs due to defects in the medium, such as discontinuities, obstructions or sharp bends.

mode: **(1)** For an *electromagnetic wave* in a *transmission medium* it is the *path* followed through the medium. **(2)** The field pattern permitted for an electromagnetic wave propagating within a *waveguide.* **(3)** An averaging technique, the mode of a number of items is the one which appears the most often. Therefore for numbers 1, 3, 3, 3, 5, 5, 6, 8, 9, 9 the mode is 3.

modem: Equipment which is used to send a *digital signal* over conventional *telephone lines.* Since these lines have been designed to carry *analogue signals,* at *Voice Frequencies (VF)* of 300 Hz to 3.4 kHz, the modem must convert digital signal into an *analogue signal,* send it down the telephone lines, and then convert it into a digital signal at the receiving end (see Figure M.10). Modem is a derived from a combination of *modulation* and *demodulation.*

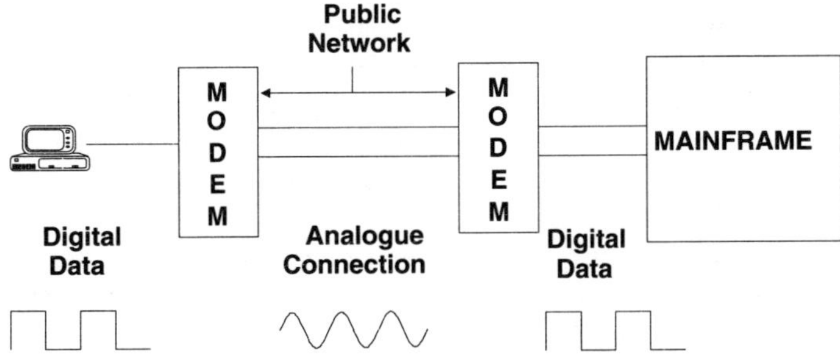

Figure M.10 Modem in an analogue network

modem data: Generally refers to the *voiceband date* which is transmitted and received by a *modem*.

modem eliminator: A device which replaces the two *modems* at the end of a *line* but still performs their function of allowing data interchange between two *Data Terminal Equipment (DTEs)*.

modem pool: A group, or pool, of *modems* which can be shared by many users, usually by dialling these from their *terminals*. This clearly saves costs by sharing a few modems amongst a relatively large number of users.

Modification of Final Judgement (MFJ): Ruling in 1982 by the District Court for the District of Columbia that the regional Bell system had to be split off from AT&T. The long distance services of the Bell system, the Bell Laboratories and Western Electric were to be retained by AT&T. Competition was subsequently introduced in the long distance *network*, but the regional network remained a monopoly. On 1 January 1984 the MFJ ruling became effective, resulting in the AT&T *divestiture*. See also *Telecommunications Act*.

Modified Alternate Mark Inversion (MAMI): An *Alternate Mark Inversion (AMI) line code* in which deliberate violations of the code are introduced, for example to maintain *synchronisation* of the *signal*.

modulate: To vary a parameter of a *carrier signal* in order to convey *information*.

modulated carrier wave: A *carrier signal* in which one or more parameters have been varied such that it contains *information*, which can be obtained by a process of *demodulation*.

modulating signal: The *signal* which is used to *modulate* the *carrier signal* and so impart its *information* to it.

modulation: The process used to *modulate* a *carrier signal* in order for it to carry *information*. There are several different methods used for this, such as *Amplitude Modulation (AM)*, *Frequency Modulation (FM)*, *phase modulation*, *angle modulation*, etc.

modulation depth: The amount or extent of *modulation*, especially applied to *Amplitude Modulation (AM)*.

modulation frequency: The *frequency* of the *modulating signal*.

modulation index: In *angle modulation* it is given by the ratio of the *frequency offset* of the *carrier signal*, following *modulation*, to the *frequency* of the modulating signal.

modulation rate: The rate at which changes occur in the *modulated carrier wave*, measured as the reciprocal of the time interval between two significant points on its *waveform*.

modulator: A device which can *modulate* a *carrier signal*.

modulo: The number of possible states, usually of a *counter*.

modulo-n: The number of events (n) which can occur before another event must takes place. For example, in a *transmission* system n *frames* or packets can be transmitted before an *Acknowledgement (ACK)* is required; or n frames or packets which can be counted in the *network* before the *counter* is reset to zero; etc.

monitor: (1) To observe or record events which occur without interfering with them. **(2)** Device used for recording and displaying events, usually on a *Visual Display Unit (VDU)*.

monitor station: A *station* on a *Local Area Network (LAN)* which *monitors* it and carries out supervisory functions, such as removing damaged *packets* from the LAN.

monochromatic radiation: *Electromagnetic radiation* which is made of a single *frequency* or a single *wavelength*. This cannot be effectively achieved in practice so the term is also applied to radiation which is within a very narrow *frequency band*.

monochrome: Refers to a device which has a single colour.

monochrome display: A display which has a single colour, i.e. black and white rather than colour.

monopole: A *dipole antenna* which essential contains half its element, as shown in Figure M.11, and fed against a *ground* plane. (See also Figure D.17(a)). The image of the monopole appears below the ground since this plane acts as a mirror. The *gain* of a monopole is twice that of a dipole and it has a *radiation pattern* above the ground plane which is the same as that of a dipole.

monomode fibre: The same as *singlemode fibre*.

Morse code: A *code*, developed by Samuel Morse for use in *telegraphy* in which *characters* are represented by a combination of dots (short duration pulses) and dashes (longer duration pulses), obtained by depressing a key and so making and breaking an electrical circuit.

Morse key: The device used to generate *Morse code*, usually consisting of a spring loaded key which is pressed down and released to make and break an electrical circuit.

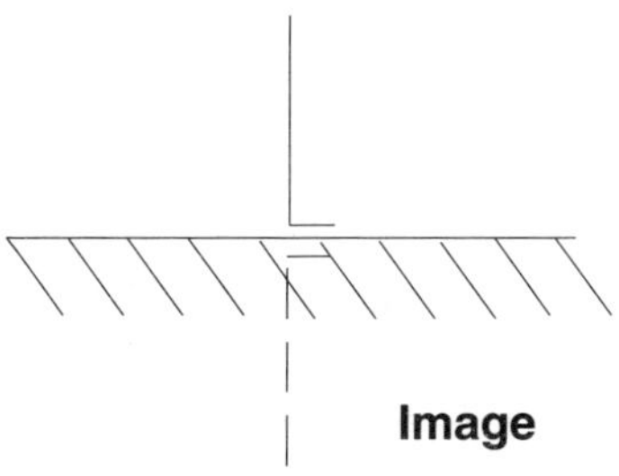

Figure M.11 A monopole

Morse telegraphy: *Telegraphy* which uses *Morse code* for communications.

Most Significant Bit (MSB): In a *binary number* it is the *bit* which represents the highest value. See also *Least Significant Bit (LSB).*

motherboard: In an electronic equipment it is the main printed circuit board to which other boards are attached.

Motion Adaptive Colour Plus (MACP): An improved *luminance/chrominance separation system used in enhanced television transmission* systems, such as *PALplus*, to receive a picture with reduced cross-luminance and cross-colour impairments.

Motion Picture Experts Group (MPEG): A specialist group, formerly a Working Group of *ISO/IEC/JTCI*, which began work in 1988 on standards for digital compression of *video* and *audio signals*. These standards are named after the group, i.e. *MPEG.*

MOTIS: *Message Oriented Text Interchange Standard.*

mouse: A hand-held device which is used to control the position of a *cursor* on a *Visual Display Unit (VDU).*

m out of n code: A *binary code* which has n *digit* words in which m digits are identical.

moving coil receiver: Receiver used in a *telephone handset*, as shown in Figure M.12. A diaphragm, made of plastic and formed into a cone, is attached to a coil wound on a former which is suspended within a magnetic field. The *voice* current from the line passes through this coil, which moves in the magnetic field and vibrates the diaphragm. This reproduces the voice *signals.*

MPDS: *Multipoint Distribution Service.*

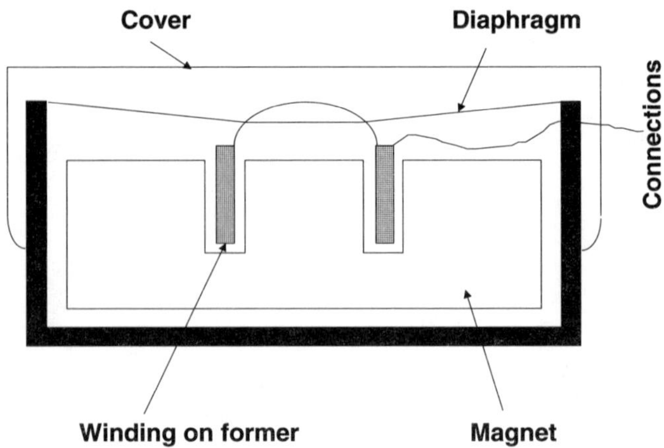

Figure M.12 Moving coil receiver

MPEG (1) A set of standards for *image compression*. MPEG-1 was developed in 1993, and adopted as *ISO/IEC* 11172, which defines a *video coding* standard for *digital* storage media, such as CD-ROM. MPEG-2 was published in 1994 and adopted as ISO/IEC 13818-1/2/3. It defines a standard for coding interlaced pictures and is suitable for high quality applications. It is the basis of the *Digital Video Broadcasting (DVB)* programme. MPEG-3 was not implemented and MPEG-4 is currently being developed. **(2)** *Motion Picture Experts Group.*

MPEG Syntactic Description Language (MSDL): Part of *MPEG*-4, it is a language to allow customised encoders to instruct receivers how to configure a matching decoder. It has many parallels with the *Java programming language.*

MPG: *Microwave Pulse Generator.*

MPOA: *Multi-Protocol Over ATM.*

MPT: *Ministry of Posts and Telecommunications.*

MPT1327 trunked radio: A *Trunked Mobile Radio* system which in UK uses the *frequency band* between 174 MHz and 225 MHz. Also referred to as Band III radio since this part of the spectrum was previous allocated to Band III radio systems. Figure M.13 shows a sample MPT1327 *address code word* structure.

MRNS: *Maritime Radionavigation Service.*

MS: *Message Store.*

MSAP: Mini Slotted Alternating Priorities.

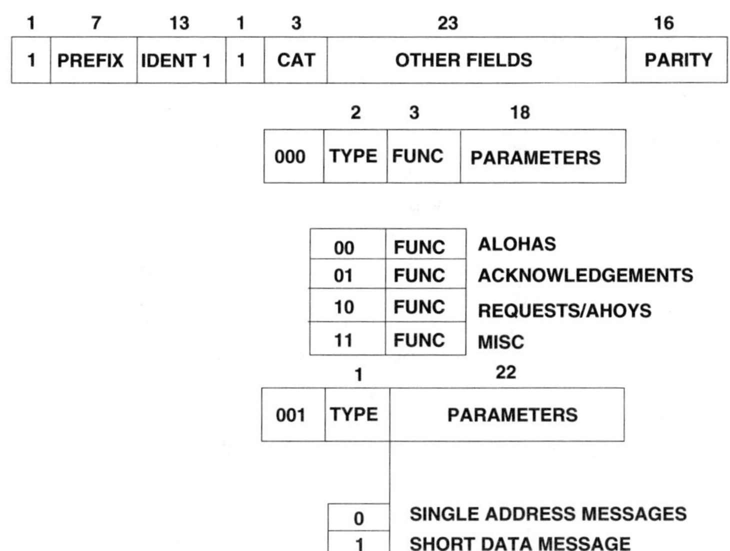

Figure M.13 MPT1327 trunking system address codeword structure

MSB: *Most Significant Bit.*
MSC: *Main Switching Centre* or *Mobile Switching Centre.*
MSI: *Medium Scale Integration.*
MSISDN: *Mobile Station ISDN number.*
MSO: *Multiple System Operator.*
MSS: *Mobile Satellite Service.*
MTA: *Message Transfer Agent.*
MTBF: *Mean Time Between Failures.*
MTBO: *Mean Time Between Outages.*
MTP: *Message Transfer Part.*
MTS: *Message Telephone Service.*
MTSO: *Mobile Telephone Switching Office.*
MTSR: *Mean Time to Service Restoration.*
MTTF: *Mean Time To Failure.*
MTTR: *Mean Time To Repair.*
MU: *Message Unit.*
MUF: *Maximum Usable Frequency.*
μ-law: *Algorithm* used in North America and Japan for *Pulse Code Modulation (PCM)* of *Pulse Amplitude Modulation* samples of *signals*. It is used in the conversion of *analogue signals*, such as *voice*, to *digital signals* prior to *transmission*.
MULDEM: *Multiplexer-Demultiplexer.*
multi-access: The facility which allows several *users* to be able to use the same resource on the *network* at the same time.
multi-address calling: A *network* feature which allows a *message* to be *addressed* and sent to a group of recipients at the same time, the message being delivered simultaneously or sequentially to each.
Multi-Carrier Modulation (MCM): A *transmission* method, which is a form of *Frequency Division Multiplexing (FDM)*, in which the *data* to be transmitted is formed by *interleaving* into several *bit streams* and these are used to *modulate* one or more *carrier signals*.
multicast: The *transmission* technique which allows a single *station* on a *network* to send *data* simultaneously to a group of other stations on the same network.
multicast address: The single *address* used by a *station* when it *multicasts* a *message*. Usually this address will be connected with a group of users who are related to each other, such as a *Closed User Group (CUG)*.
Multicast Backbone (MBONE): An addressing scheme used on the *Internet* which allows *multicasting* of *information* to all *users* who request it.
multicasting: The simultaneous *transmission* of *data* to a selected group of recipients on the *network*.
multicast packet: A *packet* which is *addressed* to and delivered to a group of *nodes* on the *network* at the same time.

multichannel: The use of a common *channel* by several *users*, using techniques such as *Frequency Division Multiplexing (FDM)* or *Time Division Multiplexing (TDM)*.

Multichannel Multipoint Distribution Service (MMDS): Also known as 'wireless cable' it has been used for many years for delivery of television to homes using *microwave* radio systems. The system can carry 33 *analogue video channels*, in the *frequency spectrum* between 2 GHz and 3 GHz, at a *bandwidth* of 500 MHz, and the number of channels is continually increasing. *Line of Sight (LoS)* operation is required, the *antenna* being located on the *subscriber's* roof and fed from the service centre, the *signals* then being taken into the house over *coaxial cable*.

multidrop: Generally refers to a system where several *nodes* share a common *transmission* facility.

multidrop line: A *channel* or *line* connecting several *nodes* where the *transmission* from each node is controlled centrally using *polling*.

multidrop network: A *network* which consists of a single *transmission path* so that transmission occur through intermediate nodes, control being exercised by end nodes.

multi-exchange call: A *call* which passes through more than one *exchange*, e.g. a *trunk call*. See also *local call*.

multiframe: A group of consecutive *frames* which form the *transmission* system, the position of each frame being determined by the *multiframe alignment signal*. Also known as *superframe*.

multiframe alignment signal: A *signal* used for *timing* in *multiframe transmission* systems.

Multifrequency (MF) dialling: The same as *Dual Tone Multifrequency signalling (DTMF)*.

Multifrequency (MF) signalling: The same as *Dual Tone Multifrequency signalling (DTMF)*.

Multifrequency (MF) tone signalling: The same as *Dual Tone Multifrequency signalling (DTMF)*.

Multi-Level Multi-Access protocol (MLMA): A *multiple access* technique which uses *reservation slots* to avoid *collisions*, as illustrated in Figure M.14 for 100 *users*. Suppose that users numbered 45, 49, 61 and 66 wish to transmit *data*. They will first *broadcast* their most significant *station* number in *frame* A. The system now knows that any number of 20 users in the 40s and 60s group wish to transmit. In frame C the 40s users are invited to place their reservations, indicating their second digit. Stations 45 and 49 so so and are therefore positively identified. Similarly in frame D users 61 and 66 identify themselves by their second digit. Each user keeps track of how many reservations have been made and the queue length.

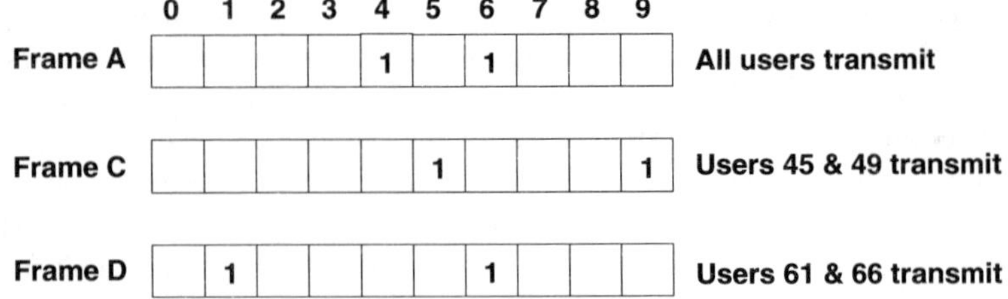

Figure M.14 Multi-Level Multi-Access protocol

Multi-Level Precedence and Preemption (MLPP): A priority allocation method in which *calls* are allocated one of several levels of precedence or priority. These are then handled in different ways by the *network*. If resources are limited then calls with higher priority can preempt calls of lower priority and take control of these resources.

Multi-line Hunt Group (MLHG): The capability within a *switching* system for connecting a *call* to another *number* within the group if the number called is *busy*.

Multi-Link Procedure (MLP): Part of the *ITU-T Recommendation* X.25 (see *X Series*), the MLP distributes and manages *packets* in a *multi-link transmission* system.

multi-link transmission: *Transmission* which makes use of more than one *link*. For example, in a *Packet Switched Network (PSN)* a *message* can be delivered by different *packets*, which form the same message, taking different *paths* or *links* through the *network*.

multimedia: The use of different type of sources, such as *voice*, *sound*, *text* and *image*, to convey *information*, as illustrated in Figure M.15. See also *Interactive Multimedia (IMM)*.

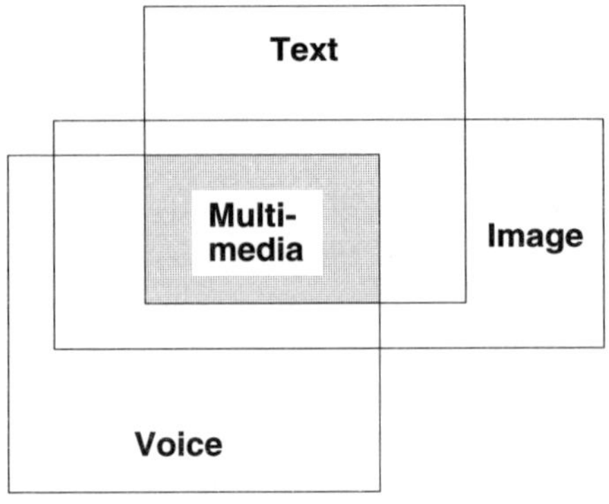

Figure M.15 Simplified illustration of multimedia services

multimedia call: A *call* which uses more than one form of media (i.e. it uses *multimedia*) for conveying *information.*

Multimedia Internet Mail Extension (MIME): Functionality provided for *messages* sent via *electronic mail (e-mail)* on the *Internet* which indicates to the *receiving terminal* which application to use in order to read the attachments to the message. Also know as Multipurpose Internet Mail Extension.

multimode distortion: The *distortion* which results from *multimode transmission*, caused by the differences in *group delay* of the different propagating modes.

multimode fibre: *Optical fibre* having a relatively large *core* diameter, compared to *singlemode fibre*, which allows *multimode transmission*

multimode transmission: *Transmission* of *electromagnetic waves* which have two or more modes, for example different *frequency* or *phase*.

multi-NAM: The ability for a *mobile phone* in a *cellular radio system* to have more than one *Numerical Assignment Module (NAM)*, i.e. two or more *telephone numbers* on different cellular systems, with the ability to choose between them, depending, for example, on *Quality of Service (QoS)* or geographical location.

multi-party connection: The facility which allows three or more *subscribers* to take part in the same *call*. See also *conference call.*

multipath effect: The effect produced in radio systems when the radio signal from the transmitter reaches the receiver via several different *paths*, e.g. due to direct transmission, reflection off nearby objects, diffraction through the atmosphere, etc. Figure M.16, for example, shows this effect for a *microwave* system where a ray refracted through the atmosphere reaches the receiver along with the direct ray from the transmitter. This causes *interference* between the signals from the various paths, resulting in *distortion* and variation in signal strength, as the

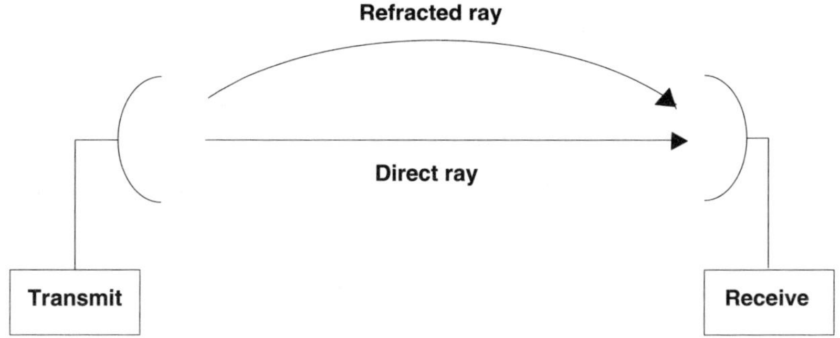

Figure M.16 Multipath effect

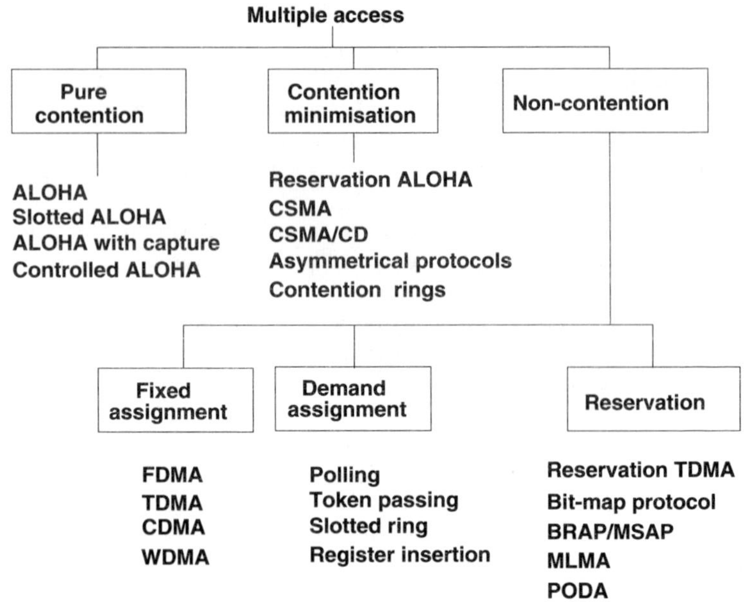

Figure M.17 Classification of multiple access techniques

signals either cancel or reinforce each other depending on the path taken. This is known as *multipath fading*.

multipath fading: See *multipath effect*.

multiple access: The ability of multiple *users* to access and share a common resource on a *network*. Many different techniques have been developed for this and they can be classified as in Figure M.17, depending on the level of *contention* allowed in the system.

multiple address message: A *message* which is to be sent to more than one destination.

multiple beam: The use of a single reflector and an array of feeds which allows an *antenna*, for example on a *satellite*, to produce multiple beams on the Earth, as shown in Figure M.18.

multiple homing: The connection of a terminal to more than one *local exchange*, usually using different *access lines* and a different *telephone number* for each exchange.

multiple media transmission: *Transmission* which uses more than one type of *transmission medium* to complete the *call*, for example, copper *twisted pair wire* for the *local loop*, *fibre optic cable* in the *trunk network* for connection to *Earth stations*, and *satellite links* for transmission to remote locations.

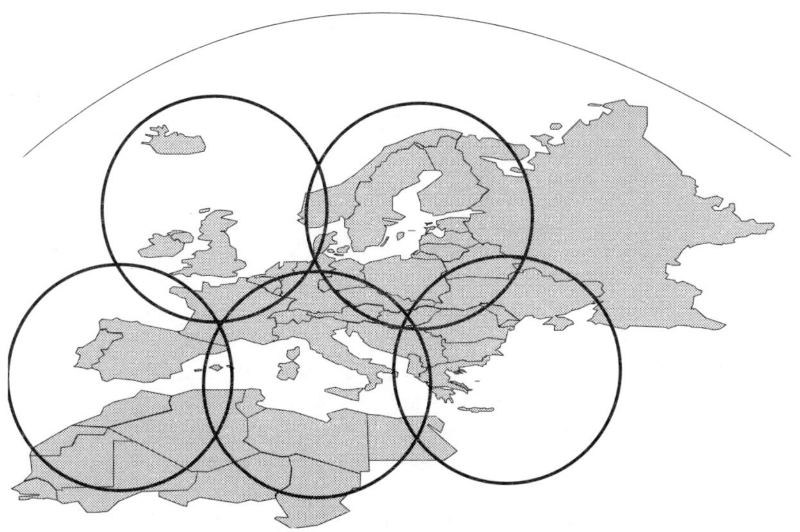

Figure M.18 Multiple spot beams generated from a satellite

multiple party conference: Same as *multi-party connection.*
multiple response code: See *correlative code.*
multiple routeing: The process of *routeing* a *message* to more than one destination, these being specified in the *message header*.
multiple spot scanning: The *scanning* process in a *facsimile* equipment where two or more *scanning spots* are used, each covering its part of the document, the total picture being built up by combining their outputs.
Multiple Sub-Nyquist Sampling Encoding (MUSE): A family of *bandwidth* reduction and *signal compression* techniques primarily used for the distribution of *High Definition Television (HDTV)* via *satellite*, such as in *Direct Broadcast Satellite (DBS)* systems.
Multiple System Operator (MSO): Term used to describe a company which has bought several smaller independent operators and has therefore also acquired their operating licences.
Multiplexed Analogue Component (MAC): A family of standards used for *analogue video transmission* in applications such as *Direct Broadcasting by Satellite (DBS)* and within studios.
multiplex: To transmit two or more *signals* over the same *transmission channel*.
multiplex aggregate bit rate: The *bit rate* in a *multiplex* system, which is equal to the sum of the bit rates of the individual *signals* being multiplexed plus *overhead bits* required for the process.

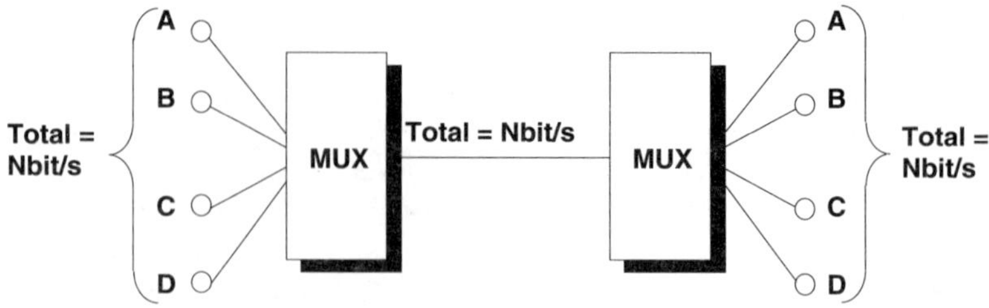

Figure M.19 Basic model of a multiplexer

multiplexed channel: A *channel* which is capable of carrying the *transmission* from two or more *users* at the same time.

multiplexer (MUX): Equipment which receives the *signals* from several low speed *channels* and combines them for *transmission* over a common high speed channel, as in Figure M.19.

multiplexer-demultiplexer (MULDEM): Equipment which can perform both *multiplexing* and *demultiplexing*.

multiplexer mountain: A term used to refer to the problem, common in systems using the *Plesiochronous Digital Hierarchy (PDH)*, in which a *signal* which is being transmitted at a high *data rate* must go through several levels of *multiplexer-demultiplexer (MULDEM)* in order to obtain a lower data rate signal. This is illustrated in Figure M.20 where a 140 Mbit/s signal must pass through multiplexers at 34–140, 8–34, and 2–8 in order to arrive at a 2 Mbit/s signal for a user. Following this the whole process must be repeated to arrive back at the 140 Mbit/s signal for onward transmission. See also *drop and insert*.

multiplex hierarchy: Generally refers to the hierarchy of *channels* used for combining lower *bit rate signals* into higher bit rates. For *Frequency*

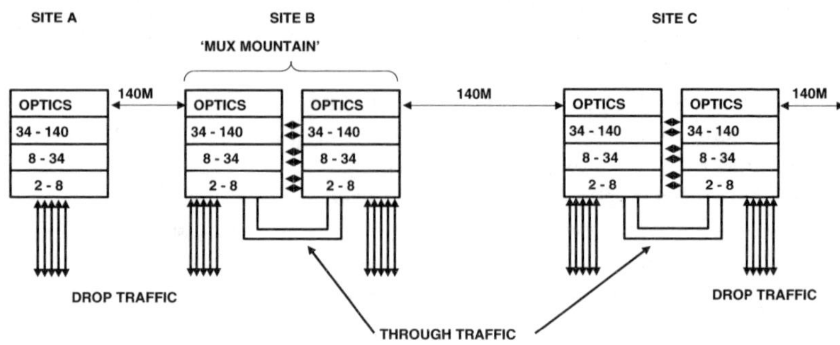

Figure M.20 Illustration of a multiplexer mountain

Division Multiplexing (FDM) this consists of 12 channels forming a group, 5 groups a *supergroup* and 5 or 10 supergroups a *mastergroup.*

multiplexing: The process of combining together several *signals* from lower speed *channels* into a higher speed channel. Several techniques exist for this, such as *Frequency Division Multiplexing (FDM), Time Division Multiplexing (TDM)* and *Wavelength Division Multiplexing (WDM).*

multipoint circuit: A *circuit* connecting three or more *nodes.* See also *multidrop.*

Multipoint Control Unit (MCU): A control unit, normally embedded in the *network,* which allows communications between more than two simultaneous *users.* Examples are units used to control audio conferencing and *videoconferencing.*

Multipoint Distribution Service (MPDS): A *microwave* radio service, in the USA, which allows one way communications between a fixed transmitting *station* and several fixed receiving stations, spread over different geographical locations.

multipoint line: Same as *multidrop line.*

Multipoint Video Distribution Service (MVDS): A *Wireless Local Loop (WLL)* technology developed in the UK. It operates in the 40.5 GHz to 42.5 GHz *frequency band* and is very similar in operation to the *Local Multipoint Distribution Service (LMDS).*

multiport repeater: A *repeater* with several inputs and outputs, the *signal* at any input appearing on every output. Commonly used to connect several *Ethernet* segments together.

multiprocessing: The simultaneous processing of several tasks by several Central Processor Units (CPU) in equipment such as a computer. This is different from *multitasking.*

Multi-Protocol Over ATM (MPOA): A *Local Area Network (LAN)* emulation system developed by the *ATM Forum.* It provides switched *routeing* functionality within an ATM environment and gives the *network layer* three intelligence.

multi-satellite link: A radio *link* connecting two *Earth stations* and involving two or more *satellites,* with links directly between these satellites, if required.

multi-server network: A *Local Area Network (LAN)* which has two or more *file servers* connected to it. *Users* on the network can obtain *information* from any of the servers to which they are connected.

multi-state signalling: *Signalling* in which more than two states are used, represented, for example, by differences in *signal amplitude.* See also *two-state signalling.*

multitasking: The ability for a computer system to manage several tasks at the same time, i.e. have several applications open simultaneously.

Only one Central Processor Unit (CPU) is used so processing can only be carried out one at a time. See also *multiprocessing*.

Murray code: *Code* used for *telex transmission* in which all *characters* are made from five units, each taking the same time to transmit. It is the basis of the *International Alphabet No. 2* produced by the *ITU-T*.

MUSE: *Multiple Sub-Nyquist Sampling Encoding*.

mutually synchronised network: A *network* in which each *clock* in the system exerts a level of influence on the *timing* or *synchronisation* of the *transmission* and reception of *data*. Also known as a *democratic network* as opposed to a *despotic network*. **MUX:** *Multiplexer*.

MVDS: *Multipoint Video Distribution Service*.

N

NA: *Numerical Aperture.*

NADF: *North American Directory Forum.*

nadir: The point which is diametrically opposite to the *zenith* and constitutes the lowest point of all.

nailed-up circuit: A *circuit* which is set up through a *switching* system to provide a semipermanent connection between two points.

NAK: *Negative Acknowledgement.*

NAM: *Numerical Assignment Module.*

name management: An aspect of *configuration management*, it allows the *user* symbolically to name and refer to resources on the *network*. Several techniques exist for this, the most popular being *white pages* and *yellow pages*.

NAMPS: *Narrowband Advanced Mobile Phone System.*

nanometre: One billionth of a metre.

nanosecond: One billionth of a second.

NANP: *North American Numbering Plan.*

narrowband: A loose term, usually used to mean 'not *broadband*'. Can be taken to imply sub-*voice frequencies*.

Narrowband Advanced Mobile Phone System (NAMPS): An enhancement to the *Advanced Mobile Phone System (AMPS)*, proposed by Motorolla Inc., which provides analogue *voice* processing with digital *signalling*, so that it can bridge to digital *cellular radio systems*.

Narrowband Integrated Services Digital Network (N-ISDN): *Integrated Services Digital Network (ISDN)* which provides two separate 64 kbit/s *channels* and one 16 kbit/s channel, making a total *capacity* of 144 kbit/s. Also referred to as *2B+D*.

narrowband modem: A *modem* whose output is within the *capacity* of a *narrowband channel*, i.e. about 4 kHz.

narrowband signal: A *signal* whose *bandwidth* is below that of a *voice channel*, i.e. 4 kHz.

n-ary code: A *code* which has n states, the value of n being greater than 1.

NATA: *North American Telecommunications Association.*

National Bureau of Standards (NBS): See *National Institute of Standards and Technology (NIST)*.

National Destination Code (NDC): Part of the *ISDN numbering* scheme, the National Destination Code, together with the subscriber's number, form the National ISDN number, as shown in Figure I.16. NDC is therefore equivalent to an *area code* in a conventional *telephone number*.

national frequency allocation: The allocation of *frequency bands* by national governments for use by services within their national boundaries. Clearly this needs to be coordinated with the international frequency allocations, and with those of neighbouring countries, in order to prevent *interference*.

National Institute of Standards and Technology (NIST): Formerly known as the *National Bureau of Standards (NBS)*, in 1968 it was given the responsibility for helping the US Federal Government make effective use of its vast base of computer and *Information Technology (IT)* equipment, and for developing *Federal Information Processing Standards (FIPS)*. Many FIPSs have been developed and published. The best known example is that on *data encryption* (FIPS 46), which has been adopted by the *American National Standards Institute (ANSI)* as X3.92.

national numbering: The *telephone numbering* scheme used within a country. Since the *ITU-T* has specified that the maximum number of *digits* for an *international number* is not to exceed 12, the maximum number of digits which a national number can have is 12 minus the number of digits in its *country code*. A national number consists of three parts: the *area code*, the *exchange code* and the *subscriber's local exchange* number. An additional prefix is also used for *trunk calls* and *international calls*. For example, in the UK the international prefix is 00. The first 0 routes the call to the *trunk exchange* and the second to the international *gateway switching centre*.

National Science Foundation Network (NSFNet): The civilian equivalent of the *ARPANET* system, which was created in the USA in 1980. It is primarily used by the academic community and operates over the *Internet Protocol (IP)*.

National Telecommunications and Information Administration (NTIA): A division of the US Department of Commerce which has the responsibility for advising the US government on *telecommunications* and related matters, including telecommunications standards.

national telecommunications network: The *Public Switched Telephone Network (PSTN)* which provides services within a country, from the *local loop* to international *lines*. Figure C.16, for example, shows the UK national *telecommunications network* which consists of a hierarchy of interconnected networks.

National Television Standards Code (NTSC): See *National Television Standards Committee*.

National Television Standards Committee (NTSC): The committee set up to develop the standard for the North American television *broadcast* system, the standard being named after the committee, i.e. the *National Television Standards Code (NTSC)*. It operates at 60 Hz and uses 525 picture lines. See also *PAL* and *SECAM*.

NBH: *Network Busy Hour.*

NBS: *National Bureau of Standards.*

NCC: *Network Control Centre.*

NCTE: *Network Channel Terminating Equipment.*

NDC: *National Destination Code.*

NDSF: *Non-Dispersion Shifted Fibre.*

NE: *Network Element.*

near-end crosstalk (NEXT): *Crosstalk* (i.e. the unwanted transfer of *signals* from one *link* to another) which occurs in the *transmission medium* at a point close to the transmitting equipment. See also *far-end crosstalk.*

near-end echo: The *echo* which occurs when a *signal* on a *transmission line* passes from a *two wire circuit* (the *local line* from the *subscriber*) to the *four wire circuit* (the *trunk line* leading to the *exchange*). See also *far-end echo.*

near field sound: Term used in the design of a *noise cancelling microphone*. Near field sound is caused by *sound waves* originating from a source close to the microphone. See also *far field sound.*

Near Video On Demand (NVOD): *Service* which allows a *subscriber* to watch public *video* programmes at almost any time. See *Video On Demand (VOD).*

NEF: *Network Element Function.*

Negative Acknowledgement (NAK): In *data transmission* this *message* is sent by the *receiving terminal* to the *transmitting terminal* to indicate that the previous message contained an error, and requesting a *retransmission.*

negative bit stuffing: See *negative justification.*

negative justification: The opposite process to *justification* or *bit stuffing*, i.e. the deletion of *bits* from a *frame* in order to reduce its *bit rate*. Also called *negative bit stuffing.*

negative temperature coefficient: A parameter which decreases as the temperature is increased.

n-entity: A concept used in the *OSI Basic Reference Model* (see Figure L.2) in which the n-entity in any *layer* is the active element within the *n-layer* in a real system. It interacts with the entities of the layers above and below and performs a set of defined functions. n can refer to any one of the *seven layers* of the model.

NEP: *Noise Equivalent Power.*

neper: A measure of the *gain, loss* or comparison of relative values between two *signals*. It is equal to $Np = \log_e V_1/V_2$ where V_1 and $_2$ are magnitudes of the two signals being compared.

NET: *Norme Europeene de Telecommunications.*

Netscape: The name of a popular browser for use with the *World Wide Web (WWW)*, which is supplied by Netscape Communications Corp.

network: A collection of *nodes* which are connected together by means of *transmission paths* and which can include a variety of functional units, such as *switching centres*, *repeaters*, etc. Different *transmission medium* can be used, such as *optical fibre*, *microwave* radio, *satellite links*, etc. and it can have several *network topologies*. Many forms of *networks* exist, such as the *Local Area Network (LAN)*, *Wide Area Network (WAN)*, *private network*, etc.

network architecture: The overall plan of the *network* indicating the position of the elements, the means of communications between them, the procedures to be followed and formats to be used, etc.

network attachment: The connection of equipment to a *Public Switched Telephone Network (PSTN)*. Only approved equipment can be so attached.

network backbone: See *backbone*.

Network Busy Hour (NBH): The *busy hour* calculated for the whole *network*.

Network Channel Terminating Equipment (NCTE): *Customer Premises Equipment (CPE)* which is used to convert *signals* from the *subscriber's* equipment to a form required by the *Public Switched Telephone Network (PSTN)*. This includes functions such as *line conditioning* and *equalisation* and the performance of *loopback tests*.

Network Control Centre (NCC): The same as *Network Management Centre*.

network control channel: The *channel* used to pass *information* relating to the *network* to the *Network Management Centre (NMC)* and to pass control commands to the nodes from this Centre. The channel does not carry any user related information.

network control phase: The periods during which network control *signals* are being exchanged between the *network* and a *Date Terminal Equipment (DTE)*, such as would occur during *call establishment*, *information* transfer, and *call disestablishment*.

Network Element (NE): Generic term used to include any active element on the *network*, such as a *telephone*, *switching equipment*, *transmission* equipment, etc.

Network Element Function (NEF): Part of the *Telecommunications Management Network (TMN)* it provides the interface between the TMN and the *Network Element (NE)*. See Figure T.9.

network facilities: The services available to a *user* of the *network*. See also *facility request*.

network hierarchy: (1) The levels of *transmission* systems and *exchanges* which make up a *network*. For the example of the *national telecommuni-*

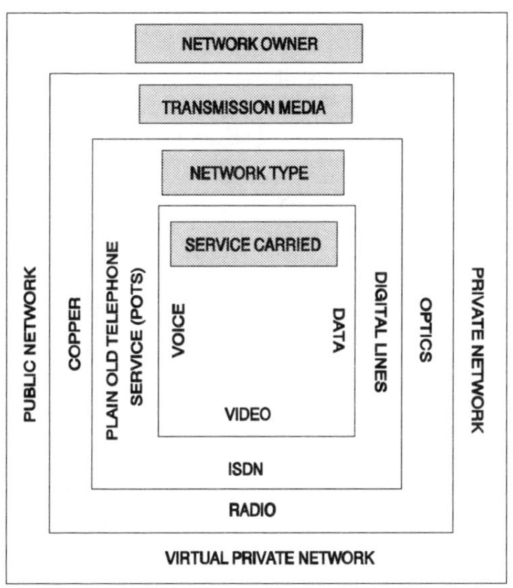

Figure N.1 The hierarchical structure of a network

cations network, shown in Figure N.1, this consists of the *local network*, the *trunk network* and the international network. **(2)** The combinations which exist within a network, from a *user* perspective, as shown in Figure N.1. These are: the services carried by the network, such as *voice*, *data* and *video*; the type of network, such as *POTS*, *ISDN* or digital lines; the *transmission medium* used, such as copper, *fibre optics* or radio; and the network owner, such as whether it is a *private network*, a public network, such as the *PSTN*, or a *Virtual Private Network (VPN)*.

network interface: The physical point at which equipment connects to a *network*. Usually this interface point refers to that connecting *Customer Premises Equipment (CPE)* to a public or *private network*.

Network Interface Unit (NIU): The unit which forms the *network interface* and often performs ancillary functions, such as *protocol conversion*, *data buffering*, protection, etc.

Network Inward Dialling (NID): The same as *Direct Inward Dialling (DID)*. See also *Network Outward Dialling (NOD)*.

Network Layer: The third *layer* of the *OSI Basic Reference Model*. Its prime aim is to provide a transparent path for the *transmission* of *data* between the *Transport Layers* within the communicating systems. It handles *addresses* and *routeing* of *circuits*. Both *connection mode transmission* and *connectionless mode transmission* have been used in the Network Layer. The Network Service Definition is provided by *ISO*

8348 and a number of standards are available for the various *protocols* and configurations that can be used.

network management: The overall control and management of a *network* to ensure its efficient operation. Several *network management functions* are performed.

Network Management Centre (NMC): The central location where overall management of the *network* occurs. This centre carries out all the *network management functions*.

Network Management Forum (NM Forum): Established in 1988 by a group of eight equipment suppliers and *PTOs* it now has about 160 members worldwide. Its aim is to accelerate the implementation of *network management* standards and to demonstrate their use. The Forum adopts international standards and draws up its own *functional standards* or *profiles* where these do not exist. In June 1990 the Forum published its Release 1 specification, which has subsequently been updated. It also publishes technical reports, which are guidelines on Forum strategies and activities. It operates Showcase programmes, which are usually shows within international trade exhibitions, where members can demonstrate products and interoperability.

network management functions: The functions required from a *network manager* have been classified by *ISO* into five groups: *fault management*, *configuration management*, *performance management*, *security management* and *accounting management*, as shown in Figure N.2. These can be considered to occur at several *network management levels* such as element, network and service.

network management hierarchy: The structuring of *network managers* within an overall *network*. This can be considered in several ways. For example the hierarchy could be on a geographical basis, e.g. local

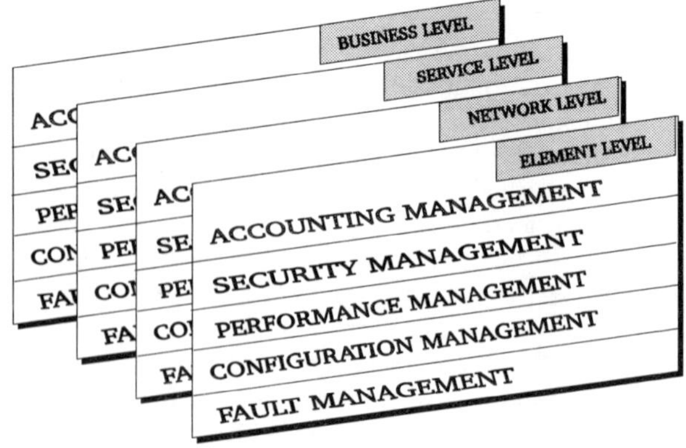

Figure N.2 Network management functional structure

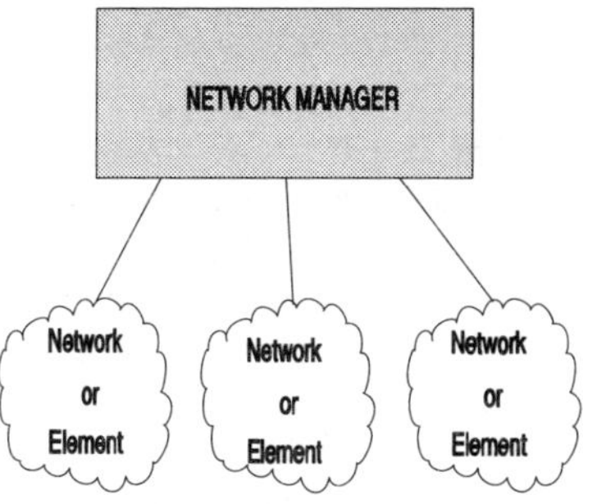

Figure N.3 Centralised network management structure

managers, metropolitan managers, national managers and international managers, the span of control and complexity of equipment managed increasing from the local to the international level. Figures N.3, N.4 and N.5 show alternative hierarchical structuring of network management. In the centralised management structure of Figure N.3 all *network management functions* are carried out at one central location. Although simple and low cost, and suitable for small networks, spread over a limited geographical area, this solution becomes difficult to operate where *information* has to be transported over long distances. The tree structure of Figure N.4 provides a distributed management hierarchy in which the level 1 or element level manager carries out all the local servicing of alarms and disaster recovery, feeding back only major alerts to the higher level (level 2) manager, and so on up the chain. The *data processing system* has been moved closer to the *Network Element (NE)* that generates the data, which enables faster response to faults and avoids

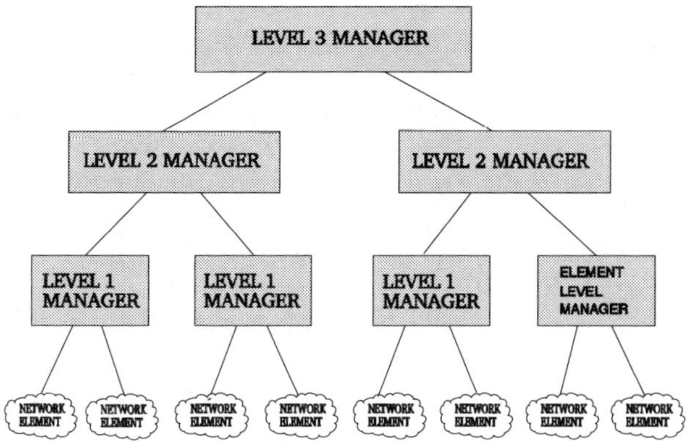

Figure N.4 Tree hierarchy of network managers

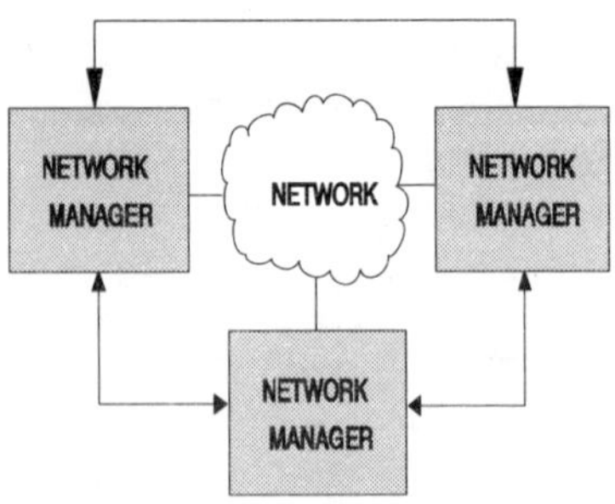

Figure N.5 Network manager peer-to-peer structure

the problem of a single failure of the central manager bringing down the whole network. The peer-to-peer structure, shown in Figure N.5, divides the network into segments, which can then be controlled by separate but interlinked managers. This structure is often used when resilience is important. In these circumstances each manager has total jurisdiction over its own segment of the network, but maintains a monitoring brief over other segments. If one of the managers were to fail then its neighbour would step in and assume temporary responsibility over this additional part of the network.

network management levels: Four levels, or layers, of *network management* are generally considered to exist: element, network, service and business, as indicated in Figure N.3. The element level is primarily concerned with day-to-day management of the network. It deals with faults as they occur and makes planned changes to the configurations. The network level maintains a network view of the system and carries out network wide tasks, such as setting up a service path between two elements on the network. It is concerned with items such as *traffic* congestion and *throughput*. The service level manages the *services* which are carried over the network, for example an X.25 service (see *X Series*), and it deals with faults which result in end customer complaints. At the business level the plans for the entire network are formulated, after consideration of the present and future aims of the organisation. It takes a longer term view in considering items such as *capacity* and growth, and costs in terms of inventory, operations and damage to the company and its reputation caused by downtime or congestion. Questions like the level of *network resilience* needed also have to be answered.

network management point: Same as *Network Management Centre (NMC).*

network management signals: The *signals*, passing in and out of the *Network Management Centre (NMC)*, which are primarily concerned with *network management* rather than the management of other functions, such as *call management* or *traffic* management.

network manager: (1) The *hardware* and *software*, usually one or more powerful computers, which carries out the *network management functions*. **(2)** The person responsible for efficient operation of the network.

network node: See *node*.

Network-Node Interface (NNI): The interface between the *network* and a *node* on the network, i.e. it is internal to the network and is not presented to the *user* of the network. See also *User-Node Interface (UNI)*.

Network Operating System (NOS): *Software* which provides *users* with access to the resources on a *Local Area Network (LAN)*, such as printers and files. It also manages the problems of security on the network and the use of these resources by several users.

network operator: Same as *network manager*.

Network Outward Dialling (NOD): Same as *Direct Outward Dialling (DOD)*. See also *Network Inward Dialling (NID)*.

network resilience: The ability of a *network* to continue operation in spite of failures of some of its elements. For example, network resilience can be obtained by the use of *alternate path routeing*, in which *traffic* is sent by an alternate route if the main route has failed. *Redundancy* can also be built into the network, where certain equipment are on *standby* and only come into operation to take over the role of a failed component.

network security: The actions taken to protect a *network* against unauthorised access, which includes access to gain unauthorised *information* or to cause damage to the network or interfere with the *traffic* carried.

Network Service Access Point (NSAP): See *Service Access Point (SAP)*.

Network Terminal Number (NTN): Part of an international number for a *Public Data Network (PDN)* as specified in *ITU-T Recommendation* X.121. This number consists of a *Data Network Identification Code (DNIC)*, of four *digits*, and a ten digit Network Terminal Number. The NTN is used for in-country *routeing* and can have any suitable format although ITU-T has recommended that the last two digits is a sub-address for use by the *subscriber*.

Network Terminating Interface (NTI): The point which marks where the responsibility for maintenance and service of the *network* by the service provider begins and ends.

Network Terminating Unit (NTU): A device which connects *terminals* to a *network*. It may carry out functions such as *protocol conversion*, and may have facilities to indicate activities such as the progression of *calls*.

Network Termination Type 1 (NT1): The interface between an *Integrated Services Digital Network (ISDN)* and the *terminal* equipment on the *subscriber's* premises. The NT1 functional grouping includes functions equivalent to *layer* 1 of the *OSI Basic Reference Model* and are associated with the termination of the network. It interfaces at the *T reference point*. (See Figure U.5.)

Network Termination Type 2 (NT2): In the *Integrated Services Digital Network (ISDN)* it is the functional group which includes functions equivalent to *layer* 1 and higher layers of the *OSI Basic Reference Model*. A *PABX*, *Local Area Network (LAN)* and *terminal* controller are all examples of equipment, or combinations of equipment, which provide NT2 functions. The NT2 is between the *S reference point* and the *T reference point* (see Figure U.5).

network topology: The physical and logical arrangement of *nodes* on a network. Several such arrangements are possible, some of these being illustrated in Figure N.6. The *point-to-point network* has a direct *link* connecting two nodes. In the *star network* the nodes radiate from a central control point. This is usually seen in configurations such as lines terminating within a central controller or switch. It is well suited to carry out a *polling* mechanism from the central location to outlying nodes.

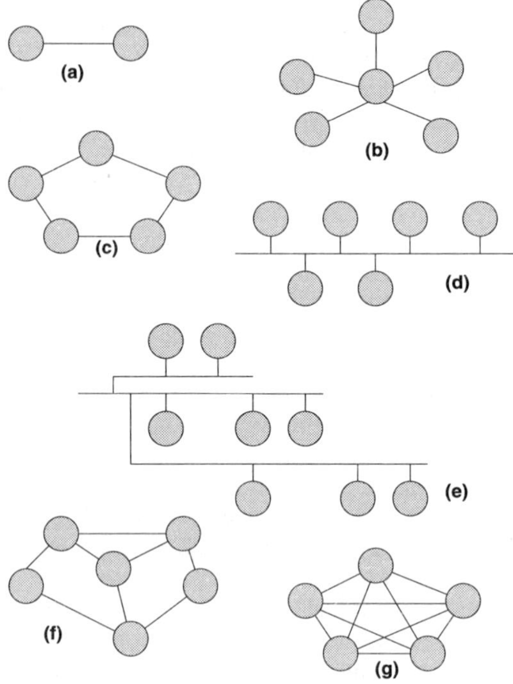

Figure N.6 Network topologies: (a) point-to-point; (b) star; (c) ring or loop; (d) bus; (e) tree; (f) mesh; (g) fully interconnected

Ring networks or *loop networks* are often preferred since they are not dependent on a single control site, and can generate *traffic* in either direction around the ring, so providing alternate *transmission paths* and therefore *network resilience*. The *bus topology* provides simultaneous access to a central *bus* by several nodes and this is the most commonly used structure within a *Local Area Network (LAN)*. *Tree networks* can be considered to be generalised forms of bus structures and have similar characteristics. In a *mesh network* the nodes are connected together but a link does not exist from every node to every other node, as is the case in a *fully interconnected network*.

Network User Identity (NUI): A combination of *user* name and password which is given to *subscribers* of a *Value Added Network Service (VANS)* in order to allow them to access the service.

neural network: A theoretical *network* which is programmed and conditioned to learn from experience so that its performance can approximate to that of the real network which it is simulating.

neutral telegraph system: A method of transmitting *signals* in *telegraphy* in which a *mark* is represented by current flow and a *space* by the absence of current.

New Licensed Operator (NLO): A *Public Telecommunications Operator (PTO)* who is licensed by an appropriate authority to provide telecommunications services, usually in competition with the incumbent *PTT*.

NEXT: *Near-end crosstalk.*

NF: *Noise Figure.*

n-function: The function performed by an *n-entity* in the *OSI Basic Reference Model.*

NID: *Network Inward Dialling.*

Nippon Denshin Denwa Kabushiki Kaish Law: One of the two laws, introduced on 1 April 1985, which started the *deregulation* and *liberalisation* process in Japan. The other law was the *Telecommunications Business Law*. These ended the domestic and international monopoly of NTT and KDD and structured NTT as a private corporation.

N-ISDN: *Narrowband Integrated Services Digital Network.*

NIST: *National Institute of Standards and Technology.*

NIU: *Network Interface Unit.*

b-key lockout: A feature of *keyboards* which prevents an output *signal* when two or more keys are depressed simultaneously. See also *n-key rollover*.

n-key rollover: A feature of *keyboards* which enables an operator to depress keys in rapid sequence. See also *N-key lockout*.

n-layer: The nth *layer* in the *OSI Basic Reference Model*, where n can refer to any one of the *seven layers*. See Figure L.2.

NLO: *New Licensed Operator.*

NMC: *Network Management Centre.*
NM Forum: *Network Management Forum.*
NMT: *Nordic Mobile Telephone system.*
NNI: *Network-Node Interface.*
no-break power supply: A power supply which can continue to provide power without any interruption even if the input supply to it fails. Also known as an *Uninterruptible Power Supply (UPS).*
NOD: *Network Outward Dialling.*
node: A point of interconnection to a *network* which may contain units or *stations* which can carry out a variety of functions. It may be a simple *terminal* or a complex *switching centre*. The node may also be polled.
node address: The *address* of a *node* on the *network*. Every node must have at least one unique node address. Also known as a node number.
node number: See *node address.*
node-to-network protocol: A *protocol* which controls the *transmission* of *information* between adjacent *nodes* in a *network*. An *end-to-end protocol*, on the other hand, controls the flow from sender to receiver, through several intermediate nodes.
noise: Unwanted random *signals* which cause *interference* with the signal being transmitted, resulting in degradation and *errors*. The noise signals may be electrical, electromagnetic or acoustic. See also *acoustic noise*, *background noise*, *impulse noise*, *intermodulation noise*, *Johnson noise*, *partition noise*, *shot noise*, *thermal noise*, and *white noise*.
noise bandwidth: See *effective noise bandwidth.*
noise cancelling microphone: A *microphone* which has been designed to compensate for the presence of acoustic noise. It has two sound ports which give access to the front and back of the microphone diaphragm. It reacts to the difference in sound pressure on the diaphragm, this being made up of the *near field sound* and the *far field sound*. The noise cancellation of the microphone is given by the difference between the near field and far field measurements.
noise current: The current generated in a *circuit* by the presence of a *noise* voltage.
noise diode: A semiconductor diode which can be used as a *noise* source. The output is mainly *shot noise*, in a broad *frequency band* up to about 5 MHz, which may be adjusted by varying the input current to the diode.
Noise Equivalent Power (NEP): The *radiant power* which, under specified conditions, produces a *Signal to Noise Ratio (SNR)* of one when measured by a *photodetector*, i.e. it is the amount of radiant power which produces a signal equivalent to the noise level.
noise factor: A measure of the noise introduced by a system, it is given by the ratio of the *Signal to Noise Ratio (SNR)* at the input of the system to the Signal to Noise Ratio at the output from the system. So if these are

equal to SNR_i and SNR_o then the noise factor (F) is given by $F = SNR_i/SNR_o$.

Noise Figure (NF): The Noise Figure is equivalent to the *noise factor* stated in *decibels*. Therefore if the *Signal to Noise Ratio (SNR)* at the input and output to a system are SNR_i and SNR_o respectively, then the Noise Figure (NF) is give by $NF = 10\log SNR_i - 10\log SNR_o$. The Noise Figure is therefore the number of decibels by which the *signal* is degraded by the system.

noise level: The *noise power*, usually when compared to a reference value, and measured in *decibels*.

noise margin: The maximum amplitude of the *noise signal* which can be applied to a system without its performance being adversely affected.

noise meter: An instrument which is used to measure noise. Instruments are available for measurement of the different types of noise, both electrical and acoustic.

noise power: The total power contained in the *noise signal*.

Noise Power Density (NPD): The *noise power* per *hertz*, i.e. the noise power in a *bandwidth* of one hertz.

noise resistance: (1) The ability of a *circuit* or a system to prevent or reduce the occurrence of *noise*. **(2)** In *thermal noise* the hypothetical equivalent resistance which has the same effect as the noise.

noise suppressor: A device which eliminates or reduces the *noise* in a system.

noise temperature: The temperature of a source of *thermal noise*, in degrees Kelvin, which gives the same *noise power* as the device being considered.

noise voltage: Voltage which is generated by the *noise signal* and causes *interference* within the *circuit* being considered.

noise weighting: System, produced by *regulatory bodies*, which defines the effect of *noise* on a *telephone user*, taking into account the *frequency* and *amplitude* of the noise source and the *attenuation* characteristics of the telephone *handset*.

nomadic communications service: Same as *Personal Communications Service (PCS)*.

nominal black: In a visual display system, such as television and *facsimile*, it is the *signal* corresponding to the blackest area which can be transmitted. See also *nominal white*.

nominal linewidth: In a display system, such as *facsimile* or a *Visual Display Unit (VDU)*, it is the average distance between the centres of adjacent *scanning* lines.

nominal white: In a visual display system, such as television and *facsimile*, it is the *signal* corresponding to the whitest area which can be transmitted. See also *nominal black*.

non-associated signalling: *Signalling* in which the *signals* for a group of *circuits* is transmitted on a separate *channel* reserved for signalling. See *Common Channel Signalling (CCS)* and *Channel Associated Signalling (CAS).*

non-blocking switch: A switch, such as a *PABX*, which has enough *switching capacity* to cater for the maximum demand so that all *calls* can be connected through it.

non-contention multiple access: *Multiple access* systems in which there is no *contention* for use of the *transmission channel.* Examples are *R-TDMA*, *BRAP*, *MLMA* and *PODA*.

non-directional antenna: An *antenna* which is used for the *transmission* and reception of *broadcast radio waves* almost evenly in any direction. See also *directional antenna*.

Non-Dispersion Shifted Fibre (NDSF): *Optical fibre*, used in legacy *transmission* systems, in which the *dispersion* zero was positioned to be between 1300 nm and 1320 nm, which was suited to transmission at 1310 nm. See also *Dispersion Shifted Fibre (DSF).*

non-disruptive test: Testing which is carried out on a system without affecting any of the *users* or *traffic* in the system.

non-linear device: A device in which the output *signal* is not a linear function of the input signal.

non-linear distortion: *Distortion* due to the *amplitude* of the output *signal* not being proportional to the input across a *frequency band*, which results in the generation of *harmonic frequencies* or *intermodulation* products.

non-linearity: See *non-linear device*.

non-periodic signal: Same as *aperiodic signal*.

non-persistent CSMA: A variation of the *Carrier Sense Multiple Access (CSMA)* technique for *multiple access* in which a *user* who is ready to send data senses the *line* and if it is free it commences *transmission.* If the *delay time* in the *transmission line* is low then there is a good chance of success. If the line is busy then the user does not continue sensing (persisting), but backs off for a random time before sensing it again, and so on. See also *p-persistent CSMA*.

Non-Return to Zero (NRZ): A *signal* represented by a sequence of *digits* in which two states are used, High and Low (1 and 0). The signal remains in this state for the whole duration of the *clock* period, i.e. it does not return to zero in-between.

non-synchronous network: Same as *asynchronous network.*

non-uniform quantisation: *Quantisation* which uses unequal intervals of quantisation. See also *uniform quantisation*.

Nordic Mobile Telephone system (NMT): The analogue *cellular radio system* developed jointly by the *PTTs* of Sweden, Norway, Denmark and Finland. It was designed to operate in the 450 MHz *frequency band* and

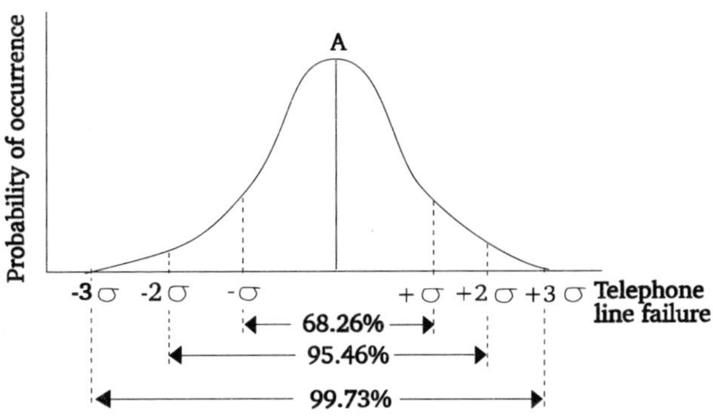

Figure N.7 The normal curve

was later adapted to also use the 900 MHz band. It saw commercial service in 1981, before *AMPS*.

normal distribution: *Probability distribution* in which the *probability* of an event, such as, for example, the failure of a *telephone line*, follows the curve shown in Figure N.7. The curve shows that most of the events occur close to the *mean* value and this is usually the case in practice. In this curve σ represents the *standard deviation* so that, for example, 68.26% of the values lie between two standard deviations points.

Norme Europeene de Telecommunications (NET): An *ETSI* standard which is expected to be applied by all Member States of the *European Union (EU)*. NETs are approved by weighted voting within *TRAC*. At the end of 1991 the procedure for the development of NETs was revised and some of the NETs are now known as *Common Technical Regulations (CTR)*.

North American Directory Forum (NADF): A Forum, consisting of a consortium of *electronic mail (e-mail)* service providers, which was formed to develop a messaging directory based on the *ITU-T Recommendation* X.500 (see *X Series*).

North American Numbering Plan (NANP): The *telephone numbering* scheme administered by *Bellcore*.

North American Telecommunications Association (NATA): A trade association of manufacturers and distributors of telephone equipment and services, it was founded in 1970.

NOS: *Network Operating System.*

notch antenna: An *antenna* with a *radiation pattern* similar to that of a *dipole*, but which is constructed by cutting a notch or slot in its surface.

notched noise: *Noise* which is distributed over a wide *frequency range* but which has one or more narrow bands of *frequency* removed from it. This noise signal is normally generated for testing purposes.

NP: *Number Portability.*
NPA: *Numbering Plan Area.*
NPD: *Noise Power Density.*
NRZ: *Non-Return to Zero.*
NSAP: *Network Service Access Point.*
NSFNet: *National Science Foundation Network.*
NTI: *Network Terminating Interface.*
NTIA: *National Telecommunications and Information Administration.*
NTN: *Network Terminal Number.*
NT1: *Network Termination Type 1.*
NTSC: *National Television Standards Committee* or *National Television Standards Code.*
NT2: *Network Termination Type 2.*
NTU: *Network Terminating Unit.*
NUI: *Network User Identity.*

null character: **(1)** An *ASCII character* (NUL) which can be inserted and removed from a sequence of characters without affecting its meaning. Primarily used as a fill character. **(2)** A character used to allow time for other operations on a *network*, such as the adjustment of a printer. Also called an *idle character.*

null modem: A device which emulates a *Data Circuit- terminating Equipment (DCE)* and so connects two *Data Terminal Equipment (DTE)* directly together. The term is usually applied to the wired RS-232 cable which allows two DTEs to communicate directly with each other.

number: In the context of *telecommunications* it is a *code* used to denote the *address* of a *subscriber*, such as the number of a *telephone.*

numbering plan: The plan used to allocate a *number* to each *subscriber* so that this number provides a unique *address* on the *network* whether it is on the *PSTN*, a *Local Area Network (LAN)* or a *Wide Area Network (WAN)*. The numbering plan is closely related to the *billing* plan.

Numbering Plan Area (NPA): The plan used to allocate *area codes* on the *Public Switched Telephone Network (PSTN).*

Number Portability (NP): The facility for retaining the same *telephone number* under changing circumstances. There are three aspects of Number Portability, known as *Local Number Portability (LNP)* in the USA: (i) Portability when moving from one *service* provider or *network* to another. This is important, especially to businesses, since changes in number would mean that all stationery would need to be changed. (ii) Portability when the *subscriber* moves from one physical location to another. This presents a problem since the number is now associated with the subscriber rather than the subscriber's location. (iii) Portability when changing from one type of service to another, such as from *POTS* to *ISDN*, or from TACS to *GSM*, etc.

Number 7: See *Signalling System No. 7 (SS7).*

Numerical Aperture (NA): A measure of the angle of acceptance of *electromagnetic waves* entering a *transmission medium*. If the rays enter at too steep an angle then they will not be propagated by *total internal reflection*. The Numerical Aperture is half the apex angle of the light *acceptance cone* (see Figure A.1) and is given by $NA = (n_1^2 - n_2^2)^{1/2}$ where n_1 is the maximum *refractive index* of the *core* and n_2 is the minimum refractive index of the *cladding* boundary.

Numerical Assignment Module (NAM): A module in a *mobile phone*, usually operating on the *cellular radio system*, which stores its *number*.

numeric pager: A pager which can only display numeric values. Typically this would be 12 or 24 *digits* so the *caller* can send a *telephone number* which is to be contacted by the pager wearer. See also *alphanumeric pager*, *tone pager* and *two-way page*.

NVOD: *Near Video On Demand.*

Nyquist rate: See *Nyquist sampling rate.*

Nyquist sampling rate: To convert an *analogue signal* into a *digital signal* the analogue signal is sampled by a periodic *pulse train* to convert it from a continuous to a discrete time. If the analogue signal has a *bandwidth* of B, then, in order to be able to reproduce the original signal, the minimum sampling rate or frequency must be f where $f = 2B$. This rate is known as the Nyquist sampling rate, as defined by the *Nyquist theorem.*

Nyquist theorem: The theorem which defines the *Nyquist sampling rate.*

Nyquist transmission rate: The maximum rate of *transmission* of pulses over a *transmission channel* (equal to twice the *bandwidth* of the channel) without the risk of *Intersymbol Interference (II).*

O

OA: *Office Automation.*
OADM: *Optical Add-Drop Multiplexer.*
OA&M: *Operations, Administration and Maintenance.*
O&M: *Operations and Maintenance.*
OAM: *Operations, Administration and Maintenance.*
OATS: *Open Area Test Site.*
object code: Computer programme which can be executed by a computer, usually produced as an output from a *compiler* or an *assembler*.
OC: *Optical Carrier.*
OCC: *Other Common Carrier.*
OCR: *Optical Character Recognition.*
octal: A number system which uses the base of eight. For example, Table O.1 gives the decimal and octal equivalents for numbers up to decimal 15.
octet: An association or grouping of eight *binary digits* or *bits*. Similar to a *byte*.
octet timing signal: The *timing signal* used to identify the first *bit* of an *octet*, in a *transmission* consisting of a continuous sequence of octets.
ODA: *Office Document Architecture* or *Open Document Architecture.*
odd parity: A method of *parity checking*, for *error detection* in *transmissions*, in which a *parity bit* is added to the *data block* or *character* such

Table O.1 Octal numbers

Octal	*Decimal*	*Octal*	Decimal
0	0	10	8
1	1	11	9
2	2	12	10
3	3	13	11
4	4	14	12
5	5	15	13
6	6	16	14
7	7	17	15

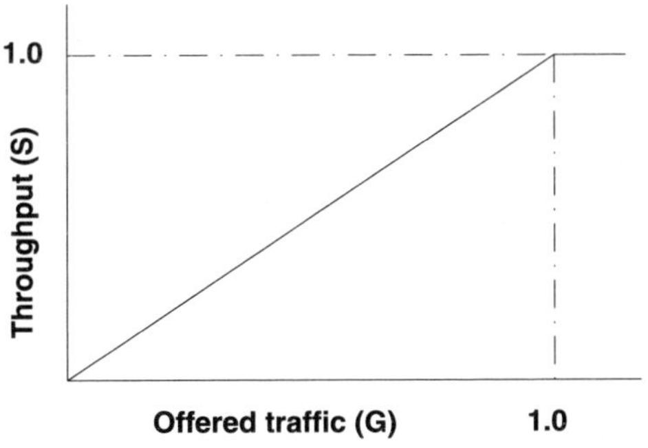

Figure O.1 Idealised throughput to offered traffic curve

that the total number of logical 1s in the block, including the parity bit, always equals an odd number.

ODP: *Open Distributed Processing* or *Originator Detection Pattern.*

OEM: *Original Equipment Manufacturer.*

OFDM: *Orthogonal Frequency Division Multiplex.*

offered traffic: The demand placed on a system by *users*, measured as the sum of all the *traffic* demands made by them. In an idealised situation the offered traffic will be directly related to the *throughput*, as in Figure O.1, all the traffic being handled by the system. In reality, at high levels of offered traffic, some *congestion* will occur and some of the traffic will be delayed or lost.

off-hook: Refers to the state in which the *handset* of a *telephone* is lifted so that the *gravity switch* in the instrument is activated. This is part of the *call setup* and *call disestablisment* process. When the *calling terminal* goes off-hook it signals the *exchange* that a call is to be made and it sends a *ringing tone* to the *called terminal.* When the *called terminal* goes off-hook the exchange is notified and it stops the ringing tone. See also *on-hook.*

off-hook signal: A *signal* sent to the *exchange* when a *telephone* goes *off-hook.* It can indicate a request for *service* or a *line*, or a *busy* condition.

office: In *telecommunications* it generally refers to a *Central Office (CO).* See also *class of office.*

Office Automation (OA): A generic term used to describe the use of technology to improve the productivity within a business. For example, the use of a *Local Area Network (LAN)*, *electronic mail*, *Intranet*, etc.

office class: The *class of office* which existed before *divestiture*, as illustrated in Figure C.18. Class 1 is the Regional Centre (RC), Class 2 the Sectional Centre (SC), Class 3 the Primary Centre (PC), Class 4 the Toll Centre (TC) and Class 5 the End Office (EO).

Office Document Architecture (ODA): *ISO* standard for the *transmission* of content and layout of a document and of *multimedia* documents. Also known as *Open Document Architecture*.

Office of Telecommunications (OFTEL): The UK *telecommunications watchdog*. It was set up in April 1984 by the *Telecommunications Act*, which established the regulatory framework within the UK. Written as Oftel.

off-line: Generally refers to equipment which is connected to a *transmission line* but is temporarily not taking part in any communications on the line. When in this state it cannot receive or send any *data*.

off-net calling: *Calls* which normally originate in a *private network* but need to go into another network, such as the *Public Switched Telephone Network (PSTN)*, in order to reach the *called terminal*. Also known as off-network calling.

off-peak: Refers to the times when *traffic* on a system is not at its peak value, such as during early morning or late evening when businesses are closed. See also *busy hour*.

Off-Premises Extension (OPX): A *terminal* which is located on another premises to the main telephone system, such as a *PABX*, but is connected to it by a *line* and can use all its facilities. Also known as Off-Premises Station (OPS).

offset antenna: An *antenna* which is front fed and in which the *antenna feed* is not directly on the axis of the reflector. It is offset to a side to reduce *signal* blocking.

off-the-air: Refers to the state where a *radio transmitter* is not shut down but is not transmitting any *signals* either. See also *on-the-air*.

off-the-air monitoring: The operation where a *radio transmitter* monitors some of its own *signals*, or that of other transmitters in the vicinity, to determine the quality of the *transmissions*.

OFS: *Operational Fixed Service*.

OFTEL: *Office of Telecommunications*.

ogive: A *data* presentation technique, as illustrated in Figure O.2, which shows the number of *users* for a *telecommunications service* classified by age. This should be compared to the *histogram* of Figure H.5. In a histogram the areas of the rectangles represent the frequencies in the different groups. Ogives show the cumulative frequency occurrences above or below a given value. From this curve it is possible to read off the total number of users above or below a specific age.

OIC: *Optical Integrated Circuit*.

OIF: *Optical Interworking Forum*.

OIW: *OSI Implementors' Workshop*.

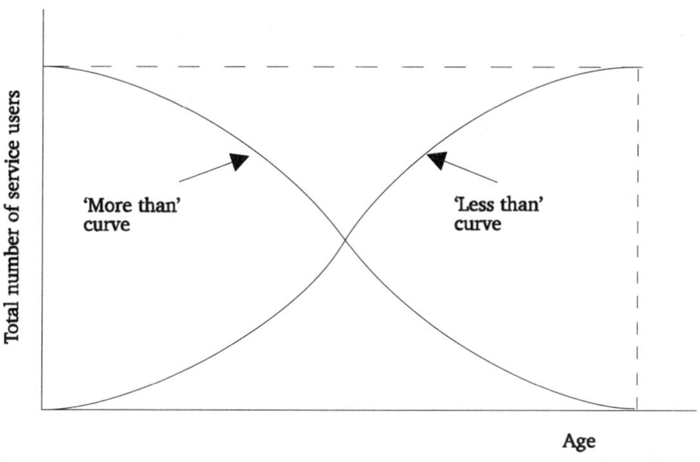

Figure O.2 An ogive

oligarchic network: A *network* which has been *synchronised* in such a manner that a few of the *clocks* on the network control the *timing* of all the other clocks.

OLR: *Overall Loudness Rating.*

OLT: *Optical Line Terminal.*

OLTU: *Optical Line Terminating Unit.*

Omega navigational system: A global radio navigational system which used the *signals* from eight *transmitting stations*, deployed globally, to provide positional information.

omnidirectional antenna: An *antenna* whose emitted and received *radiation patterns* are nondirectional.

OMNIPoint: A set of specifications for *network management*, produced by the *Network Management Forum (NMF)* in collaboration with other interest groups. These standards are primarily concerned with *service* management and are intended to address the business related problems faced by service providers.

ONA: *Open Network Architecture.*

ones complement: The ones complement of a *binary* numeral can be found by reversing the state of its *bits* in every position. Therefore the ones complement of 01110010 is 10001101.

ones density: A requirement is placed on *binary data* carried within the *PSTN* that there cannot be more than seven consecutive zeros in a *binary string*. This is because a binary one is needed for maintaining *timing* of *repeaters* and other clocking devices. Therefore a binary one is inserted if seven zeros are detected at any time in the *data transmission*.

One Stop Billing (OSB): The facility provided by a *service* provider where a single bill is given to the *subscriber* for a variety of services used, often on a global basis by large multinational corporations.

One Stop Shopping (OSS): A concept in which all the *services* required by a *subscriber*, such as a large multinational corporation, are provided by a single service provider. It is also closely related to *One Stop Billing (OSB)*. This is also known as *bundling*, where various services are provided through a single operator with one end bill, e.g. *local calls*, *long-distance calls*, *cable TV*, *video*, *Internet* access, etc. OSS clearly requires co-operation between *PTOs* in different countries for it to be successful on an international scale. To this end OSS concepts were defined by *CEPT's Commercial Action Committee (CAC)* in 1986 and in September 1989 most of the CEPT countries signed a Memorandum of Understanding (MoU) on the provision of OSS.

one-to-many call: A *call* in which a *subscriber* in one location is connected to several other subscribers in different locations. See also one-to-one call and *many-to-many call*.

one-to-one call: A *call* which occurs between a single *subscriber* in one location and a single subscriber in another location. See also *one-to-many call* and a *many-to-many call*.

one-way trunk: A *trunk*, usually connecting two *Central Offices (CO)*, or a Central Office and a *PABX*, in which the *traffic* always originates from one direction only. During the *call* traffic can flow in both directions, but the call is always originated from one end of the trunk, i.e. *seizing* of the trunk occurs at one end only.

on-hook: The condition which exists when a *telephone handset* is replaced in its cradle and the device is inactive except for the bell which can ring if a *call* is received. Going on-hook after a call is also a form of *signalling* to tell the *exchange* to commence *call disestablishment*. See also *off-hook*.

ONI: *Operator Number Identification*.

on-line: The state where a *node* is connected to the *network* and is actively communicating with it.

on-line processing: The state which exists when *users* are connected directly to a processing unit and can carry out processes as required, i.e. it is a *real-time system*. See also *batch processing*.

on-line system: Same as *real-time system*.

on-net calling: A *call* in which the *calling terminal* and the *called terminal* are both located on the same *private network*. Also known as on-network calling. See also *off-net calling*.

On-Off Keying (OOK): Same as *Amplitude Shift Keying (ASK)*.

ONP: *Open Network Provision*.

on-premises wiring: *Cables*, both copper and *optical fibre*, which are situated on a *subscriber's* premises and are owned by the subscriber. They may form part of a *Local Area Network (LAN)* and may also connect to the *Public Switched Telephone Network (PSTN)*.

on-site paging: A *radio paging* system which operates on a premises or a small campus. It is normally owned and operated by the same organisation. The area covered can be quite large, as in Figure O.3 where two remote sites belonging to the same organisation are operated with the same on-site paging system, being connected together via a *tie line*. On-site paging systems can vary in size from a dozen or so *pagers* to well over a thousand pagers. See also *wide area paging*.

ONT: *Optical Network Termination*.

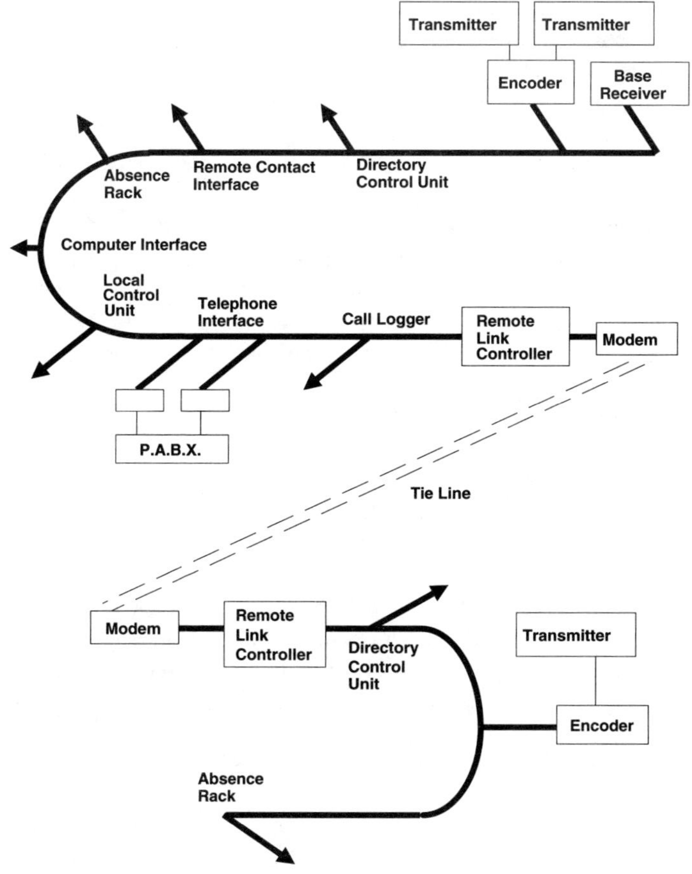

Figure O.3 On-site paging system

on-the-air: Refers to a radio *transmitting station* which is transmitting at least a *carrier signal*. This carrier may be *modulated* with the *information*. See also *off-the-air*.

ONU: *Optical Network Unit*.

OOF: *Out Of Frame*.

OOK: *On-Off Keying*.

OPC: *Originating Point Code*.

open air transmission: *Transmission* in which the *transmission medium* is the atmosphere. This is the normal transmission mode for *radio waves*.

Open Area Test Site (OATS): Refers to an open *field* site used for measurements of *radiated emission*, related to *Electromagnetic Compatibility (EMC)* tests. The open area site minimises errors due to reflections, as can occur in a chamber.

Open Distributed Processing (ODP): A reference model from the *JTC1* which defines a set of standards to enable distributed processing, i.e. the ability to carry out processing on equipment supplied by several different manufacturers.

Open Document Architecture (ODA): See *Office Document Architecture (ODA)*.

Open Network Architecture (ONA): Introduced by the *FCC* in 1987, as part of its *Computer Enquiry* III, it specifies that the public *network* needs to be open to all users, and it aims to do this by ensuring that network operators provide *equal access* to *service* providers, i.e. it must treat all operators equally, which includes its own internally owned companies and external companies.

Open Network Provision (ONP): An *EC* Council Decision of February 1990. Its aim is to provide open access to harmonised *services* across Europe. It specifies the conditions for a *network* which is open to *users* and service providers. It has three components: technical interfaces, usage and supply conditions, and *traffic* principles. ONP covers a wide range of applications, such as *leased lines* (introduced in 1992), *ISDN*, *voice telephony*, *packet switching*, mobile services (including *paging*), *telex*, *Intelligent Networks (IN)*, and *broadband*. In 1996 *Directorate General* XIII published the Draft Amending Directive to the ONP Voice Telephony Directive, proposing that mobile operators be drawn into the same regulatory framework as wireline operators, i.e. contributing to the funding of universal service (see *universal service obligation*) and go to cost-based pricing.

open numbering plan: An older *numbering plan* which does not have a fixed number of *digits* in the *telephone number*. This was primarily used when the *network* contained *step-by-step exchanges*, the number of digits for a *call* being determined by the number of *switching* stages it was routed through. Modern systems use a *closed numbering plan*, with a fixed number of digits irrespective of the distance of the call.

Open Shortest Path First (OSPF): A *routeing protocol*, used in large *TCP/IP* networks, such as the *Internet*. It is based on a link state *algorithm* which contains a table showing the best route for each direct link from one *router* to another in the *network*. The algorithm used sorts the table to determine the best, or shortest, route and this is used for any *transmission*. However, in the event that this fails backup routes are used, hence the term Open Shortest Path First. See also *Routeing Information Protocol (RIP)*.

Open Software Foundation (OSF): A private consortium of equipment suppliers formed to set common standards for *open systems*, including operating systems such as *UNIX*.

open system: Computer and communications systems which can communicate with each other using internationally recognised standards based *protocols*.

open system architecture: A *network architecture* which enables *open system* communications. Also refers to the *Open System Interconnect (OSI)*.

Open System Interconnect (OSI): In 1978 the *International Standards Organisation (ISO)* established a programme, called the Open Systems Interconnect, to define an architecture and set of standards to serve as the basis for *open systems*. The *OSI Basic Reference Model* was published as ISO 7498.

Open System Interconnect Environment (OSIE): The environment in which *open systems* operate when they are using the *Open System Interconnect (OSI)* definitions.

Open System Interconnect (OSI) service definition: A definition of the services provided within each of the *layers* of the *OSI Basic Reference Model*.

open system interface: The situation where the interface between different pieces of equipment is such that they can communicate with each other, e.g. they use a common standard, such as *Open System Interconnect (OSI)*.

open wire: *Transmission medium* consisting of pairs of bare copper wire which are supported on insulators mounted above the ground.

operating environment: The total system, which normally includes the *operating system* and other *user* application *software*, the *databases*, and the *telecommunications access* method.

operating system: A computer programme which resides permanently in the computer memory and manages all its basic tasks, such as running applications, accessing memory, controlling printers, etc.

operating time: The time between the instance that a request is made for an action and the instance when the action is carried out. For example, in a *call setup*, it is the time needed for *seizing* the *line*, *dialling* the *number*, the elapsed time before the connection is made, and finally negotiating the setup with the *terminal* at the other end.

Operational Fixed Service (OFS): *FCC* regulated *service* which is provided by fixed *microwave* equipment.

Operations, Administration and Maintenance (OA&M or OAM): The functions which are carried out within a *network management* system. Often used to describe a *network manager*. Also called *Operations and Maintenance (O&M)*.

Operations and Maintenance (O&M): See *Operations, Administration and Maintenance*.

Operations Support System (OSS): The *hardware* and *software* which is used to manage an operation, such as a business. Commonly used to refer to the *network management* systems which provide overall control of all the network functions, at the network, service and business levels. (See *network management functions* and Figure N.3.)

Operations System (OS): Part of the *Telecommunications Management Network (TMN)*, as in Figures T.4 and T.10. Same functionality as *Operations Support System (OSS)*.

Operations Systems Function (OSF): Part of the *TMN functional architecture* (see Figure T.9), the OSF processes information to support and control the realisation of various telecommunication management activities. Many types of OSFs are realisable, depending on the TMN, and can range from business, customer, *service*, *network* and basic levels of abstraction. Business OSFs are concerned with the management and coordination of the total business or enterprise. Service OSFs usually provide the interface to customers and are primarily involved in service aspects of the network. The network OSFs cover the realisation of the network, based on TNM application functions, by communicating with the basic OSFs. In small networks these basic OSFs may not be present and then the network OSFs need to communicate with the *Network Element Function (NEF)* or directly with the *Mediation Function (MF)*.

operator assisted call: A *call* which needs the involvement of a human operator, such as, for example, when making a *collect call*, a person-to-person call, etc.

Operator Number Identification (ONI): Operator used in a *Centralised Automatic Message Accounting (CAMA)* office to obtain verbally the *calling number* for *calls* originating in *offices* not equipped with *Automatic Number Identification (ANI)*.

Operator Service Provider (OSP): Company in the USA providing competitive toll *operator services*, such as for *billing* and *call* completion.

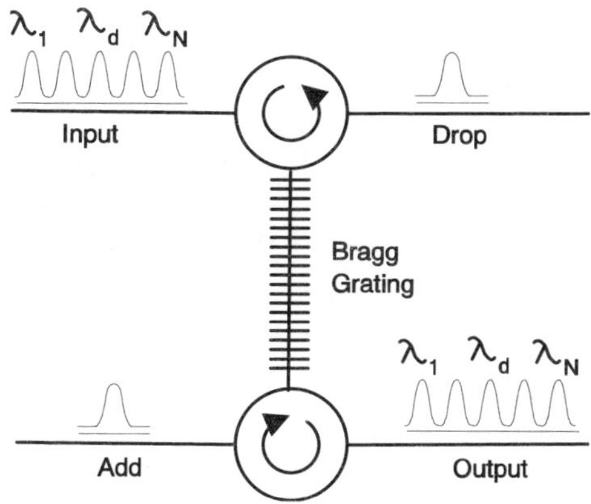

Figure O.4 Add-drop multiplexer using circulators and Bragg gratings

operator services: Generic term used to describe any telephony related *service* which is obtained from an operator. It includes *directory services* and *operator assisted calls.*

Optical Add-Drop Multiplexer (OADM): An *Add-Drop Multiplexer (ADM) which can operate using light signals.* It allows one *wavelength channel* to be separated from a group of others within a *transmission*, and for another channel to be added, either at the same wavelength or a different wavelength from the one which has been removed. Figure O.4 shows a simple OADM which uses a combination of circulators at each end of a Bragg grating, to add and drop the wavelengths.

optical amplifier: An *amplifier* which is used to increase the power of an *optical signal*. This is usually used to compensate for *transmission losses* within a *transmission medium.* See also *optical attenuator.*

optical attenuator: An *attenuator* which is used to decrease the power of an *optical signal.* See also *optical amplifier.*

optical blank: The initial casting which can be subsequently modified, such as by moulding, polishing, drawing, etc., to produce the required optical component, e.g. a lens or *optical fibre.*

optical budget: In the design of an optical *network* it is the amount of optical power *loss* which can be sustained in the various sections, such as in *optical fibres, optical connectors, optical splitters*, etc., while still maintaining the required level of performance in the overall system. For a link it could also be the difference between the output of the light source to the threshold of the *photodetector*, measured in *decibels.* Also known as *optical link budget.*

Optical Carrier (OC): Part of the *SONET* system it defines the *data transmission rate* and is equivalent to the *Synchronous Transport Mo-*

dule (STM) of the *Synchronous Digital Hierarchy (SDH)* system. The Optical Carriers are OC-1 at 51.84 Mbit/s; OC-3 (equivalent to STM-1) at 155.52 Mbit/s; OC-12 (equivalent to STM-4) at 622.08 Mbit/s; OC-48 (equivalent to STM-16) at 2488.32 Mbit/s and OC-192 (equivalent to STM-64) at 9953.28 Mbit/s.

Optical Character Recognition (OCR): The recognition and storage of written characters and words by means of a machine which optically scans the characters and then converts them into electrical *signals*.

optical combiner: A passive device which combines the optical power of *signals* from several *optical fibres* into a smaller number of fibres. See also *optical splitter*.

optical connector: Same as *Fibre Optic Connector (FOC)*.

optical cross connect (OXC): A *cross-connect* which is used to switch *optical signals* between different *paths*.

optical coupler: A device which can connect together two optical devices, such as an *optical fibre*, optical source, optical detector, etc., so that light is coupled from one device to the other.

optical disk: A very high capacity *data* storage medium in which the *binary digits* are written into its photosensitive surface by a *LASER* and it can be read off by a sensor as the disk revolves under it.

optical fibre: See *fibre*.

optical fibre cable: A *cable* assembly consisting a *optical fibres*, which act as the *transmission medium* for *optical signals*, and associated protection and strengthening members. For example Figure O.5 shows a cable

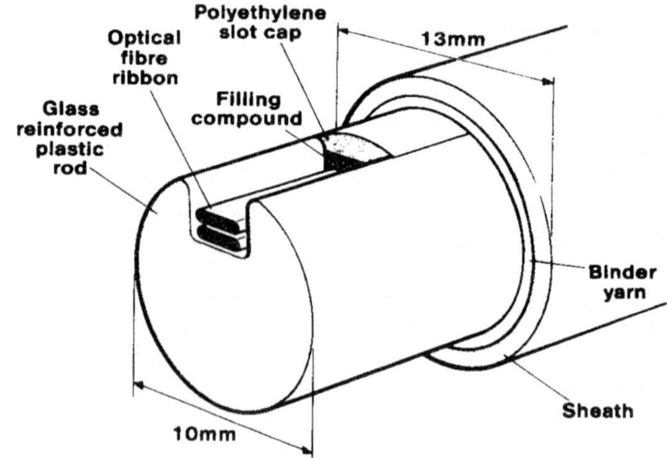

Figure O.5 Self-supporting cable

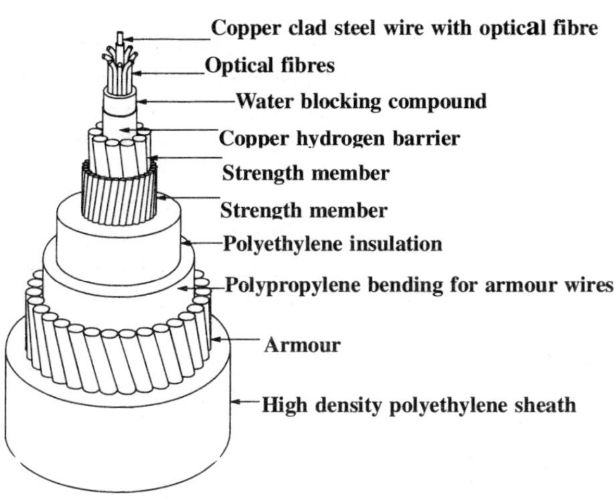

Figure O.6 Submarine optical fibre cable

construction which has been used for low fibre counts on live power transmission lines and Figure O.6 a construction which has been used for *submarine cables*. See also Figure C.1.

optical fibre dispersion: The *dispersion* which occurs when an *optical signal* travels along an *optical fibre transmission medium.*

optical fibre pigtail: Same as *fibre pigtail.*

optical fibre scattering: *Scattering* of an *optical signal* as it travels along the length of an *optical fibre*, caused by impairments in the *transmission medium.*

optical fibre splicing: The joining together of two pieces of *optical fibre* so that it forms a permanent join and there is a low level of *optical signal* loss at the join. Two main techniques exist for making this splice, *fusion splicing* and a *mechanical splice*.

optical fibre splitter: Same as *fibre optic splitter.*

optical filter: A device which lets through selected *wavelengths* and stops others, from the *optical signal* which passes through it.

Optical Integrated Circuit (OIC): A device which contains miniature optical components, such as *Light Emitting Diode (LED)*, *photodetector*, *optical waveguide*, etc., all mounted in a monolithic or hybrid integrated circuit.

optical interconnection panel: A panel which is used to interconnect individual *optical fibres*, primarily for administration purposes.

Optical Interworking Forum (OIF): An industrial forum, established by Cisco and Ciena and supported by several telecommunications equipment suppliers and users, such as AT&T, Bellcore, Hewlett-Packard, Qwest, Sprint and WorldCom. It held its inaugural meeting in Atlanta, USA, on 8 June 1998. Its main aim is to accelerate the deployment of optical *networks* by providing a forum for equipment users, manufacturers and service providers to discuss and resolve key issues and to ensure interoperability.

optical isolator: A device, inserted in the *transmission path* of an *optical signal*, which prevents *reflections* from reaching the signal source.

Optical Line Terminal (OLT): Same as *Optical Line Terminating Unit (OLTU)*.

Optical Line Terminating Unit (OLTU): Equipment which terminates an *optical fibre transmission line*, usually converting *optical signals* to electrical signals, and vice versa. It is usually situated in the *Central Office (CO)*. Also known as *Optical Line Terminal (OLT)*. See also *Optical Network Unit (ONU)*.

optical link budget: See *optical budget*.

optical network layering: A concept used in an *all optical network* for defining the communications over the network. Communications occur within a *layer* and between layers. For example, in Figure O.7 communi-

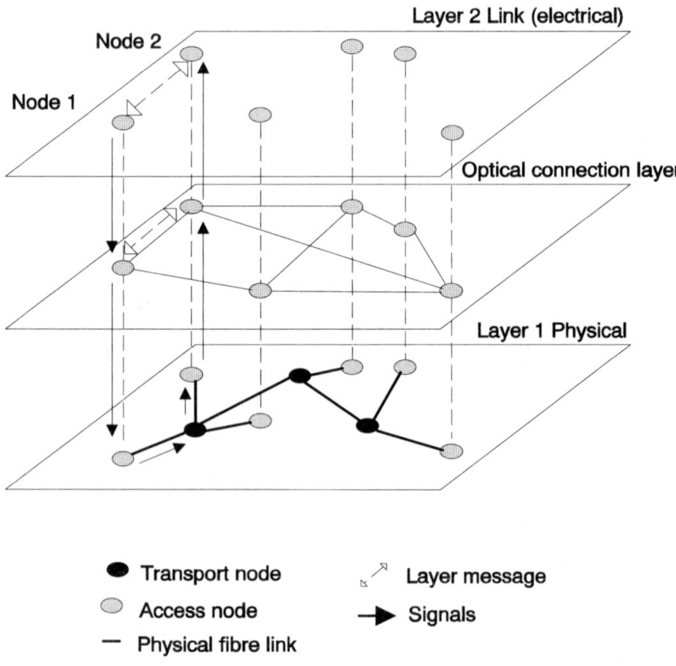

Figure O.7 Optical network layering

cations between nodes 1 and 2 in Layer 2 causes *signals* to pass to the optical connection layer and physical layer, where, for example, the *wavelengths* to be used in the communications could be selected.

Optical Network Termination (ONT): Termination point for *optical fibre access* systems.

Optical Network Unit (ONU): Equipment which terminates an *optical fibre transmission line*, usually converting *optical signals* to electrical signals, and vice versa. It is usually situated close to the *subscriber*. For a *Fibre To The Building (FTTB)* system the ONU would be located in the customer's building, whereas for *Fibre To The Curb (FTTC)* it would be located on the curb and supply several buildings.

optical power: The *radiant power* of *electromagnetic radiation*, whose *frequency band* lies in the *optical spectrum*.

optical power budget: See *optical budget*.

optical power efficiency: The ratio of the *optical power* output from a source to the input power, which is usually an electrical input.

optical power meter: Instrument used to measure the *optical power* and energy from a *narrowband signal* source, such as a *LASER*. A range of calibrated detector probes are normally available for use with the equipment, to suit the source being measured. The power is normally stated in *decibels* or milliwatts.

optical radiation: *Electromagnetic radiation* which is in the *optical spectrum*.

optical range: The distance between two points on the Earth's surface, measured in a direct *Line of Sight (LoS)*.

optical receiver: A device which can receive *information* carried in an *optical signal*. Usually it would contain an optical detector, *optical amplifier*, *equaliser*, and *signal processing* stages. A digital receiver would also include *clock extraction* and *regeneration* stages. Generally several of these stages would be implemented electronically so that the device is often called an *optoelectronic receiver*. See also *optical transmitter*.

optical repeater: A *repeater* which can handle *optical signals*. It amplifiers and, if necessary, reconstructs it for onward *transmission*.

optical scanner: A *scanner* which uses light as a *scanning beam* to illuminate the object being scanned.

optical signal: A *signal* in the *optical spectrum* which contains *optical power* and is transmitted along an *optical waveguide*.

optical spectrum: See *visible spectrum*.

optical switching: *Switching* in which *optical signals* are involved.

Optical Time Division Multiple Access (OTDMA): A variation of *Time Division Multiple Access (TDMA)* which is used in an *all optical network*. It is based on the principle of Time Division Multiple Access but uses all optical components to maximise the benefits gained from the large *transmission bandwidth* available from *singlemode fibres*.

Optical Time Division Multiplexing (OTDM): *Multiplexing* which uses *interleaving* of high *bit rate data* streams of *optical signals*.

Optical Time Domain Reflectometer (OTDR): An instrument used to measure distances in an optical *transmission medium*, such as *optical fibre*. For example, it can measure the distance to a splice, the end of a fibre, defects and breaks in the fibre, etc. It works on the principle of launching an *optical signal* into the transmission medium and then noting the time before parts of the signal are scattered and reflected back from the point being measured.

optical transmitter: A device which generates *optical signals*. It would include stages for *modulation* of the *carrier signal* and for control of the *optical power* level. See also *optical receiver*. Generally several of these stages would use electrical signals, so that it is often called an *optoelectronic transmitter*.

optical waveguide: *Waveguide* used for the *transmission* of *optical signals*. The most commonly used optical waveguide is *optical fibre*.

optimised traffic: *Traffic* in *multiplexing* systems in which many users share a common connection, the traffic being optimised by ensuring that the *transmission packets* are as full as possible whilst still maintaining the required *Grade of Service (GoS)*.

optoelectronic: Generic term used to describe devices which have both optical and electrical properties. It could, for example, generate an electrical voltage or current in response to an *optical signal* (such as in a *photodetector*) or convert an electrical input into an optical output (such as in a *LASER*).

optoelectronic detector: See *photodetector*.

optoelectronic receiver: See *optical receiver*.

optoelectronic transmitter: See *optical transmitter*.

oracle: An early *teletex transmission* system, introduced in the UK by the IBA-ITV, which was incorporated with the conventional television *signals*. See also *Ceefax*.

orbit: The path taken by an object relative to a reference. For example the path taken by a *satellite* relative to the Earth. Figure O.8 shows some examples of satellite Earth orbits. See also *Low Earth Orbit (LEO)*, *Medium Earth Orbit (MEO)*, and *Geostationary Orbit (GEO)*.

orbital inclination: The angle between the plane of the *orbit* and the equatorial plane of the Earth.

orbital perturbations: Changes in a *satellite's orbit* caused by factors such as the gravitational fields of the Sun and the Moon, the pressure of solar radiation and solar wind, etc. For a *communications satellite* these

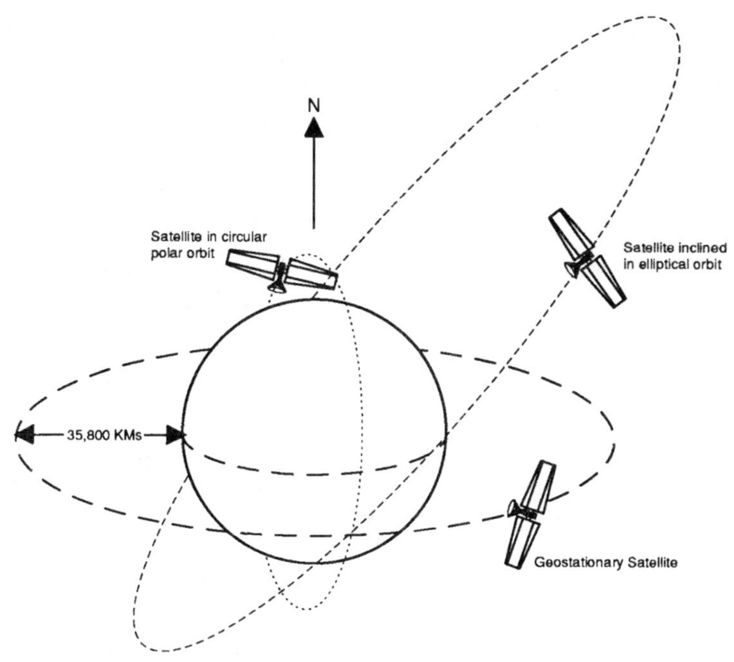

Figure O.8 Some satellite orbits

perturbations can result in *interference* with other satellites operating in the same *frequency band*.

orderwire: Same as *Engineering Orderwire (EOW)*.

orderwire message: The *message* exchanged on an *Engineering Orderwire (EOW)* between technical staff, and used for the control of the *network*.

Original Equipment Manufacturer (OEM): A term used to apply to the manufacturer and to the equipment which is sold by one organisation but is manufactured by another organisation. The seller would normally badge the equipment as its own, so that the identity of the actual manufacturer is not revealed.

originating office: The *Central Office (CO)* on the *network* from which the *call* originates and enters the system.

Originating Point Code (OPC): The *address* of the *exchange* on the *network* which is setting up a *long distance call* using *ITU-T Signalling System No. 7 (SS7)*.

originating traffic: A *terminal* which is generating *outgoing calls*. See also *terminating traffic*.

Originator Detection Pattern (ODP): A pattern of *characters*, sent by a *modem* operating on the V.42 *protocol*, to determine if the modem connecting to it is using a similar protocol.

Orthogonal Frequency Division Multiplex (OFDM): A means for *transmission* of high *data rate bit streams* by *modulation* of many *carrier signals* in parallel, each at a relatively low data rate.

OS: *Operations System.*

OSB: *One Stop Billing.*

OSF: *Open Software Foundation* or *Operations Systems Function.*

OSI: *Open System Interconnect.*

OSI Basic Reference Model: The communications model defined by the *International Standards Organisation (ISO)* as part of *Open Systems Interconnect (OSI)*. The model deals with communications between systems and not processing within *end systems*. It gives an abstract structure for the subdivision of communication functions and so forms a framework for coordination of current and future standards, rather than defining the standards themselves. Figure O.9 shows the traditional OSI *seven layer model* in which communications is taking place between two end systems, using an intermediate or *relay system*. The OSI concept

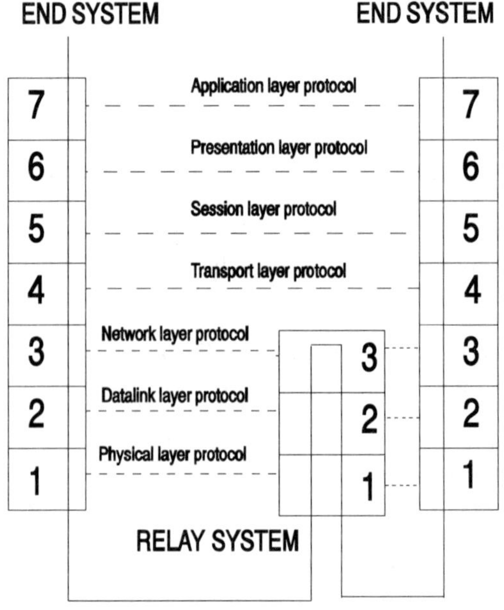

Figure O.9 OSI Basic Reference Model, showing communications between two end systems using an intermediate relay system

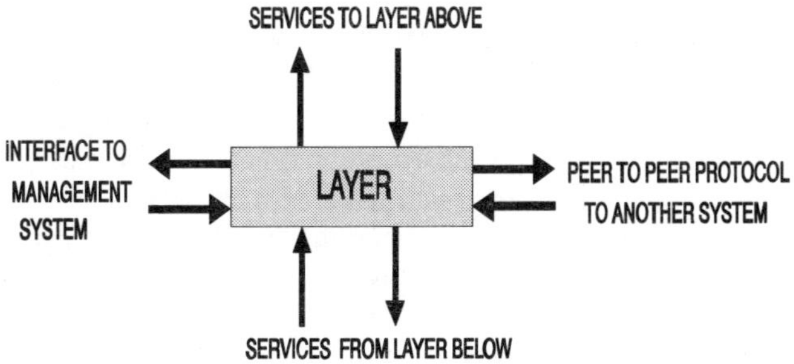

Figure O.10 Communications using services within the seven layer OSI model

assumes that exchanges of data between layers involves at least four components: the communications *path* between the systems; the systematic delivery of the *data* along this path, which requires a structured exchange of *information*; a common language (the syntax) which is understood by all parties involved in the data interchange; and the same conceptual interpretation of the data on all sides, i.e. the semantics. OSI has divided these requirements into a hierarchical structure within its reference model. The basic concept is that the functions involved in any communications can be divided into *seven layers*. Communications between layers within different systems is defined by service primitives. OSI standards define the *services* each layer provides and the service definitions are then the basis for the protocol defining how layers in one system communicate with layers at the same level within another system. Figure O.10 illustrates this concept. Layers provide services to their own users and to the layers above, using their own functions or those contained within services provided by the layer below. See also, *Application Layer*, *Presentation Layer*, *Session Layer*, *Transport Layer*, *Network Layer*, *Data Link Layer*, *Physical Layer*.

OSIE: *Open System Interconnect Environment*.

OSI Implementors' Workshop (OIW): One of the regional workshops working on *Open System Interconnect (OSI)* standards development. It has links with the *NIST* and is based in the USA. See also *European Workshop for Open Systems (EWOS)* and *Asia and Oceania Workshop for OSI Standardisation (AOWS)*.

OSP: *Operator Service Provider*.

OSPF: *Open Shortest Path First*.

OSS: *Operations Support System* or *One Stop Shopping*.

OTDM: *Optical Time Division Multiplexing*.

OTDMA: *Optical Time Division Multiple Access*.

OTDR: *Optical Time Domain Reflectometer*.

Other Common Carrier (OCC): Term used in the USA to describe *common carriers*, licensed by the *FCC*, who provide long distance communications services in competition with the incumbent common carriers. These include specialised common carriers, domestic *satellite* carriers, and domestic or international record carriers.

outage: A condition which exists when the system is non-operational so that the *user* cannot get any *services*.

outage duration: Same as *outage period*.

outage period: The period of time from the onset of *outage* to when the *service* is restored. Also known as *outage duration*.

outage probability: The probability of an *outage* occurring. See *outage ratio*.

outage ratio: The mean time, over a measurement period, when there is *outage*, measured as the sum of all the outages during the period divided by the time period. It is a measure of the *outage probability*.

outband signalling: Same as *out-of-band signalling*.

outgoing access: The feature which allows *users* on different *networks* to communicate with each other using the facilities of the networks to which they both have access.

outgoing call: A *call* which goes outside the area administered by a local switch or *PABX*.

outgoing call handling: The procedure used to handle an *outgoing call*. For example the *PABX* may need to decide on the *route* to be selected (such as in *Least Cost Routeing (LCR)*) or the *common carrier* to be used for the call.

outgoing line: The *transmission line* which carries the *message* from the *exchange* to its destination.

outgoing trunk: The *trunk line* used to carry the *outgoing call* from the *exchange*.

outgoing trunk queuing: The facility which allows a *user* to be placed in a queue when an *outgoing trunk* is *busy* and then to be called back when the trunk becomes free.

out-of-band components: *Signals* which occur within a *voice frequency bandwidth* but not within the voice channel, i.e. occupying the frequencies up to 300 Hz and from 3.4 kHz to 4.0 kHz.

out-of-band flow control: A *flow control* technique in a *Packet Switched Network (PSN)*, where the *receiving terminal* can stop accepting *data*, even after it has sent a *clear to send signal*, if the *traffic* becomes too heavy.

out-of-band signalling: *Signalling* in which the signalling information is carried on *channels* different from that used to convey the main *user information*. See also *Common Channel Signalling (CCS)*.

Out-Of-Frame (OOF): An error condition on a *T1 circuit* which exists when two or more consecutive *framing bits*, which have been received, have errors.

out-of-frame alignment: Same as *loss-of-frame alignment.*

out-of-order signal: A *signal* which is sent, in the backward direction to the *calling terminal*, to indicate that the *call* cannot be completed because the *called terminal* is not working, or because a defect exists on the *line* to it.

outside plant: Equipment which is located outside an *exchange*, i.e. between the *Main Distribution Frame* in the exchange and the entrance of another exchange or a *subscriber* premises. This includes all equipment, such as *cables*, conduits, poles, repeaters, etc.

out-slot signalling: *Signalling* which uses *timeslots* which are different from those used to transport *user data.*

outsourcing: Generally considered to be equivalent to *Facilities Management (FM)*, as used in the *Information Technology (IT)* industry. In this the organisation pays an outside company to look after the day-to-day running of all its internal IT and *telecommunications* equipment. Generally this is done so that the organisation can concentrate on its core business. The outsourcing company can also usually provide the service at a lower cost, since it has a larger pool of skilled staff that it can use to solve any *network* problems, if they arise.

outstation: (1) A *station* on a *network* which is located in a remote area, such as in a *rural communications* system, and serves a local community. The outstation is linked to a *central station* and communicates with it. See Figure C.13. **(2)** A radio station which is connected to the network but is listening rather than transmitting.

Outward Wide Area Telecommunications Service (OUT-WATS): See *WATS.*

OUT-WATS: *Outward Wide Area Telecommunications Service.*

overall loss: The end-to-end loss in a *network*, which includes the sum of all the losses, usually expressed in *decibels.*

Overall Loudness Rating (OLR): A measure of the end-to-end *transmission loss* of a *telephone circuit*, in terms of the acoustic to acoustic loss between the *sending terminal* and the *receiving terminal.* It is equal to the sum of the Send Loudness Rating (SLR), the Receive Loudness Rating (RLR) and the Circuit Loudness Rating, as shown in Figure O.11.

overfill: The situation which arises when the *optical signal* from the source is greater than that which can be handled by the detector. For example the cone of light from a source is greater than the area of the *photodetector* on which it falls.

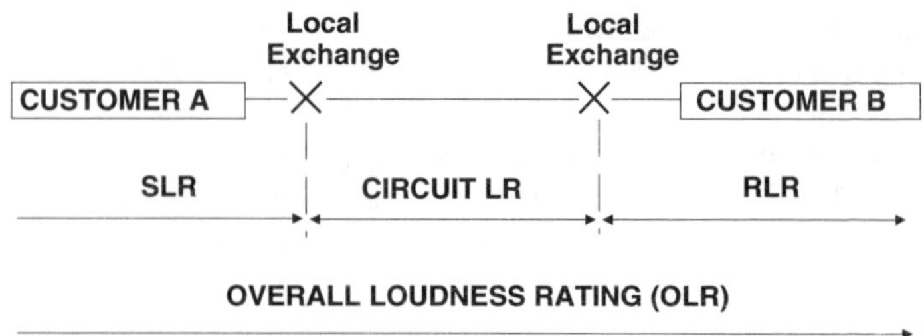

Figure O.11 Overall Loudness Rating

overflow: The situation which occurs when the *traffic* is greater than that which can be handled by the normal *circuit* and is caused to be diverted to an alternative circuit.

overflow call: A *call* which has been diverted to an alternate *circuit*, due to *overflow* on its main circuit.

overflow indicator: An audio or visual indicator which occurs when an *overflow* takes place.

overhead: In *data transmission* it is the extra data which is sent, in addition to the *user information*, for ancillary use, such as in *error detection*, *synchronisation*, etc. Retransmission of user information, which has originally been received in error, is also a form of overhead.

overhead bit: A *bit* which is an *overhead*, i.e. it is not part of the *user information*. It provides control functions, such as *error detection*, *synchronisation*, etc.

overlap: A defect in *facsimile transmission* which causes adjacent *scanning* lines to overlap.

overlay network: A *network* which is added over essentially the same route as an existing network. It does not replace it but usually carries *supplementary services*. For example, an older copper based network could be overlaid with an *optical fibre* network to provide a higher *bandwidth* service, the two working in parallel.

overlay paging: A *paging* system which operates with another form of mobile system. For example a *pager* could be used to alert a driver, who had left his car, that a *call* is waiting for him on the *Personal Mobile Radio (PMR)* system in the car.

overlay service: *Services* which are provided as an additional option to the main service.

overload: Any instance where the load placed on a system exceeds the capabilities of the system so, that it cannot perform within its specified characteristics. For example the *traffic* on a *switch* could exceed its traffic handling capabilities, so that some of the *calls* are lost.

overmodulation: *Modulation* which is greater than that needed to produce 100% modulation of the *carrier signal*, resulting in *distortion* of the demodulated output.

override: The ability of a *node*, which has been granted priority, to carry out *line seizure*, even though the line may be *busy* on another *call*.

overrun: The situation which results in *loss* of *data* during a *transmission*, due to the fact that the *transmitting terminal* is sending data faster than the *receiving terminal* can accept it.

oversampling: *Sampling* in a *Time Division Multiplexing (TDM)* system where a sample is taken of each *bit* from each *channel* more than once.

over the horizon transmission: *Data transmission* which occurs over the Earth's horizon, i.e. not in a straight line. This is done in several ways, such as by *ducting* in the Earth's atmosphere, by *tropospheric scatter* and by *meteor scatter communications*.

overtone: An integral multiple of the *frequency* of a *sinusoidal waveform*, which excludes the *fundamental frequency*.

own exchange call: Same as *local call*.

OXC: *optical cross-connect*.

P

PABX: *Private Automatic Branch Exchange.*

pacing: A method of *flow control* in which the *receiving terminal* regulates the *traffic* rate from the *sending terminal.*

packet: A collection of *bits* which are grouped into a unit, containing *user information* as well as control information, such as the *address* of the destination, *error control* information, size of the packet, etc. Packets travel over a *Packet Switched Network (PSN).*

Packet Assembler-Disassembler (PAD): A device which enables equipment not designed for operating over a *Packet Switched Network (PSN)* (X.25 network) to do so. It takes a *character* stream, such as from *asynchronous terminals*, as in Figure P.1, and converts them into *packets* for transport over the PSN. At the other end this data is again converted into a character stream for the asynchronous host. There are several standards for PADs, defined by the *ITU-T*, such as X.3, which defines how the asynchronous terminals should interface; X.28 which determines how the network messages are to be interpreted; and X.29, which negotiates parameters needed for end-to-end session compliance.

packet buffer: Part of the memory of a *packet switching exchange* or *terminal*, where *packets* are temporarily stored when received or when waiting *transmission.*

packet collision: *Collision* which occurs between *packets* in a *Packet Switched Network* or in a *multiple access* system.

packet disassembly: The process of converting the *packet* into a *message* for delivery to non-packet *terminals.* Part of a *Packet Assembler-Disassembler (PAD)* unit.

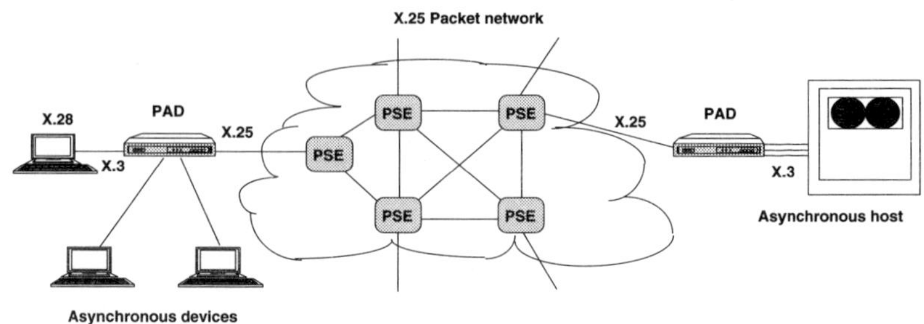

Figure P.1 A PAD in a Packet Switched Network

Packet Internet Groper (PING): A test facility available in a *TCP/IP* network, such as the *Internet*, in which a query, in the form of a *packet*, is sent to a distant processor and its presence on the *network* is determined by receiving a confirmation back.

packetised voice: The conversion of *voice* into a *digital signal* for *transmission* as *packets* over a *Packet Switched Network (PSN)*.

packet level protocol: *Protocol* which is concerned with the handling of *packets* within a *Packet Switched Network (PSN)*. It is at Level 3 of the *ITU-T Recommendation* X.25.

packet mode: The *data communications* mode which uses *packet switching* rather than some other form, such as *circuit switching* or *message switching*.

packet mode terminal: *Data Terminal Equipment (DTE)* which has the capability for handling *packets*, e.g. formatting, transmitting, receiving etc., to and from a *Packet Switched Network (PSN)*.

packet network: Same as *Packet Switched Network (PSN)*.

packet radio: A *packet mode* of *transmission* in which *radio channels* are used as the *transmission medium*.

Packet Switched Data Network (PSDN): Same as *Packet Switched Network*.

Packet Switched Exchange (PSE): The *node* or *exchange* within a *Packet Switched Network (PSN)* which is capable of carrying out all the *packet switching* functions for the network.

Packet Switched Network (PSN): A *network* in which *packet switching* is used for *data communications*.

Packet Switched Public Data Network (PSPDN): *ITU-T* terminology for a public *network* using *packet switching* for *data communications*.

packet switching: A method of *data communications* in which the *data* is formed into discrete segments, usually with their own control information, and is routed through the *network* in these envelopes, referred to as *packets*. Packets occupy a *communications channel* for a short duration, so that packets from several users can share the same channel. There are two different modes for *transmission*, known as *connection mode transmission* and *connectionless mode transmission*.

packet switching service: A public *service*, provided by a *PTO*, using *packet switching*.

packet transfer mode: The method of *data transfer* between *users*, which uses *packet switching*.

packing density: The amount of *information* which can be stored on a storage medium, such as a computer disk.

PACS: *Personal Access Communications System*.

PAD: *Packet Assembler-Disassembler*.

page: A unit of *information* which is to be handled, stored, displayed or printed as a single unit. For example, in a *videotex* system a page consists of several *frames*, each of several *characters*.

pagefax: *Facsimile apparatus* used in the publishing industry for transmission of pages of newsprint. The majority of these systems use proprietary standards. See Table F.2. See also *photofax*.

pager: A device which is carried by mobile *user* and is used to contact them by *paging*. There are several types of pagers, such as *numeric pagers*, *alphanumeric pagers* and *tone pagers*.

pager codes: The *transmission codes* used in *paging* systems. A few of these are given in Table P.1 for both one way pagers (forward) and *two way pagers* (return).

page store: A memory unit capable of storing *information* which comprises a *page*.

paging: The communications systems which allows *users* who are mobile to be contacted by sending them a short *message*. Early paging systems only enabled messages to be sent in one direction, to the pager from a central control station, but *two way pagers* are now also available. See also *on-site paging* and *wide area paging*.

paired cable: Generally refers to *cable* made from *twisted pair wire*.

paired-disparity code: *Code* in which two different conditions are used to represent *bits* or *characters*, using a set of rules. For example, in *Alternate Mark Inversion (AMI)*.

pair gain: Any system in which *subscribers* can be served by a fewer number of wire pairs (usually *twisted pair wire*) than otherwise, e.g. by *multiplexing*, etc. This includes increasing the number of channels over the existing pair of wires, as is done in *ISDN* and in *Digital Subscriber Line (DSL)*. It is measured as the ratio of the number of wire pairs needed without the pair gain system to that required with the pair gain system.

PAL: *Phase Alternation by Line*.

PAL-M: *Phase Alternation by Line-Modified*.

PALplus: Colour television system developed in Germany in 1989 to provide enhanced analogue *transmission*, with wide screen aspect ratios using existing *PAL* receivers.

PAM: *Pulse Amplitude Modulation*.

parabolic reflector antenna: A *reflector antenna* which has the *antenna feed* at its focus F, as shown in Figure P.2. The energy from the feed goes to point P on the surface and is then reflected parallel to the axis to point A. The equation for this is given by $r^2 = 4F(F-z)$. For a reflector of diameter D the equation at the edge of the reflector is given by $F/D = 1/4[\cot(\theta_o/2)]$, where F/D represents the depth of the paraboloid.

Table P.1 Some pager code formats

		POCSAG	*ERMES*	*FLEX*	*ReFLEX*	*InFLEXion*
Data speed (bps)	Forward	1200 1400	6250	1600 3200 6400	1066 3200 6400	112000 (per 50kHz data only: voice, NDA)
	Return	No Return			8000 1600 6400 9600	8000 1600 6400 9600
Modulation	Forward	2 level FSK	4 level FSK	2 level FSK 4 level FSK	2 level FSK 4 level FSK	Voice: NDA Data: NDA
	Return	No Return			FSK4	FSK4
Bandwidth	Forward	25kHz FM	25kHz FM	25kHz FM	25Gz/50kHz FM (linear)	50kHz (min) FM (linear)
	Return	No Return			12.5kHz (min)	12.5kHz (min)
Forward Error Correction	Forward	BCH (31,21) +p	BCH (30,18) +Interleaving	BCH (31,21) +p + interleaving +**	As FLEX	NDA
	Return	No Return			RS	RS
Capacity 100ch messages BHCR = 0.25	Forward	13000 in 25kHz channels	60000 in 25kHz channels	74000 in 25kHz channels	200000 in 50kHz channels	Data: NDA Voice: NDA

parabolic orbit: A *satellite orbit* which has the shape of a parabola. It is therefore not a closed orbit, as in the case of a circular orbit or an elliptical orbit.

parallel interface: An interface between devices in which the entire *character* or group of *bits* (usually eight bits or a *byte*) is transmitted at the same time using *multiple access* points. See also *serial interface.*

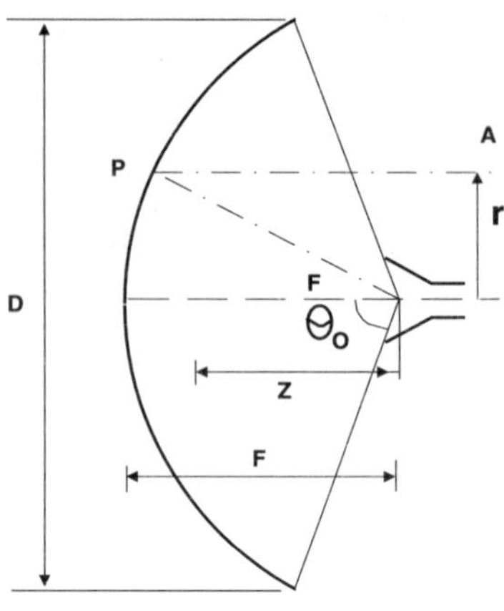

Figure P.2 Parabolic reflector antenna

parallel mode: Refers to the *parallel transmission* mode of *data transfer*.

parallel port: A *port*, usually used for *input/output* of *data* between a processing unit and its peripherals, which handles the data in *parallel mode*.

parallel to serial converter: A device which received data in the form of a *parallel transmission* and outputs this as a *serial transmission*.

parallel transmission: A method of *data transmission* in which groups of *bits* making up a *character*, usually a *byte*, are sent simultaneously over separate *transmission media*.

parity: Technique used in the *transmission* of *binary coded information* in which the total number of logical 1s which are transmitted are always maintained at an odd or an even value. See *odd parity* and *even parity check*.

parity bit: The *bit* which is set to a logical 0 or logical 1, in a *character*, during *parity* setting prior to *transmission*, and which is then checked during the *parity check* on the received *data*.

parity check: The process for *error detection* in *transmission* by adding a *bit* to each *character*, so that it has an odd or an even number of bits, and then checking for this in the received data.

parity check coding: The coding technique used during *transmission* with *parity check*.

parity error: An error which has been detected in the received *data*, when the number of *bits* in the *character* is not equal to either an even or an odd number, as in the transmitted data.

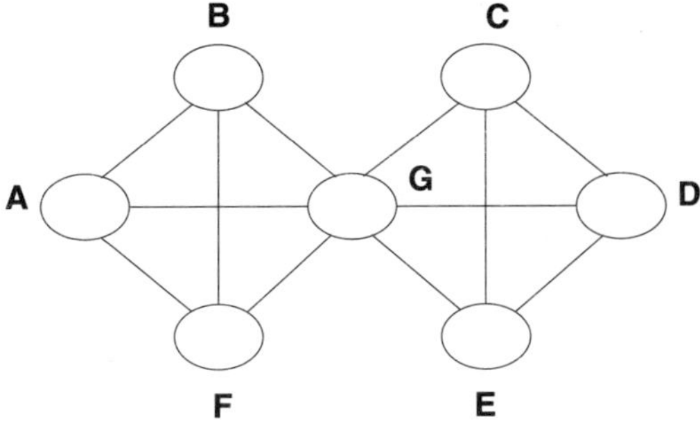

Figure P.3 Partially interconnected network

parking orbit: Refers to the first *orbit* into which a *satellite* is placed, soon after it has been launched. From this it would be moved into the *transfer orbit* and then to its final orbit.

partially interconnected network: A *network* in which a *link* occurs between many, but not all, the *nodes* on the network, as shown in Figure P.3. See also *fully interconnected network.*

partial response code: A group of *line codes* in which a controlled amount of *Intersymbol Interference* is introduced into the received *signal,* to produce a multi-level signal having the required line code characteristics. An example of a partial response code is *duobinary code.*

partial response equaliser: An *equaliser* which uses a *partial response code* to convert a *binary signal* into a three level signal.

partial response signal: The *signal* produced after it has been coded using the *partial response code.*

partitioned database: A *database* which has been split for ease of access by certain groups of *users.* For example if the users are spaced geographically apart, and each group uses a particular part of the database, these portions can be located close to the users in order to minimise the *transmission* requirements.

partition noise: A form of *noise* which occurs in multi-electrode devices, such as transistors and valves, due to the division of the current between the various elements. See also *thermal noise, shot noise, intermodulation noise* and *impulse noise.*

party line: *Telephone line* shared by two or more *subscribers.* A common *telephone number* is used, the different subscribers being differentiated by some other method such as the *ringing tone.* This is also known as *shared service,* whilst one subscriber per line is called exclusive service.

pascal: The unit used to measure sound pressure and equal to one newton per square metre.

passband: The *frequency band* which can be transmitted by a system without significant *attenuation* (usually 3 *decibels* below the maximum value). For example, the passband of a *telephone* circuit is in the *range* from 300 Hz to 3.3 kHz.

passband filter: A *filter* which allows only those *frequencies* within the *passband* to go through without appreciable attenuation. Also known as *band pass filter*.

passive communications satellite: A *communications satellite* which basically reflects *signals* between *Earth stations* or retransmits after only simple amplification and signal reshaping.

passive hub: A *hub*, used for adding *nodes* to a *network*, which is not connected to an external power source.

passive optical device: A device which can handle *signals* in the *optical range* of the *electromagnetic spectrum*, which does not require an external power source for its operation. Examples of passive optical devices are *optical fibres*, *optical splitters*, *optical attenuators* and *optical filters*.

Passive Optical Network (PON): A *Fibre In The Loop (FITL)* technology in which a single fibre is shared between several *subscribers* using *optical splitters*, which are *passive optical devices*. An example is shown in Figure P.4 in which 1:4 way splitters are used and the subscribers are connected to an *Optical Network Unit (ONU)* over *twisted pair wire* using *Digital Subscriber Line (DSL)* technology.

passive star: A *star network* in which the *node* at the centre of the star does not perform any functions on the *signals* which pass through it.

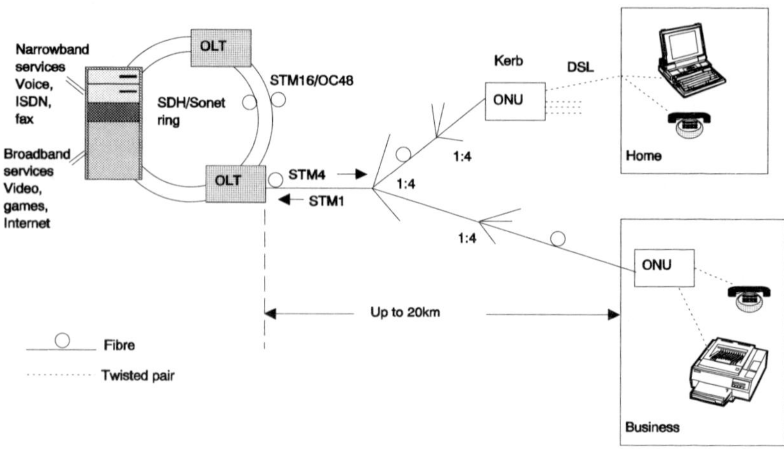

Figure P.4 Passive Optical Network

passthrough: The process of gaining access to an element on a *network* by going through another element, but not using any of its services.

password: A combination of characters which are used for *authentication* of a *user*, or a system, before it is granted access to a *network* or *service*.

patch cord: A short length of cable, with *connectors* at each end, used to make connections at a *patch panel*. Also called a patch lead. Patch cords can be copper or *fibre optic*.

patch panel: A device consisting of many terminal blocks, which is used to make temporary connections and changes between elements on a *network*, using *patch cords* for the connections. It performs a similar function to a *distribution frame* but the connections on a patch panel are intended to be temporary. There are generally four categories of patch panels: wire patching, where connections are made by individual wires; *cable* patching, using 2, 4, 6 or 8 core cable; *software* patching, where links are done by software configurations; and fibre patching, where *optical fibre* leads are used for the patch cords.

path: The communications route or *channel* between the *transmitting terminal* and the *receiving terminal*. It includes all the *network* resources between these, such as *cables*, *multiplexers*, *switching equipment*, etc.

path attenuation: The loss of power in a *signal* in the *path* between a *transmitting terminal* and a *receiving terminal*, usually measured in *decibels*.

path clearance: The term is usually used to describe the vertical distance between a radio beam, as used in *Line of Sight (LoS)* communications systems, such as *microwave*, and any obstructions on the ground. The curvature of the Earth has an effect on path clearance, as shown in Figure P.5, and this is greatest at the centre of the *path*. If d_1 and d_2 are distances in miles, then the curvature of the Earth in feet is given by $h = (d_1 d_2)/1.5K$, where K is a constant. For straight line microwave propa-

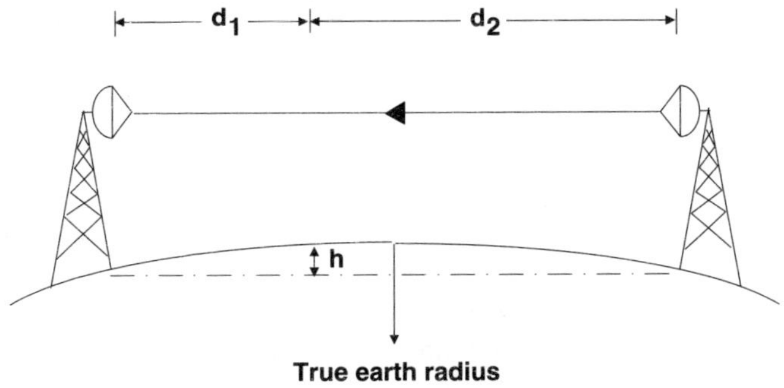

Figure P.5 Path clearance

gation K is equal to 1.0, but it can vary if atmospheric effects are taken into account, which cause the radio beam to curve.

path length: The physical length of the *path* between a *transmitting terminal* and a *receiving terminal.*

path loss: The loss of *signal power* which occurs in the *transmission path.* This occurs in all *transmission medium* and can be due to a variety of reasons. See for example, *refraction*, *scattering*, *absorption* and *fading*.

Path Overhead (POH): Part of the structure of an *SDH frame*. The Path Overhead *bytes* are combined with the *Synchronous Container* to form the *Virtual Container (VC)*, as in Figure V.7. The POH bytes allow the *network* operator to carry out several important functions, such as monitoring the *error rate* of the *transmission path.*

path trace: The *network* feature which allows the *path* taken by the *data* between a *transmitting terminal* and the *receiving terminal* to be determined.

PAX: *Private Automatic Exchange*.

payload: **(1)** Generic term used to describe the *information* part of the *data* being transmitted. **(2)** In the structure of an *ATM cell* (see Figure A.21) it is the part which contains *user* information, *signalling* or *Operations and Maintenance messages*. The payload moves transparently from one end-to-end of the *network*. **(3)** For a *satellite* it refers to the total mass which needs to be sent into *orbit* by its launch rocket systems.

Payload Type Identifier (PTI): Three *bits* which form part of the *ATM cell header*. This provides information regarding *congestion* on the *path*, the type of *ATM Adaptation Layer (AAL)* used to create the *payload* and on whether any *Operations and Maintenance* information is contained in the cell.

pay per view: A *video* service in which a one off payment is made to view it. Commonly used with *cable television* systems. See also *Video On Demand (VOD)* and *Near Video On Demand (NVOD)*.

payphone: *Telephone*, usually situated in a public place, in which the caller pays for each *call* as it is made. This can be done either by inserting money into the machine, or by use of a *pre-paid calling card*.

PBT: *Push Button Telephone*.

PBX: *Private Branch Exchange*.

PC: *Private Circuit* and *Personal Computer*.

PCI: *Protocol Control Information*.

PCM: *Pulse Code Modulation*.

PCMCIA: *Personal Computer Memory Card International Association*.

PCM-30: A *Pulse Code Modulation (PCM)* format used for *Time Division Multiplexing (TDM)* of 30 *voice* or *data channels*, along with one *frame alignment* and one *signalling* channel, onto a single *twisted pair wire*. It is specified in *ITU-T Recommendation* G.732.

PCN: *Personal Communications Network.*

PCS: *Personal Communications Service.*

PDH: *Plesiochronous Digital Hierarchy.*

PDM: *Pulse Duration Modulation.*

PDN: *Public Data Network.*

PDU: *Protocol Data Unit.*

peak busy hour: Same as *busy hour*. Also stated as peak hour.

Peak Information Rate (PIR): The maximum *bandwidth* which a *Virtual Circuit (VC)* may use in a *Frame Relay (FR)* system. Any *data* which exceeds the PIR is discarded by the *network*. See also *Committed Information Rate (CIR)*.

peak load: The highest *traffic* which the *network* is expected to carry. Usually this peak load occurs in the *busy hour*.

peak spectral power: The maximum value of the *radiance* from a light source, measured at specified *wavelengths*.

peak to average ratio: A measure of the quality of an analogue circuit. It is the ratio of the maximum value of a signal in the circuit to its time average value.

peak-to-peak: A measure of the maximum value of a parameter, given by the extremes of its waveform, as in Figure P.6. The peak value is A and the peak-to-peak value is B.

peer entity: Usually refers to a *layer* in the *OSI Basic Reference Model*, located in an *end system*, which corresponds to an equivalent layer in another end system with which it is communicating.

peer interaction: The interaction between components which are considered to be equal in their communications capabilities, such as *modems* at each end of a *link*.

peer-to-peer communications: A communication system in which items performing the same functionality communicate directly with each other. Commonly used in systems based on a *layered architecture*, where the *layers* at the same level communicate directly with each other, although they use the *services* of the layers below them for doing so.

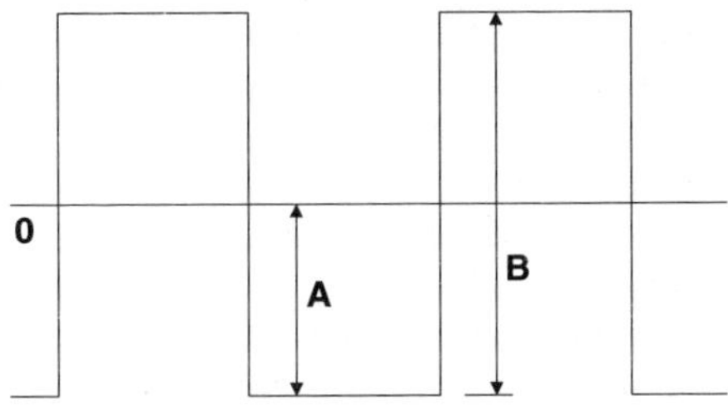

Figure P.6 Peak and peak-to-peak values

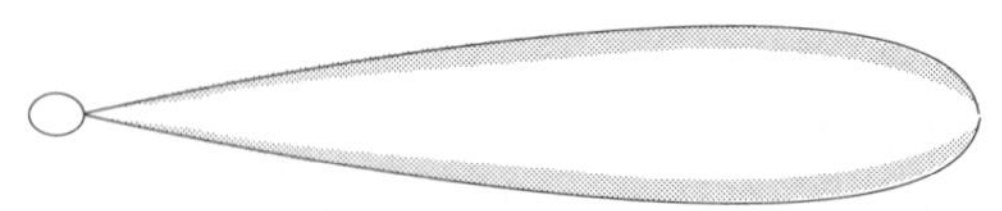

Figure P.7 Pencil beam radiation pattern

peer-to-peer network: A *network* in which *nodes* can communicate with each other, as required, without having to wait for permission from a central controller, or having to route the *data* through a *host processor.*

peg count: Measure or count of the number of calls which are made or received within a specified period of time.

pel: The smallest element in a *Visual Display Unit (VDU)* screen which is capable of being controlled individually.

pencil beam antenna: An *antenna* in which the radiation beam is confined to a narrow beam of energy, as shown in Figure P.7.

penetration: (1) In a *CATV* system it is the number of homes actually connected to the service compared to the *homes passed.* **(2)** An unauthorised access to a communications or processing facility, which is successfully carried out.

percentage overflow: The traffic *overflow* measured as a percent of the total *traffic.*

percent break: In a *dial pulse* it is the ratio of the dial open time to the sum of the open and closed times, expressed as a ratio.

performance management: One of the five *network management functions* defined by *ISO* (see Figure N.3). The aim of performance management is to monitor and improve on the performance of the *network.* It gathers statistical information to enable both long term and short term planning and predictions of short term trends. Performance monitoring requires the maintenance of logs of objects and states, so as to show trends, and the adjustment of network resources in response to trends.

performance testing: The generic term used to cover a variety of tests which are done on communications systems, to ensure that they meet end-to-end performance requirements.

periapsis: In the *orbit* of a *satellite* it is the point at which the satellite is closest to the centre of the body about which it it orbiting. See also *perigee.*

perigee: The term is usually used to describe *periapsis* when applied to a *satellite* which has an *orbit* around the *Earth.* So it is the lowest point of the satellite orbit, when it is closest to the centre of the Earth and therefore to its gravitational pull.

perigee altitude: The vertical distance from the *perigee* to a point on the surface of the Earth.

perigee motor: During the launch of a *satellite* it is the rocket which is fired during the second stage of the launch, to move the satellite out of its *parking orbit* and into its *transfer orbit.*

periodic signal: A *signal* in which the sequence of events repeats exactly after a fixed period of time. This time is known as the *periodic time* or period. If v(t) is the value of a signal at time t and v(t+T) is its value in time (t+T) then, for a periodic signal, v(t) = v(t+T) where T is known as the *periodic time*.

periodic time: See *periodic signal.*

peripheral equipment: Any equipment which forms part of a complete communications system but is not an integral part of it. For example a printer or a *Visual Display Unit (VDU).*

peripheral interface: The interface which exists on many computing and communications devices, which allows *peripheral equipment* connected to it to be changed without the need for special adapters.

per-line unit: Any device whose use is dedicated to a given *line* and is therefore not available for shared use.

permanent call tracing: The *network* feature which allows *call tracing* to be applied to all *calls* made through any part of the network, such as a *line*, *trunk* or *exchange*.

permanent copy: Same as *hard copy*.

Permanent Signal (PS): A *signal* which is generated when a *telephone* is *off-hook* and no *call* is in progress. This signal can occupy a large part of the *bandwidth* in the *switching centre*.

Permanent Virtual Circuit (PVC): A *Virtual Circuit* which is set up permanently between two *Data Terminal Equipment (DTE)* in a *Packet Switched Network (PSN)*. No *call setup* or *call clearing* is needed.

permissible interference: *Interference* which is permitted by regulations, such as by the *ITU*. See also *International Table of Frequency Allocations*.

permitted service: See *primary service*.

permutations: The number of ways in which a proportion can be chosen from a group, when sequencing within each *combination* is involved. Therefore the number of permutations of two letters out of four letters A, B, C, D is 12, i.e. AB, BA, AC, CA, AD, DA, BC, CB, BD, DB, CD, DC.

Personal Access Communications System (PACS): USA equivalent of the *Personal Handyphone System (PHS)*.

personal communications: Generic term used to describe a communications system which allows a *user* to communicate with other users, and to access other *services*, irrespective of their location, and also whilst on the move. The user is allocated a personal number which is independent of location.

Personal Communications Interface: USA equivalent of the *CT2/CIA* standard.

Personal Communications Network (PCN): Standard for a *cellular radio system* operating in the 1800 MHz *frequency band* in Europe. In the USA it is known as the *Personal Communications Service (PCS)* and operates in the 1900 MHz band.

Personal Communications Service (PCS): See *Personal Communications Network (PCN).*

Personal Computer (PC): A computer which is primarily intended for use by a single person. Several Personal Computers can be networked together and may be used to access larger computers, such as *minicomputers* or *mainframes.*

Personal Computer Memory Card International Association (PCMCIA): The association which develops standards for small cards which plug into portable and desktop equipment, to provide enhanced functions, such as additional memory or *LAN* access using radio.

personal global communications: Same as *personal communications.*

Personal Handyphone System (PHS): Japanese *Cordless Telephony (CT)* standard which makes use of the *PSTN*. Developed in Japan in 1989 under the guidance of the *Ministry of Posts and Telecommunications (MPT)*. It operates in the *frequency band* from 1895 MHz to 1918 MHz and 77 *carrier frequencies* are shared amongst three applications within this band, as shown in Figure P.8. The basic radio parameters for the system are given in Table P.2.

personal identification code: Same as *Personal Identification Number (PIN).*

Personal Identification Number (PIN): A number which is allocated to a person for *authentication* by a system. Generally this authentication is done before access is granted to a *service*, such as to use a mobile *telephone* or to withdraw money from a bank teller machine. The PIN

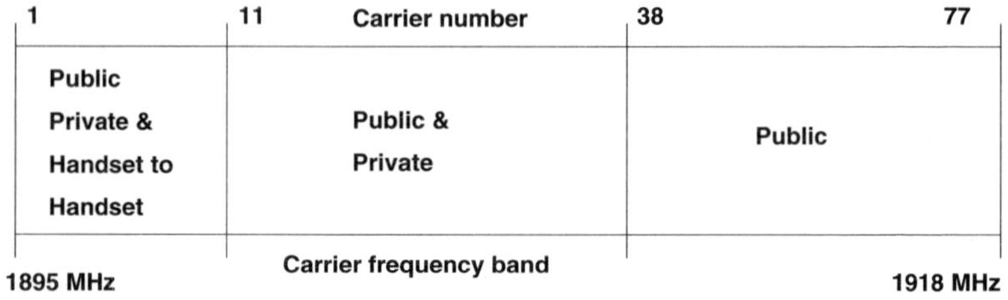

Figure P.8 PHS spectrum allocation

Table P.2 PHS radio parameters

Parameter	*Value*
Frequency band	1895 MHz to 1918 MHz
Radio channel spacing	300 kHz
Transmitted symbol rate and over-air bit-rate	192 kbaud and 384 kbit/s
RF carriers	77
Peak and mean transmitted power; base station, high power	4 W max. and 100–500 mW
Peak and mean transmitted power; base station, standard power	160 mW and 20 mW
Peak and mean transmitted power; indoor base station	80 mW and 10 mW
Modulation	π/4 QPSK
Duplex channels per frame	4
Frame period	5 ms
Bits per timeslot	244 bits
Guard time bits and time	16 or 41.7 μs
Net channel traffic bearer rate	32 kbit/s
Diversity	Base station antenna diversity

number is often used in conjunction with a card with is inserted into the machine, so as to provide double protection against fraud.

Personal Mobile Radio (PMR): A mobile radio standard primarily intended for use to control fleet vehicles, such as by the emergency services (police, fire, ambulance), taxi firms and delivery vehicles. The majority of PMR systems operate within the *frequency band* 30 MHz to 960 MHz. Also known as *Private Mobile Radio.* See also Trunked Mobile Radio.

Personal Number Service (PNS): A *network* feature which allows *subscribers* to be allocated a personal number. This single *telephone number* can be used for both fixed and mobile services. The Personal Number Service can work to a predetermined schedule, to determine where and if to contact the subscriber. Since personal numbers are not tied to a particular location they are also useful in being able to contact subscribers long after they have moved house.

personal telecommunications: Same as *personal communications.*

Personal Wireless Telecommunications: USA equivalent of the *DECT* standard.

person to person call: An *operator assisted call* which is set up between two parties. The caller asks for the *called party* by name and is only connected once this party answers the call.

peta: Prefix used to indicate a number multiplied by one thousand million million, i.e. 10^{15}. See, for example, *petahertz.*

petahertz: A measure of *frequency*, equal to 10^{15} hertz.

PFM: *Pulse Frequency Modulation.*

phase: The position of a *signal* relative to a reference point or a reference signal. Phase is usually stated in angular degrees. See also *phase shift.*

phase aligner: A system which aligns the incoming *data* to the *local clock.* It differs from *clock extraction* in that a shorter acquisition time is needed.

Phase Alternation by Line (PAL): A colour television *transmission* system, used in the UK and parts of Europe, which has a *scanning* rate of 625 lines per *frame*, and operates with primary power at 220 volts, 50 Hz. See also *NTSC* and *SECAM.*

Phase Alternation by Line-Modified (PAL-M): A modified version of *Phase Alternation by Line (PAL)* which uses a primary power of 220 volts, 50 Hz, but in which the *scanning* rate is 525 lines per *frame.*

phase-amplitude distortion: *Distortion* of a *signal* caused by variations of the *phase shift* at different *amplitudes.*

phase bandwidth: The *frequency band* over which the phase-frequency relationship of the signal is linear.

phase diagram: A diagram which shows, graphically, the *phase* relationship between two *signals.*

phased array: An arrangement of *antennas*, such as in an array of *dipoles*, in which the phase of the *signal* to each antenna element is varied in such a way as to enable the beam to be shaped and directed, without the need to physically move any of the antennas.

phase difference: The difference in *phase* between two signals. Figure P.9, for example, illustrates the phase difference between two signals having *sinusoidal waveforms* and identical parameters apart from *phase.*

phase distortion: The *distortion* of a *signal* caused by the difference in propagating times, in the *transmission medium*, of its different *frequency* components. This results in the relative *phases* of the different compo-

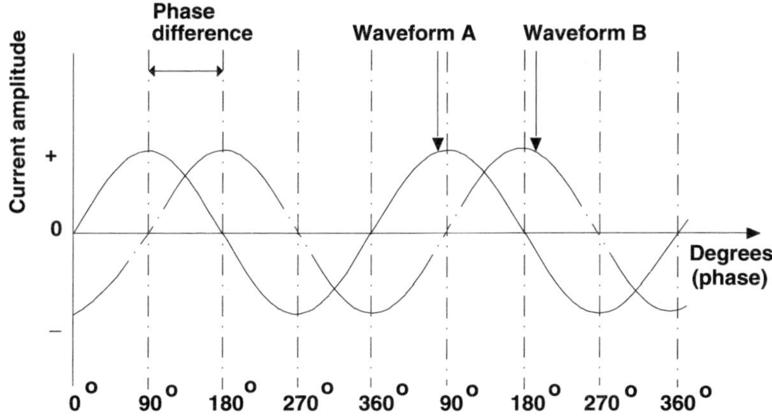

Figure P.9 Phase difference

nents of the *waveform* not being maintained. Also known as *phase-frequency distortion.*

phased orbit: In a group of *satellites* it is the *orbit* which has a defined relationship with one or more of the other satellites.

phase-frequency distortion: See *phase distortion.*

phase-frequency equaliser: A *network* or *circuit* which compensates for *phase-frequency distortion* by making the *phase shift* relatively constant over *frequency band.*

phase hit: A sudden, unwanted and significant *phase shift* of a *signal.*

phase inversion: The change in *phase* of a *signal* by 180°.

phase inversion modulation: *Modulation* in which the states are represented by *signals* which are 180° apart in *phase.*

phase jitter: The rapid and continuous variation (*jitter*) in the *phase* of a *signal.* This variation may be random or cyclic.

Phase Modulation (PM): A *modulation* technique in which the *phase* of the *carrier signal* is varied by the *modulating signal*, as in Figure P.10. See also *Amplitude Modulation (AM)* and *Frequency Modulation (FM).*

Phase Reversal Keying (PRK): *Phase Shift Keying (PSK)* in which the *phase* of the *carrier* is shifted between two states located 180° apart.

phase shift: The change in the *phase* of a *signal.*

phase shifter: The device used on each element of a *phased array*, which allows the direction of the beam from the *array antenna* to be varied rapidly, without the need for mechanical movement.

Phase Shift Keying (PSK): Technique for the *transmission* of *digital signals*, over analogue *lines*, using *phase modulation.* See also *Frequency Shift Keying (FSK)* and *Amplitude Shift Keying (ASK).*

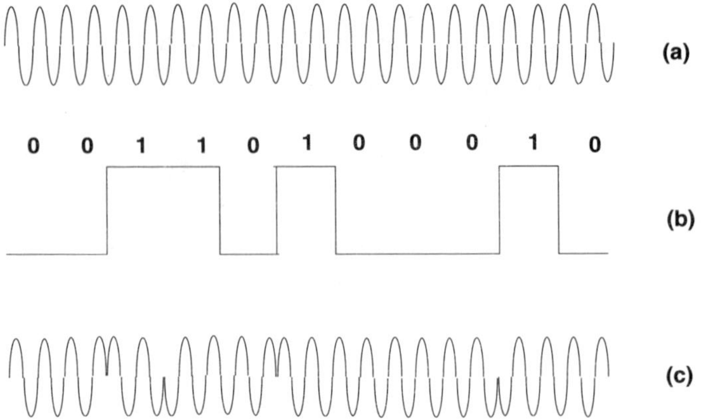

Figure P.10 Phase modulation: (a) carrier; (b) modulating waveform; (c) modulated waveform

phase velocity: For a *signal* propagating in a *transmission medium* it is the velocity with which an observer would need to move in order to make the *phase* of the signal appear to remain constant.

phase wander: The gradual change, or *wander*, in the *phase* of a *signal*.

phone mail: Same as *voice mail*.

phonetic alphabet: Spoken words which are used to represent the alphabet, to avoid confusion between letters transmitted by radio or *telephone*. These are given in Table P.3 and are usually accompanied by pronunciation guides for different languages, so that the words sound the same in all languages.

phosphorescence: The *electromagnetic radiation* from a material which has been excited by an *electron beam*. The emissions occur during the excitation and continue for a period after it has stopped, as in a television screen.

photocurrent: The electrical current generated in a *photodetector*.

photodetector: A device which converts light energy into electrical *signals*. It is generally used as a light sensor.

photodetector responsivity: A measure of the effectiveness of a *photodetector*. It is given by the ratio of the r.m.s. output current to input optical *signal*.

photodiode: A semiconductor diode which acts as a *photodetector*, the reverse current through it being equal to the strength of the incident optical *signal*.

photofax: *Facsimile apparatus* used for the transmission of full tone photographs. The size is normally below A4 and the image is reproduced on photographic paper. See Table F.2. See also *pagefax*.

Table P.3 Phonetic alphabet

Letter	*Word*	*Letter*	Word
A	Alpha	N	November
B	Bravo	O	Oscar
C	Charlie	P	Papa
D	Delta	Q	Quebec
E	Echo	R	Romeo
F	Foxtrot	S	Sierra
G	Golf	T	Tango
H	Hotel	U	Uniform
I	India	V	Victor
J	Juliette	W	Whiskey
K	Kilo	X	Xray
L	Lima	Y	Yankee
M	Mike	Z	Zulu

photograph facsimile telegraphy: *Telegraphy* in which the *facsimile signals* carry a range of tone information, to enable the reproduction of photographic images. Also known as *phototelegraphy*. See also *document facsimile telegraphy*.

photometer: An instrument which measures the level of *irradiance* in a light source.

photometric power: The power from a light source which is measured to take into account the perception of the *human eye*, whose sensitivity varies with the *wavelength* of the light.

photometric unit: Units concerned with the measurement of visible light, and with the response of the *human eye* to this radiation.

photometry: The science and technology concerned with the measurement of light detected by the *human eye*. This involves *electromagnetic radiation* in the *wavelengths* between 380 nm and 750 nm. The basic unit of power is the *lumen*. See also *radiometry*.

photon: A fundamental unit of electromagnetic energy. For an emission of frequency f the photon energy is given by hf, where h is *Planck's constant.*

photon ranging: Same as *correlation ranging.*

photopic response: The response curve of the *human eye* to *electromagnetic radiation* established by the Commission International de l'Eclairage (CIE) in 1924. This peaks at 555 nm and takes into account factors such as the sensitivity of the eye being ten times more sensitive to light in the green part of the *visible spectrum* compared to the blue part. See also *scotopic response.*

phototelegraphy: See *photograph facsimile telegraphy.*

PHS: *Personal Handyphone System.*

physical address: The *address* which identifies a *node* on a *network.*

physical channel: The *transmission medium* through which the *signals* pass, such as *fibre optics*, copper wire, etc.

physical control layer: Same as *Physical Layer.*

physical interface: Generally used to describe the *mechanical interface.* See also *electrical interface.*

Physical Layer: The lowest *layer* in the *OSI Basic Reference Model*, it interfaces to the physical *transmission medium* and converts the *data* into *signals* which are compatible with it, e.g. optical, electrical and radio. See Figure O.4.

physical link layer: Same as *Physical Layer.*

PIC: *Primary Interexchange Carrier.*

pico: A prefix used to indicate one trillionth of a unit.

picture element: The smallest unit of a picture, such as in a *facsimile* transmission or a *Video Display Unit (VDU).* Many picture elements are used to build up the total picture of the display. Also called a *pixel.*

picture frequency: The *frequency* which is generated by *scanning* an object, such as in a *Visual Display Unit (VDU)* or *facsimile* system.

pigtail: A short length of *cable*, made, for example, from copper wire or *optical fibre*, which is permanently connected at one end to a component and is free at the other end. It is used to connect the component to other devices.

pilot: A *signal* which is used for a variety of supervisory or control functions, but not for carrying any *user information.* For example in *Frequency Division Multiplexing (FDM)* systems discrete *frequencies* are added to each translated band of *traffic channels* for monitoring and *Automatic Gain Control (AGC)*, some of these being shown in Table P.4. Regulating pilots are also used to regulate the gain of *repeaters*, so as to compensate for the loss in cable systems, a few of these being shown in Table P.5. Frequency comparison pilots are used to maintain the master

Table P.4 Some reference pilots

Pilot	*Frequency*
Group reference pilot	84.080 kHz
Supergroup reference pilot	411.920 kHz
Hypergroup or mastergroup reference pilot	1552 kHz
Supermastergroup reference pilot	11096 kHz

Table P.5 Some regulating pilot frequencies

Coaxial system	*Pilot frequency*
1.3 MHz, 300 channel	1364 kHz
4 MHz, 960 channel	4092 kHz
4 MHz, 900 channel	4287 kHz
12 MHz, 2700 channel	12435 kHz
18 MHz, 3600 channel	18480 kHz
60 MHz, 10800 channel	61160 kHz

Table P.6 Some frequency comparison pilots

Coaxial system	*Pilot frequency*
1.3 MHz, 300 channel	60 kHz
4 MHz, 960 channel	60 kHz
4 MHz, 900 channel	300 kHz
12 MHz, 2700 channel	308 kHz
18 MHz, 3600 channel	564 kHz

oscillators used on *carrier* generators, and a few of these are shown in Table P.6.

pilot make busy circuit: A *circuit* which uses a *pilot* to put certain *trunks* of a *network* into a *busy* mode so that they will not handle *traffic*. This is done in specific instances, such as when an *outage* has occurred or when performance has fallen to unacceptable levels, for example due to *fading* in *radio transmission* systems. This ensures that signals will not be transmitted with major defects.

pilot tone: A *pilot* consisting of a single *frequency* in the audible *frequency range*.

PIN: *Positive-Intrinsic-Negative*.

PING: *Packet Internet Groper*.

PING packet: See *Packet Internet Groper*.

ping pong: A method of achieving *full duplex* transmission over a *half-duplex channel* by rapidly and automatically reversing the direction of *data transmission* when either end has *data* to transmit.

pinout: Generally refers to the pin configuration on a *connector* which indicates which wires in a *cable* assembly connect to other locations.

pipe: The conceptual link between devices or applications, and includes the *transmission medium*, a *buffer* between applications, etc.

PIR: *Peak Information Rate*.

pitch: The distance between adjacent entities, such as dots on a *Visual Display Unit (VDU)* or the position of a *tone* on a *frequency* scale.

pixel: See *picture element*.

PKES: *Public Key Encryption System*.

plain language: See *plain text*.

Plain Old Telephone Service (POTS): The basic *telephone service* provided over the *PSTN* and using conventional *access lines* such as *twisted pair wire*.

plain text: *Text* which has not been through *encryption*. After encryption it is known as ciphertext.

planar array: An *antenna* having all its elements in one plane. It would generally have a large *antenna aperture*.

Planck's constant: A physical constant which is equal to 6.626×10^{-34} joule-seconds.

plane polarisation: *Polarisation* of an *electromagnetic wave* in which the *field vectors* are in a plane perpendicular to the direction of propagation.

plant: A generic term used to describe all the equipment used to provide *telecommunication services*. This includes both *outside plant* and *inside plant* equipment.

PLAR: *Private Line Automatic Ringdown*.

Plastic Clad Silica (PCS): *Optical fibre* which has a glass *core* but plastic *cladding*. See also *plastic fibre*.

plastic fibre: *Optical fibre* in which the *core* and *cladding* are both made from plastic. See also *Plastic Clad Silica (PCS).*

platform: Usually refers to the basic *hardware* and *software* elements (e.g. computer, *operating system, relational database*) on which an *application programme* is built.

plenum cable: *Cable* which is located in the plenum, i.e. the space above the suspended ceiling used in modern office buildings. The cable is designed to be flame retardant and have low smoke emission, in case of fire.

Plenary Assembly: Part of the *ITU*. In the old ITU structure Plenary Assemblies for the *CCITT* (now *ITU-T*) and *CCIR* (now *ITU-R*) used to meet every four years to approve standards produced by their Study Groups. This process was changed in the 1988 IXth Plenary Assembly meeting in Melbourne, Australia, to speed up the standards making process by enabling standards to be approved by majority voting, by correspondence.

Plenipotentiary: Part of the *International Telecommunications Union (ITU)* structure (see Figure I.9), the Plenipotentiary is held approximately every seven years. It is the highest authority of the ITU, responsible for setting its policies and for revisions to its constitution.

Plesiochronous Digital Hierarchy (PDH): Standard for *plesiochronous transmission*, part of the *digital hierarchy* developed by the *ITU-T* and widely used around the world. Table D.4 shows the two main systems in use. See also *Synchronous Digital Hierarchy (SDH).*

plesiochronous transmission: *Transmission* in which nominal *synchronisation* exists between two or more *signals* being transmitted, but over the long term there could be variations since their *timing* is not derived from a common *master clock*. See also *asynchronous transmission, anisochronous system, mesochronous signal, isochronous transmission* and *synchronous transmission.*

PLM: *Pulse Length Modulation.*

plug: A *male connector* which is used to connect wires into a *jack.*

plugboard: A panel containing arrays of *jacks* into which *plugs* are inserted to set up connections between equipment.

PM: *Phase Modulation* or *Pulse Modulation.*

PMR: *Personal Mobile Radio* or *Private Mobile Radio.*

PNS: *Personal Number Service.*

POCSAG: *Post Office Code Standardisation Advisory Group.*

PODA: *Priority Oriented Demand Assignment.*

POH: *Path Overhead.*

POI: *Point of Interface.*

pointer: **(1)** *Data* in a computer programme which indicates the location of further *information.* **(2)** In a *Visual Display Unit (VDU)* a manually

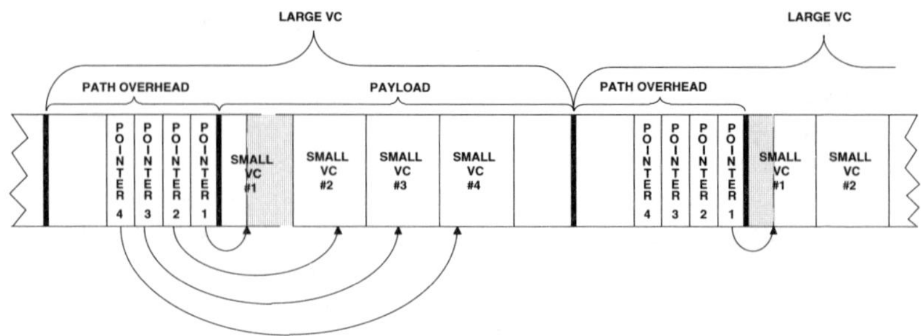

Figure P.11 Use of pointers in SDH frames

operated symbol used to highlight portions of the display on the screen. **(3)** Used within the *Synchronous Digital Hierarchy (SDH)* to indicate the start of information within a *frame*. For example pointers are used when nesting a smaller *Virtual Container (VC)* into a larger one, as in Figure P.11.

Point Of Interface (POI): The point which represents the interface between the *networks* belonging to different *telecommunication* operators, and so defines the technical parameters, test points and the demarcation of responsibility. In the USA, for example, it represents the interface point between the *Local Access and Transport Area (LATA)* and *inter-LATA* functions.

Point Of Presence (POP): Terminology established after *divestiture* to indicate the point at which a local telephone operator terminates a *subscriber line* before it is connected to *leased lines* or long distance *PSTN* lines. For example, in the USA it represents the point in the *Local Access and Transport Area (LATA)* from which a *Local Exchange Carrier (LEC)* connects to an *Interexchange Carrier (IEC)*.

Point Of Sale (POS): Generally refers to the *terminal* (POS terminal) located in retail outlets which records customer purchases, often with bar code readers, and issues receipts. It would generally be connected on-line to credit card checking centres.

Point Of Service (POS): The point at which the *call* moves from the *network* belonging to a local company to one belonging to a long distance company.

Point Of Termination (POT): The physical point at which the *network* enters the *subscriber* premises and the responsibility of the telecommunications operator ends.

point size: The size of a *character* to be printed or displayed, which is specified in units called points. One point equals 1/72 of an inch.

point source: A source of *electromagnetic radiation* which is so small or so distant from the observation point that the radiation appears to come from a planar surface.

point-to-multipoint: A *transmission* system in which *data* goes from one *terminal* to several other terminals.

point-to-point: A *transmission* system in which *data* goes from one *terminal* to another terminal only. See also point-to-multipoint.

Point-to-Point Protocol (PPP): The *protocol* used to transport *Transmission Control Protocol/Internet Protocol (TCP/IP)* over the *Public Switched Telephone Network (PSTN)*. It is used to allow *subscriber* access to a node belonging to an *Internet Service Provider (ISP)*, using the *Public Switched Telephone Network (PSTN)*.

point-to-point network: A *network* in which all *lines* between *nodes* are *point-to-point* and no intervening nodes are involved. See *network topology* and Figure N.7.

Poisson distribution: A method for calculating *probability distribution* for discrete events. It is used when the number of times an event occurs and does not occur cannot both be defined. For example, one can state the number of times a telephone circuit failed over a given period of time, but not the number of times when it did not fail. Poisson distribution can be calculated using probability charts, as in the example of Figure P.12. This gives the probability that an event will occur at least m times when the mean (or expected) value is known.

polar code: *Line code* using two balanced levels, positive and negative, to represent *binary data*. Also called *Non-Return to Zero (NRZ)*.

polar diagram: Generally refers to a plot for a source, such as a light or *radio wave* emitter, showing how the relative intensity of the emitted *signal* varies with angular displacement.

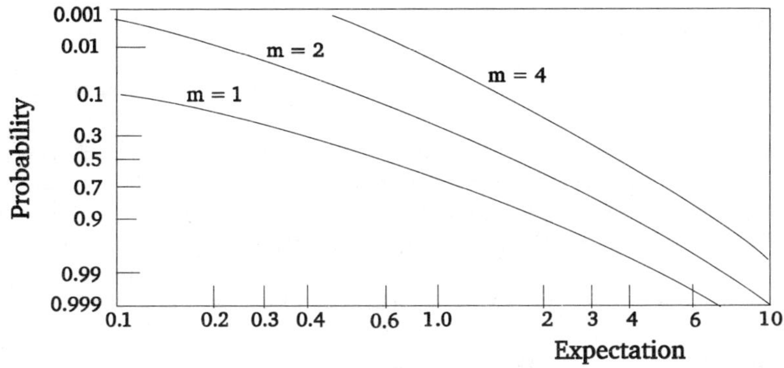

Figure P.12 Poisson probability paper

polarisation: The variation of direction and *amplitude* of an *electromagnetic wave* with time. See *linear polarisation*, *circular polarisation* and *elliptical polarisation.*

polarisation modulation: *Modulation* in which the *polarisation* of the *carrier* is varied by the *modulating signal.*

polarised light: Light which has been through *polarisation.*

polarity: The condition in which two opposite situations occur. For example positive and negative polarity from a power source.

polar operation: Operation of a system in which different *polarities* are used to indicate the different states. For example a current or voltage of one polarity is used to indicate the first state and a current or voltage with equal and opposite polarity is used to indicate the second state.

polar orbit: A *Geostationary Satellite Orbit (GSO)* which passes over the North and South Poles of the Earth.

poll: The process in which a central *node* asks an associated node whether it has any *data* for the central node.

poll character: The *character*, or sequence of characters, which is sent by the central *node* during its *poll.*

poll cycle: In a *polling* system it is the time taken by the central *node* to go around and *poll* all the nodes to which it is connected.

polling: A method for controlling multiple *nodes* on a *network*. Each node in turn is interrogated by a central station to see if they have any *data* to transmit. It they do then they commence *transmission* until their *buffer* is empty. At the end of this they send a 'complete' message and the next node in turn is *polled.* If a user does not have any data to transmit it simply returns the 'complete' message. See also *hub polling* and *roll-call polling.*

polling delay: The time interval between two consecutive *polls* of a node by a master node.

PON: *Passive Optical Network.*

POP: *Point Of Presence.*

port: The physical access point into and out of an electrical equipment or *network.*

portability: The ability to move from one system to another. For example, to move a *software* system from one computer to another; to move equipment from one location to another; to move a *telephone number* from one physical location to another (see *Number Portability*), etc.

port selector: A device which enables more *data lines* to be handled by a single port, by switching between these lines. Also known as a data PBX.

POS: *Point Of Sale.*

POSI: *Promoting conference for OSI.*

positive acknowledgement: An *error control* technique in which an *acknowledgement* is sent by the *receiving terminal* to the *sending terminal* to state that the *information* has been correctly received. See also *Negative Acknowledgement (NAK).*

positive feedback: System in which some of the output *signal* from the device is fed back to the input of the device in such a way that the fed back signal reinforces the original input signal.

Positive-Intrinsic-Negative (PIN): A type of *photodiode* in which the junction consists of three semiconductor materials: one doped positively, one not doped at all (intrinsic) and one doped negatively. It is commonly used as a *signal* detector in *fibre optic* systems.

positive justification: Same as *justification.*

positive temperature coefficient: A parameter which increases as the temperature is increased. See also *negative temperature coefficient.*

POSP: Private Off-Site Paging.

possible crosstalk: *Crosstalk* components which are noted to exist but which do not effect the *user* at the point being considered, although they may effect the user at some other point.

POST: *Power On Self Test.*

postal address: The *address* which is used for messages sent and received by post, which would primarily be for *electronic mail.*

Postal, Telegraph and Telephone (PTT): Legacy term, usually referring to the *telecommunications* authority within a country, often a publicly owned body. The term is also loosely used to describe any large *common carrier.*

POS terminal: See *Point Of Sale.*

post office: Location where *messages* are stored in an *electronic mail* system and are subsequently distributed.

Post Office Code Standardisation Advisory Group (POCSAG): UK developed standard for *digital paging*, named after the study group which developed it. Adopted by the *ITU-T* as Radio Paging Code RCP No. 1.

POT: *Point Of Termination.*

POTS: *Plain Old Telephone Service.*

power balancing: The process of adjusting the power in various parts of a system to achieve optimum performance. For example: in a radio system, adjusting the power in various transmitters for optimum reception and to avoid *interference* between individual transmitters; in a *satellite* system adjusting the amount of available power between the different functions which need to be carried out by the satellite.

power budget: The amount of power loss which a system can tolerate whilst still maintaining its performance within specified limits. Generally this refers to the *signal power* so that it is equal to the difference between the power from the transmitter and the power sensitivity of the receiver.

power density: *Signal* power within a specified *frequency band*, expressed in watts per hertz.

power down: The process of turning off an equipment (removing power). Usually this involves following a specified procedure in order to avoid damaging the system. See also *power up*.

power level: The amount of *signal* power at any point, stated as a ratio to a reference value. It is usually given in *decibels* and if the reference value is 1 milliwatt it is in *dBm*.

power limited: A system whose performance (such as *traffic* handling capability) is limited by the amount of *signal* power available.

power line: A communications system in which *radio frequency signals* are transmitted over the cables, usually in a building, which carry the main *AC* power.

power margin: The unexpected losses which need to be allowed for in the calculation of a *power budget* for a system. This would take into account all the components making up the system, but would not included losses which can be predicted, for example due to ageing, *coupling loss*, etc.

Power On Self Test (POST): Ability of an equipment to automatically go through a series of tests, to check for correct functioning of its main elements, when it is powered up.

power plant: Generally refers to all the equipment which is used to supply power to a location, such as a *Central Office (CO)*. This would include batteries, motor-generator set, *Uninterruptable Power Supply (UPS)*, protection equipment, etc.

power up: The process of turning on an equipment (applying power to it) following a set procedure in order to avoid damaging it. See also *power down*.

p-persistent CSMA: A *Carrier Sense Multiple Access (CSMA)* technique which uses slotted *channels*. When a *user* has *data* to transmit it senses the channel and if it is free it transmits with *probability* p (i.e. it delays *transmission* to the next slot with probability 1–p). If the next slot is idle then the user will again transmit with probability p, and so on. If the slot is not free at any time then the user waits a random time and then tries again.

PPM: *Pulse Phase Modulation* or *Pulse Position Modulation*.

PPP: *Point-to-Point Protocol*.

PPS: *Precise Positioning Service*.

PRA: *Primary Rate Access*.

PRBS: *Pseudorandom Binary Sequence*.

preamble sequence: A short sequence of *bits*, sent before the main body of the *message*, usually to condition the *receiving terminal*, for example, by ensuring *synchronisation*, *error control*, etc.

precedence: The level of urgency attached to a *message*, which indicates to the *receiving terminal* the order in which incoming *data* is to be handled.

precipitation attenuation: *Attenuation* of an *electromagnetic wave* as it passes through the atmosphere containing precipitation, such as rain, hail, sleet, snow and fog. This attenuation can be due to several causes, such as *scattering*, *reflection*, *refraction* and *absorption.*

Precise Positioning Service (PPS): A very accurate positioning service, using the *Global Positioning System (GPS)*, which is available to authorised *users* only, such as the military.

predictive dialling: Automatic *telephone dialling* method, used in *call centres* and using *Computer Telephony Integration (CTI)*. In this the system predicts when an agent is likely to become free and dials the next *telephone number*, connecting the agent to it when the *called party* answers. At the same time details of the called party are displayed on a screen in front of the agent. See also *preview dialling*.

pre-emphasis: The process of increasing the magnitude of certain *frequencies* in the *signal*, compared to other frequencies, in order to compensate for other changes in the system.

pre-emption: The process of interrupting a lower priority *transmission* in order to let a higher priority *message* through.

preferred call forwarding: The feature which allows a *user* to apply *call forwarding* to a selected group of *telephone numbers*.

prefix: *Information* inserted in front of the main body of *data* which qualifies the information which follows. For example, in a *PABX* a 9 before a number would indicate a call on the *PSTN*, and the *country code* inserted before the main number, when making an *international call*, indicates the country in which the *called party* is located.

preliminary call: A *call*, made in a radio transmission system, to set up communications between the calling and called *stations*, prior to the main body of the communications.

premium rate service: Additional *services*, available for an extra fee, over those provided by the normal *telephone* service. Examples are *Calling Line Identification (CLI)* and *call forwarding*.

pre-paid calling card: A card which is programmed to allow a level of *calls* to be made from a *payphone*, depending on the amount of money paid by the *subscriber* for the card. The card is inserted into the *payphone* every time a call is made and it automatically deducts the amount of money used up.

Presentation Layer: The sixth *layer* of the *OSI Basic Reference Model*, which enables the *Application Layer* to interpret the *data* being exchanged between two *open systems*.

preventive maintenance: Periodic inspection and maintenance of equipment, which ensures that potential problems are found before they occur and cause a major *service* disruption.

preview dialling: Term used in a *call management system* in which the *Computer Telephony Integration (CTI)* system predicts when an agent is likely to become free and presents the agent with information on the next client to be called. The agent, however, is given the choice whether to call the number or not. By pressing a single key the agent can confirm the choice and the *telephone number* is dialled by the system. See also *predictive dialling*.

PRF: *Pulse Repetition Frequency*.

PRI: *Primary Rate Interface* or *Pulse Repetition Interval*.

price capping: Term used when the *telecommunications watchdog* sets the maximum price which can be charged by a *Public Telephone Operator (PTO)* for a *service*. This is normally done when there is insufficient competition in the market to regulate prices.

primary access: Same as *Primary Rate Access (PRA)*.

primary centre: A class 3 office (see *class of office*) which links together *toll centres* and can also act as a toll centre for local *end offices*.

primary channel: The *channel* which acts as the main *path* for *transmission* between *terminals*. See also *secondary channel*.

primary colours: Red, green and blue, which can be combined to create the other colour.

primary group: Generally refers to the lowest level of a *multiplex hierarchy*. For example, the primary group for a *Pulse Code Modulation* system contains either 24 *channels* (in Japan and USA) or 30 channels (in Europe).

Primary Interexchange Carrier (PIC): The long-distance *carrier* to which the *subscriber* is normally routed in an *equal access* area.

Primary Rate Access (PRA): An *Integrated Services Digital Network (ISDN)* interface standard, contained within *ITU-T Recommendation* I.431. It is based on existing primary *Pulse Code Modulation (PCM)* formats. It therefore has two transmission rates, 1.544 Mbit/s in the USA and Japan and 2.048 Mbit/s in Europe, as shown in Figure P.13. See also *Basic Rate Access*.

Primary Rate Interface (PRI): Same as *Primary Rate Access*.

primary rate ISDN: Same as *Primary Rate Access (PRA)*.

primary route: The *path* which is normally used between a *transmitting terminal* and a *receiving terminal*. This it the path which is first attempted during *call setup* and alternative paths are only tried if the primary route is not available. See also *secondary route*.

primary service: Service definition contained within the *International Table of Frequency Allocations*. It refers to a *frequency band* which is

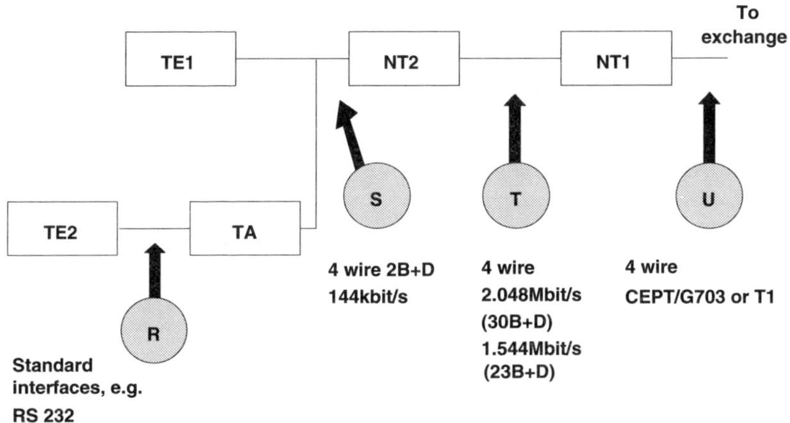

Figure P.13 ISDN Primary Rate Access

shared amongst several *services*. In this instance those services defined as a primary service or *permitted service* have priority over those which are a *secondary service*. Also, when new frequency plans are prepared, primary services will have a prior choice of frequencies over a permitted service.

primary trunk switching centre: *ITU-T* terminology for a *Group Switching Centre (GSC)* or a class 4 office (see *class of office*).

priority facility: The *network* feature which enables certain *users* or type of *traffic* to be given priority or *precedence* over other users or types of traffic.

priority indicator: A *code* in the *message header* which indicates the priority to be attached to the message and therefore determines the sequence in which messages are transmitted.

Priority Oriented Demand Assignment (PODA): A *multiple access protocol* in which *channels* are divided into *frames* and each frame has *data* and *reservation slots*. *Users* make reservations for a number of slots and are allocated a priority for *transmission* on *data* slots. A central station then allocates the data slots to users according to their priority and demand. See also *FPODA* and *CPODA*.

Private Automatic Branch Exchange (PABX): A privately owned and privately operated *switch* which interconnects *telephones* within an organisation, often using *private lines*, and also connects into the *Public Switched Telephone Network (PSTN)*. Almost all of the functions within the PABX are carried out automatically, without operator intervention. Figure P.14 shows the main elements of a PABX.

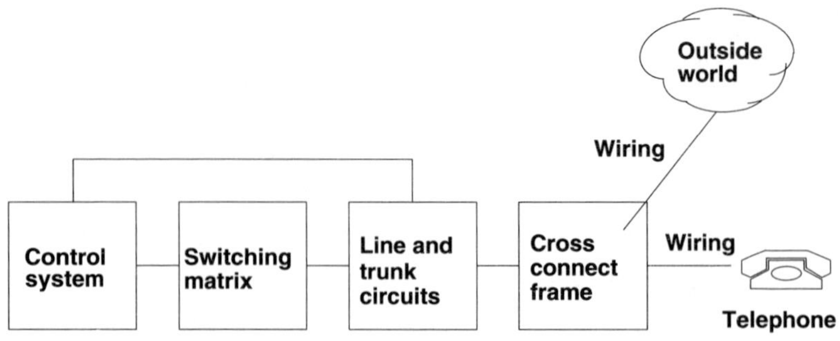

Figure P.14 Main elements of a PABX

Private Automatic Exchange (PAX): A privately owned and privately operated *switch* which provides *telephone* services to users within an organisation, but does not connect into the *PSTN*.

Private Branch Exchange (PBX): Same as a *PABX* but the amount of automatic operation is severely limited, requiring frequent operator intervention. Often the term PBX is used to mean PABX.

Private Circuit (PC): A *circuit* or *line* which is designed for the exclusive use of a person or organisation. Generally there is no access via the *PSTN*.

private key encryption system: A *data encryption* system which uses a single *key* to both encrypt and decrypt *messages*. The key is known to a *user* or a group of users. Example of a private key encryption system is the *Data Encryption Standard (DES)*.

private leased circuit: See *leased circuit*.

private line: See *Private Circuit (PC)*.

Private Line Automatic Ringdown (PLAR): A *leased line* connecting two single *terminals* such that when either one of the terminals goes *off-hook* the other terminal *rings*.

Private Mobile Radio (PMR): See *Personal Mobile Radio*.

private network: Usually refers to a *network* which is owned and operated by a private organisation. However, the term is also used to refer to a network containing *leased lines* from a *Public Telephone Operator (PTO)*, but which are dedicated to the exclusive use of the organisation.

Private Off-Site Paging (POSP): Privately run *on-site paging* system which is operated on a larger area (up to about 10 square miles). The *frequency* chosen often needs to be adjusted to prevent *interference* with other on-site systems in the area.

private wide area paging: Generic term to cover private *paging* systems which operate over a wide geographical area. Examples are *overlay paging*, *revertive paging* and *Private Off-Site Paging (POSP)*.

PRK: *Phase Reversal Keying.*

probability: If an event A occurs n times out of a total of m cases then the probability of occurrence is stated to be P(A) = n/m. Probability varies between 0 (will never occur) to 1 (certain to occur). If P(A) is the probability of an event occurring then 1–P(A) is the probability of it not occurring.

probability distribution: The spread of *probability* of occurrence. Examples are *binomial distribution*, *Poisson distribution*, *normal distribution*, *exponential distribution* and *Weibull distribution.*

probability of failure: A measure of the reliability of a system, it is the *probability* that it will fail over a period of time.

probing: Technique used in *polling* systems in which groups of *users* are polled at the same time, to find the users who are ready for *transmission.* It is similar to the *adaptive tree walk protocol.*

procedural model: The model which defines the actions which are to be taken to achieve the desired objectives, and the results of these actions.

procedure sign (prosign): A combination of one or more letters or *characters* which are used to represent frequently used statements within a communications system, such as an instruction, request for information, etc. Procedure signs are most often used to convey *messages.* See also *procedure word.*

procedure word (proword): In *voice* radio based systems a word which is used to convey a wider procedural meaning, for procedures which are carried out frequently. Procedural words are used most often to carry out communications operations. See also *procedure sign.*

proceed-to-select: The *signal* which a *calling terminal* receives during the *call establishment* phase, to inform it that its *call request* signal has been recognised and it can now transmit the *address* of the *called terminal.*

proceed-to-send: The *signal* sent to the *calling terminal* to state that the *network* is now ready to receive the *calling signal.*

product modulator: A *balanced modulator* which is used to eliminate the *carrier* and generate only *sidebands.* The modulator acts as a switch which multiplies the *baseband* by a quasi square wave carrier. Its output contains upper and lower sidebands and a baseband filter is used to remove all components apart from the required sidebands, as shown in Figure P.15.

profile: See *functional standard.*

programming language: A language which is designed to be used to write computer instructions, and which can be understood and interpreted by the computer. See also *High Level Language (HLL)* and *low level language.*

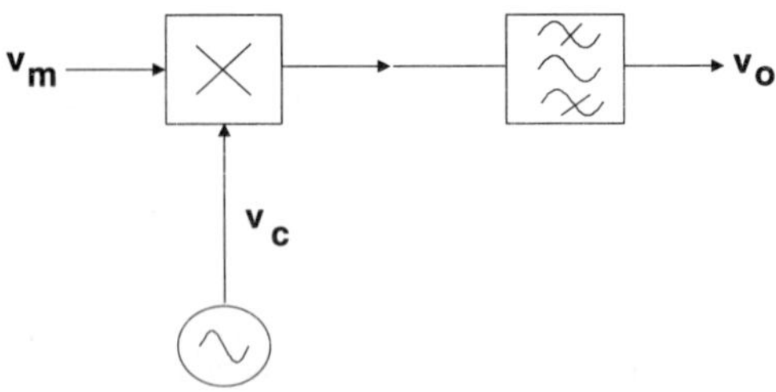

Figure P.15 Product modulator

progressive conference call: A *conference call* in which participants can be added as the call goes along, existing participants having to wait each time a new one is added.

progress tone: The *tone* received by *subscribers* informing them of the progress of their *call*. Examples are *dial tone*, *busy tone*, etc.

Promoting conference for OSI (POSI): One of the two bodies responsible within Japan for promoting *Open System Interconnect (OSI)* standards, the other being the *Asia and Oceanic Workshop for OSI Standardisation (AOWS)*. POSI was formed in 1985 with the prime aim of furthering the use of OSI by international cooperation. The founder members were Oki Electric Industries, Toshiba, Nippon Telegraph and Telephone (NTT), Mitsubishi Electric, Hitachi, NEC and Fujitsu.

prompt: (1) A message, often in the form of an audible or visual signal, to inform the *user* that the system (e.g. computer) has finished an action or is waiting for a user instruction. **(2)** Spoken words, often produced by a *voice processing system*, which guides the user into the use of a facility or service.

propagation coefficient: For a *cable* having a resistance of R, in ohms per unit length, leakage of G, in siemens per unit length, inductance of L, in henries per unit length and capacitance of C, in farads per unit length, the propagation coefficient P is given by $P = [(R+j\omega L)(G+j\omega C)]^{1/2}$. For any *propagation medium* the propagation coefficient can be represented by a real and imaginary term. The real part of the propagation coefficient gives the *attenuation* of the medium, in nepers per unit length, and is called the attenuation coefficient. The imaginary part is called the phase coefficient.

propagation constant: Same as *propagation coefficient.*

propagation delay: The time taken for a *signal* to travel a specified distance though the *propagation medium.*

propagation medium: Same as *transmission medium.*

propagation mode: The method of transmission in the *propagation medium.* For example, in *radio transmission* this mode can be by *ground waves*, *sky waves*, *scattering*, etc.

propagation velocity: The speed at which a *signal* travels through the *propagation medium.*

propagation unavailability: Term used in *microwave transmission* systems to describe the *outage* condition, for example, due to excessive *fading.*

Proportional Traffic Distribution Facility (PTDF): Facility for *routeing* of international *traffic* in countries having more than one *gateway switching centre.* The PTDF ensures that the national *network* shares outgoing traffic evenly amongst these gateways.

protection switching: The facility to recover from a failure by switching to another *transmission path.* For example, 1+1 protection switching provides a standby path which can be switched to if the main one fails or deteriorates, e.g. due to *signal fading* or high errors. In N+M protection switching the N paths are protected by M alternative paths.

protocol: Rules, usually defined by a standards making body, for carrying out a specific function, such as exchange of *information* between two systems, *synchronisation, error control,* etc.

protocol analyser: Test instrument used for analysis of the *protocol* used by communicating devices, such as those on a *Local Area Network (LAN).*

Protocol Control Information (PCI): Part of the *Protocol Data Unit (PDU)* which coordinates the information exchange between entities of a *layer,* using the services provided by the next lower layer in an *open system.*

protocol conversion: To convert the interpretation of a *message* from one *protocol* to another, usually so that it can be understood by equipment, working on different protocols, which are communicating with each other.

protocol converter: A device which carries out *protocol conversion.*

Protocol Data Unit (PDU): *Data* unit which is passed between peer *layers* in the *OSI Basic Reference Model,* containing *information* relating to *protocols,* such as *address* and *control. The PDU contains the Protocol Control Information (PCI)* and the Service Data Unit (SDU) functions, as shown in Figure P.16.

protocol stack: Usually refers to the *software layers* used for communications between systems, covering all the layers of the *protocol* involved. Examples are the *seven layer model* OSI protocol stack, and the protocol stack used within the *Transport Control Protocol/Internet Protocol (TCP/IP).*

protocol testing: Tests which are carried out to ensure that the *protocol* being used meets requirements, as specified by the standards making bodies.

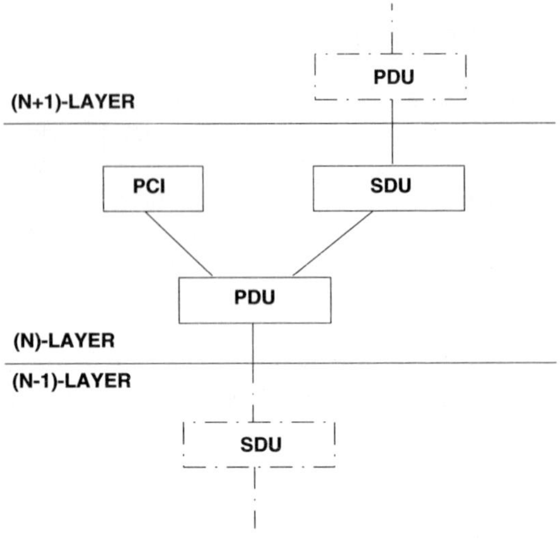

Figure P.16 Data Units

provisioning: Refers to the process of getting the installed equipment operational, so that *service* can be provided to a *user*.

proximity switch: A switch which operates when it is within a set distance of an object, but with no mechanical contact between the switch and the object being sensed.

proxy agent: An entity which provides *network management* for a device by sitting in front of it and relaying information from a remote network manager. A proxy agent has no inherent network management capabilities. For example, Figure P.17 shows the use of a proxy agent to

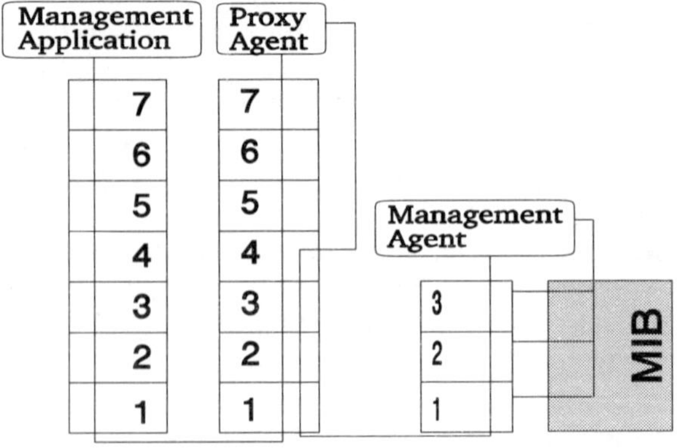

Figure P.17 Use of a proxy agent

control a device which does not have a full *seven layer protocol stack*. The proxy agent uses the lower layer, in the example shown layer 3, for communications.

PRR: *Pulse Repetition Rate.*

PS: *Permanent Signal.*

PSC: *Public Service Commission.*

PSDN: *Packet Switched Data Network.*

PSE: *Packet Switching Exchange.*

pseudocode: A *code* which needs to be interpreted before it can be acted on by a processor.

pseudo duplex: A technique for presenting *full duplex* capabilities to a *user*, whilst still operating in *half duplex* mode. *Data* is *buffered* to prevent loss.

pseudo-random: A sequence which repeats itself to some extent, but is sufficiently random to pass accepted statistical tests for randomness.

Pseudo-Random Binary Sequence (PRBS): A sequence of *binary digits* which is *pseudo-random.*

pseudo-random noise generator: A *noise* generator obtained by setting the generator to a very long periodic sequence, so that the output stream appears to be random even though it repeats at intervals.

PSK: *Phase Shift Keying.*

PSN: *Packet Switched Network* or *Public Switched Network.*

PSN call: A *call* made over a *Packet Switched Network (PSN).* The phases involved in the call are shown in Figure P.18.

psophometer: An instrument which is used for measuring *channel noise* and includes a *weighting network.*

psophometric line weighting: Different *noise frequencies* have different effects on a listener using a *telephone*. To cater for this a *line noise weighting* curve was developed by the *ITU-T*, as illustrated in Figure P.19. This is one of the weightings used in a *psophometer.*

psophometric weighting: See *psophometric line weighting.*

PSPN: *Public Switched Packet Network.*

PSPDN: *Packet Switched Public Data Network.*

PSTN: *Public Switched Telephone Network.*

psychoacoustic: Terminology use to describe the interaction between the characteristics of sound and their interpretation by the human brain. See, for example, *cocktail party effect*, *Haas effect* and *echo.*

PTDF: *Proportional Traffic Distribution Facility.*

PTI: *Payload Type Identifier.*

PTM: *Pulse Time Modulation.*

PTO: *Public Telephone Operator.*

PTT: *Postal, Telegraph and Telephone.*

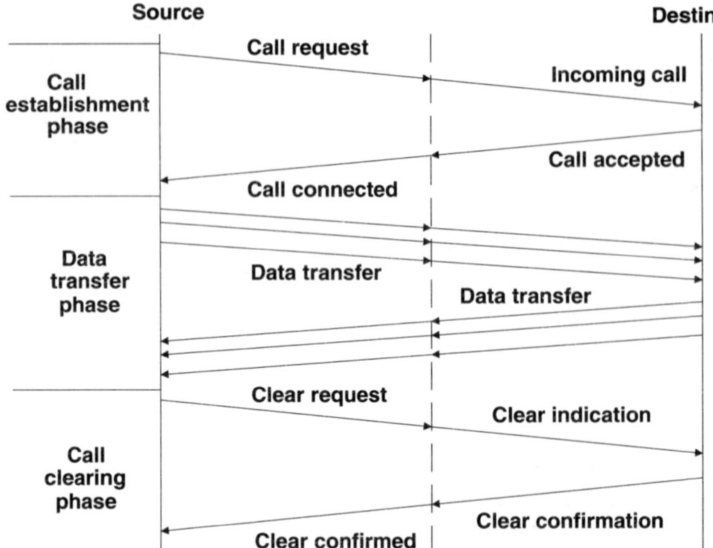

Figure P.18 Phases of a PSN call

Public Data Network (PDN): A *data network* which is operated by a *PTT*, *common carrier* or a *Recognised Private Operating Agency (RPOA)*, specifically for *data communications*. It may use *packet switching* or some other technology. See also *Packet Switched Public Data Network (PSPDN)*.

public data transmission service: The *service* operated over the *Public Data Network (PDN)*.

public dial-up port: *Ports* on a *network* which are available to authorised *users* as required, by access over the *Public Switched Telephone Network (PSTN)*.

Public Key Encryption System (PKES): An *encryption* system which uses two *encryption keys*, a public key (which is known to everyone) and a private key (which is only known to the user who receives the message). The private key is used for decryption of the message.

public message: A *message* which can be seen and read by anyone who has access to the system on which it is transmitted.

public network: Generally refers to a *network* owned and operated by a licensed telecommunications operator, providing *PSTN services* to the public.

Public Service Commission (PSC): *Regulatory body* in the USA who has responsibility for regulating *telecommunications services* at the state level. Also known as the *Public Utility Commission (PUC)*.

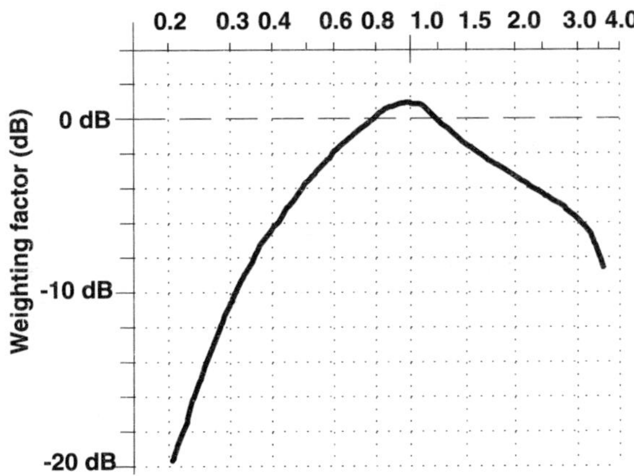

Figure P.19 Psophometric weighting curve

Public Switched Network (PSN): Same as *Public Switched Telephone Network (PSTN).*

Public Switched Packet Network (PSPN): Same as *Packet Switched Public Data Network (PSPDN).*

Public Switched Telephone Network (PSTN): *Public network* which provides *dial-up access* and has been primarily designed to carry *voice traffic*. It represents the major part of the networks in most of the world.

public telegraphy network: A *telegraph network* which is operated by a *common carrier* for *public* use.

public telephone network: Same as *Public Switched Telephone Network (PSTN).*

Public Telephone Operator (PTO): Generic term used to describe any operator who offers *telecommunications services* to the general public. PTOs would generally own their *network* infrastructure.

Public Utility Commission (PUC): Same as *Public Service Commission (PSC).*

PUC: *Public Utility Commission.*

pulse: A *waveform* having a brief transition to a state different from its usual state, as shown in Figure P.20.

Pulse Amplitude Modulation (PAM): A *sampling* technique, as shown in Figure P.21, in which a *baseband signal* is sampled by the *pulse train*, to give an output which consist of the pulse train with amplitude equal to that of the baseband signal at the sample time. These waveforms are shown in Figure P.22. Pulse Amplitude Modulation forms the basis of *Pulse Code Modulation (PCM)* which is widely used in *digital tele-*

Figure P.20 Illustration of a pulse

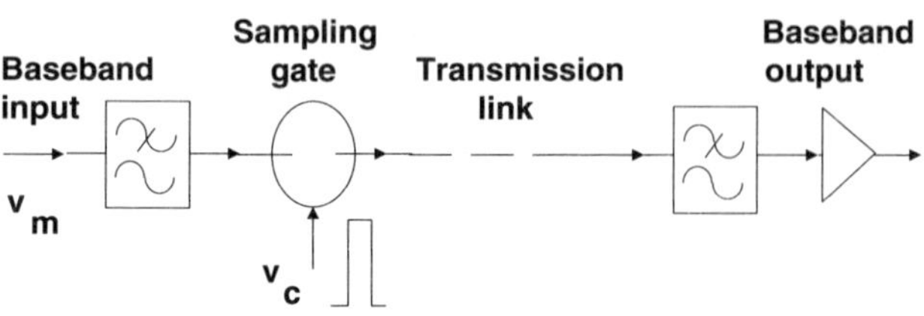

Figure P.21 Basic Pulse Amplitude Modulation system

phony, the sampling rate (*Pulse Repetition Rate (PRR)*) normally equalling 8000 times per second.

Pulse Code Modulation (PCM): An *ITU-T* standard for converting an *analogue signal*, such as *voice*, into a *digital signal*. In this technique *Pulse Amplitude Modulation (PAM)* is used to sample the analogue signal, as in Figure P.23, and each sample is then *encoded* into eight bits in the *Analogue to Digital Converter (ADC)*. At a sample rate of 8000 per second this gives a *transmission rate* of 64 kbit/s. At the receiving end a *Digital to Analogue Converter (DAC)* is used for *decoding*.

pulse compression: The process of reducing the *pulse duration*.

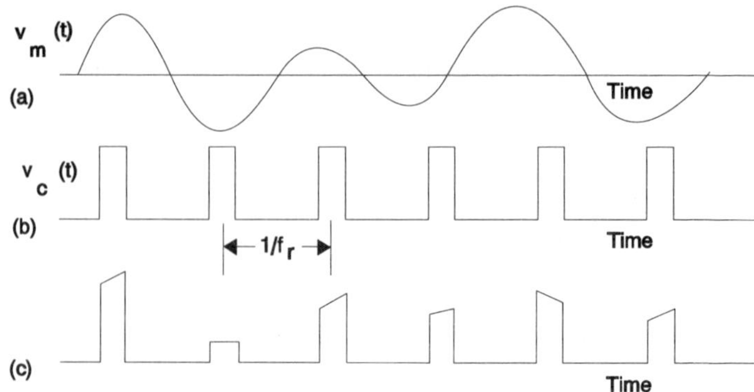

Figure P.22 Pulse Amplitude Modulation: (a) baseband signal; (b) unmodulated pulse train; (c) modulated pulse train

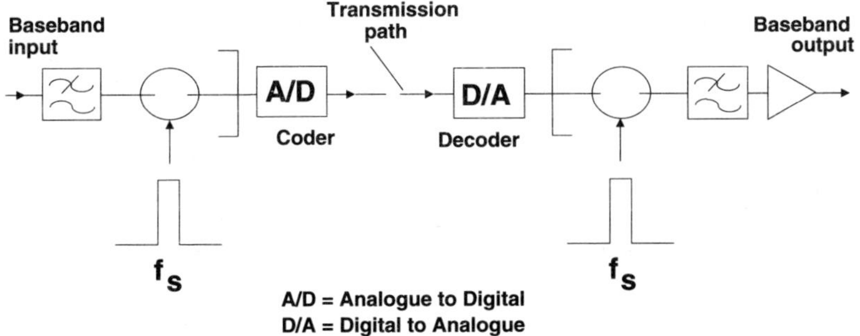

Figure P.23 Pulse Code Modulation

pulse dispersion: The increase in the *pulse duration* (i.e. *dispersion*) as it moves through the *transmission medium.*

pulse duration: The period of time between the start and end of a *pulse*, as in Figure P.20.

pulse duration distortion: The *distortion* of a *pulse*, measured as the difference between the *pulse duration* at two different points in the *transmission medium.*

Pulse Duration Modulation (PDM): *Modulation* technique in which the *pulse duration* of a pulsed *carrier* is varied by the *modulating signal.*

pulse duty factor: The ratio of the *pulse duration* to the *Pulse Repetition Interval (PRI).*

pulse fall time: The *fall time* of a *pulse*. See Figure F.2.

Pulse Frequency Modulation (PFM): *Modulation* technique in which the *Pulse Repetition Rate (PRR)* of a pulsed *carrier* is varied by the *modulating signal.*

Pulse Length Modulation (PLM): Same as *Pulse Duration Modulation (PDM).*

Pulse Modulation (PM): *Modulation* in which some characteristic of a *pulse* is varied by the *modulating signal.* See, for example, *Pulse Duration Modulation (PDM)*, *Pulse Frequency Modulation (PFM)*, and *Pulse Phase Modulation (PPM).*

Pulse Phase Modulation (PPM): *Pulse Modulation (PM)* technique in which the *phase* of a pulse is varied rather than its duration or *amplitude*. See also *Pulse Amplitude Modulation (PAM)* and *Pulse Duration Modulation (PDM).*

Pulse Position Modulation (PPM): Same as *Pulse Phase Modulation (PPM).*

pulse rise time: The *rise time* of a *pulse*. See Figure F.2.

Pulse Repetition Interval (PRI): The time between one pulse and the next, as in Figure P.20.

Pulse Repetition Rate (PRR): The number of *pulses* in a unit period of time, such as a second. It is therefore measured in pulses per second.

Pulse Repetition Frequency (PRF): Same as *Pulse Repetition Rate (PRR).*

pulse stuffing: Insertion of *pulses* into a *pulse train* to achieve some objective, such as *synchronisation*, etc.

Pulse Time Modulation (PTM): The generic term used to describe any *modulation* technique which effects the time relationship of a *pulse*. See, for example, *Pulse Frequency Modulation (PFM)*, *Pulse Phase Modulation (PPM)*, and *Pulse Duration Modulation.*

pulse train: A large number of *pulses* which occur in succession and have similar characteristics.

pulse width: Same as *pulse duration.*

Pulse Width Modulation (PWM): Same as *Pulse Duration Modulation (PDM).*

pure ALOHA: Same as *ALOHA.*

pure contention: Same as *contention.*

push button dialling: Same as *Multifrequency (MF) dialling.*

Push Button Telephone (PBT): A *telephone* which uses *push button dialling.*

push to talk: A *telephone* or *radio communications* system in which the operator needs to keep a key pressed during speech. Generally the communications mode is *simplex.*

PVC: *Permanent Virtual Circuit.*

PWM: *Pulse Width Modulation.*

Q

Q Adaptor Function (QAF): Part of the *TMN functional architecture* (see Figure T.9), the QAF provides the translation between a *Telecommunications Management Network (TMN)* interface and a proprietary non-TMN interface, this latter part being outside the TMN, as shown in Figure T.9. The QAF is therefore used to connect a *Network Element (NE)* which does not support a TMN interface to a TMN.

QAF: *Q Adaptor Function.*

QAM: *Quadrature Amplitude Modulation.*

QCIF: *Quarter Common Intermediate Format.*

QDU: *Quantising Distortion Unit.*

Q Interface Adapter (QA): Part of the *TMN physical architecture* (see Figure T.10), it is the device which a *Network Element (NE)* without a TMN interface connects to. A Q Interface Adapter may incorporate a *mediation function.*

QoS: *Quality of Service.*

QPSK: *Quadrature Phase Shift Keying.*

QSAM: *Quadrature Sideband Amplitude Modulation.*

Q Series: *ITU-T Recommendations* for the *Integrated Services Digital Network (ISDN)*, some of these being summarised in Table Q.1. *Broadband ISDN* aspects are covered under SG11, as shown in Figure Q.1.

Q-switch: A device used to control the output from a *LASER*, which increases the power level by preventing any output until the energy level has reached a specified level, and then shortens the *pulse duration* so that the energy is concentrated into a shorter period.

Q-switched LASER: A *LASER* whose output is controlled by a *Q-switch.*

quad: Term used to describe a *cable* with two pairs of *twisted pair wire*, the two pairs then being twisted together. One pair is used to carry the *telephone* conversation and *signalling*, the other pair acting as a spare, or providing a second *line* for conversations and signalling, or carrying power to ancillary functions in the *handset*.

Quadrature Amplitude Modulation (QAM): *Modulation* technique which combines *Amplitude Modulation (AM)* and *Phase Modulation (PM)*, so as to increase the number of states available. This enables higher *data rates* to be transmitted whilst keeping to the same *bandwidth* of the *transmission medium*. For example, *ITU-T Recommendations* V.26 and V.27 enable four state QAM to produce *Quadrature Phase Shift Keying (QPSK)* to provide a data rate of 2.4 kbit/s over *telephone lines*, with a *signalling* rate of only 1200 bauds. Sixteen state QAM (ITU-T Recommendation V.22 bis) provides 2.4 kbit/s with a signalling rate of

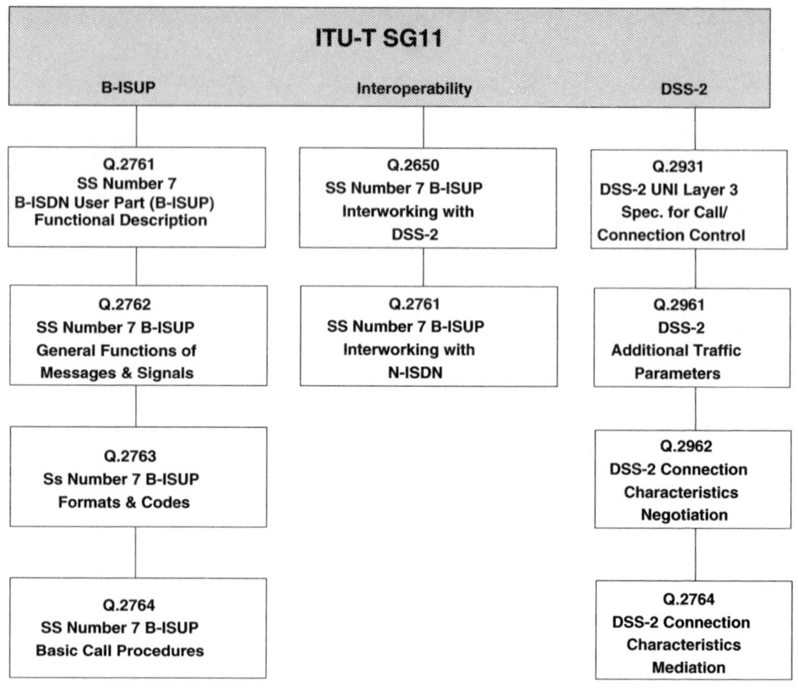

Figure Q.1 ITU-T Recommendations, Q Series

only 600 bauds. Also known as *Quadrature Sideband Amplitude Modulation (QSAM).*

Quadrature Phase Shift Keying (QPSK): *Phase Shift Keying (PSK)* in which four different phase states, usually 90° apart, are used for every period of the *carrier signal.*

Quadrature Sideband Amplitude Modulation (QSAM): See *Quadrature Amplitude Modulation (QAM).*

qualifier bit: Specified in *ITU-T Recommendation* X.29 relating to a *Packet Assembler-Disassembler (PAD)* in a *Packet Switched Network (PSN)*, it is the *bit* in the *packet header* which defines whether more than one level of *information* is contained in the *data* and whether this is control information.

Quality of Service (QoS): A generic term used to define the quality of a *telecommunications* system. It may be measured subjectively, i.e. the quality of a *telephone call* experienced by a *user*, or it can be defined by parameters such as *Bit Error Ratio (BER)*, probability of *call blocking*, number of *lost calls*, etc. See also *Grade of Service (GoS).*

Table Q.1 Some Q Series ITU-T Recommendations

Recommendation	*Description*
Q.60	Management services methodology
Q.513	OAM exchange interfaces
Q.750	Overview of management within Signalling System No. 7
Q.751	Managed Objects within Signalling System No. 7
Q.752	Monitoring and measurement within Signalling System No. 7
Q.753	Management functions within Signalling System No. 7
Q.810	Information model for switching and signalling
Q.811	Q3 interface lower layer protocol
Q.812	Q3 interface upper layer protocol
Q.921	Frame format at the Data Link Layer
Q.931	Signalling protocol in the D channel
Q.941	Management protocol for user-network interface

quantisation: The process in which an *analogue signal* is sampled and this sample is assigned a *digital* value. The values form non-overlapping groups, or *quantisation levels*, and the sampled signal is assigned to one of these groups. Quantisation is used in *Pulse Code Modulation (PCM)*. The analogue signal is subsequently re-created from the sampled values. See also *A-law* and *μ-law*.

quantisation distortion: The *distortion* introduced into the *analogue signal*, following the process of *quantisation*, measured as the difference between the analogue signals before and after quantisation.

quantisation distortion power: The power available in the distortion component of the *signal* which has occurred due to *quantisation*.

quantisation error: The error which occurs during the process of *quantisation* in which there are insufficient *quantisation levels* to represent the *analogue signal* exactly. It is given by the difference between the *digital* or encoded value of the signal and the actual value of the signal at that point. Also known as *quantisation noise*.

quantisation interval: The difference between two adjacent *quantisation levels*.

quantisation level: The number of discrete digital values, or groups, which can be used to represent the *amplitude* of the *analogue signal* at the *sampling* point during *quantisation*. For example if four bits are used for this the number of levels is 15 and with seven bits it is 127. The greater the quantisation level the lower the *quantisation error*.

quantisation noise: Same as *quantisation error*.

quantiser: A device which carries out *quantisation*, i.e. assigns digital values to correspond to the *amplitude* of the *analogue signal* at each *sampling* point.

Quantising Distortion Unit (QDU): It is a measure of the *distortion* introduced into an *analogue signal* when it is converted to a *digital signal*, using *Pulse Code Modulation (PCM)*, and then back again to an analogue signal. *ITU-T Recommendation* G.113 defines one Quantisation Distortion Unit as distortion resulting from a single 8 *bit A-law* or *μ-law* PCM signal. QDUs are cumulative and are added in the case that several PCM conversions take place in tandem.

Quarter Common Intermediate Format (QCIF): Specified in *ITU-T Recommendation* H.261 *digital video coding*, it has half as many *pixels* in each direction compared to the *Common Intermediate Format (CIF)*, i.e. 144 lines of 176 coded *pels*. It is suitable for use as a downgraded version of CIF in applications such as low *bit rate transmission* and small screen displays.

quartile: The division of a series of numbers into four parts or quartiles.

quartile deviation: The *dispersion* of a number from the average of a series of numbers can be found by dividing the series into *quartiles* and then stating the deviation as the interquartile range, i.e. the difference between the first and third quartile numbers, or the quartile deviation, which is half this value.

quasi-analogue signal: A *digital signal* which has been modified so as to resemble an *analogue signal*, for example for *transmission* over an analogue *line*.

quasi-synchronous operation: An operating mode which is almost, but not completely, synchronous.

query: **(1)** The *signal* sent by a *master station* to other *nodes* on a *network*, asking for specific *information*. **(2)** The process of searching for and retrieving *data* from a *database*.

query language: A computer programming *language* designed for carrying actions such as a *query*.

queue: A collection of items arranged in a sequence. For example, a queue of *nodes* on a *network* waiting to transmit *data* during *polling*, or a queue of *data* in a *buffer* waiting for *transmission* over a *busy line*.

queue discipline: The procedure used to determine the sequence in which the *queue* will be served. For example *service* can be on a basis of first in first served.

queueing: The process of delaying an action until a more favourable situation occurs. For example *calls* may be queued until *trunks* become free.

queueing delay: The delay caused to an action being completed because it is held in a *queue*. For example, in the case of a *call*, it would be the time between completion of *signalling* by the *calling terminal* to the instance when the *ringing* occurs at the *called terminal*.

queueing theory: The analytical study of the behaviour of *queues*, widely used in *telecommunications* to predict the performance of a *network* under *traffic* conditions.

quiescent phase: The phase in an operation where little activity is taking place. For example, during *call setup* it is the period during which the *network* and *receiving terminals* are indicating their status for *transmission*.

QWERTY: The layout of a standard *keyboard*. The first line of letters on the keyboard spell out the word qwerty.

R

RACE: *Research and development in Advanced Communications technologies in Europe.*

raceway: A structure made from metal or plastic which is used to locate electrical and communications wiring within a building. Usually this would be located under a raised floor or in the space above a false ceiling.

rack: A metal, open frame, structure into which equipment is mounted and *cables* run to connect to the equipment. Also referred to as a cabinet.

rack mounted equipment: Equipment which has been mechanically designed to be mounted in a *rack.*

RAD: *Rapid Applications Development.*

radar: Radar stands for RAdio Detection And Ranging. It is equipment which is used to measure the location, distance and movement of distant objects. This is done by transmitting a pulsed *radio frequency signal* at the object and recording the characteristics of the signal which is reflected back from the object.

radar beacon (RACON): A device which is used in radionavigation systems. It provides a pulsed *radio frequency signal*, either automatically or in response to an interrogation signal, which can be received by another object and provides it with location information.

radial distribution: Generally refers to a *network* in which wiring runs from a central point outwards to other *nodes*. See also *star network.*

radian: A unit of measure for angles. One radian (rad) equals the angle which is covered by a circular arc having the same length as the radius of its circle.

radiance: The *radiant power* at a point, measured per unit solid angle, per unit projected area. It is expressed in watts per square centimetre per *steradian.*

radiant energy: Energy which is transferred via *electromagnetic waves*. It is measured in joules and is the time integral of *radiant power.*

radiant excitance: The total *radiant power* per unit surface area, measured in watts per square centimetre.

radiant flux: Same as *radiant power.*

radiant intensity: The *radiant power* per unit solid angle measured in a given direction. It is given in watts per *steradian.*

radiant power: The time rate of flow of *radiant energy* from an *electromagnetic radiation* source. It is measured in joules per second or watts. Also known as *radiant flux.*

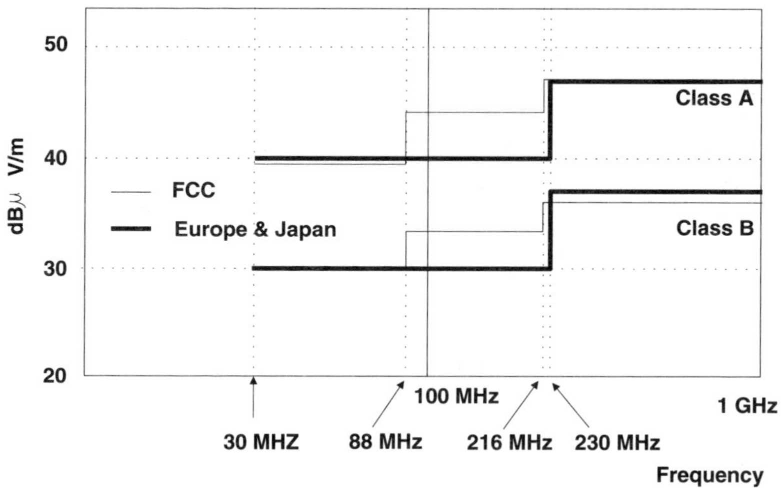

Figure R.1 Radiated emission limits

radiated emission: The *electromagnetic radiation* which is transmitted through the atmosphere rather than conducted through a solid material, such as wire. Limits are set on the maximum level of these emissions, to avoid undue *radiated interference*, by curves such as in Figure R.1. See also *conducted emission.*

radiated interference: *Radio Frequency Interference (RFI)* caused by *radiated emissions.* See also *conducted interference.*

radiating cable: Same as *leaky cable.*

radiation hardened: Equipment is said to be radiation hardened if it can recover normal operation within a reasonable period following an *Electromagnetic Pulse (EMP).*

radiation pattern: The diagram which provides a plot of the *irradiance* from an *antenna* along different directions. See also *Horizontal Radiation Pattern (HRP)* and *Vertical Radiation Pattern (VRP).*

radio astronomy: Astronomy based on the analysis of *radio waves* from *space.*

radio channel: A *frequency band* which is wide enough for use in *radiocommunications.*

radio channelling plan: A *frequency allocation plan* which divides the total *frequency band* into *radio channels*. For example, in Figure R.2 the frequency band has been divided into two sub-bands, A and B. Each of these has been further divided into channels, 1 to 20. Each of these channels has a *bandwidth* which allows a single uni-directional *link* for

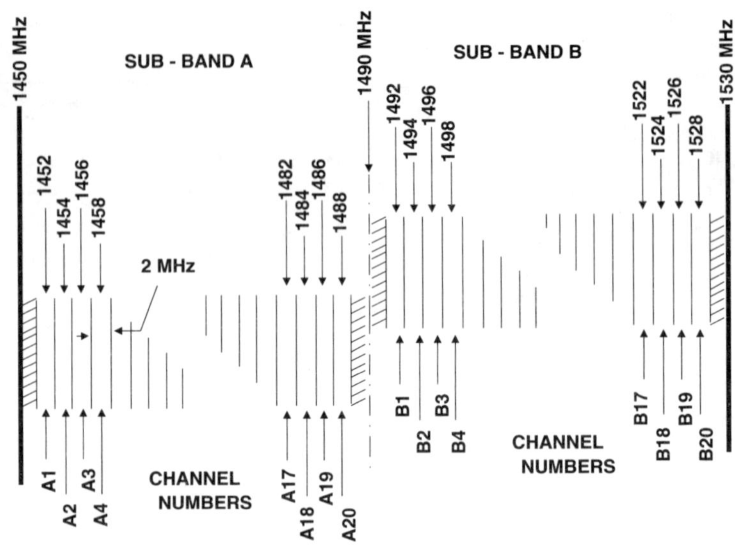

Figure R.2 Radio channelling plan

transmission, so two of the channels (e.g. A1 and B1) would be needed for a bi-directional link.

Radio Common Carrier (RCC): A *common carrier* who provides *radiocommunications services*.

radiocommunications: Generic term used to cover any form of communications which occurs using *radio waves* and operating within the *radio frequency spectrum*.

Radiocommunications Advisory Group (RAG): Part of the *ITU-R* (see Figure I.10), it monitors and provides guidance to the ITU-R Study Groups, as well as undertaking other tasks, such as recommending actions to be taken to increase cooperation with other organisations and advising the Director of the *Radiocommunications Bureau*.

Radiocommunications Assembly: Part of the organisation of the *ITU-R* (see Figure I.10) it contains the Study Groups which carry out the standardisation development work within the ITU-R.

Radiocommunications Bureau: Part of the *ITU-R* organisation (see Figure I.10) the Radiocommunications Bureau is run by a Director who is responsible for organising and coordinating the work of the ITU-R. It provides all the administrative and technical support to the Conferences and Study Groups, applies the provisions of the *Radio Regulations*, coordinates the preparation and publication of all documents, and records and registers frequency assignments and orbital characteristics of space services, as well as maintaining the *Master International Frequency Register.*

Radiocommunications Sector: Part of the organisation of the *ITU*, it is known as the *ITU-R* and carries out all the work in the field of *radiocommunications* for the ITU. See Figure I.9.

radio coverage: The geographical area within which a *radio transmitter* can provide *signals* of sufficient strength so that they can be effectively picked up by a *receiving station.*

radio coverage diagram: A diagram, usually in the form of a *polar diagram*, which shows the effective *radio coverage* from an *antenna.*

Radio Direction Finding (RDF): The technique of finding the direction in which a *transmitting station* is located by reception and analysis of its *radio transmissions.*

radio fix: The technique for locating the position of a *transmitting station* by taking several *radio direction finding* measurements, the location being where these intersect.

Radio Frequency (RF): *Electromagnetic radiation* which covers a *frequency band* from 3 kHz to 300 GHz.

radio frequency assignment: The allocation of particular *radio frequencies*, or a *radio frequency band*, to a *station* for its use. This assignment is normally granted by a *regulatory body* and is closely controlled to prevent *interference* between adjacent *transmitting stations.*

radio frequency band: The *radio frequency spectrum* is divided into a series of radio frequency bands, as shown in Figure R.3, each band covering a specified *frequency range.*

radio frequency carrier wave: A *carrier signal* within a *radio frequency band.*

Radio Frequency Interference (RFI): *Interference* which is caused by *signals* within the *radio frequency spectrum.* See also *conducted interference* and *radiated interference.*

radio frequency signal: A *signal* within the *radio frequency spectrum* which is used for *transmission* through the atmosphere.

radio frequency spectrum: The *electromagnetic spectrum* covering the range from 3 kHz to 300 GHz, as shown in Figure R.3. This is also given in Table R.1.

radio horizon: For an *antenna* it represents the horizon at which *radio waves* transmitted from the antenna are at a tangent to the Earth's surface.

Radio In The Loop (RITL): Same as *Wireless In the Local Loop (WILL).*

Radio Link Protocol (RLP): A Layer 2 *protocol*, specified within the *Global System for Mobile communications (GSM)*, for *error correction* between the *mobile phone* and the *network*, in the GSM non-transparent *service.*

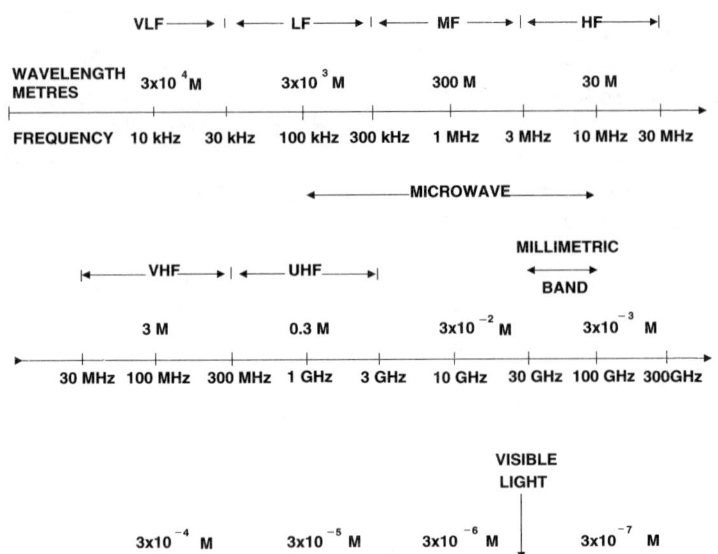

Figure R.3 Radio frequency bands

Table R.1 Radio frequency bands

Band name	*Band number*	*Frequency range*
Very Low Frequency (VLF)	4	3–30 kHz
Low Frequency (LF)	5	30–300 kHz
Medium Frequency (MF)	6	300–3000 kHz
High Frequency (HF)	7	3–30 MHz
Very High Frequency (VHF)	8	30–300 MHz
Ultra High Frequency (UHF)	9	300–3000 MHz
Super High Frequency (SHF)	10	3–30 GHz
Extremely High Frequency (EHF)	11	30-300 GHz
Tremendously High Frequency (THF)	12	300–3000 GHz

radiometric units: The measurement units used within *radiometry*, such as *radiance*, *radiant intensity*, *irradiance*, *radiant power* and *radiant exitance*.

radiometry: The measurement of *electromagnetic radiation* from all the *wavelengths* within the *optical spectrum*, not taking into account any human perceptions. The ideal detector for radiometric measurements is one with a flat response with *wavelength*. See also *photometry*.

radio modem: A *modem* which can operate over a *radio channel*, for example by the use of *cellular radio systems* or *packet radio*.

radio paging: Same as *paging*.

radio propagation: The *transmission* or propagation of *radio waves* through the atmosphere. Radio propagation is effected by several factors, such as *fading*, *multipath effects*, *reflections*, *diffraction*, etc.

radio range: The *range*, measured as the maximum distance from a *radio transmitter*, at which the *radio signal* still has sufficient strength to be correctly received.

Radio Regulations: Document published and maintained by the *ITU-R*. Central to the Radio Regulations is the *International Table of Frequency Allocations*.

radio relay: A *relay station* which handles *radio frequency signals*.

Radio Resource management (RR): Used for *signalling* within *General System for Mobile communications (GSM)*, the Radio Resource management is concerned with managing the *logical channels*. This includes *paging*, *hand-off*, channel assignments, etc. See also *Connection Management (CM)* and *Mobility Management (MM)*.

radio telegraph: The *transmission* of *telegraph code* using a *radio frequency carrier wave*.

radio telephone: A *telephone* which can operate using *radio frequency signals*.

radio transmission: *Transmission* which occurs using *radio frequency signals*. This can take many forms, both *point-to-point* and *broadcast*.

radio transmitter: Equipment which carries out *radio transmissions*.

radio wave: *Electromagnetic waves* within the *radio frequency spectrum* which are propagated through the atmosphere.

RADSL: *Rate Adaptive Digital Subscriber Line*.

RAG: *Radiocommunications Advisory Group*.

RAID: *Redundant Array of Inexpensive Disks*.

rain attenuation: *Attenuation* of *radio frequency signals* due to *absorption* and *scattering* by rain.

rain barrel effect: Sound *distortion* on a *telephone line* caused by over-*equalisation*. The effect is the same as when talking into a partially filled rain barrel.

raised cosine: A *pulse* shape which provides low *Intersymbol Interference*. This is shown in Figure I.14. The waveform within the magnitude of the Nyquist pulse envelope and has half the peak *amplitude* at time T/2.

raised cosine channel: A *transmission channel* having no *Intersymbol Interference* at the sample times of adjacent *signalling* intervals.

R-ALOHA: *Reservation ALOHA*.

RAM: *Random Access Memory*.

Raman effect: The non-linear *scattering* of a *photon* by a molecule in the *transmission medium*. This has the effect of producing multiple *wavelengths* of *electromagnetic radiation* from a source with a narrow band of wavelengths.

Raman scattering: See *Raman effect*.

Random Access Memory (RAM): Storage device in which information can be stored and read, the time to access any location in the store being the same.

random error: Error in *data* in which any *digit* is as likely to have the *error* as any other digit.

random noise: Same as *white noise*.

random routeing: *Routeing* strategy in which a number of *paths* are tried at random until a free path is found.

random signal: A *signal* in which variations in parameters, such as *amplitude*, *frequency* and *phase*, occur in an unpredictable manner. It is not possible to predict the future value of the signal based on previous *data*. Also known as *stochastic signal*. See also *deterministic signal*.

range: **(1)** The maximum distance which a *signal* can be effectively transmitted and received. **(2)** The difference between two values, such as two numbers.

ranging: The process of determining the *range* or distance from a reference point to the measurement point. For example ranging is used to determine the range of an *Optical Networking Unit (ONU)* before it can be connected into an existing *Passive Optical Network (PON)*. Several techniques exist for this, such as *photon ranging* and *correlation ranging*.

RAP: *Remote Access Point*.

Rapid Applications Development (RAD): Technique used to quickly develop a new system, which concentrates on delivering 80% of the functionality in 20% of the time.

RARP: *Reverse Address Resolution Protocol*.

raster: The *scanning* of an image, such as on a *Visual Display Unit (VDU)* screen or a *facsimile* document, by means of a *modulated* light beam or *electron beam*. This is done to write the image (*pixels*) or to read it by recording the reflected beam.

raster image: A *bit* mapped image, consisting of *pixels*.

raster scan: The process in which a *raster beam* moves from side to side along a display region, so covering the whole area. See also *raster*.

Rate Adaptive Digital Subscriber Line (RADSL): Also abbreviated to RDSL. A *Digital Subscriber Line (DSL)* technology in which the RADSL *modems* automatically and dynamically adjust their *transmission* speed to match the quality of the *local loop*. This means that it is possible to serve a wider spectrum of *subscribers* although the speed to some of them may be reduced.

Rayleigh criterion: The criterion used to determine if the surface of a ground is rough or smooth, when considering the effect of *reflections* in *electromagnetic wave* propagation. If the angle between the reflected ray and the ground plane (θ) is small then the Rayleigh criterion states that the surface can be treated as being smooth, for a *signal wavelength* of λ, if the undulations on the surface are no greater than $\lambda/(16\theta)$.

Rayleigh fading: *Fading* which is primarily due to *multipath effects*. It is also known as fast fading. This can be predicted mathematically by the Rayleigh distribution function, as in Figure R.4.

Rayleigh scattering: *Scattering* caused in a *transmission medium* due to minute imperfections in the structure of the material. This normally sets the lower limit of the *attenuation*. It varies as the reciprocal of the fourth power of the *signal wavelength*.

RBER: *Residual Bit Error Ratio*.

RBL: *Re-Broadcasting Link*.

RBR: *Re-Broadcasting Receiver*.

RBS: *Robbed-Bit Signalling*.

RBOC: *Regional Bell Operating Company*.

RCC: *Radio Common Carrier*.

RCU: *Remote Concentrator Unit*.

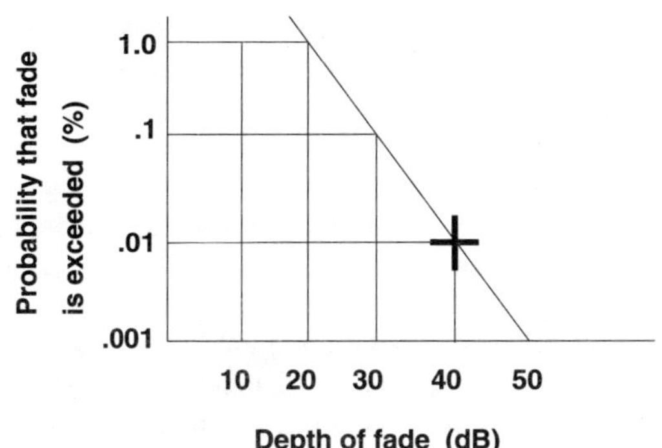

Figure R.4 Rayleigh fading

RD: *Received Data.*
RDF: *Radio Direction Finding.*
RDN: *Relative Distinguishing Name.*
RDSL: *Rate adaptive Digital Subscriber Line.*
RDT: *Request Data Transfer.*

READ coding: *Code* developed by the *ITU-T* and used for *data compression* in *facsimile transmission.* It is a RElative ADdress (or READ) technique in which the new transmitted line to be coded is based on the previous line.

read only: Relates to *information* which is available for reading only. It cannot be modified or copied.

Read Only Memory (ROM): Storage device in which the stored *data* is *read only.*

ready-for-data signal: A *signal* sent by a *Data Circuit-terminating Equipment (DCE),* such as a *modem* to its *Date Terminal Equipment (DTE)* to indicate that contact has been made with the DCE (e.g. modem) at the remote end. The DTEs at the two ends can now begin communications.

ready state: The operational mode in which a *Data Circuit-terminating Equipment (DCE)* and its *Data Terminal Equipment (DTE)* are ready to accept a *call,* i.e. they are not engaged on any other activity.

ready to send: A *signal* sent by a *Data Circuit-terminating Equipment (DCE)* to its *Data Terminal Equipment (DTE)* to indicate that it has made contact with the remote DCE and that *data transmission* between the DTE and remote DTE can take place.

Real Time Network Routeing (RTNR): *Dynamic routeing* strategy for *traffic* implemented using a proprietary protocol by AT&T.

real-time system: A system in which interchange of *data* takes place in *interactive mode.* The response to a *message* is fast enough to affect the subsequent message. See *interactive system.*

reasonableness check: Check made on *data* received by, or transmitted from, a system to ensure that its value conforms to certain predefined criteria. It is therefore a method for *error detection* and is commonly used in *real-time systems.*

Re-Broadcasting Link (RBL): *Video* and *radio frequency transmission links* used to feed *information* from a main *transmitter* to a lower power *relay station,* or between high power main transmitters forming a regional *network.*

Re-Broadcasting Receiver (RBR): High grade receivers and transmitters used in a *Re-Broadcasting Link (RBL).* Incoming *signals* are usually converted to an intermediate *frequency* and then to the required *channel* frequency.

recall: Facility which enables a *dial tone* to be obtained, or the operator to be contacted, after a *call* has been established.

RECC: *Reverse Control Channel.*

Received Data (RD): Reference to the *data signal* received in an *RS-232 transmission* from the *Data Circuit-terminating Equipment (DCE)* to the *Data Terminal Equipment (DTE).*

Received Line Signal Detector (RLSD): A *signal* in *RS-232 transmission* which informs the *Data Terminal Equipment (DTE)*, attached to a *Data Circuit-terminating Equipment (DCE)*, such as a *modem*, that communication is occurring with the distant DCE.

received noise power: The absolute value of the power contained in the *noise* which occurs at the receiving end of a *transmission.*

Received Signal Level (RSL): The strength of a *signal* at the *receiving terminal*, within a specified *bandwidth.* It is usually measured in *decibels* relative to one milliwatt.

Receive Loudness Rating (RLR): One of the divisions of the *Overall Loudness Rating (OLR)*, as shown in Figure O.11.

receive only: A *terminal* which can receive *data* but cannot transmit it, or an operating mode in which only *signal* reception can occur. An example is a printer which receives data and produces a *hard copy* output.

receiver: A device which receives a *data transmission* and interprets and modifies this before presenting it to a *user.*

receiver clock: The *clock* within a *receiver* which provides *timing* for its operations, often ensuring *synchronisation* between the receiver and the *signals* from the *transmitting terminal.*

receiver off-hook tone: A *tone* which is sent by the *Central Office (CO)* to alert *subscribers* when their *telephone* has been *off-hook* for greater than a prescribed time and no connection has been made.

receiver threshold: A measure of the maximum amount of *noise* which the *receiver* can tolerate without significant performance degradation. It is given by the maximum *noise power* to *signal carrier power* which can be tolerated at the input of the receiver whilst still providing an acceptable *Signal to Noise Ratio (SNR)* at its output.

receiving antenna: See *antenna.*

receiving Earth station: An *Earth station* which receives *signals* from a *satellite.* It may then distribute this to other *terminals* over the Earth or send signals to a *transmitting Earth station* for communication to a satellite.

receiving sensitivity: One of the parameters used to determine the performance of a *telephone.* It is the ratio of the sound pressure level in an *artificial ear* to the voltage applied to the terminating impedance of the *exchange* feed bridge. It is designed to have a flat *frequency response* between 300 Hz and 3.4 kHz, with a gradual fall of below 300 Hz and a sharper fall above 3.4 kHz, as shown by the typical curve of Figure R.5. See also *sending sensitivity.*

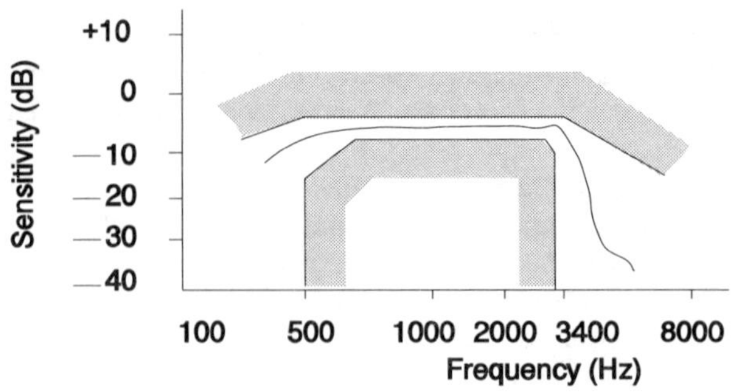

Figure R.5 Receiving sensitivity

receiving station: A *station* which acts as a *receiver*.

receiving terminal: A *terminal* which acts as a *receiver*.

recipient network: In *Number Portability* it is the *network* to which the *number* had been ported. See also *donor network*.

Recognised Private Operating Agency (RPOA): A private organisation who operates a telecommunications system and who has been recognised by the *ITU*. An RPOA can attend ITU meetings but cannot generally vote at these meetings. Examples are BT and Mercury in the UK and Nippon Telephone and Telegraph Corporation (NTT) in Japan.

Recommendations: (1) Standards published by the *ITU*. See *ITU Recommendations*. **(2)** One of the instruments available within the *European Community (EC)*, under the *Treaty of Rome*. A Recommendation has no legal power, being a recommendation for a course of action. It may be issued by the *European Commission* or the *European Council*.

reconfiguration: The process of changing the arrangement of the components within a system or *network*.

record: A group of related items which are combined to form a single unit of *information*. A record generally consist of several *fields*. For example a record could contain the name of a *terminal*, its make, *software* content, and *address*.

recorded information service: A *service* which allows *subscribers* to gain access to *databases* of *information*, on items such as the weather, sports results, stock market prices, etc., using the *Public Switched Telephone Network (PSTN)*.

recorder warning tone: A *tone* which is applied to the *line* during a *telephone call* if one of the parties is recording the conversation, to indicate that the call is being recorded.

record locking: A technique for maintaining the integrity of a *record* by enabling only one *user* at a time to access and change it. When one user accesses the record all other users are locked out.

Record Separator (RS): A *character* which is used to define the separation between *records*.

recoverable error: An *error* following which *error recovery* can be implemented.

recovery: The action of bringing a system back to its normal operational state after a failure or an abnormal event.

recovery procedure: The procedure followed, in the event of a fault, to recover the system back to the state which existed before the fault occurred. This could involve, for example, automatically writing all current data onto memory, while the system is failing, and then restoring this from memory when the fault has been corrected.

recovery time: The time needed for a system to return to its normal operating state following an abnormal event or a failure.

Red Alarm: An *alarm signal* which is generated in a *T1* system if an *error*, such as *Out Of Frame*, persists for more than 2.5 seconds.

RED-BLACK concept: The procedure, used for secure communications systems, in which RED *traffic* (i.e. classified *information* carried in *plain language*) is carried by different equipment from BLACK traffic (i.e. unclassified information or classified information which has been through *encryption*).

red box: An illegal device which connects into a *payphone* and simulates the tones made when a coin is put into the coin box. This tricks the system, or operator, into thinking that the correct amount of money has been deposited for the *call*.

Red-Green-Blue (RGB): The three different colour *video signals* which are transmitted in a display system.

redirected call: A *call* which has been sent to an alternative *address* on the *network*, primarily because the original address was not available.

redundancy: (1) A *network* arrangement in which more elements are provided than that needed for its operation, the additional elements coming into operation in the event of failure of the main elements. See also *standby* and *hot standby*. **(2)** Extra *data*, such as *characters*, *bits* and *bytes* which may be added to the transmitted *information* to carry out auxiliary functions, such as *error detection*, *synchronisation*, etc. This additional data can be removed without affecting the sense of the main *message*.

redundancy checking: Technique used for *error detection* and *error correction* in *data transmission*, in which additional data is appended to the *signal*. This is used at the *receiving terminal* to determine if an *error* has occurred and, in some systems, to correct for this error. The additional data is then discarded.

Redundant Array of Inexpensive Disks (RAID): Computer storage system using several disk drives in a redundant configuration.

redundant bits: Additional *bits* which are added to the *data transmission* for *redundancy*.

redundant code: A *code* in which the *information* being transmitted is represented by more *signal* elements than that needed to transmit the information.

reed electronic exchange: An older form of *telephone exchange* which used *reed relays* as the main switching element.

reed relay: An electronic component consisting of two metal iron reeds enclosed in a sealed glass tube and surrounded by a coil. Current in the coil causes the reeds to move and either make or break the contact.

Reed-Solomon code: A subclass of the *Bose Chaudhure Hocquengherm code*.

re-encryption: *Encryption* of a *message* which has already been encrypted, without first going through *decryption*.

reference antenna: An *antenna* which is used to calibrate the *radiation pattern* from other antennas. It may be real or theoretical.

reference clock: Same as *master clock*.

reference model: Model used to define the behaviour of *telecommunications* systems and specified by standards making bodies, such as the *ITU*. There is, for example, a reference model for the *Integrated Services Digital Network (ISDN)* and one for the *Telecommunications Management Network (TMN)*, both these also defining *reference points*.

reference noise: A *noise power* level which is chosen as a reference to measure other *noise signals* against. See also *noise weighting*.

reference point: Abstract interface point between devices, specified in *reference models*. For example, Figure R.6 shows the reference points in the *ITU-T* reference model for the *Integrated Services Digital Network (ISDN)*, these being the *R reference point*, the *S reference point* and the

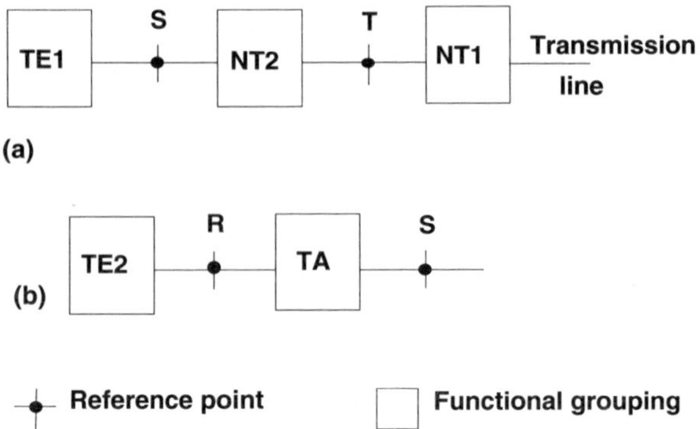

Figure R.6 ISDN reference points

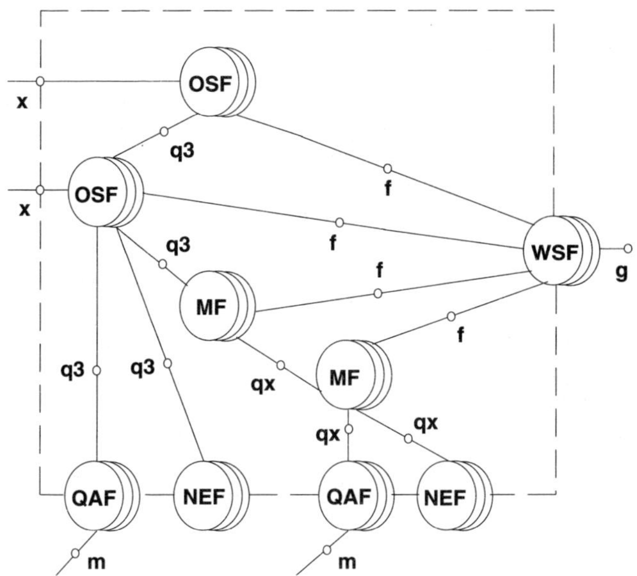

Figure R.7 TMN reference points

T reference point. Figure R.7 shows the reference points for the *TMN functional architecture*.

reference system: A system against which the performance of other systems are measured. It may be real or theoretical.

refile: **(1)** Conversion of *traffic* which has been generated using one set of procedures into traffic which uses another set of procedures. Usually this is done to transfer the *message* over a different *network*. **(2)** The principle used by a *Public Telephone Operator (PTO)* to overcome the differences in *accounting rates* between operators. This is illustrated in Figure R.8. In this it is assumed that Country 1 pays Country 3 an accounting rate of \$1.0 per minute, but it has negotiated a rate of \$0.1 per minute with

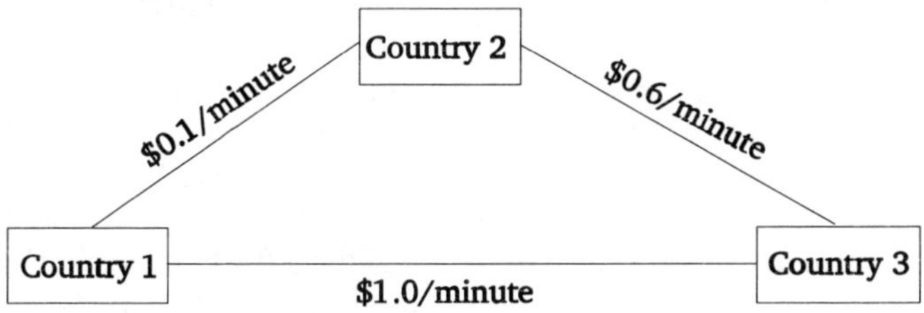

Figure R.8 Illustration of refile

Country 2 who has, in turn negotiated a rate of $0.6 per minute with Country 3. Clearly Country 1 will send its traffic to Country 3 via Country 2, a process known as refile.

refile message: The *message* which is to be *refiled.*

reflectance: A measure of *reflection*, it is the ratio of the reflected light power to incident light power.

reflection: The abrupt change in the direction of *transmission* of an *electromagnetic wave*, such as light or *radio waves*, as it moves from one *transmission medium* to another, the wave being returned back to the originating medium.

reflection density: Parameter given by the logarithm to the base ten of the reciprocal of the *reflectance*.

reflector antenna: An *antenna* which depends on *reflection* of the *signals* from the *antenna feed*. See, for example, *parabolic reflector antenna* and Figure P.2.

ReFLEX: A development of the *FLEX paging protocol*, introduced Motorola Inc. in September 1995. It allows the *pager* to transmit a short, low power, *signal* in *acknowledgement* of *messages* received.

refraction: The bending of an *electromagnetic wave* as it passes from one *transmission medium* to another having a different *refractive index*.

refractive index: A measure of the amount of *refraction* in a *transmission medium*. It is given by the *propagation velocity* of an *electromagnetic wave* in vacuum to that in the medium. The higher the refractive index of the medium the greater the refraction of the electromagnetic wave.

refractive index profile: In a *transmission medium*, such as *optical fibre*, it is the variation of the *refractive index* across the cross section of the medium. See, for example, *step index fibre* and *graded index fibre*.

reframe time: The time between receipt of a *Frame Alignment Signal (FAS)* at the *Data Terminal Equipment (DTE)* and the time when alignment occurs.

refresh: The process of repeating an operation in order to maintain the *data*. For example, in a *Visual Display Unit (VDU)* the image is continually written on the screen so that it remains visible. In some memory systems the data also needs to be periodically refreshed in order to remain in the store.

refresh rate: The rate at which the *data* needs to be *refreshed* in order to be stable. For a *Visual Display Unit (VDU)*, for example, this is 50 times a second in the UK and 60 times per second in the USA.

regeneration: **(1)** The process in which a *digital signal* is restored to its original characteristics, so overcoming the effects of *distortion* and *noise*. **(2)** Sometimes used to describe *refresh*. Regeneration cannot be used on *analogue signals*.

regeneration rate: The rate at which the *data* goes through *regeneration*.

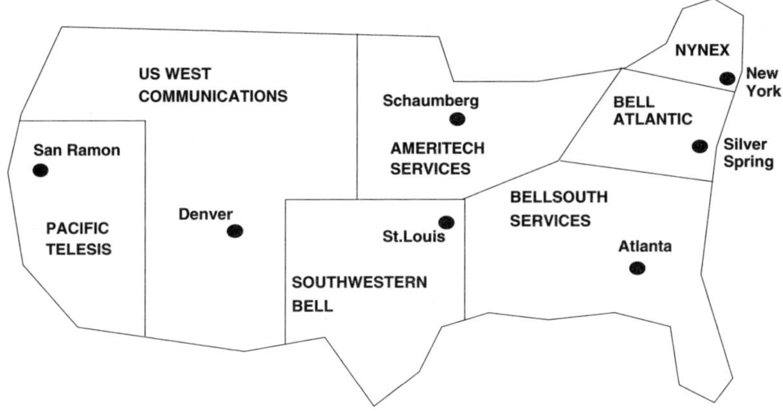

Figure R.9 Regional Bell Operating Companies (following MFJ)

regenerative repeater: A *repeater* which carries out *regeneration* of *digital signals*.

regenerator: A device in which *regeneration* of a *digital signal* occurs.

Regional Bell Holding Company (RBHC): Sometimes used to refer to a *Regional Bell Operating Company (RBOC)*.

Regional Bell Operating Company (RBOC): Following the *divestiture* of AT&T in January 1984, the twenty-two local operating companies of the Bell system were formed into seven regional holding companies, referred to as the Regional Bell Operating Companies or 'Baby Bells'. These are shown in Figure R.9.

regional centre: A class 1 office (see *class of office*) in the USA connecting together *sectional centres*.

regional exchange: Same as *regional centre*.

Regional Radiocommunications Conference (RRC): Part of the *ITU-R* organisation (see Figure I.10), the RRC is held either by an ITU Region or a group of countries who have a mandate to develop an agreement on a particular subject. These conferences do not have the power to modify the *Radio Regulations* and the output from them are only binding on the countries which take part in the agreement.

Regional Workshop Co-ordinating Committee (RWS-CC): Committee set up by the *European Workshop on Open Systems (EWOS)*, the *Asia Oceania Workshop for Open Systems (AOW)* and the *OSI Implementors' Workshop (OIW)* to monitor their joint work programme and activities.

register: A device which is generally used for the temporary storage of *data*. For example, a register is used to receive and store *dial pulses*, and when *dialling* has been completed it is forwarded to control the switch. See also *register insertion ring* and Figure R.10.

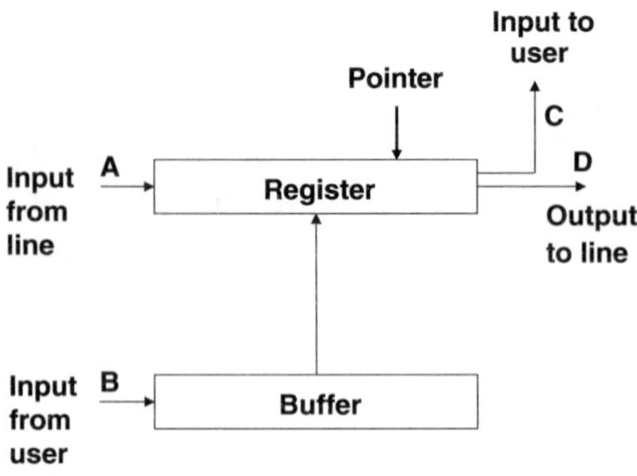

Figure R.10 Interface to a register insertion network

Registered Jack (RJ): Plugs uses with *telephone* and *data communications* equipment, which have been registered with the *Federal Communications Commission (FCC)*. There are a large number of different types, with different number of wires, pins and shapes, for use in various applications. Examples are RJ11, RJ45, etc.

register insertion ring: A *multiple access* technique, as illustrated in Figure R.10. *Data* from the *line* feeds into the *register* and its position is located by a *pointer*. As the register fills up the pointer moves to indicate the level of *traffic* on the line. When a *user* wishes to transmit *data* this is moved into the *buffer*. A check is then done to see if there is sufficient room in the register, and if there is the data is loaded from the buffer into the register and is clocked out onto the line. If there is no room in the register the user must wait, so *contention* is avoided.

register translator: Part of the older *strowger exchange*, it provided a *dialling* translation function, so helping *users*.

registration programme: Programme by the *Federal Communications Commission (FCC)* in the USA which required equipment which is to be connected to the *Public Switched Telephone Network (PSTN)* to be registered with the FCC. This requires that the equipment has been tested to ensure that it will not damage the PSTN.

regression: A method for establishing a mathematical relationship between two variables. Several equations may be used to determine this relationship, the most common being that of a straight line. Figure R.11, for example, shows the number of defective public *telephones* which were reported at seven instances in time. These are seen to lie on an approximate straight line, AB, the equation of the line being given by $y = mx + c$, where x is the independent variable, y is the dependent variable, m is the slope of the line and c is its intercept on the y axis.

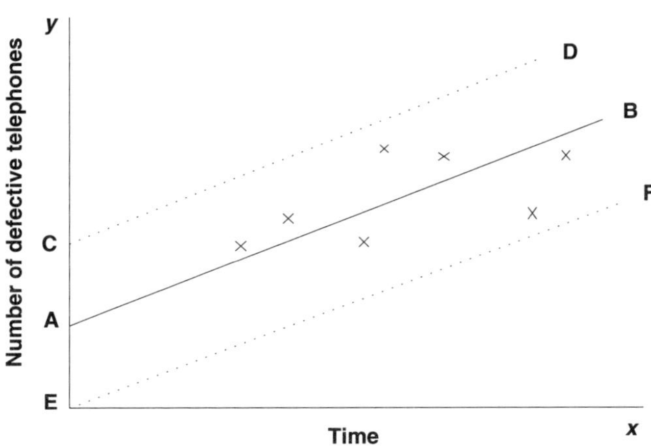

Figure R.11 A scatter diagram

regression testing: Selective retesting of equipment, which has already been tested, to ensure that changes made to the unit has not introduced faults into parts which were not directly modified.

Regular Pulse Excitation Long Term Prediction (RPE-LTP): A variant of *Linear Predictive Coding (LPC)*, specified by *ETSI* and used in *GSM cellular radio systems*. It is a *speech* coder with long term prediction, using speech blocks of 20 ms and operating at 13 kbit/s with an additional 3 kbit/s for *error control*.

regulations: (1) One of the legal instruments available under the *Treaty of Rome* to enforce *European Community (EC)* rules. Regulations are mainly issued by the *European Council (EC)*. They are legally binding in their entity on all Member States. **(2)** Directives issued by a *regulatory body*, usually within a country, which needs to be followed by all operators within that country.

regulatory body: Same as *telecommunications watchdog*.

relational database: A *database* in which *data* is stored and retrieved according to the relationships between items.

Relative Distinguishing Name (RDN): In a *Management Information Tree (MIT)* it is the name of an object relative to any other superior object below the *root*. For example, in Figure M.5 object D has an RDN of D, relative to C and an RDN of CD, relative to B.

relative error: Relative error is the *absolute error* represented as a percentage of the actual value. So, if the actual number is 36845.615, and it is represented to the nearest hundred by 36800, then the absolute error is 36845.615–36800 = 45.615 and the relative error is 45.615/36845.615 = 0.12%. It is written as 36800 ± 0.12%.

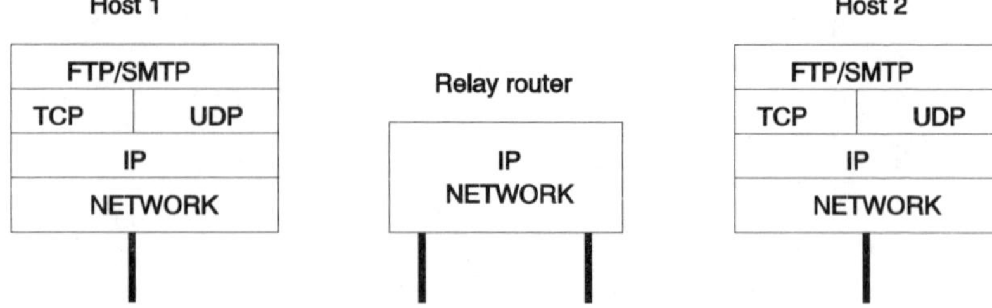

Figure R.12 Use of a relay router

Relative Intensity Noise (RIN): This is the *noise* generated by the *signal* source itself, before the signal is launched into the *transmission medium.*

relative transmission level: The ratio of the power of the test *tone signal* at a given point to that at any other point chosen as a reference.

relay: (1) An electromagnetic device with contacts which can open or close under control of a electrical voltage or current, and which controls the operation of another circuit. **(2)** A *relay station.* See also *relay system.*

relay router: A *router* which acts as an intermediate router between two host routers, such as on the *Internet* using the *Internet Protocol (IP)* (see Figure R.12). The relay router forms part of both *networks* and has two *IP addresses*, one for each network.

relay satellite: A *satellite* which operates as a *relay station.*

relay station: An intermediate *station* which receives *signals* from one station on the *network* and passes it on, without significant modifications, to another station. None of the information received is intended for the relay station. See also *intermediate exchange.*

relay system: Part of the *OSI Basic Reference Model*, see Figure O.4, it provides an intermediate *relay* function between two *end systems.* Only the lower three layers of the model are impacted, the *Physical Layer*, the *Data Link Layer* and the *Network Layer.*

release: (1) The ending of a *call*, usually by pressing the release button (RLS) on the *telephone handset.* **(2)** Authorising the release of a secure *message.*

release time: The time between voltage or current being removed from the *relay* coil and the relay contacts returning to their de-energised state (which may be open or closed).

reliability: The *probability* that a system or *service* will continue to function within specified performance limits over a period of time. See also *reliability of service.*

reliability of service: A measure of the *reliability* of a *service.*

reliability target: The targets for *reliability* set for different systems and subsystems within a *network*, in order for the whole network to meet an overall reliability figure. These targets are normally specified as the *probability of failure* over a specified period of time.

Reliable Transfer Service Element (RTSE): Part of the *Application Layer* standards within the *OSI Basic Reference Model.* These elements ensure reliable communications, protecting against failures and informing the *sending terminal* when an *error* occurs.

remote access: (1) Access by a remote *node* on a *network* with central processing systems. **(2)** The PABX feature which allows a user on a remote site to dial into the PABX and use its features.

Remote Access Point (RAP): Equipment which is located in remote areas and provides a high speed access to a *network* for *users* in those locations who would not otherwise have access to the network. Examples are *concentrators* and *multiplexers*.

remote alarm: An alarm which is generated on a distant point on the *network.*

remote base station: A *base station* which is located remote from the operator's console base station, in order to gain *radio coverage* in *VHF/UHF* systems. See also *communal base station.*

remote bridge: A *bridge* which has the same functionality as a *local bridge*, or *bridge*, except that it can be used to link *Local Area Networks (LAN)* to a *Wide Area Network (WAN)*, as shown in Figure R.13. Two identical bridges are needed at each end of the *link* which gives these bridges the alternate name of half bridge.

remote call forwarding: A *network* feature which enables a *subscriber* to have a telephone number which is located in a local *PABX* or *exchange*. Calls made to this number by local users will be at local rates and all calls will be forwarded to a distant number specified by the subscriber, e.g. to the subscriber's headquarters.

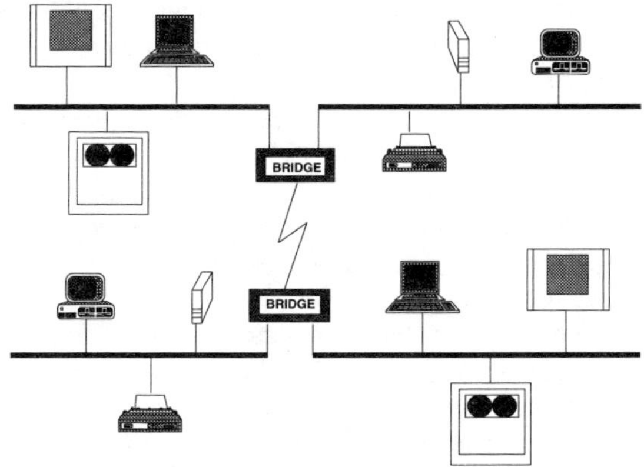

Figure R.13 LAN interconnection using remote bridges

remote concentrator: A *concentrator* which is located close to remote *subscribers* and is used to reduce the number of lines fed from this point into the main *exchange*.

Remote Concentrator Unit (RCU): The unit which acts as the *remote concentrator*.

Remote Monitoring (RMON): An enhancement to the *Simple Network Management Protocol (SNMP)* which was introduced in 1992 and has subsequently been updated by the *Internet Engineering Task Force (IETF)* to RMON v.2 (RFC 1271, 1513). Although basically a *Management Information Base (MIB)* the RMON specification defines *network* monitoring functions and interfaces which allows communications between remote monitors and SNMP based managers. These monitors listen to *traffic* on the network, compiling *data* on parameters such as performance and *throughput*, and produce reports and error analysis.

Remote Operation Service Element (ROSE): Part of the *Application Layer* of the *OSI Basic Reference Model*. It has functions for supporting remote operation in a distributed *open system* environment. It has the ability to request that a remote operation be performed on another open system, a response or an *error* being returned.

remote orderwire: An *orderwire* which has been extended to a point where it is more convenient for operations and maintenance staff.

repeat dialling: A feature of a *PABX* or an *exchange* which allows the last *number* dialled to be automatically redialled if an earlier attempt resulted in the number being *busy*.

repeater: The simplest form of interworking device, which operates at the *Physical Layer*. It can extend the *range* of a *Local Area Network (LAN)* by collecting the *signal* and then repeating it on another identical LAN. However repeaters introduce other imperfections into the system, such as *propagation delay*, so they cannot be used without due consideration. Figure R.14 indicates the location of repeaters within a building. See also *regenerative repeater*.

repeatered fibre optic link: A *fibre optic transmission medium* which incorporates *repeaters*.

repeaterless fibre optic link: A *fibre optic transmission medium* which does not include any *repeaters*. This may be for several reasons, such as the short distance not requiring it, or the loss of *bandwidth*, or increase in *attenuation*, not being acceptable.

replicated database: The system in which one master *database* is used to create and update several other databases which are located at remote regions, close to users.

Request Data Transfer (RDT): A *signal* sent by a *Data Terminal Equipment (DTE)* to its *Data Circuit-terminating Equipment (DCE)*, asking for a connection to be set up.

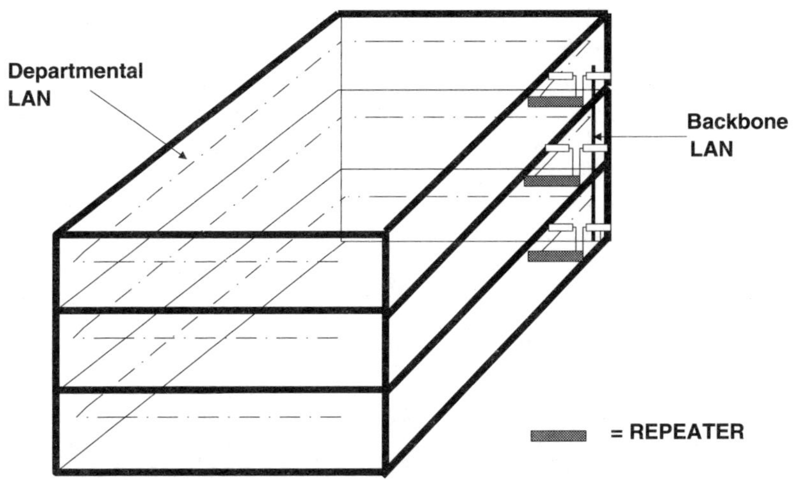

Figure R.14 Location of repeaters within a building

Request For Comment (RFC): Request made by a standards making body for comments on a proposed standard before it is published. For example, the *Internet Activities Board (IAB)* published three RFCs in August 1988 to define *SNMP* and these have now been issued by the *Internet Engineering Task Force (IETF)*. RFCs form a definition of the *Internet* standards.

request repeat signal: A *signal*, sent by a *receiving terminal* to the *transmitting terminal*, asking for a *retransmission*.

Request To Send (RTS): The *signal*, sent by a *Data Terminal Equipment (DTE)* to its *Data Circuit-terminating Equipment (DCE)*, indicating that it wishes to transmit *data* to a remote DTE. The DCE will reply with a *ready to send* signal when a connection to the remote DCE is set up.

rerun point: In a control programme it is the point at which all the *data* needed to rerun the programme is available.

resale carrier: A *common carrier* who buys spare capacity and services, wholesale, from a larger common carrier and then sells these on to the general public on a retail basis. See also *value added service provider*.

Research and development in Advanced Communication technologies in Europe (RACE): A major *European Community (EC)* initiative for the introduction of *Integrated Broadband Communications (IBC)*. It included two phases, covering research, field trials and installation, with *optical fibre links* connecting EC capital cities, and first commercial service by 1996. (See Figure R.15.) The RACE 1 funding was set at ECU 550 million and for Phase 2 it was ECU 489 million.

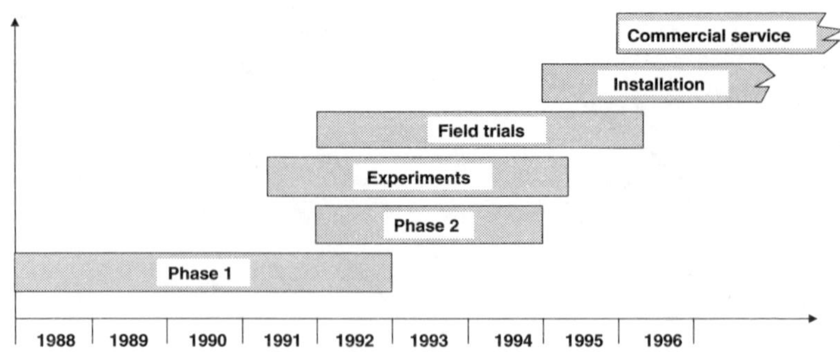

Figure R.15 The RACE programme leading to IBC service

Reseaux IP Europeens (RIPE): Body responsible for allocating *Internet* numbers in Europe and is equivalent to *Internic* which is responsible for USA. See also *Internet Assigned Numbers Authority (IANA)*.

reseller: A generic term used to describe any person or organisation who buys equipment or services and then resells it without making any changes to it. See also *Value Added Reseller (VAR)*.

reservation: Usually refers to the process, within a *multiple access* technique, where a *terminal* wishing to transmit *data* first reserves part of the *capacity* of the *transmission medium*. Example is *Reservation ALOHA (R-ALOHA)*.

Reservation ALOHA (R-ALOHA): A *multiple access* technique which is similar to *ALOHA* but uses some form of *reservation* of *capacity* as a means of minimising *contention*. Several techniques can be used for these reservations, such as *explicit reservation* and *implicit reservation ALOHA*.

reservation slot: *Timeslots* within a *transmission frame* which have been introduced for the express purpose of providing a *capacity reservation* mechanism for multiple users, so minimising or avoiding *contention* in *multiple assess* systems. See, for example, *Reservation ALOHA (R-ALOHA)* and *Reservation TDMA (R-TDMA)*.

Reservation TDMA (R-TDMA): Reservation TDMA is similar to the *Reservation ALOHA (R-ALOHA)* except that *Time Division Multiple Access (TDMA)* is used to access the *reservation slots*. The *transmission channel* is divided into N *reservation slots* and kN *data* slots per *frame*, as in Figure R.16, where N is the number of *users*. Each user is allocated one data slot within each of the k subgroups, for its use, and one reservations slot. Reservations are made in the reservation slot and now there is no *contention* since no other user can transmit in it. If the owner of a data slot does not wish to use it then it is assigned to other ready users, usually on a round robin basis.

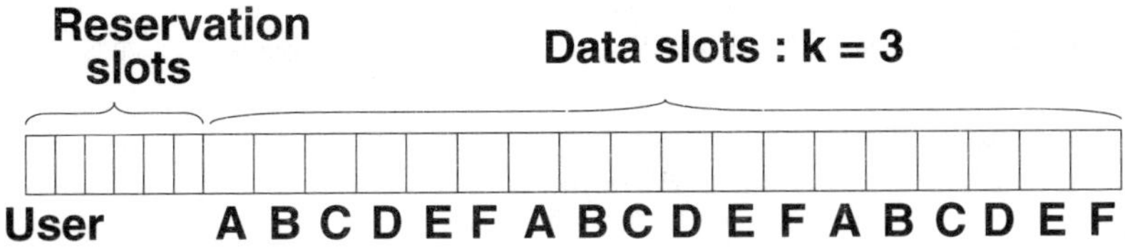

Reservation slots

Data slots : k = 3

User A B C D E F A B C D E F A B C D E F

Figure R.16 Frame structure of R-TDMA

residential terminal: A *terminal* which is used primarily in a residential rather than a business environment, i.e. a lower level of *traffic volume* is sent and received.

Residual Bit Error Ratio (RBER): A measure of *transmission* quality, specified in *ITU-T Recommendation* 594-1. It is the *Bit Error Ratio (BER)* for 'good' *frames*. It is given by the ratio of the number of *bits* of *data* which have been incorrectly received, but not subsequently detected or corrected, to the total number of bits sent.

resolution: Generally used as a measure of the amount of detail which can be shown on a display screen, such as a television or *Visual Display Unit (VDU)*, or on a printed page, such as from a printer or *facsimile*. It is measured as the number of dots (*pixels*) per unit area.

response time: In a *real-time system* it is the delay between the end of a *message* and the beginning of the response to that message.

restoration priority: The priority which is allocated to restoring the *services* to the different *users*, following a long period of service disruption, such as due to *overload* or an *outage*.

restricted access: Generally refers to the situation where access to certain types of *services* is limited to certain class of *users*, usually determined by the limitations of the *switching equipment*.

retransmission: The process in which the *sending terminal* resends the original *message* because it has not been correctly received by the *receiving terminal*. For example, by an *Acknowledgement (ACK)* not being received back by the sending terminal, or by a *Negative Acknowledgement (NAK)* being received back.

retrieval services: Part of the *service classification* specified by the *ITU-T*. Retrieval services is part of the *interactive services* group and includes the other half of the *messaging services* in which retrieval of *information* is required, such as from a *database*. Examples of retrieval services are *voice mail*, *facsimile* and *video*.

retrograde orbit: Refers to the *satellite orbit* in which the projection of the satellite onto the Earth moves in an opposite direction to that of the rotation of the Earth.

return loss: Difference, measured in *decibels*, between the energy in the incident *signal* and that of the signal following *reflection*, at a signal reflecting point.

return path: Often used to describe the *channel* which carries *information* back to the *sending terminal*, to inform it regarding the progress of the *call*.

Return to Zero (RZ): A method of *encoding binary data* in which the signal level always returns to zero between each encoded *bit*.

Return to Zero (RZ) binary code: See *Return to Zero*.

Reverse Address Resolution Protocol (RARP): *Internet Protocol (IP)* used by a system to determine its Internet *address* at start up.

reverse battery signalling: *Signalling* in which the battery connection and the *ground* connection are reversed on the loop tip and ring. This provides the *off-hook signal*.

reverse channel: A *data channel*, usually with limited *bandwidth*, which allows control *information* to pass from the *receiving terminal* to the *sending terminal*.

Reverse Control Channel (RECC): The *transmission channel* used between the mobile and *base station* in *trunked mobile radio*.

Reverse Path Forwarding and pruning (RPF): Technique used for *multicasting* in large *networks*, such as the *Internet*. In this the *information* is sent to several destinations, who then use a map of the network to send a copy to their neighbours, but only if they are on a route which is the shortest back to the original source. Also receivers can inform the sender if they are interested in the material so that next time only information of interest is sent to them.

reverse video: A technique for highlighting parts of an image on a screen by making the foreground the same colour as the rest of the background, and the background the same as the rest of the foreground.

revertive paging: An addition to *Personal Mobile Radio (PMR) in which paging messages* are received by a car radio and then forwarded on to a *pager* being carried by the driver (who may have temporarily left the vehicle).

RF: *Radio Frequency*.

RFC: *Request For Comment*.

RF carrier wave: A *carrier wave* which is within the *Radio Frequency* range.

RF filter: A *filter* which has been designed to let through *frequencies* within the *Radio Frequency range*, and to reject frequencies above and below this.

RFI: *Radio Frequency Interference*.

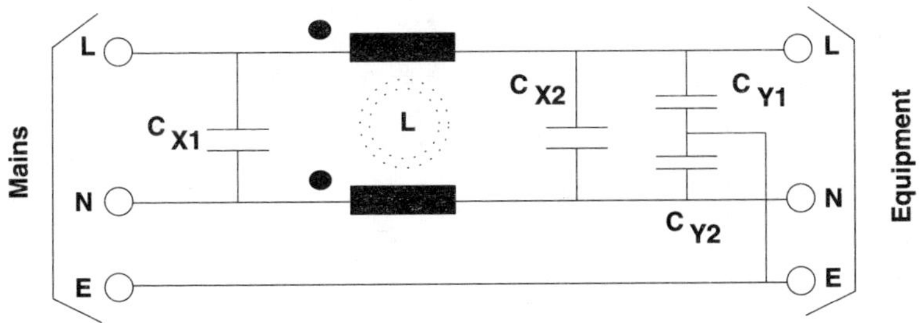

Figure R.17 A mains RFI filter

RFI filter: *Filter* used to suppress *Radio Frequency Interference (RFI)*. There are many constructions used, Figure R.17 showing a typical mains filter circuit.

RF modem: A *radio modem* operating in the *Radio Frequency (RF) range*.

RGB: *Red-Green-Blue*.

RGB gun: The *electron gun*, used in a *Visual Display Unit (VDU)*, to activate the phosphors on the screen and so produce the colours needed by a combination of *Red-Green-Blue*.

rhombic antenna: An *antenna* which has wire radiators, whose shape encloses the sides of a rhombus.

ribbon cable: A *cable* in which the individual wires are laid side by side, so that the cable has an overall dimension which is flat instead of round.

Right To Use (RTU): Fee charged for use of *software*, such as on a switch or *PABX*.

RIN: *Relative Intensity Noise*.

ring: **(1)** The *ringing* sound in a *telephone*. **(2)** A *ring network*.

ringaround: The situation where a *call* is routed back through an *exchange* which is already in the process of clearing the same call.

ringback signal: The *signal* which is fed back to a *calling terminal* to indicate that the *called terminal* is *ringing*. This normally takes the form of an interrupted *audio tone* in the caller's earpiece.

ringdown: *Signalling*, usually operating between *terminals* on the same tie line, in which an *AC signal* is automatically sent down the line, to ring a bell, operate a lamp, etc., if one terminal goes *off-hook*.

ringdown modem: A *modem* which is programmed to always *ring* a preset *terminal*.

ringdown circuit: See *ringdown*.

ringing: *Signalling* in a *telephone network* in which an *AC signal* is sent down the *line* to ring the called *telephone handset*. The ringing signal consists of a *sinusoidal waveform* with its axis shifted by the *exchange* battery (–50 volts), as shown in Figure R.18. The signal is applied in a cycle of short on periods and longer off periods.

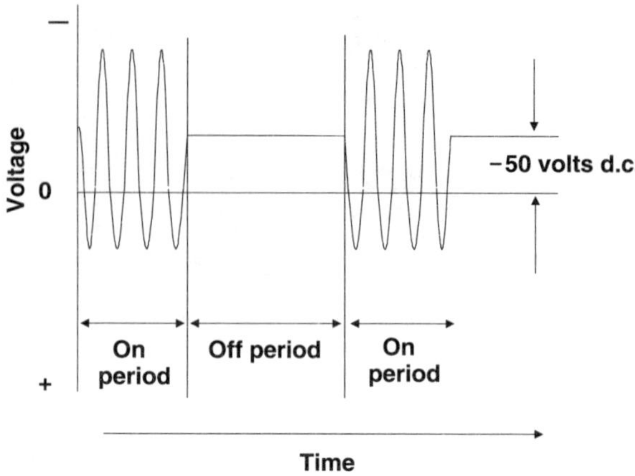

Figure R.18 Typical ringing waveform

ringing tone: The *tone*, received by the *calling terminal* from the called *exchange*, to indicate that the *called terminal* is being rung.

ring latency: In a *ring network* it is the time needed for the *signal* to propagate once around the ring.

ring network: Same as *ring topology* and *loop network*.

ring off: *Signal*, sent to a switchboard, to indicate completion of a *call*.

ring topology: *Network topology* in which the *transmission medium* is arranged in a closed loop and signals from one *node* to another can only be transmitted by going past intermediate nodes, as in Figure R.19. See also Figure N.7.

ring trip: The *signal* sent to the *exchange* to stop *ringing* once the *called terminal* answers the *call*.

R interface point: Same as *R reference point*.

RIP: *Routeing Information Protocol*.

RIPE: *Reseaux IP Europeens*.

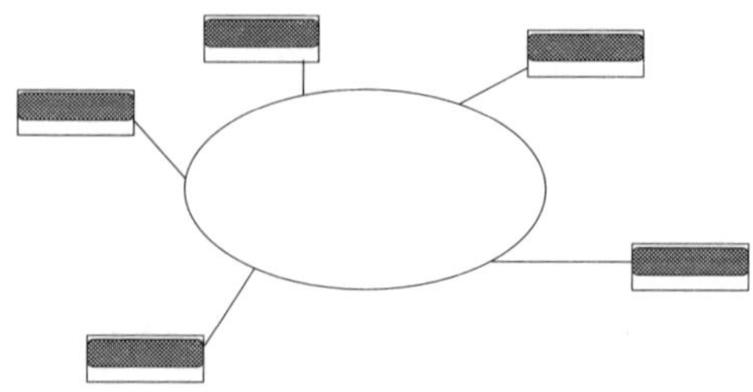

Figure R.19 Ring topology

ripple voltage: An *AC* component of voltage which is superimposed on to the *DC* voltage, usually as a result of imperfect filtering following rectification to convert AC to DC.

riser: Vertical shafts within buildings which are used to carry *cables* providing various *services* between floors of the building.

riser cable: *Cable* made from copper or *optical fibre* which is used to carry *telecommunications*, power and other services between floors of a building, being located in a *riser*.

rise time: The time needed for the *waveform*, such as a *pulse*, to rise from 10% to 90% of its maximum value. See Figure F.2 and *fall time*.

RITL: *Radio In The Loop.*

RJ: *Registered Jack.*

RLP: *Radio Link Protocol.*

RLR: *Receive Loudness Rating.*

RLSD: *Received Line Signal Detector.*

RMON: *Remote Monitoring.*

roaming: Ability of a mobile communications device (such as a telephone on a *cellular radio system*) to switch from one radio transmitter to another, so that the *call* being made is uninterrupted as the caller moves between geographical locations.

roaming number: Temporary *telephone numbers* allocated to *subscribers* as they move between geographical areas (usually by the *Home Location Register (HLR)*) so that *incoming calls* can be routed to them.

Robbed Bit Signalling (RBS): *Signalling* technique, used in North America, in which the least significant *bit* from the *PCM* is overwritten by signalling information every 750 μs.

rocking armature receiver: Receiver used in a *telephone handset*, it is similar in construction to the rocking armature transmitter shown in Figure R.20. As *AC signal* in the coil moves the armature which vibrates the diaphragm and produces the sound in the listener's ear.

rocking armature transmitter: Transmitter used in a *telephone handset*, as shown in Figure R.20. Sound on the diaphragm causes the armature to rock which induces an *AC signal* in the coil, which represents the original sound.

rod vision: The elements of the *human eye* which dominate in poor lighting conditions, when the eye's sensitivity has shifted towards the blue end of the *electromagnetic spectrum*. Figure R.21 shows the spread of rod vision and *cone vision*.

roll-call polling: *Polling* used, for example, for *multiple access* control, in which a *central station* polls each *user* in turn, as shown in Figure R.22. When a user receives a poll it puts its *message* on to the common line and the addressed user retrieves it. When all messages have been com-

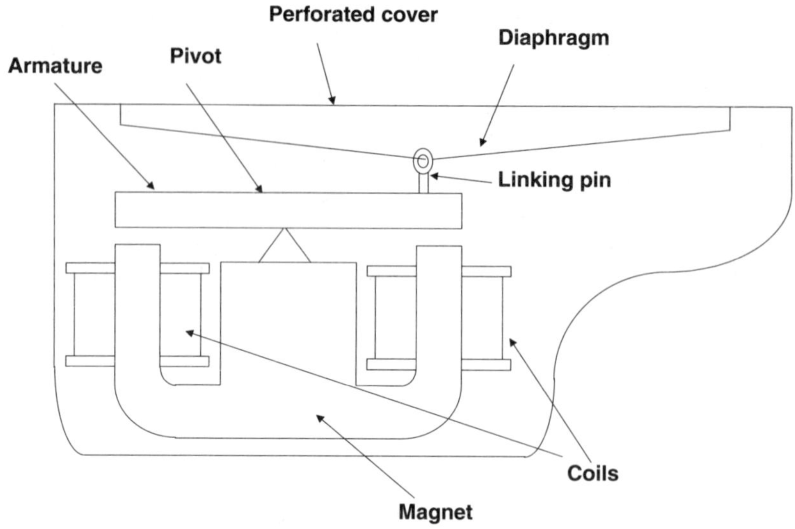

Figure R.20 Rocking armature transmitter

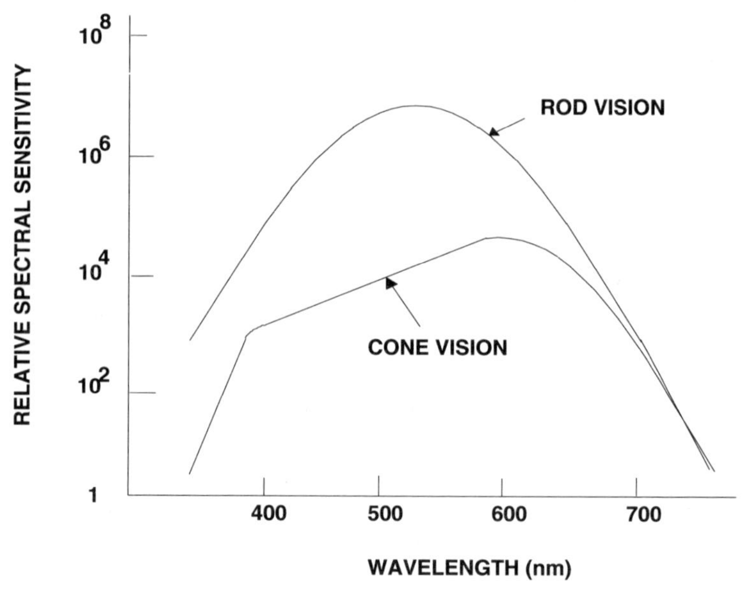

Figure R.21 Cone and rod vision

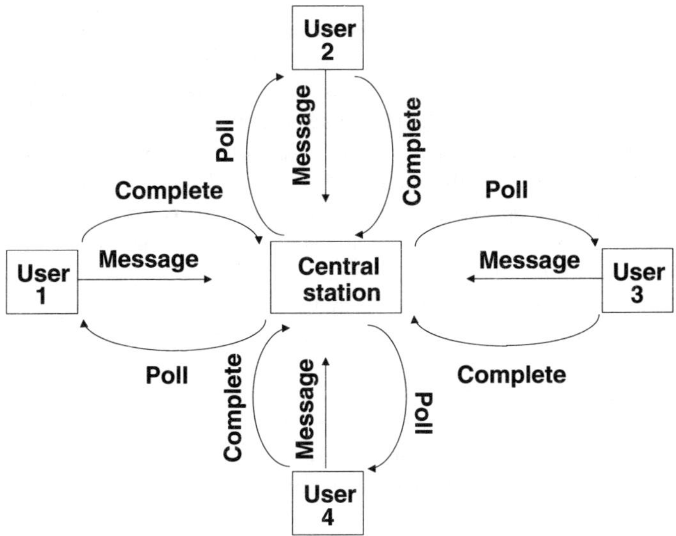

Figure R.22 Roll call polling

pleted the user sends a 'complete' message to the central station, which then polls the next user in its planned sequence.

ROM: *Read Only Memory.*

root: The base or head of a structure. For example the root directory is the base directory structure of a disk. The root of a *Management Information Tree (MIT)* is the head of the tree, as shown in Figure M.5.

root bridge: Part of the *Spanning Tree algorithm* it is the *bridge* which is allocated the highest priority of all the bridges on the *network*, the network and routes being configured around it.

ROSE: *Remote Operation Service Element.*

rotary dial: Part of the *telephone handset* it is the device which is wound up and released to generate *dial pulses* corresponding to the *telephone number.*

round trip: The distance covered by a *signal* from the *sending terminal* to the *receiving terminal* and back again.

round-trip delay: The time taken for a *round trip.*

route: The *path* taken by a *signal* from a *transmitting terminal* to a *receiving terminal.*

route addressing: *Addressing* in a *Packet Switched Network (PSN)* in which the address of any endpoint varies according to who is sending the *data*. For example, in the network of Figure C.4 A would send information to B along route 7, 6, 3, 1 which is its route address for B, whereas C would send it along 2, 1 which is C's route address for B.

route dialling: Setting up of a *route* through a *network* in which the *address* of each *switching centre* along the route is set up.

route diversity: The provision of two or more geographically separate *routes* between two points. This is used for security reasons, since the same *data* can be sent over these diverse routes with the confidence that a failure on any one of the routes will still enable the data to be received.

routeing: The technique used for sending a *message* from a *source address* to a *destination address*. This includes determining the *route* to be taken and the *transmission* mode to be used.

routeing diagram: A map, usually held within a *router*, which shows the preferred *routes* through a *network*. See also *routeing plan* and *routeing table*.

routeing indicator: Part of the *message header*, it provides information regarding the *routeing* of the message.

Routeing Information Protocol (RIP): *Routeing algorithm* for use within the *Internet* covering *networks* which are within a single *span of control* (*Autonomous System (AS)*). It is based on a distance vector *algorithm*.

routeing plan: The plan used for *routeing* through a *network*. See also *routeing diagram* and *routeing table*.

routeing table: A table, stored within *packet switches*, which indicates the *route* out from the switch for any *node* on the *network*. For example, of the network shown in Figure R.23, the routeing table for switch 2 is given by Table R.2 and that for switch 3 is given by Table R.3.

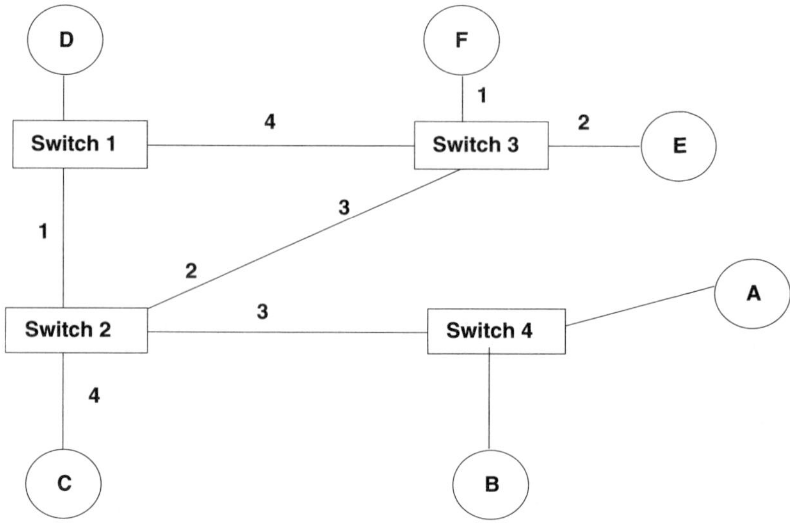

Figure R.23 Routeing example

Table R.2 Routeing table for switch 2 of Figure R.23

End point device 2 address	*Link*
A	3
B	3
C	4
D	1
E	2
F	2

Table R.3 Routeing table for switch 3 of Figure R.23

End point device 3 address	*Link*
A	3
B	3
C	3
D	4
E	2
F	1

router: A device which has the ability to choose *routes* along which to send *data* in a *network*. This is usually done based on *protocols* and *addresses*. Routers are sophisticated devices, operating at the *Network Layer* of the *OSI Basic Reference Model*. They can create logically extended networks, consisting of separate subnetworks, as shown in Figure R.24.

RPE-LTP: *Regular Pulse Excitation Long Term Prediction*.

RPOA: *Recognised Private Operating Agency*.

RR: *Radio Resource management*.

RRC: *Regional Radiocommunications Conference*.

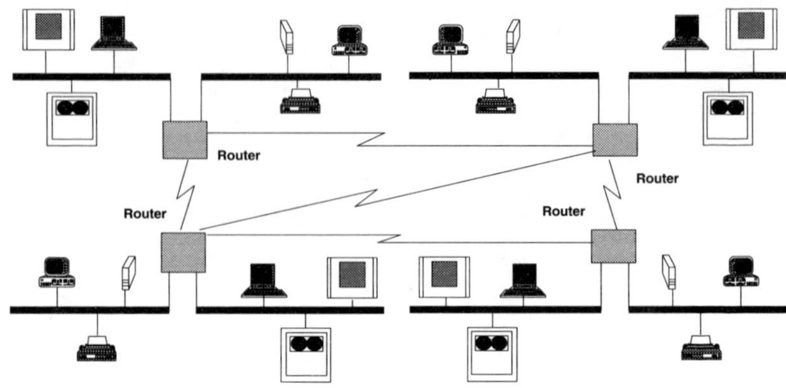

Figure R.24 An extended network using routers

R reference point: Also known as the *R interface point* (see Figure R.6). The R reference point forms part of the *ISDN* reference model and enables equipment without a recognised interface to be connected to an ISDN *network*.
RS: *Record Separator.*
RSL: *Received Signal Level.*
RS standards: Recommended Standards by the *Electronic Industries Association (EIA)* covering *data communications* interfaces. A few of these are given in Table R.4.
R-TDMA: *Reservation TDMA.*
RTNR: *Real Time Network Routeing.*

Table R.4 Some RS standards

Standard	*Description*
RS-232	Electrical and mechanical characteristics for interfacing between data processing devices, such as computers, terminals, modems, etc.
RS-422	Intended for use with RS-449
RS-423	Extends the length of RS-232 and is used in conjunction with RS-422.
RS-449	37 pin higher speed transmission between DTE and DCE.

RTS: *Request To Send.*
RTSE: *Reliable Transfer Service Element.*
RTU: *Right To Use.*
rural communications: *Telecommunications* systems which need to operate in areas of sparse population. Often it covers hostile terrain with poor road access, so high equipment reliability is important and maintenance must be easy to carry out. These systems primarily use *VHF* and *UHF* radio bands and *satellite*.
RWS-CC: *Regional Workshop Coordinating Committee.*
RZ: *Return to Zero.*

S

SA: *Standards Australia.*

SABM: *Set Asynchronous Balanced Mode.*

SAC: *Standards Advisory Committee.*

Safety Extra Low Voltage (SELV): A voltage, defined in IEC 950, which is non-hazardous even when occurring between an equipment (such as a computer) with a single fault and an attachment (such as a *modem*).

Sagnac fibre optic sensor: A sensor which uses a rotating optical fibre loop and mirrors to measure the angular velocity and angular displacement of light beams.

sample and hold: An analogue sampler used in *Pulse Code Modulation (PCM)* to sample *analogue signals* and hold them until the next sampling instant.

Sample Rate Converter (SRC): A device which converts between different sample rates, as used for recording *digital audio signals.*

sampling: (1) Taking a random selection of a small number of items from a much larger population and carrying out tests on this sample to determine the characteristics of the total population. **(2)** The process of taking periodic readings of a *signal*, at regular intervals. These samples may be further coded, as in *Pulse Code Modulation (PCM).*

sampling rate: The number of samples (see *sampling*) taken in a unit of time.

sampling time: The time between two consecutive samples (see *sampling*). It is the reciprocal of the *sampling rate.*

SANZ: *Standards Association of New Zealand.*

SAP: *Service Access Point.*

SAPI: *Service Access Point Identifier.*

SAT: *Supervisory Audio Tone.*

satellite: A body that revolves around another body or around the centre of gravity of a system containing more than one body.

satellite channel: A communications *path* involving the use of a *communications satellite* with associated *Earth stations.*

satellite Earth station: See *Earth station.*

satellite footprint: The pattern of *satellite signal* illumination on the surface of the Earth. Complex satellite *antenna* design is often adopted to maximise the overlap of the satellite footprint with the desired geographical area.

satellite frequency allocations: The *frequency bands* which have been allocated for use with satellites. For example, Table S.1 shows the bands allocated for fixed satellite service, both for the *uplink* and *downlink.*

Table S.1 Fixed satellite service frequency allocations

Uplinks (GHz)	*Downlinks (GHz)*
5.85–6.425	3.625–4.2
6.725–7.025	4.5–4.8
12.75–13.25	10.7–10.95 and 11.2–11.45
14.0–14.5	10.95–11.2 and 11.45–11.7 11.7–12.2 and 12.5–12.75
27.5–29.5	17.7–19.7
29.5–30.0	19.7–20.2

satellite link: A *radio channel* which usually forms part of a *data transmission* channel and comprises an *uplink* between a transmitting *Earth station* and a *satellite*, and a *downlink* between the satellite and a receiving Earth station.

Satellite Master Antenna Television (SMATV): *ETSI* standard ETS 300 473 for *digital broadcasting* systems for television, sound and *data services*. It make use of *Earth stations* to receive and distribute programmes to local communities, such as large hotels, apartment blocks, institution sites, etc.

satellite orbit: The *path* repeatedly travelled by a *satellite*. Each orbit has a period (*satellite period*) which remains approximately fixed. The most common orbit used by communications satellites is the *geostationary orbit*, although elliptical orbits with a high altitude coverage, and inclined elliptical orbits with short orbital periods, are both used.

satellite paging: The use of communications *paths* via *satellite uplinks* and *downlinks* in *paging networks*.

satellite period: The time taken for a *satellite* to complete one *orbit*. This period (T) is expressed in hours and is a function of the radius of the Earth (r = 3913 km), the altitude of the satellite (A in km). If K is a constant (equal to 396613.52 km^3/s^2) then: $T = \frac{\pi}{43200}\sqrt{\frac{(r+A)^3}{K}}$

satellite relay: A *repeater* (active or passive), attached to a *satellite* in *geostationary orbit*, which amplifies and relays communications *signals* between an Earth based *transmitter* and an Earth based *receiver*.

satellite station: A centre which communicates with one or more artificial *communications satellites* in Earth *orbit*.

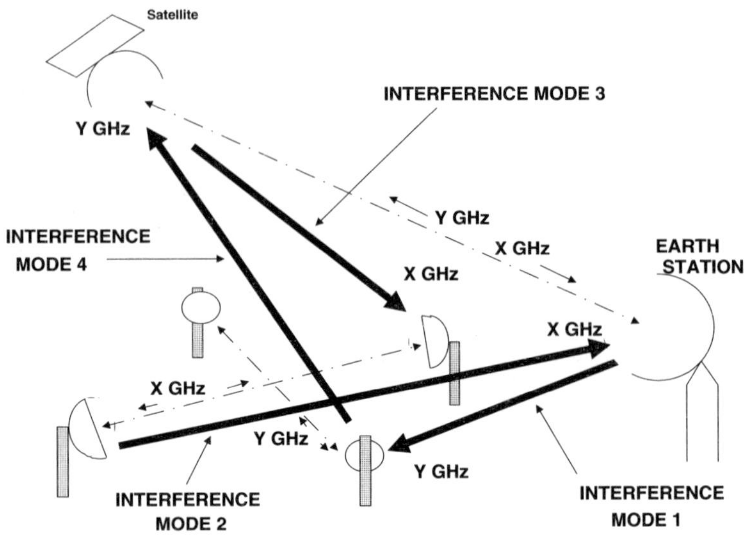

Figure S.1 Interference modes between satellite and terrestrial links

satellite telephone: A portable *telephone* that can send and receive *signals* via *satellites*.

satellite transmitters: *Radio transmitters* used on *satellites*. *Interference* can arise between adjacent satellites using Earth stations located close to each other and between satellite and terrestrial *links*, as shown in Figure S.1. In this the wanted *signal paths* are shown as broken lines.

Satellite Wide Area Network (SWAN): A *Wide Area Network (WAN)* which incorporates the use of *data transmission channels* between *Earth stations* and *satellites*, as well as using terrestrial *links*.

saturation: (1) The situation of a *telecommunications* component when it reaches its maximum *traffic capacity*, equivalent to one *erlang* per component. **(2)** When applied to linear electronic devices, saturation is the point where the output of the device deviates from being linear with respect to the input.

saturation routeing: *Call routeing* technique in which all possible *paths* are tried on a *broadcast* basis, the best path being selected following *Acknowledgement* from the *terminal* at the end of each path.

saturation signalling: *Signalling* technique in which a *route* is found from the *transmitting terminal* to the *receiving terminal* by interrogating all the intermediate *switching centres* until the end *terminal* is found and a *ringing tone* is applied to it.

save and repeat: A feature of a *telephone handset* or *PABX* which allows a number to be temporarily stored and then dialled as required by

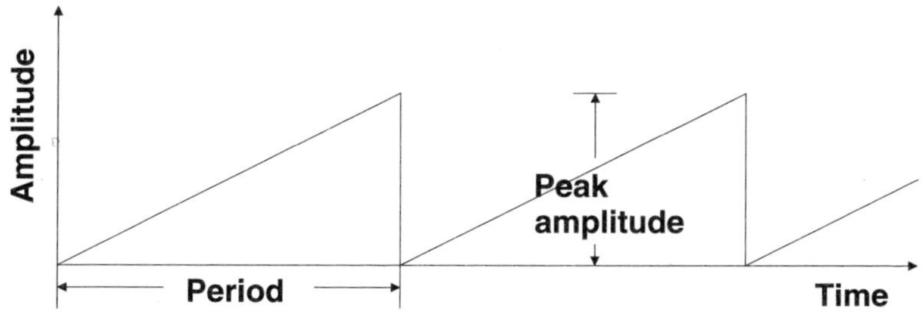

Figure S.2 Sawtooth waveform

pressing a single key on the telephone *keypad*. See also *last number redial*.

sawtooth waveform: A *waveform* which has the shape of the teeth on a saw, as in Figure S.2.

SBR: *Statistical Bit Rate*.

SBS: *Simulated Brillouin Scattering*.

scalability: An attribute of an item, such as a communications *network*, which allows it to grow either physically by the addition of groups of *nodes*, or in *capacity*, for example by upgrading some of the existing nodes. The cost for this is also proportional to the growth and does not increase exponentially.

scalar: A quantity or variable is scalar if it has magnitude but does not have direction, e.g. mass, time, *phase* or volume. See also *vector*.

scalar feed horn: A *feed horn* used with an *antenna* which has a symmetrical *radiation pattern* which is independent of *polarisation*.

scan: The process of *scanning*.

scanner: A device that methodically senses each part of a two or three dimensional object and generates an *analogue signal* or *digital signal* corresponding to the scanned object. Scanners are essential for the functioning of equipment such as a *facsimile* machine or a *Visual Display Unit (VDU)*.

scanning: **(1)** The periodic examination of *telecommunications traffic* activity to check for the need for further processing. **(2)** Methodical analysis of the attributes of a telecommunications input, such as the colours and densities of picture elements, of the *amplitude* and *frequency* of an *audio signal*. The results can then be converted into coded signals to be transmitted along *data communications channels* and to be displayed, for example on a *facsimile* equipment or a *Visual Display Unit (VDU)*. **(3)** Sweeping a *radar* beam through a specified region in order to detect possible signals.

scanning beam: An *electron beam* used in systems, such as television cameras, to translate the light entering the camera into a series of electrical *signals*. The beam scans a surface which has been electrically charged according to the light intensity which is incident on it.

scanning line rate: The rate at which the *scanning beam*, used in systems such as *facsimile* equipment or a *Visual Display Unit (VDU)*, crosses an imaginary line that is perpendicular to the *scanning* direction.

scanning period: The time it takes the *scanning beam*, used in systems such as *facsimile* equipment and a *Visual Display Unit (VDU)*, to cross an equivalent point on two consecutive scanned lines. The scanning period is the reciprocal of the *scanning line rate*.

scanning rate: The linear speed of the *scanning spot* over the scanned material, such as a *facsimile* document or a *Visual Display Unit (VDU)*.

scanning spot: The area on the material being scanned, for example by a transmitting *facsimile* machine or a *Visual Display Unit (VDU)*, which is illuminated at any instant. Also the area of the blank document being covered by the *receiving terminal* at any one instant.

scatter diagram: A graph in which most points are scattered within statistical error of a postulated line or curve. (See Figure R.11.)

scattering: The loss of *signal* due to irregularities in the *transmission medium*. For example, it is the loss of *light wave* signal, used for *data transmission*, along *optical fibres*. Scattering is caused by effects which include the variation in the density of the *cable*, known as Mie scattering, and the intrinsic molecular structure of the material making up the cable, called *Rayleigh scattering*.

SCC: *Standards Council of Canada* or *Specialised Common Carrier*.

SCCP: *Signalling Connection Control Part*.

SCE: *Service Creation Environment*.

Scientific and Industrial Organisation (SIO): One class of member of the *ITU*. Although they can attend meetings, such organisations cannot generally vote. There are about one hundred and fifty organisations, including IBM, Nortel and Siemens.

SCL: *Supervisory Control Language*.

SCM: *Subcarrier Multiplexing*.

scotopic response: The response of the *human eye* under low lighting conditions, when the spectral response shifts towards lower *wavelengths* compared to the eye's *photopic response*.

SCP: *Service Control Point*.

SCR: *Sustained Cell Rate*.

scrambler: A device used for *scrambling* a *signal*.

scrambling: The process of reversibly altering a *telecommunications signal* in order to make it unintelligible to all *receiving terminals* which are not aware of the scrambling method used. Some processes used in scrambling include substituting, inverting and displacing the *signal*, but not *encryption*, which would offer greater security.

scratchpad: A temporary store used within a processing device, such as a computer or a *telephone handset*. For example, scratchpads are used within *cellular radio systems* to temporarily store *telephone numbers* during a *call*.

Screened Foiled Twisted Pair (S-FTP): An *Unshielded Twisted Pair (UTP) cable* with an additional overall braided screen around the foil and pairs of the copper wire. It has excellent *EMC* performance.

Screened Twisted Pair (STP): A *cable* consisting of twisted pair wire with a foil around it to protect it from *Radio Frequency Interference (RFI)*. Also known as *Shielded Twisted Pair*.

screening router: A *router* that operates according to predetermined criteria in order to let through or to block certain *traffic*, e.g. based on the *network address*. It is often used as the component of a *firewall*.

screen refresh rate: See *refresh rate*.

scrolling: The process of continuously moving the content of a *Visual Display Unit (VDU)* vertically or horizontally.

SCSI: *Small Computer System Interface*.

SCVF: *Single Channel Voice Frequency*.

SDCCH: *Standalone Dedicated Control Channel*.

SDH: *Synchronous Digital Hierarchy*.

SDH frame: The *frame* structure used within the *Synchronous Digital Hierarchy (SDH)*, as shown in Figure S.3. This frame has been designed to transport 64 kbit/s *channels*, or any higher rate which is an integer multiple of this basic rate. The frame repeats at intervals of 125 μs.

SDH multiplexing structure: The *multiplexing* system used within the *Synchronous Digital Hierarchy (SDH)* to obtain higher *bit rate transmissions*, as shown in Figure S.4.

SDLC: *Synchronous Data Link Control*.

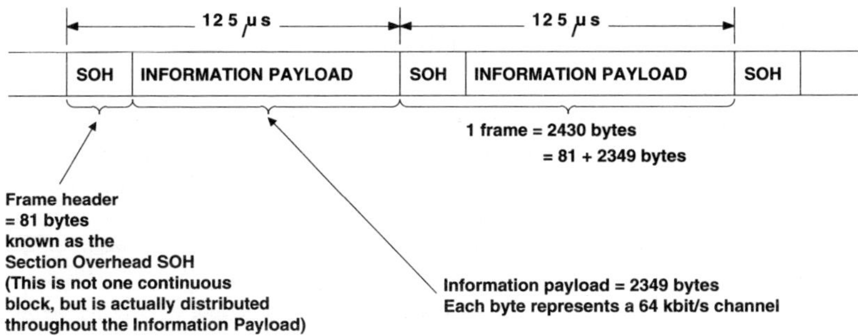

Figure S.3 Basis of SDH frame structure

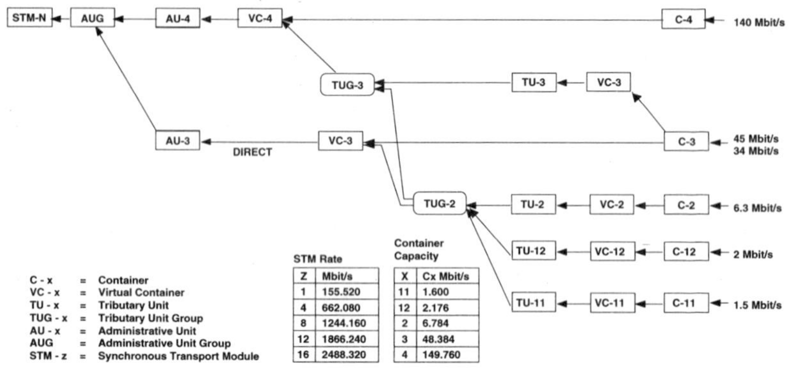

Figure S.4 SDH multiplexing structure

SDM: *Space Division Multiplexing.*

SDN: *Software Defined Network.*

SDR: *Speaker Dependent Recognition.*

SDSL: *Symmetrical Digital Subscriber Line.*

SDU: *Service Data Unit.*

SDV: *Switched Digital Video.*

SEC: *Single Error Correction.*

SECAM: Systeme Electronique Couleur Avec Memoire. A colour television *broadcast* system, developed in France, which uses 625 lines per *frame* and 50 Hz at 220 volts primary power. The main distinguishing feature of SECAM is that the two colour difference *signals* are transmitted sequentially rather than simultaneously, as in the case with *PAL* and *NTSC.*

secondary calling: Generally used to refer to a form of private *wide area paging*, to supplement public mobile radio systems, where the *user* of the *mobile phone* is paged following receipt by the radio of their paging *address* from the *base station.* See also *revertive paging.*

secondary centre: An *exchange* on the *Public Switched Telephone Network (PSTN)* which links the *tertiary trunk exchange network* to the *trunk circuits* of *regional exchanges.*

secondary channel: The *channel*, in a system where two channels share the same interface, which has the lower *data signalling rate capacity.* The secondary channel is often used to obtain *diagnostic information* and perform tests on the system, without interfering with the performance of the *primary channel.*

secondary characters: The *characters* found in the *figures shift* of a *telegraph code.*

secondary route: An alternative *route* taken by *data* in a *Packet Switched Network (PSN)* when the *primary route* is unavailable.

secondary service: Refers to the *service* which has been allocated a *frequency band* within the *ITU International Table of Frequency Allocations*, but has been designated to be a secondary service. Secondary services have the lowest priority. They must not cause harmful *interference* with a *primary service* or a *permitted service* and they cannot claim protection from interference from these services.

secondary station: Part of a communications *network*, the secondary *station* is a remote station which carries out functions, such as unbalanced *link* level operations, under control of a primary station. It receives commands and generates responses.

secondary trunk exchange: An *exchange* on the *Public Switched Telephone Network (PSTN)* which is connected to other *trunk exchanges* and handles major *routeing*.

sectional centre: A *control centre*, of a *Public Telephone Operator (PTO)*, linking *primary centres*.

sectorisation: A technique for adding *capacity* to a *cellular radio system* by using several *directional antennae*, each with its own *channel* set, at the *base station*.

secure-access card: A *smart card* that allows the *user* access to *telecommunications* or other *services*. Secure-access cards frequently rely for their *security* on a further identifying signature, such as a *Personal Identification Number (PIN)* that has to be input by the user.

secure line: An optical or wire *telecommunications data transmission path* which carries *information* coded in a form which would be indecipherable to unauthorised interceptors. A secure line could also be one that is physically inaccessible to unauthorised people, or is protected from environmental damage.

secure telephone unit: A *telecommunications terminal* that has been approved by the US government as being secure enough to transmit or receive sensitive *voice*, *data* or *facsimile messages*.

secure transmission: The *transmission* of *telecommunications data* which can only be interpreted by predetermined recipients. For example, the transmission of *binary coded* telecommunications sequences in *spread spectrum* systems, where the *messages* encoded in the sequences can only be interpreted by users with the proper *key* for the spread spectrum code sequence generator.

security: Generic term used to describe the techniques for preventing unauthorised access to *data*. It may involve several techniques, such as *encryption*.

security management: One of the five groups of *network management functions* specified by *ISO*, and given in Figure N.2. Security management includes the *authentication* and authorisation of access to the *network* by *users*. Although shown as a separate function, security management cuts across several other functional areas, for example the need to limit access to users for certain *configuration management* functions. Security management therefore determines who may do what in controlling a network. It is usual to provide security access in a layered structure, the upper *layers* being able to perform all the functions available to the lower layers.

SED: *Single Error Detection.*

segment: **(1)** A continuous part of a *network*, such as a segment of a *bus network*. **(2)** A *block* of 64 *bytes*, as used in a *Packet Switched Network.*

segmentation: The process involved in breaking a larger *network* or *message* into smaller portions.

seizing: The act of accessing a *circuit* or *channel*, or making it *busy* so that it is not available to other *users*.

seizing signal: A *signal* used by the calling end of a *trunk* or *line* to request a *service* from another part of the *telecommunications* system.

selection: The process of choosing the *telecommunications path* or destination *terminal* for a *message*.

selection digits: The *telephone number* of the receiver of a *call*, expressed in *signal* elements.

selection signal: The *characters* in a facility request that define the *network* facilities required to complete the communications *path* and the *address* of the *receiving terminal*.

selective calling: The process in which a transmitting *node* can determine which of the other nodes on the *network* are able to receive its *message*.

selective fading: The *fading* of radio *signals* in which one or more of the individual *frequency* components of the signal fades independently of the others.

selective ringing: The *ringing* of only the desired *terminal* out of several terminals attached to a common *line*, such as a *party line*.

selectivity: The ability of a *receiving terminal* to pick up the desired *signal* whilst rejecting other adjacent signals.

selector: An electromagnetic switch which selects *lines* in *Strowger exchanges*.

self clocking: Same as *self timing*.

Self-Phase Modulation (SPM): The broadening of a *pulse* of light as it travels through the *transmission medium*, caused by the *intensity* dependence of the *refractive index* of the medium.

self-supporting cable: Generally refers to *optical fibre* communications *cable* for use on high voltage power line pylons, capable of being installed whilst power is present. See, for example, Figure O.5.

self-test: A *diagnostic* test performed on itself by an item of communications terminal equipment, to assess the performance of the equipment. The test may be administered and analysed by a specialised *network* diagnostic controller.

self timing: Generally refers to the ability of a piece of communications equipment to extract the *clock* from the incoming *data signal*, so that a separate *timing channel* is not required.

SELV: *Safety Extra Low Voltage*.

semaphore: **(1)** A mechanism for *synchronisation* of concurrent processes in computer or *telecommunications* systems, to prevent several processes from entering mutual critical stages at the same time. Semaphores work by *signalling* when specific actions have been completed. **(2)** A semaphore is also an early method of sending optical *telegraph signals* by the use of devices such as fires, flags, shutters and wooden arms.

semiautomatic switching: *Telephone switching equipment* in which attendants convert oral instructions given by *users* into a form that can then be used by automatic switching systems.

Semiconductor Optical Amplifier (SOA): *Optical amplifier* constructed from semiconductor material. The *gain* of the device is achieved in an identical way to that of a *LASER*.

semiduplex: A *transmission* system which is *simplex* at one end and *duplex* at the other.

sender: **(1)** A device, found in an *exchange*, that controls the *transmission* of *signals* along the correct communications *path*. **(2)** A *transmitting terminal*.

sending sensitivity: A measure of the *transmission* performance of an individual *telephone*, defined as the ratio of the voltage across the terminating impedance of the *exchange* feed bridge and the sound pressure level injected into the *microphone*. This ratio usually varies over the *frequency range*.

Senior Officials Group on Telecommunications (SOGT): A group comprising the European Ministers of *Telecommunications* and Industry which meets every six weeks under the chairmanship of the Director General of DG XIII (see Table C.1). A subcommittee of SOGT is the *Analysis and Forecasting Group (GAP)*.

separate channel signalling: A system of *data transmission* over several *channels* where one channel is dedicated to the transmission of control *signals* related to the data carried over the other channels.

sequential: Several events are said to be sequential if they are in chronological sequence but do not overlap.

sequential scanning: *Scanning* in which each line is scanned *sequentially*.

sequential tone signalling: The use of *sequentially* transmitted *tones*, each tone uniquely corresponding to a *digit*, to send communications *data*

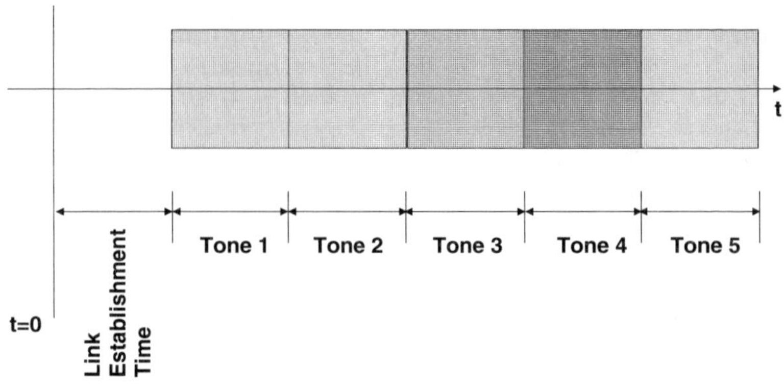

Figure S.5 Sequential tone signalling

over *radio channels*, avoiding the effects of Rayleigh fading, as shown in Figure S.5. The internationally agreed tones are given in Table S.2, for the EEA, *ITU-R* and *ITU-T*.

serial: Events are said to be serial if they occur, or are processed, *sequentially*. Serial elements may be given sequential alphanumeric designations for the purposes of control or planning.

serial access: The *sequential transmission* of *data* into or out of a *telecommunications* device or storage medium. Generally this would occur one *bit* at a time, as in the transmission of an *ASCII character*, each character being bound by a *start bit*, and a *parity bit* and *stop bit*.

serial communications: See *serial access*.

serial attribute coding: A system of storing *character* attribute *codes* as non-displayable characters in the page store. This results in spaces in the *text* at the beginning of each sequence with changed attributes. These spaces mostly coincide with the spaces between the words so are often not visible. Serial attribute coding is used in many *teletext* systems.

serial bit stream: A *sequential* series of *binary digits* which, taken together, often provide the *code* for specific *information*. See also *bit stream*.

serial interface: Interface to equipment which is used for *serial access*. See also *parallel interface*.

serial mode: Refers to the *serial transmission* mode of *data* transfer in which the *bits* that make up the data are sent one after another along the same *transmission path*, rather than simultaneously along different paths. See also *parallel mode*.

serial transmission: See *serial access*. It is the most common mode of *data transmission*, the bits of data being sent *sequentially* along one *transmission path*. See also *parallel transmission*.

Table S.2 Sequential tone signalling frequencies

Standard	*EEA*	*ITU-R*	*ITU-T*
Digit 1	1124	1124	697
Digit 2	1197	1197	770
Digit 3	1275	1275	852
Digit 4	1358	1358	941
Digit 5	1446	1446	1209
Digit 6	1540	1540	1335
Digit 7	1647	1640	1477
Digit 8	1747	1747	1633
Digit 9	1860	1860	1800
Digit 0	1981	1981	400
Repeat	2110	2110	2300
Tone length (ms)	40	100	100

server: A *node* on a *network* which allows other nodes on the network to access it and make use of its resources. Usually servers are located on a *Local Area Network (LAN)* and provide *services* to workstations and processing units on the same LAN.

service: Functions provided by one *node* on the *network* to another. See also *service classifications*.

Service Access Point (SAP): The geographical location at which physical access to a communications *service* is provided, or the part of a communications system which provides the *user* with service in person, or via automated equipment. In an *OSI Basic Reference Model* the Service Access Point is the conceptual port between an (N)-layer and an (N+1)-layer. (See Figure S.6.)

Service Access Point Identifier (SAPI): Used as part of the *ISDN* the SAPI is used to identify the *service* that the *signalling frame* is intended for. Two different types of *terminals* on the same *bus* can use different SAPIs for each type of service, so the *network* has the option of handling

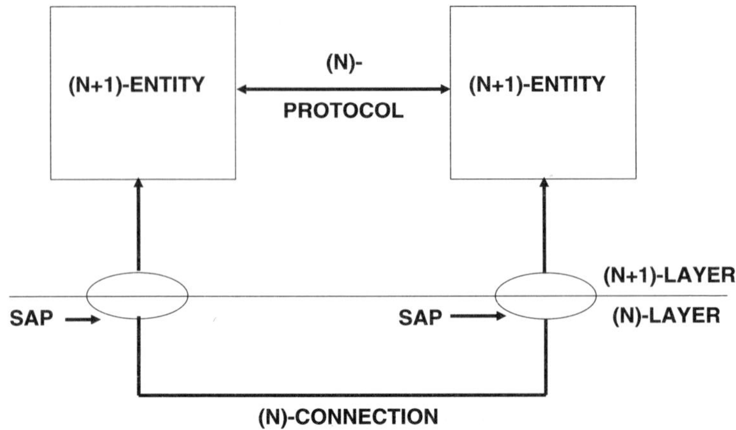

Figure S.6 Service Access Point

the signalling associated with the different services in separate modules. The value of the SAPI is fixed for a given service.

service area: Generally describes the geographical area over which a *service provider* operates. See, for example, Local Access and Transport Area (LATA).

service classification: A classification system, adopted by the *ITU-T*, which recognises the two main groups of *services* as being *interactive services* (transmission in both directions) and *distribution services* (transmission in one direction at a time). Each group is then further sub-classified as shown in Figure S.7.

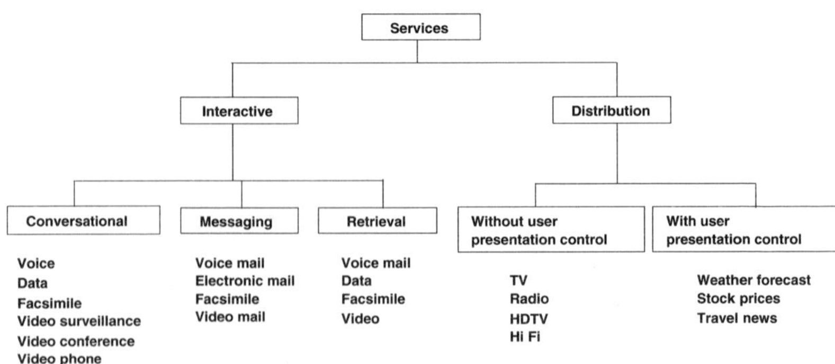

Figure S.7 Service classification

Service Control Point (SCP): A component of an *Intelligent Network (IN)* which contains the intelligence of the network, including all the actions needed to service a *call*. It will access service *databases* which may be located remote to the *Service Switching Point (SSP)*.

Service Creation Environment (SCE): A component of an *Intelligent Network (IN)*, comprising a *software* development system which allows rapid changes to existing communications *services* and the equally fast creation of new ones.

Service Data Unit (SDU): The data unit concept used within the *OSI Basic Reference Model*. The SDU is the data which is passed to the (N)-layer from the (N+1)-layer for forwarding to the peer (N+1)-entity in the destination system, as in Figure P.16. Therefore the (N)-layer transfers the SDU transparently.

Service Level Agreement (SLA): An agreement between a *service provider* and a *service user* which specifies certain performance factors, such as *availability of service*, *throughput*, etc.

service provider: (1) A *Public Telephone Operator (PTO)* who provides a *service* for profit. **(2)** In the *ISO Basic Reference Model* the service provided by the lower layers to a *service user*. See Figure L.2.

Service Management Point (SMP): A component of an *Intelligent Network (IN)* which manages all functions of the *network* and the *services* it provides, including the interface to the network operator. Examples are *performance management*, *configuration management*, etc.

Service Switching Point (SSP): The main access point into an *Intelligent Network (IN)*. The Service Switching Point determines which *service* is required and its location, as well as any other information, before passing the *call* on to the *Service Control Point (SCP)*.

service user: The *user* of a *service* provided by a *service provider*.

SES: *Severely Errored Seconds*.

SESR: *Severely Errored Second Ratio*.

session: The period of time during which two *terminals* on a *network* are connected together by a *transmission path*, to allow communications to occur between them.

Session Layer: The fifth *layer* within the *OSI Basic Reference Model* (see Figure O.9) which controls the exchange of *information* between the *Presentation Layer* within different *open systems*. It does this by establishing connections, *synchronisation*, disconnection, *flow control* to prevent flooding, etc. The standard within the Session Layer has different classes or subsets which provide options to enable the lower layers to be matched to the needs of the upper layers, depending on the application.

Set Asynchronous Balanced Mode (SABM): Command sent in *LAPB* to establish a *DCE/DTE link*. This is confirmed when the other end issues an *Acknowledgement (ACK)*.

settlement rate: The sum of money paid by one *telecommunications* operator to another if it generates much more *traffic* than it receives. The settlement rate is equal to 50% of the *accounting rate* on the difference in traffic from the other operator.

settling time: The time needed for the *frequency* of a multifrequency device to change to, and stabilise at, a new operating frequency.

set top box: See *television set top box.*

setup: The process of preparing a computing or other control or *data processing system* in order for it to perform a particular job.

seven hundred service: A *telephone service* which allows *subscribers* to receive *calls* at various locations via a single *telephone number*, when the calls are made using the same *common carrier.*

seven layers: Refers to the *layers* within the *OSI Basic Reference Model*, as shown in Figure O.9.

seven layer model: Same as *OSI Basic Reference Model.*

Severely Errored Second (SES): A definition of *error* occurrence from *ITU-T Recommendation* G.821, which states that fewer than 0.2% of one second intervals should have a *Bit Error Ratio (BER)* worse than 10^{-3}.

Severely Errored Second Ratio (SESR): The ratio of *Severely Errored Second (SES)* to the total *available time.*

S-FTP: *Screened Foiled Twisted Pair.*

SGMP: *Simple Gateway Monitoring Protocol.*

shadow fading: A *fading* effect accounting for some path loss in *radio wave* propagation in *cellular radio systems.* Shadow fading is the fading caused by the type of terrain, such as large buildings in an urban environment, trees in a rural area, or water.

Shannon's law: In 1948 Shannon extended the *Nyquist theorem* to the case of a *channel* experiencing *random noise*. The work of Shannon sets out the fundamental rules within which communications engineers must operate, as well as allowing them to check the efficiency of a *code.* Shannon's law gives the theoretical maximum *data rate* at which error free *digits* can be transmitted over a *bandwidth* limited channel in the presence of *noise*. If C is the channel *capacity*, B the bandwidth in *hertz*, and SNR the *Signal to Noise Ratio*, then the law states that $C = B \log_2 (1+SNR)$.

shaped beam: The *radiation pattern* from an *antenna*, such as on a *satellite*, which has been shaped so that its projection profile on the surface of the Earth matches the desired contours of the area to be covered, such as countries, continents, etc. The production of the shaped beams can be achieved by either using multiple overlapping beams, or by physically changing the shape of the surface of the beam reflector.

shared access: Access in which several devices on the *network* use the same *transmission medium.* If this is on a *Local Area Network (LAN)* some form of *multiple access technique* needs to be used. See also *party line*.

shared service: The use of a common *service* by several *users*. Also sometimes used to refer to *shared access* and *party line*.

shared tenant service: The provision of shared *telecommunications services* by several tenants in a building or complex. For example several tenants could share a large *PABX* since any one would not have the *traffic* to economically utilise its full *capacity*.

SHF: *Superhigh Frequency*.

shield: The metallic shield which surrounds the insulated conductors in shielded *cables*. The shield reduces stray electrical fields and protects operators.

Shielded Screened Twisted Pair (S-STP): A 100 ohm copper communications *cable* in which each individual pair of wires has a separate *shield* and then the whole is encased in an overall screen. The cable is capable of *data rates* up to 600 MHz.

Shielded Twisted Pair (STP): A pair of communications wires that are individually insulated, twisted together and encased in a *cable* that has a *shield* to prevent *Electromagnetic Interference*, so providing virtually *noise* free *transmission* and a physically protected *line*.

shielding: The use of *shields* to protect a *circuit* from the effects of external *Electromagnetic Interference* or *Radio Frequency Interference (RFI)*, or to prevent the same interference from escaping from the circuit. The term also applies to the material used to perform the shielding.

shift character: A *character code* that allocates a specific meaning to following codes, until negated by a further shift character, for example the *letters shift* character in the *International Alphabet No. 2*.

Shift In (SI): A *character code* that indicates that the following codes have the default meaning specified in the code set in use.

shift keying: The switching from one state to another, such as, for example, in *Frequency Shift Keying (FSK)*, *Phase Shift Keying (PSK)*, etc.

Shift Out (SO): A *character code* that indicates that the following codes have a meaning other than the default meaning specified in the code set in use.

short break supply: A duplicated power supply which can be used with only a slight interruption of the power if the main power supply fails. See also *no-break power supply*.

short dipole antenna: A *dipole antenna* having a short length in which the current distribution is linearly tapering from a maximum value at the centre to zero at the tips, as shown in Figure S.8.

short haul: Generally refers to a communications system which is served by a single *switching centre* and which operates over distances of less than about 20 km, without use of any long distance *trunks*.

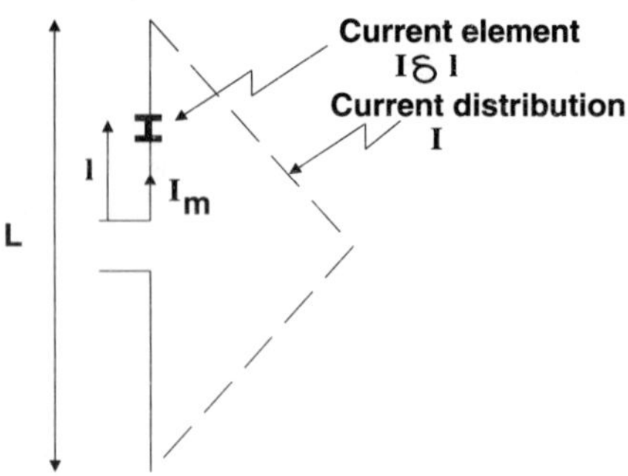

Figure S.8 Short dipole

short haul modem: A *modem*, connecting a *telecommunications terminal* to a *transmission line*, which is only suitable for use over relatively short distances, such as in-house systems rather than over the *Public Switched Telephone Network (PSTN)*.

Short Message Service (SMS): The ability to transmit a short *message*, in the region of 160 *characters*, and to be able to store this in the *handset*, such as used in a *cellular radio system*, for later retrieval.

short monopole antenna: For a vertical monopole on a perfectly conducting plane (Figure S.9) the current distribution is approximately sinusoidal and the *radiation pattern* above the plane is identical to that of a *dipole* of length 2H. For a short monopole the current distribution tapers linearly from a maximum at the base to zero at the top. For a short

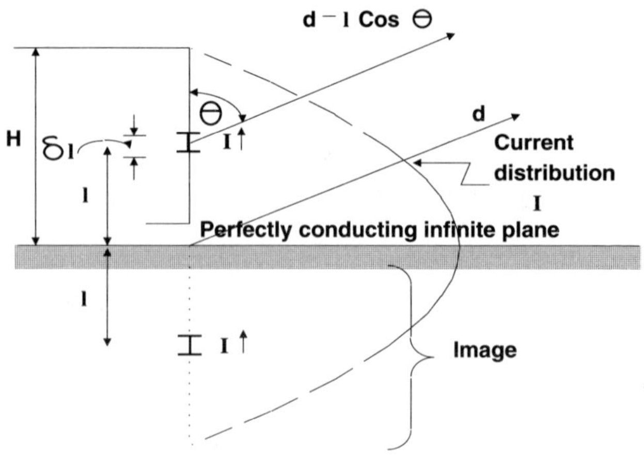

Figure S.9 Vertical monopole over a ground plane

monopole (H 0.1 λ) the *radiation pattern* is identical to that of a *short dipole*. However the power level is half that of a short dipole of length 2H, with the same current at its centre, since power can only be radiated above the reflecting plane.

shot change: A sudden change of a *video* image.

shot noise: *Noise* which occurs in an active device, e.g. a valve, transistor, etc. It is due to the random variation of velocity of electron movement and its magnitude is proportional to the square of the *Direct Current (DC)* through the device.

SI: *Shift In.*

sideband: The *frequency band* resulting from *modulation*, which is located above or below the *carrier signal.*

side lobe: The off-axis response of an *antenna* that is in a different direction to, and smaller than, its *main lobe.*

sideral day: The mean period of one Earth rotation, equal to 23 hours, 56 minutes, 4.095 seconds. The sideral day can also be defined as the time between two successive transits of a fixed star past a given meridian on Earth.

sidetone: The *transmission* of sound from the *microphone* to the earpiece of the same *telephone*. This results in a combination of *background noise* and the *voice* of the speaker on the telephone. The sidetone is useful as it allows the user to hear their own voice.

Sidetone Masking Rating (STMR): A parameter used to describe the talking effects of *sidetone* (as different from the *background noise* effects). It is a function of the characteristics of the *telephone* and the *local loop*, as in Figure S.10. If SLR and RLR are the Send Loudness Rating and the Receive Loudness Rating of the telephone set, A is the weighted sidetone balance *return loss*, then STMR (in *decibels*) = SLR + RLR + (A–1).

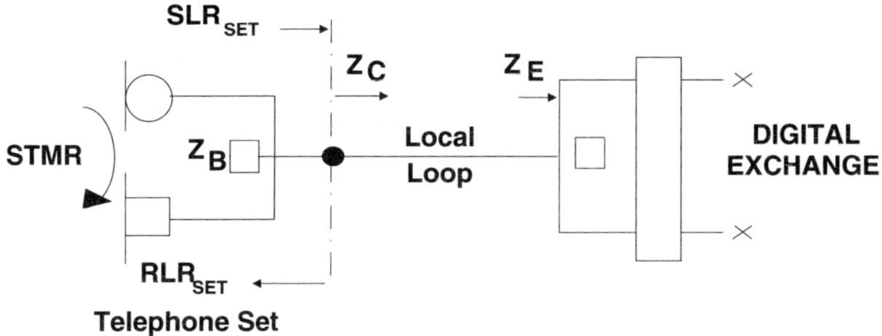

Figure S.10 Sidetone influences

sigma: A set of *telephone* wires that are treated as a unit for purposes including the establishment of electrical current balance and the computation of *noise* levels.

signal: Electrical or *electromagnetic waves* which are used to convey *information.*

signal bandwidth: The *bandwidth* or *frequency range* needed to send a *signal* accurately over a communications *channel.* The bandwidth for some signals is significantly less than that needed for others. For example, a 4 kHz bandwidth is needed for the *transmission* of *voice* signals, but up to 5 MHz is needed for *video* signals.

signal carrier: Same as a *carrier signal.*

signal conditioning: Modifications made to a *signal,* such as amplifying it, to make it more suitable for handling in future stages in the process, such as carrying *information.*

signal converter: A device which changes *signals* from one form into another (e.g. *digital signal* to *analogue signal*) for handling in subsequent stages (e.g. a *modem* which is used to transmit on an analogue *line*).

signal distance: See *Hamming distance.*

signal distortion: The *distortion* or unspecified and undesired changes in the input-output characteristics of a *signal,* such as its *amplitude distortion, frequency distortion* or *phase distortion.*

signal element: Generally refers to one *pulse* which forms part of an overall *signal.*

signal frequency network: A *network* in which the *nodes* use the same *frequency* for *transmission,* with a suitable *guard band* between them.

signal frequency shift: The numerical difference between the *frequencies* of the *data signals* corresponding to the black signal and the white signal in *facsimile transmission.*

signal ground: The common *ground* used for all *signals* except for *frame* ground.

signal level: The strength of a *signal.* It can be stated in absolute units (e.g. voltage or power) or relative to the signal strength at some other point.

signal level compensator: An *Automatic Gain Control (AGC)* used in wire communications systems to adjust the incoming *signal level* prior to it entering the *receiving terminal.*

signalling: The use of *signals* for controlling communications systems, such as establishing and terminating connections and managing *networks.* Also refers to the actual *transmission* of communications *data* and the data itself.

Signalling Connection Control Part (SCCP): Part of the *ITU-T Recommendation, Signalling System No. 7,* it assists with transfer of *messages* by providing additional *routeing* and management functions.

signalling interworking: The provision of a subsystem that can recognise and convert between two different *signalling* systems during the modification or replacement of a communications *network* involving the change of signalling standard.

signalling plan: A plan showing the *signalling* system which is to be used in various parts of a *network*.

signalling rate: The rate, usually in *bits* per second, at which a *telecommunications* device can transmit or receive a *data signal*.

Signalling System No. 7 (SS7): An *ITU-T* defined *signalling* system for *Common Channel Signalling* designed for use with *digital circuits* and based on *Signalling System No. 6*. The system defines the format and content of *packets* on the *D channel*, as well as providing internal control and *network* intelligence.

Signalling System No. 6 (SS6): An *ITU-T* defined *signalling* system. It was the first *Common Channel Signalling* system to be standardised and operates on *analogue circuits* using 2.4 kbit/s or 4.8 kbit/s *modems*.

Signalling Terminal Equipment (STE): *ITU-T Recommendation* for a *gateway* which supports the interworking between discrete *Packet Switched Networks (PSN)*, as shown in Figure S.11. Recommendation X.75 supports X.25 to provide such a gateway in special nodes known as Signalling Terminal Equipment. They provide connectivity services but operate at the first three layers of the *OSI Basic Reference Model*.

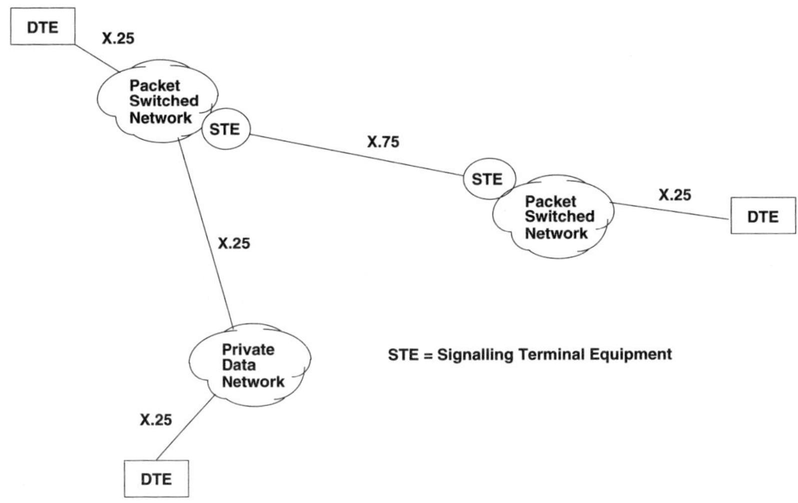

Figure S.11 Interworking via X.25 and X.75

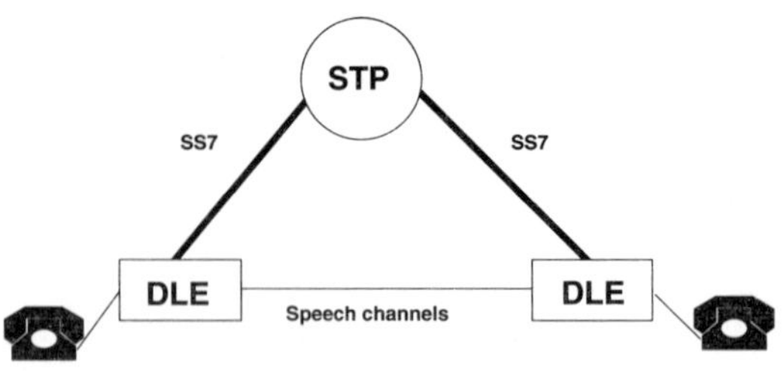

DLE Digital Local Exchange
SS7 Signalling System No. 7
STP Signalling Transfer Point

Figure S.12 Signalling Transfer Point

Signalling Transfer Point (STP): A *packet* switch, in the *ITU-T Signalling System No. 7 (SS7)*, which enables different *routeing* of *signalling messages* to that carrying *telephony traffic*, as in Figure S.12. The STP ensures reliable transfer of SS7 messages under fault conditions.

signal phase: See *phase*.

signal power: The *signal strength*, measured in *decibels*.

signal processing: The transformation of communications *signals* into other forms, such as new wave shapes, coding arrangements or power levels, using mainly electronic or optical devices.

signal return circuit: A *return path* for a *signal* from the load back to the signal source.

signal sampling: See *sampling*.

signal to crosstalk ratio: The ratio, in *decibels*, of the powers of the wanted *signal* and the unwanted signal from another *channel* (*crosstalk*) at any specific point in a *circuit*.

signal-to-listener echo ratio: The ratio of the *signal power* to the power of signals echoed back to the *transmitting terminal*, for example by changes in the characteristics of the *transmission channel*.

Signal to Noise Ratio (SNR): The ratio of the *signal power* to the *noise power*, or the signal *amplitude* to the noise amplitude, at any point in a communications system, expressed in *decibels*. The Signal to Noise Ratio can also be used to express the ratio of wanted relevant information to unwanted irrelevant information passing through a point.

Signal to Quantisation Noise Ratio (SQNR): The ratio of the desired *signal power* to the power of *quantisation noise*. This noise arises in *Pulse Code Modulation (PCM)* systems during the analogue to digital interconversion of *audio signals*.

signature: The *baseband signal* received from a *source* which is used to locate, and help to recognise and identify the source. A signature of an *electromagnetic wave*, that has been reflected from or transmitted through an object, comprises those parts of the wave which give *information* about the object.

signature analysis: The analysis of a *signature* in order to produce *information* about the signature *source* or the object which has reflected or transmitted the *electromagnetic wave*.

sign bit: A *bit* that indicates whether the number it refers to is positive or negative.

significant condition number: The number of different significant conditions that characterise a *signal code* used in communications systems. For example, 2 in the *binary Non-Return to Zero (NRZ)*, 3 in *Alternate Mark Inversion (AMI)*, etc.

significant market power: A term which can be applied to large telecommunications operators within Europe by the *telecommunications watchdog*. This places an obligation on these operators to follow certain conditions in dealing with other operators, especially competitors, e.g. *carrier preselection*.

SIM: *Subscriber Identity Module*.

Simple Gateway Monitoring Protocol (SGMP): The first demonstration, in 1987, of a system to monitor *TCP/IP networks* by managing the *routers* and *gateways* that interconnected widely spaced networks. This developed into the *Simple Network Management Protocol (SNMP)*.

Simple Mail Transfer Protocol (SMTP): The *Transmission Control Protocol/Internet Protocol (TCP/IP)* that manages the transfer of *electronic mail* on the *Internet*.

Simple Network Management Protocol (SNMP): A *protocol* which is widely used to manage networks employing the *Transmission Control Protocol/Internet Protocol (TCP/IP)* as the transport mechanism. It has three main elements which are defined in *Request For Comment (RFC)* issued by the *Internet Activities Board (IAB)*. These are: the *Structure of Management Information (SMI)*, specified in RFC 1065, upgraded to RFC 1155; The *Management Information Base (MIB)*, published as RFC 1066, upgraded to RFC 1156; and SNMP itself, published as RFC 1067, upgraded to RFC 1157. Whereas SMI and MIB documents define the set of *managed objects*, the SNMP document defines the means for getting *information* between management *stations* and managed objects. SNMP usually uses the *User Datagram Protocol (UDP)*. Figure S.13 shows communications using SNMP and Figure S.14 gives the structure of an SNMP protocol.

simplex: A communications system which has one arrangement or state, such as the ability for *transmission* to occur in one direction only, or the use of a single wire or *carrier frequency*.

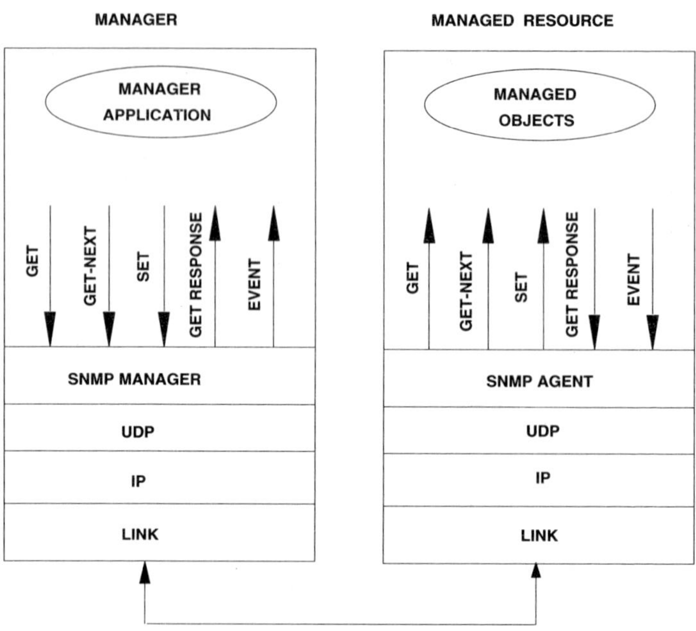

Figure S.13 Communications using SNMP

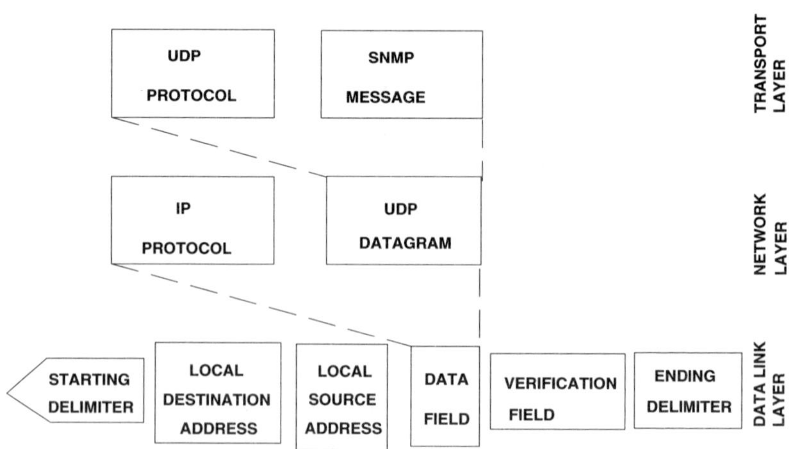

Figure S.14 Structure of an SNMP protocol

Simulated Brillouin Scattering (SBS): A non-linearity effect in the conduction of light in *optical fibres*, which can be considered as a *scattering* of a *photon* to a lower energy, with the excess energy being carried away by a photon.

Simulated Raman Scattering (SRS): A non-linearity effect in the conduction of light in *optical fibres* arising from the *scattering* of a *photon* by a molecule in the amorphous silica that makes up the optical fibre. The scattering results in a transition state change by the silicon molecule, and the loss of part of the photon energy.

sine wave: Same as *sinusoidal waveform.*

singing: An unwanted, self-sustained, *audio frequency* oscillation occurring in a *circuit.*

singing margin: The difference between the point at which *singing* occurs and the *gain* of the device effected.

singing path: The *path*, which is usually a loop, in a *circuit* around which *singing* occurs. See Figure F.9.

singing point: The threshold *gain* point of a *circuit* beyond which *singing* will occur.

single band: Communications equipment which can operate in only one *frequency band.*

single bit error: A *data transmission error* resulting from the inversion of a single *bit*, such as the replacement of a 1 by a 0 or a 0 by a 1.

single call sign calling: The use of subordinate *station call signs* in the establishment and exchange of *messages* in *radio telephone* and *radio telegraph* operation.

Single Channel Voice Frequency (SCVF): A modification to the *telex signalling* system to improve the performance of the telex *network.* Adopted by *ITU-T Recommendation* R.20. It uses the following frequency allocations: telex *exchange* to telex machine, Space 0 = 1180 Hz and Mark 1 = 980 Hz; telex machine to telex exchange, Space 0 = 1850 Hz and Mark 1 = 1650 Hz. This signalling system also enabled telex machines to operate at speeds up to 300 *baud.*

single current circuit: A *circuit* over which *data transmission* occurs as a *digital signal* by applying a positive voltage directly to the *transmission line* to represent a 1 and zero voltage for a 0. This contrasts with a *double current circuit* where positive and negative voltage differences are used.

single current signalling: A system of *signalling* in *telegraphy* in which the direct voltage from the 120 volt *exchange* feed is applied to the *line* in the *mark* condition but not in the *space* condition, as in Figure S.15.

Single Error Correction (SEC): An *error correcting code* which can correct for single errors which can occur during *transmission.*

Single Error Detection (SED): An *error detecting code* which can detect single errors which occur during *transmission.*

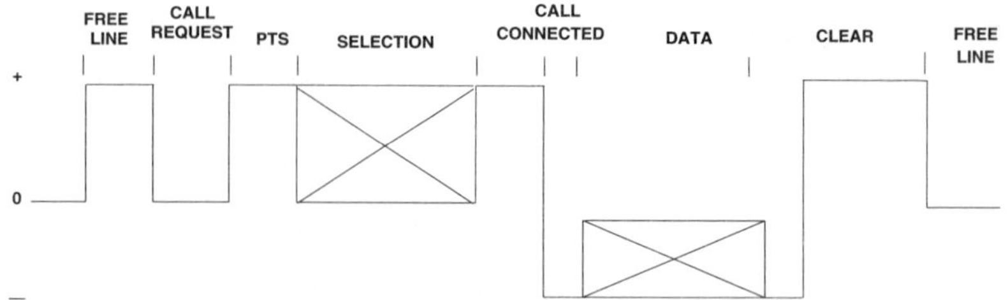

Figure S.15 Single current call progression

Single European Act: Treaty signed by members of the *European Community* on 17 and 28 February 1986, which amended the earlier treaties of Paris and Rome and defined the goal to be reached as a single European market by 31 December 1992.

single frequency signalling: The use of only a single *frequency* to send *dialling*, control and administration *signals* over a communications *line*. The choice of line *signalling tone* is important since it should not interfere with the *speech* band, and a frequency of 2280 Hz is generally used. The system operates on a forward and backward signalling basis, as in Figure S.16. Signals are sent as an application or removal of the 2280 Hz tone, i.e. tone on and tone off modes.

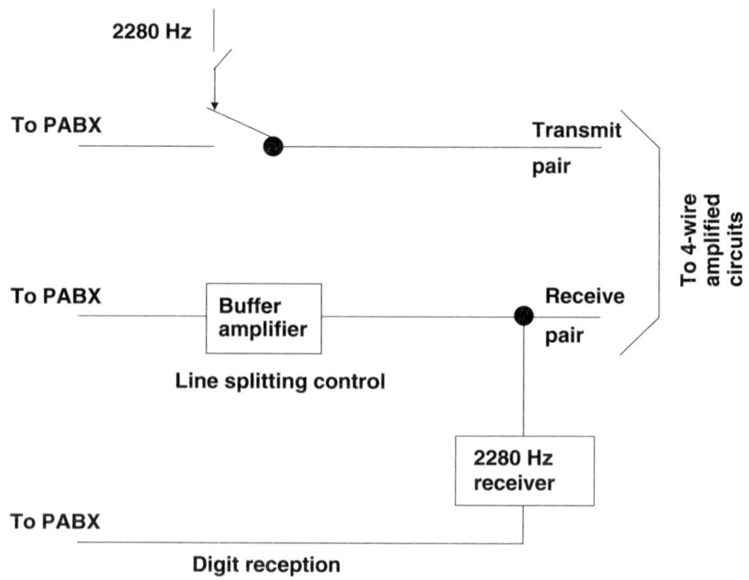

Figure S.16 Single frequency signalling

single harmonic: A *sine wave* with only the *fundamental frequency.*

single harmonic distortion: The ratio of the power of a given harmonic in a *signal* to the power of the *fundamental frequency*, expressed in *decibels* or as a percentage.

single line repeater: A *repeater* which amplifies a telegraph *signal* by using a series-connected pair of cross-coupled polar relays.

singlemode fibre: *Optical fibre* which has a very small *core* diameter and can only accommodate one path of light.

single mode launching: Coupling of an *electromagnetic wave* into a *transmission medium* so that only a single propagation mode is accommodated.

single sideband modulation: A form of *amplitude modulation* in which the *carrier wave* and one of the *sidebands* resulting from the modulation process are suppressed, and only the single remaining sideband is transmitted. This economises on the use of *bandwidth* and power, and is primarily used in carrier telephony and *High Frequency (HF)* radio.

single sideband transmission: *Transmission* method in which one of the *sidebands* (*upper sideband* or *lower sideband*), resulting from the process of *modulation*, is removed prior to transmission.

Single Sideband Suppressed Carrier (SSBSC): *Amplitude modulation* technique in which the power of the *carrier signal* and of either the *upper sideband* or *lower sideband*, is suppressed, as shown in Figure S.17.

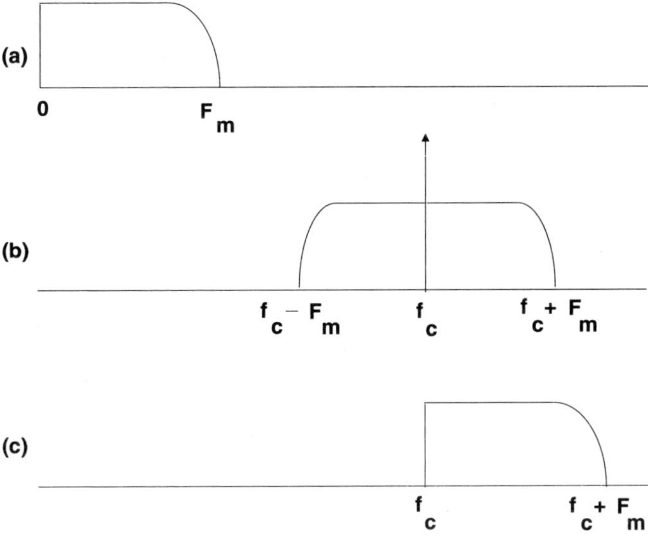

Figure S.17 Amplitude Modualtion spectra: (a) baseband signal; (b) simple Amplitude Modulation; (c) Single Sideband Suppressed Carrier

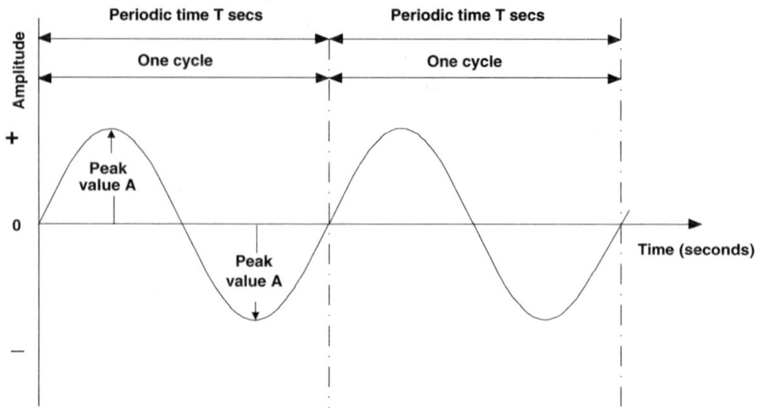

Figure S.18 Sinusoidal waveform

sink: The *receiving terminal* in a communications system, or more generally, an absorber of energy. See also *source*.

S interface: The interface at the *R reference point* of an *ISDN* reference model, as shown in Figure R.6.

sinusoidal waveform: A *waveform* that corresponds to the equation $x = A \sin \omega t$, where A is the peak value of the amplitude (see Figure S.18), and ω is the angular velocity. Also $\omega = 2\pi f$, where f is the *frequency* (equal to the reciprocal of the *periodic time*) and t is the time in seconds.

SIO: *Scientific and Industrial Organisation.*

SIR: *Speaker Independent Recognition.*

SITA: *Societe Internationale de Telecommunications Aeronautique.*

skew: **(1)** the difference in arrival time of *bits* during *parallel data transmission.* **(2)** The angular deviation of the received *frame* from rectangularity in *facsimile* transmission, caused by faults in the *synchronisation* of the *scanner* and the recorder. **(3)** An *electromagnetic wave* in a *waveguide* which is not parallel to its axis.

skewness: **(1)** A measure of the offset between two *signals.* **(2)** A measure of the deviation of a set of numbers from a symmetrical distribution, such as shown in Figure S.19. There are several mathematical ways for expressing skewness. They all give a measure of the deviation between the mean, *median* and the *mode*, and they are usually stated in relative terms. Examples are (mean–mode)/(*standard deviation*) and 3(mean–median)/(standard deviation).

skew ray: A ray of light which travels in an *optical fibre* by means of *total internal reflection* but does not travel parallel to the fibre axis, instead adopting a helical path.

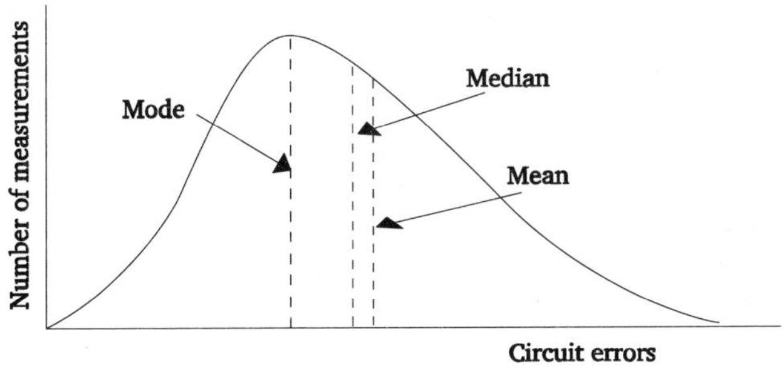

Figure S.19 Illustration of skewness

skin antenna: An *antenna* which is sited on the outside surface, or skin, of another object, such as an aircraft, spaceship, ship mast, etc.

skin effect: The effect in which as the *frequency* of the *signal* is increased the current and magnetic fields rise closer to the surface of the *transmission medium.*

skip distance: The distance between the transmitter of a *radio wave* and the closest point that the wave touches the Earth after having been reflected from the *ionosphere*.

sky noise: The *noise* received by a communications *antenna* from sources other than the Earth (i.e. from the sky) and from space-based objects made on Earth.

sky wave: A radio frequency communications wave that is transmitted from an Earth bound transmitter, but is reflected from the *ionosphere* at some point in its path. This can happen several times (multi-hop), as shown in Figure S.20. The layers of the ionosphere are illustrated in Figure E.4.

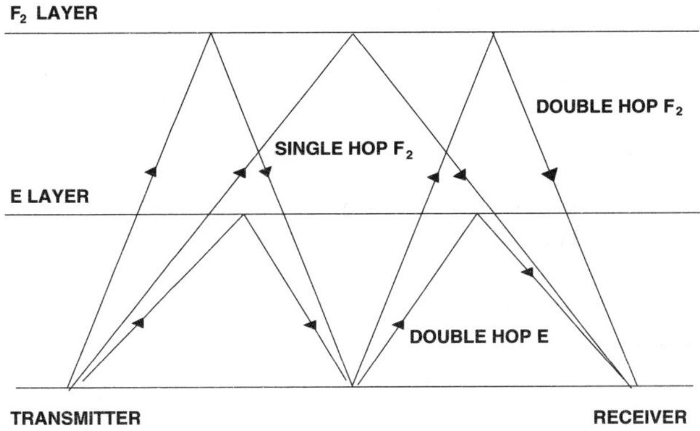

Figure S.20 Multi-hop sky wave paths

SLA: *Service Level Agreement.*

slamming: Term used, especially in North America, where a *subscriber's* account is moved from one long distance operator to another without the subscriber's full understanding of the implications or their consent.

slave: See *master-slave.*

slave station: A *station* which acts as a *slave* to a *master station.*

sliding window: A communications system in which the *window size* (see Figure W.3) can be varied to allow the *transmitting terminal* to send a number of units of *information* before an *Acknowledgement (ACK)* is received.

slip: A defect in *transmission* which causes one or more *bits* to be lost or extra ones to be inserted.

slot: **(1)** A unit of time in which *data transmission* can occur, such as a slot in the *frame* of a *TDM* system. See *timeslot.* **(2)** Mechanical position in a shelf into which a printed circuit board can slide.

slot antenna: An *antenna* consisting of a slot in the surface of a conductor or *waveguide.*

slotted ALOHA: A form of *ALOHA packet transmission* system in which a *terminal* is only able to transmit *data* during specified *timeslots*, which are *synchronised* for all *receiving stations.*

slotted ring: A *network* with a *ring topology* in which *data timeslots* of a fixed length circulate in one direction around the ring, and *transmitting terminals* are only allowed to place data in free slots, which then circulate until they reach the *receiving terminal*, where they are removed, again freeing the slot for reuse.

SMAE: *System Management Application Entity.*

Small Computer System Interface (SCSI): *Hardware* and *software* standard for the attachment of *peripheral equipment* to processing devices.

Small Scale Integration (SSI): An integrated circuit having a density of ten circuits or less on a single silicon chip. See also *Medium Scale Integration (MSI)* and *Large Scale Integration (LSI).*

smart antenna: A processor controlled *phased array antenna* with no moving parts, that can be programmed to automatically change its parameters, such as beam *frequency* and *polarisation.*

smart card: A plastic card, the size of a credit card, with an embedded integrated circuit, which can be used in various secure applications, such as communications system access, financial transactions, and building and computer access. This card is often used in conjunction with a *Personal Identification Number (PIN).*

SMASE: *System Management Application Service Element.*

SMATV: *Satellite Master Antenna Television.*

SMDS: *Switched Multimegabit Data Service.*

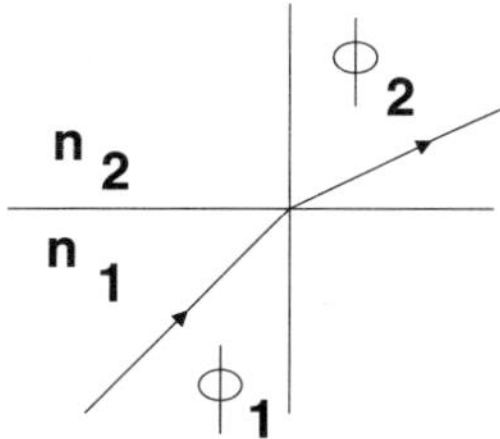

Figure S.21 Illustration for Snell's law

SMF: *System Management Function.*
SMFA: *System Management Functional Area.*
SMI: *Structure of Management Information.*
SMP: *Service Management Point.*
SMS: *Short Message Service.*
SMTP: *Simple Mail Transfer Protocol.*
SNA: *System Network Architecture.*
Snell's law: A law relating the angle of incidence, of a light or other *electromagnetic radiation*, on a *refractive index* boundary, to the angles of *reflection* and *refraction*. For example, in Figure S.21 if n_1 and n_2 are the refractive indices of the two material, then $n_1 \sin \varphi_1 = n_2 \sin \varphi_2$.
SNMP: *Simple Network Management Protocol.*
snow: The effect of *random noise* in a *Video Display Unit (VDU)*, such as a television or radar screen, which appears as a uniform distribution of fixed or moving spots.
SNR: *Signal to Noise Ratio.*
SO: *Shift Out.*
SOA: *Semiconductor Optical Amplifier.*
Societe Internationale de Telecommunication Aeronautique (SITA): The international body which controls the *data communications network* used by many of the world's airlines.
soft copy: A non-permanent image which is displayed on a device such as a *Cathode Ray Tube (CRT)* or computer screen. The soft copy is often stored in the computer as *software*.
soft decision decoding: A decoding system in which the input to the *decoder* is the sample *analogue signal*. This is usually digitised before input to the decoder. It provides extra *coding gain*, compared to *hard decision decoding*, typically of about 2 dB.
software: A set of programmes and procedures loaded into a read-write memory for execution, which are used to operate communications or other computer based systems.
Software Defined Network (SDN): The facility provided by *common carriers* which enables part of the *Public Switched Telephone Network*

(PSTN) to be configured so as to act as a *Virtual Private Network (VPN)* for a *subscriber*.

software interface: The *protocol* which defines the *interface* between two computers, allowing them to successfully interact with each other.

SOGT: *Senior Officials Group on Telecommunications.*

SOH: *Start Of Heading.*

solar noise: *Noise* received from the Sun, in the form of *electromagnetic radiation*, which can cause *interference* with *radio communications*, television, *microwave* and *light wave* communications systems.

solar wind: A plasma, given out by the Sun, which can penetrate the Earth's magnetic field to cause *interference* with *radio communications* systems and damage electronic components in susceptible equipment, such as radar and power supplies.

solitron: A *pulse* of *electromagnetic radiation* having a special shape, power level and spectral distribution, such that it can travel long distances in the *transmission medium* without appreciable *dispersion.*

SOM: *Start Of Message.*

SONET: *Synchronous Optical Network.*

sound bandwidth: The range of *frequencies* between 20 Hz and 20 kHz which is normally occupied by electrical *signals* corresponding to *sound waves.*

sound carrier: The part of a television *broadcast* wave which carries the *audio signal* but not the *video signal.*

sound-powered telephone: A *telephone handset* which derives the power needed for the operations involved in *voice transmission* from the power contained in the *voice signal*, but may use an external power source for *ringing.*

sound programme: The *data signals* which comprise a radio programme, the sound component of a television programme, or other *audio signals* for *broadcast.*

sound wave: An energy *wave* transmitted through a *transmission medium* as a result of vibrations of the particles of the medium parallel to the direction of travel. Sound waves used for communications are usually in the *audio frequency band* of 20 Hz to 20 kHz.

source: **(1)** The device from which *electromagnetic waves* originate, e.g. a light source. **(2)** The place where a *message* being transmitted originates, e.g. a *transmitting terminal*. See also *sink.*

source address: The *address* of the *source* of a *message*, usually carried in the *header* of the message.

source coding: One aspect of *data* coding in the *transmission channel* of a digital communications system prior to transmission. It generally involves the translation of the data provided by the *source* into a form more suitable for further processing.

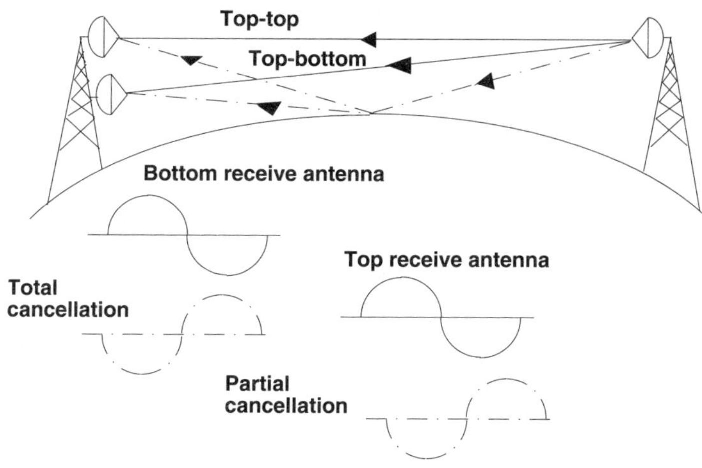

Figure S.22 Space diversity

space: (1) The name given to one of two possible *line* conditions, the other being *mark*. Combinations of these make up *characters*, such as in *alphabetic telegraphy*. Space usually represents a *binary* 0 in *data communications* and mark represents a binary 1. **(2)** The time interval between code elements in *Morse code*.

space diversity: A method of *data transmission* in which *fading* and other *signal* degrading effects are minimised by the use of two or more separate and independent *transmission paths* to transmit the same data, which are then compared and recombined upon reception. Figure S.22, for example, shows this used in a *microwave* transmission system and Figure S.23 shows the arrangement of the two *antenna* heads. Each antenna is connected to a separate receiver and the output from the receivers provide the combined output signal. Cancellation on one of the antennas may occur, due to fading, etc., but this is unlikely to occur on both paths.

Space Division Multiplexing (SDM): *Multiplexing* which uses spatially separated conductors or *waveguides* to provide separate communications *channels*.

space division switching: A mode of *analogue switching* in which a separate physical *path* is provided for each *call* and is held continuously for the exclusive use of that call.

space division system: *Multiplexing* systems in which the multiplexing is by techniques other than *Time Division Multiplexing (TDM)*, e.g. by *Frequency Division Multiplexing (FDM)*.

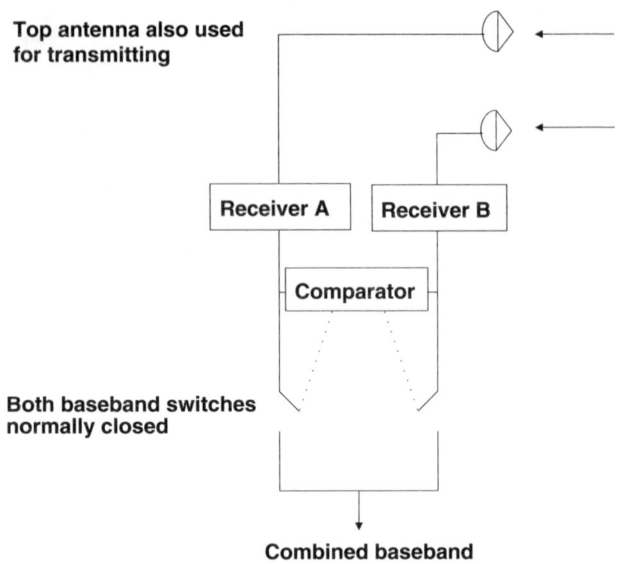

Figure S.23 Space diversity receivers

space Earth propagation: The *transmission* of *radio frequency signals* from a *satellite* to an *Earth station*. One of the major considerations in space Earth propagation is to identify and minimise the sources of power loss.

space switched system: A system capable of *switching* communications *signals* that have been through *Frequency Division Multiplexing (FDM)* rather than *Time Division Multiplexing (TDM)*.

space switching: The act of *switching* in an *exchange* which deals with signals which have been through separate paths or through *Frequency Division Multiplexing (FDM)* rather than *Time Division Multiplexing (TDM)*, or *timeslot* interchange.

Space Time Space (STS): A *switching* process used in some *exchanges* in which the *traffic* is first passed through *space switching* then through *time switching* and finally through *space switching*. See also *Time Space Time*.

space waves: *UHF* and *VHF* short range radio communications waves that are propagated both directly through the atmosphere and using *reflection* from the ground.

Spanning Tree algorithm: *IEEE* 802.1 standard which overcomes the problem in complex *networks* of *packets* perpetually circulating around the network (*broadcast storm*). The Spanning Tree algorithm determines a *root bridge* for the network, which is allocated the highest priority and the network is then configured around this. Selected routes are automatically disabled so that a *loop network* is converted into a *tree network* with only one active path from one *node* to any other node.

span of control: For a *node* in a *network* it generally refers to the number of other devices or nodes controlled by that node.

spatial redundancy: A type of *redundancy* exploited by the MPEG-2 (see *MPEG*) *image compression* system, in which adjacent *pixels* which are not independent are correlated with neighbouring pixels.

SPC: *Stored Programme Control.*

Speaker Dependent Recognition (SDR): A system of *speech recognition* which requires training to the *voice* of a caller before particular words can be identified.

Speaker Independent Recognition (SIR): A system of *speech recognition* which is independent of the particular speaker using the system. Several qualities of the spoken words are analysed and compared to a series of pre-prepared templates for each word. The templates are normally constructed by taking samples from between 100 and 200 different accents speaking the word to be recognised.

speaker verification: An *access security* system where callers are asked to speak a word or phrase which is then compared to the same word or phrase spoken previously by the same caller during an enrolment session. Also called *voice printing*.

Specialised Common Carrier (SCC): A *common carrier* which is specialised in terms of the market it serves or the *services* it provides. See also *value added service provider*.

Specialised Satellite Service Operator (SSSO): Licences granted to six operators in the UK, on October 1998, to provide *services* via *satellite*.

Special Telecommunications Action for Regional development (STAR): A programme introduced by the *European Community (EC)* in October 1986, to cover the period from 1987 to 1991, which made available a fund of money to improve the provision of advanced *telecommunications services* to peripheral regions of the Community.

spectral absorptance: The absorptance of *electromagnetic radiation* within the *transmission medium* through which it is propagating. This varies according to the *wavelength* of the radiation.

spectral width: For an *electromagnetic wave*, such as *light wave* or *radio wave*, it is the *wavelength* interval at which the *power level* is greater than its maximum value by a specified amount.

spectroradiometer: An instrument which detects and analyses the *wavelength* components of *electromagnetic radiation*. The spectroradiometer comprises several standard sections: input optics, a dispersing element such as a diffraction grating, a detector, and *signal processing* equipment.

spectrum envelope: The total pattern of *signals* generated by the *modulation* of an *information* carrying signal into a *carrier signal*. The exact shape of the spectrum envelope would depend on the shape of the information carrying signal and carrier signal, as well as the type of modulation technique used.

spectrum signature: The particular mix of *radio frequencies* detected from a source or bouncing off a specific object. The frequencies may or may not be produced as a deliberate output of the source. The mix and pattern of frequencies can often act as an identifying signature for that source.

speech: A communications *signal* which represents the sound of human *voice*. Same as *voice signal*.

speech channel: A *channel* designed to carry the *bandwidth* used for human *speech*, usually from 300 Hz to 3.3 kHz. See *voiceband*.

speech circuit: A *circuit* designed to carry *data signals* representing the *bandwidth* used for human *speech*, usually 4 kHz.

speech communication: The successful *transmission* and *switching* of *data signals* representing the human *speech*.

speech compression: Same as *voice compression*.

speech interpolation: See *analogue speech interpolation* and *Digital Speech Interpolation (DSI)*.

speech path: A *transmission path* in a *telephone network* which carries human *speech*, i.e. has a *bandwidth* of up to 4 kHz.

speech recognition: Generic term used to describe communications and processing systems which are able to recognise the human *speech* and take appropriate actions. See also *Speaker Dependent Recognition (SDR)* and *Speaker Independent Recognition (SIR)*.

speech scrambling: The conversion of *speech signals* into a form ready for *transmission*, that would be unintelligible if intercepted and overheard by an unauthorised person.

speech traffic: Communications *traffic* which represents human *speech* in either digital or analogue form, using a *bandwidth* of about 4 kHz.

speed calling: A communications *service* in which a short *code* dialled into the *terminal* would instruct the terminal or attached switch or *station* to automatically dial a much longer preset *telephone number*.

speed dialling: *Dialling* at a speed greater than 10 pulses per second. Often used to refer to *speed calling*.

spill forward: The process of sending the *address* of a *call* to a preceding *Central Office (CO)* so that it can take over control of the call.

spill-over traffic: *Traffic* which cannot be handled by the basic *route* of a switched system during *peak load* so it is transmitted along an alternative *path*.

spin: **(1)** Mechanical rotation of a body about its axis, such as experienced by a *satellite* which is in its *orbit*. **(2)** The movement of *subscribers* from one *service* level to another, e.g. taking up more *Cable TV (CATV) channels*.

splice: The join between two pieces of *transmission medium* created by *splicing*.

splicing: The process of joining two pieces of *transmission medium* to create a longer *path*. The *splice* should be done so as to minimise disruption to the *signal* being transmitted. Several techniques can be used for splicing in *optical fibres*. *Mechanical splicing* couples the two ends together under pressure, *fusion splicing* uses heat to melt and join the ends together, and bonded splicing uses epoxy resin to glue the ends together. See *fibre optic splicing*.

split screen: A *Visual Display Unit (VDU)* or other display screen which is divided into two or more separate areas. These could be used to display different files, different parts of the same file, several outputs from different closed circuit TV cameras, etc. For example, Figure S.24 shows the used of a split screen in *videophony* and *videoconferencing terminals*, where the aspect ratio of the display is considerably increased.

split-screen transmission: *Data transmission* over a *network* in which only the parts of the data which are changing are sent over the *line*. Also known as partial transmission.

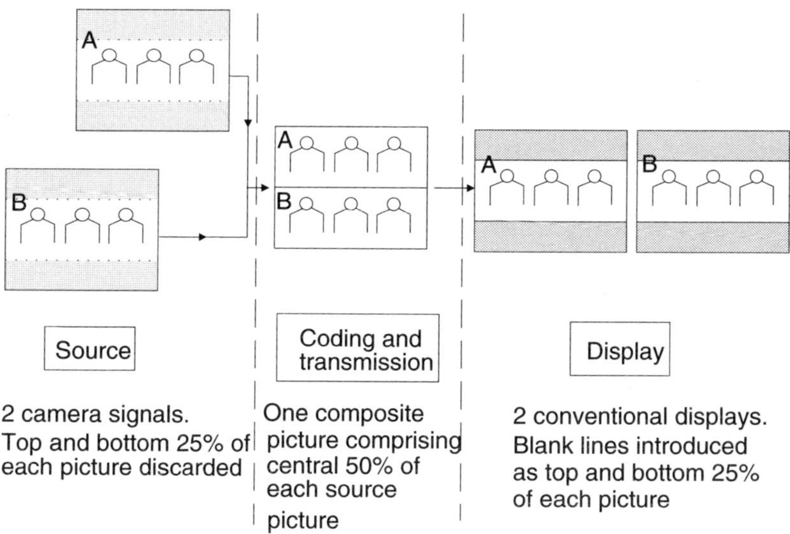

Figure S.24 Split screen

splitter: A device used to split the *transmission medium* so the *signal* can be shared between various *users*, and the signal *loss* at the splitting point is low. Splitters are used, for example, in a *Passive Optical Network (PON).*

SPM: *Subscriber Private Metering* or *Self-Phase Modulation.*

spoof: The act of deliberately inducing a component or *user* of a communications system to perform incorrectly.

spot beam: A narrow *radio communications* beam formed from a *satellite* based *antenna*, which illuminates a limited area of the Earth's surface.

spot beam antenna: An *antenna* used on a *satellite* to produce a *radio communications* beam covering a small region of the Earth's surface.

spread spectrum: The *transmission* technique in which the *signal* is spread over a wide range of *frequencies.* This makes it easier to avoid *interference* arising from a *jamming signal* or due to *multipath effects.* It also makes it difficult to detect the signal, for covert operations, and to demodulate it, to achieve privacy. Several techniques are used for spread spectrum communications, such as *Direct Sequence Spread Spectrum (DSSS)* and *Frequency Hopping Spread Spectrum (FHSS).*

Spread Spectrum Multiple Access (SSMA): A *multiple access* technique, similar to *Code Division Multiple Access (CDMA).*

spurious emission: The emission of *electromagnetic radiation* which is outside the essential *bandwidth* or is at a level which could be reduced without affecting the *transmission* of *information.*

SQL: *Structured Query Language.*

SQNR: *Signal to Quantisation Noise Ratio.*

square law: See *inverse square law.*

square waveform: A *waveform* which has a rectangular shape and represents two significant conditions, *mark* and *space* (Figure S.25). The waveform can be represented mathematically as the sum of the *fundamental frequency*, f, and all the odd *harmonic frequencies* (2n+1)f, where n tends to infinity.

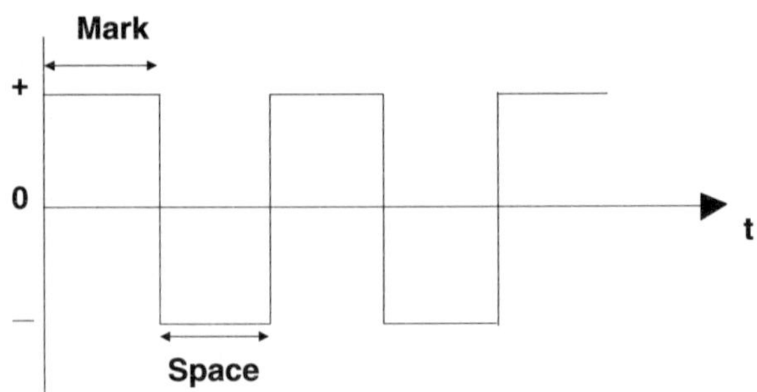

Figure S.25 Square waveform

squelch: An electronic *circuit* which suppresses the *audio signal* output of a receiver, and removes undesired lower input signals in the absence of a sufficiently strong desired input signal.

SRC: Sample Rate Converter.

S reference point: One of the *reference points* of the *ISDN* reference model, as shown in Figure R.6. The S reference point provides the interface between *Terminal Equipment 1 (TE1)* (which is equipment which complies with the ISDN *User-Network Interface (UNI)*) and the *Network Termination Type 2 (NT2)*, which includes functions equivalent to several layers on the *OSI Basic Reference Model*. Examples of NT2 equipment is a *PABX*, *Local Area Network (LAN)*, etc.

SRS: *Simulated Raman Scattering.*

SSBSC: *Single Sideband Suppressed Carrier.*

SSI: *Small Scale Integration.*

SSMA: *Spread Spectrum Multiple Access.*

SSP: *Service Switching Point.*

SS7: *Signalling System No. 7.*

SS6: *Signalling System No. 6.*

SSSO: *Specialised Satellite Service Operator.*

S-STP: *Shielded Screened Twisted Pair.*

stability: The extent to which a parameter varies with relation to some other factor, such as time, temperature, etc.

staggercast: See *Near Video On Demand (NVOD).*

standalone: A programme or device that operates independently of another programme or device.

standalone mode: See *local mode.*

Standalone Dedicated Control Channel (SDCCH): One of the four basic categories of control channels, specified by *ETSI*, for use within the *GSM*. See also *Broadcast Control Channel (BCCH)*, *Common Control Channel (CCCH)* and *Associated Control Channel (ACCH).*

standard code set: A set of internationally agreed and approved *codes* for use in communications *networks.*

standard deviation: A measure of the *dispersion* of a set of numbers from the average. It is calculated by squaring the deviations of the numbers from the *arithmetic mean*, so eliminating their sign, adding the numbers together, taking their mean and then the square root of the mean.

standard frequency and time signal: A radio *signal* which is *broadcast* by government operated radio *stations* in several countries at specific times and *frequencies.* Also the *carrier frequency* and time signal is emitted as described in *ITU-R Recommendation* 460.

standard interface: *Interface* defined in models developed by the standards making bodies. See, for example, *S interface.*

Standards Advisory Committee (SAC): Committee that forms part of the wireline exchange trade association in the USA, the *Exchange Carriers Standards Association (ECSA)*. The committee is responsible to the *ANSI* for the *TI Committee*.

Standards Association of New Zealand (SANZ): An association which is independent of government and develops standards for use within New Zealand, including standards relating to a wide range of *telecommunications* goods and *services*. It is also responsible for testing to the standards, and marking goods which conform to the standards.

Standards Australia (SA): The leading standards setting body in Australia and the Australian representative on international standards bodies, such as *ISO* and *IEC*.

Standards Council of Canada (SCC): A Canadian government body, founded in 1970, which approves standards produced by other organisations into a National Standards System. It encourages voluntary standardisation in areas not covered by law.

standard test finger: A metal device, shaped like but slightly smaller than an average human finger, which is used to define which parts of an electrical system are accessible, and whether the voltage that can be touched is within the safety margin.

Standard Text Markup Language (STML): A *protocol* developed by the printing industry to standardise the *interface* between various different versions of document creation *software* and intelligent printing equipment. This protocol has been extended to the *Hypertext Markup Language (HTML)*.

standby: A dormant status of operation for a communications system, computer, power supply, etc. in which the system is not in use but is maintained ready for use at very short notice. See also *hot standby*.

standby plant: Equipment which is maintained on *standby*.

standing charge: The standard periodic charge made by communications *service* providers to their *subscribers*, which may include hire of a *terminal*, rent of a *line*, allocation of a *telephone number*, etc. This charge is made irrespective of whether the facility is used or not.

standing wave: A wave that forms in a *transmission line*, resonant cavity, or vibrating string, with fixed ends, by the combination of two or more waves. A series of maxima (antinodes) and minima (nodes) occur along the wave (Figure S.26), with the distance between equivalent points along the wave being equal to half the *wavelength*.

Standing Wave Ratio (SWR): The ratio of the *amplitude* of the *standing wave* at an antinode to the amplitude of the same wave at a node in a uniform *transmission line*. The ratio is given by (1+r)/(1–r), where r is the *reflection* coefficient.

STAR: *Special Telecommunications Action for Regional development*.

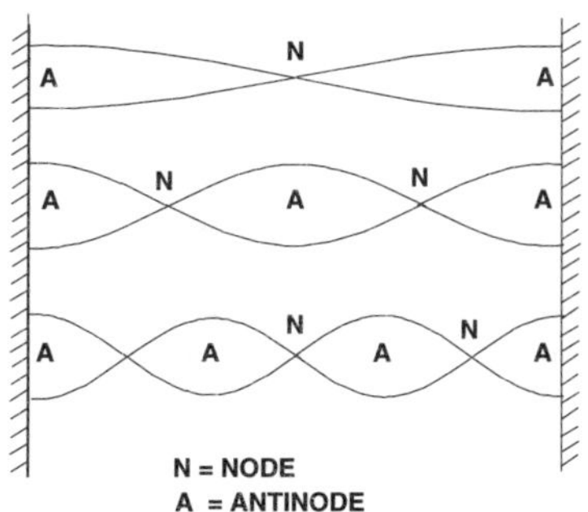

Figure S.26 Standing waves

star connection: The connection of several *terminals* by individual *cables* connecting each terminal to a central distribution box or star coupler.

star network: A *network topology* in which each *terminal* is directly connected to a central *node* which handles all the communications between *users*. See *star topology*.

star shaped ring: A communications *Local Area Network (LAN)* topology in which dedicated cabling is used to connect each workstation to a *wiring closet*. See Figure S.27.

start bit: A *binary digit* or *bit* at the start of a series of bits which represents a *character* or some other similar *information*. It is used in *asynchronous transmission* and enables *synchronisation* to occur at the *receiving terminal*. See also *stop bit*.

start code: A *code* signifying the beginning of a *character*, or other unit of *information*, in a transmission sequence of *bits* representing one *character*.

Start Of Heading (SOH): A *code* signifying that the succeeding group of *characters* is a *header* containing routeing or other control *information* for the attached text.

Start Of Message (SOM): A *signal* which indicates to other *nodes* on the *network* that the *addresses* of the nodes to which the *message* is being

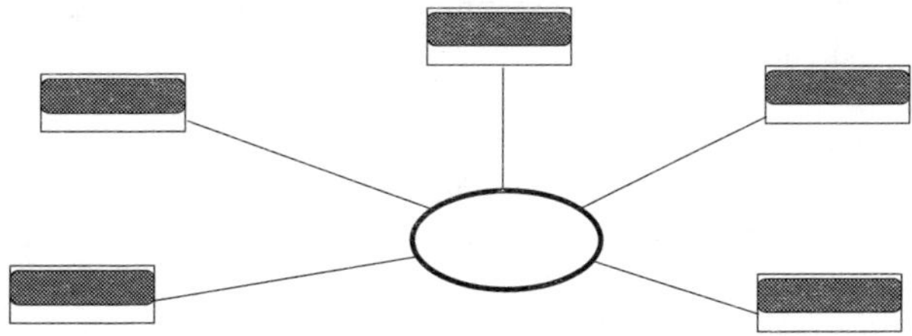

Figure S.27 Star shaped ring

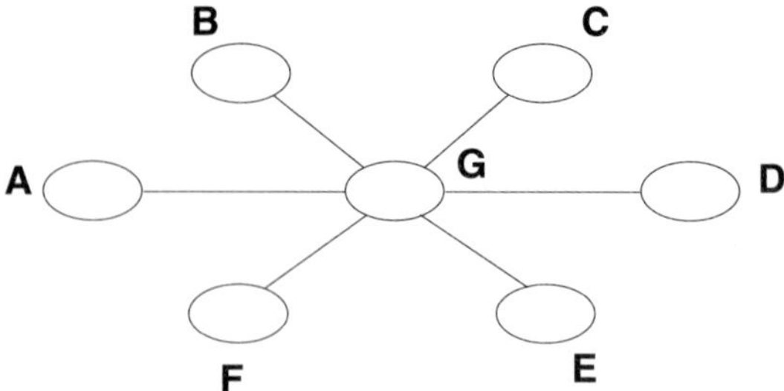

Figure S.28 Star topology

sent is to follow. It also activates automatic *message switching* equipment or aids in other automatic switching systems.

Start of Text (STX): A *transmission control character* which signals the start of *text*, i.e. the main body of the *message* following the *header*.

star topology: A communications *Local Area Network (LAN)* topology in which the networked devices are connected directly via dedicated cabling to a central network device (Figure S.28). This device normally acts as the controller for the network.

start pulse: A coded *pulse* used for *synchronisation* of the *transmitting terminal* and *receiving terminal* in an *asynchronous transmission* system. The start pulse is transmitted at the start of every *character*. See also *stop pulse*.

start-stop signal: *Signal*, in the form of *pulses*, which is used by a *transmitting terminal* to indicate the beginning and end of a group of pulses which represent a *character* or similar *information*.

start-stop transmission: A method of transmitting *information characters* in an *asynchronous transmission* system. Each character is preceded by a *start code*, which synchronises the *receiving terminal* to enable it to decode the received *data*, and is ended by a *stop code*.

state diagram: A diagram which uses nodes to indicate the internal states of a system, and branches that indicate the transitions from one state (or node) to another. State diagrams may also incorporate a temporal element by indicating state changes.

static: The *interference* in communications *signals* caused by natural electrical atmospheric disturbances.

static routeing: Same as *fixed routeing*.

station: A collection of communications equipment connected to a *network*, with a unique *address*, and capable of sending and/or receiving *data*.

station battery: A battery located within a *station* which provides *DC* power for all communications requirements, including radio and *tele-*

phone equipment, and may also provide emergency power for lighting and other facilities.

station clock: The *clock* or other *timing* device in a *station* which provides the timing reference and timing *signals*, controlling communications equipment which require *synchronisation*.

station keeping: The control of a *satellite* to ensure it maintains the desired *orbit*.

Statistical Bit Rate (SBR): One of the four *ATM Transfer Capabilities* defined by the *ITU-T*. It uses *statistical multiplexing* to efficiently handle *bursty traffic* with known throughput requirements. It corresponds to *Variable Bit Rate (VBR)* specified by the *ATM Forum*.

statistical multiplexer: A *multiplexer* which carries out *statistical multiplexing*. Also known as a statmux.

statistical multiplexing: A form of *multiplexing* in which *channels* are allocated on a statistical basis, which may take one of several forms, such as allocating channels according to probability of need based on previous usage. This system reduces high speed channel *idle time*. See *statistical Time Division Multiplexing*.

statistical Time Division Multiplexing: A *Time Division Multiplexing (TDM)* system which concentrates *data* from many *channels* on to fewer channels. It is possible that the aggregate input *traffic* can temporarily exceed the available output *capacity*, and in this situation the data can be temporarily stored in buffers, as shown in Figure S.29. The system works

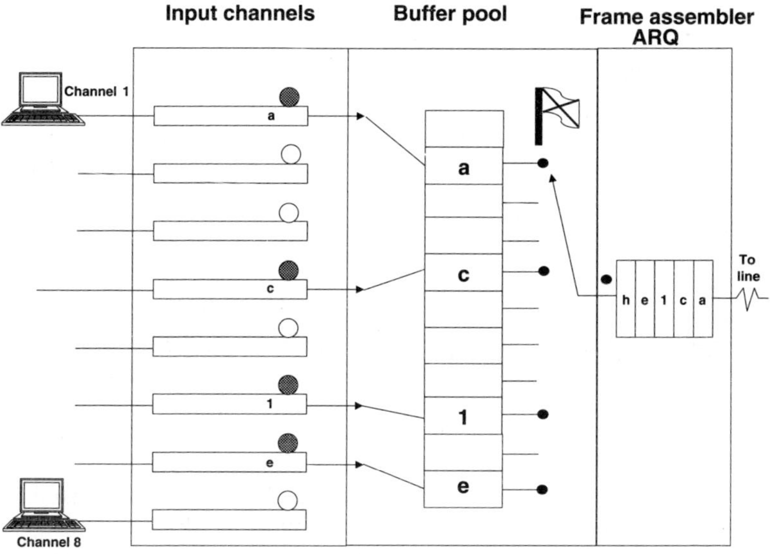

Figure S.29 Simplified operation of a statistical multiplexer

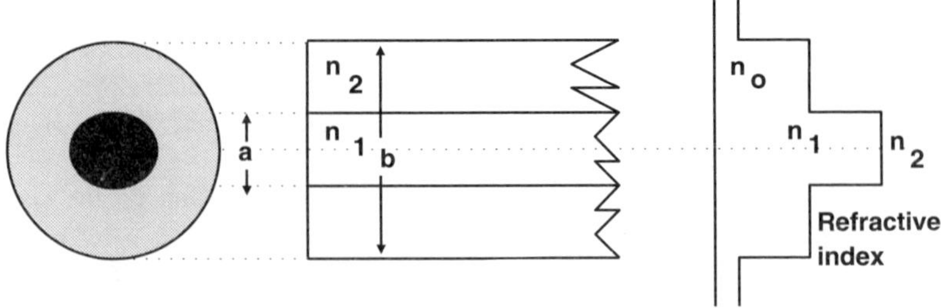

Figure S.30 Step index fibre

by scanning the input buffers and transferring any data available there into a buffer queue waiting to be assembled and transmitted as a *frame*. This system can only handle *asynchronous transmission*.

statmux: See *statistical multiplexer*.

STD: *Subscriber Trunk Dialling*.

STDM: *Synchronous Time Division Multiplexing*.

STE: *Signalling Terminal Equipment*.

step-by-step exchange: See *Strowger exchange*.

step-index fibre: *Optical fibre* in which the *refractive index* changes rapidly (step change) between the *core* and the *cladding*, as shown in Figure S.30. See also *graded index fibre*.

steradian: A unit of solid angle equal to the solid angle projected at the centre of a sphere by an area equal to the sphere radius squared on the surface of the sphere. There are 4π steradians in each sphere.

stimulated emission: The emission of *electromagnetic radiation* which occurs when the internal energy level of a quantum system drops to a lower energy level, caused, for example, by the presence of *radiant energy* of a suitable *frequency*.

STM: *Synchronous Transport Module* or *Synchronous Transfer Mode*.

STML: *Standard Text Markup Language*.

STMR: *Sidetone Masking Rating*.

stocastic signal: See *random signal*.

stop and wait ARQ: A method of *error correction* in the *OSI* model of a *Packet Data Network (PDN)* in which an *Acknowledgement (ACK)* or a *Negative Acknowledgement (NAK)* is sent to the *transmitting terminal* and the *data* is retransmitted if it has not been correctly received. See also *go back N ARQ*.

stop band: The *frequency band* which is stopped and cannot be transmitted through a *filter*, *telephone* or other communications device.

stop bit: A *binary digit* or *bit* at the end of a series of bits which represent a *character* or other similar *information*. The stop bit is used in *asyn-*

chronous transmission to indicate the end of the character and usually returns the *line* back to the *idle state*. See also *start bit*.

stop pulse: A *pulse* which occurs at the end of a *character* during *asynchronous transmission*, to indicate the end of the character. See also *start pulse*.

stop-start traffic: A form of communications *traffic* in which many short *packets* travel in one direction with slightly longer packets travelling in the reverse direction. This is often associated with short turn-around and *transit delays*, and is a characteristic of *asynchronous transmission*.

store and forward switching: A *switching centre* which temporarily stores *data* until it can be forwarded to its destination.

store and forward system: A communications system which incorporates *exchanges* or *nodes* that are capable of storing *data* and then releasing them for *transmission* when a clear *transmission path* becomes available. Such systems allow data operation to continue even when the transmission path is temporarily not available.

store and forward transmission: The *transmission* of *data* via one or more *switching centres* which temporarily store and then re-transmit the data, possibly after some manipulation.

Stored Programme Control (SPC): The centralised control of the operation of an analogue or digital *exchange* using stored *software* programmes on dedicated *hardware*. The tasks performed by the controller could include interpreting *signals* received from *subscribers*, setting up *switching paths* through the exchange, and determining which *messages* are sent to *subscribers* and other exchanges. This has the advantage that facility changes can be made centrally and the variety of hardware can be limited. The device which runs the software is often known as the *Stored Programme Controller* and the exchange is known by the same name.

Stored Programme Controller (SPC): See *Stored Programme Control*.

STP: *Shielded Twisted Pair* or *Signalling Transfer Point*.

strata graph: A graphical method of presenting *data* which shows how the total value is split amongst its constituent parts. For example, Figure S.31 shows that the total revenue obtained by a *PTO* steadily increases with time, but that only *services* B and D have growth, whilst service C is reducing and may eventually become unprofitable.

stratosphere: The portion of the *Earth's atmosphere* that extends from between 10 to 13 km to about 50 km above the Earth's surface, depending on season, latitude and time of day (see Figure E.4). The stratosphere is situated between the *troposphere* and the *ionosphere* and has a nearly constant vertical temperature.

stream cipher: A practical *encryption* system in which a stream of *plaintext* symbols is *enciphered* each symbol at a time. The symbols are

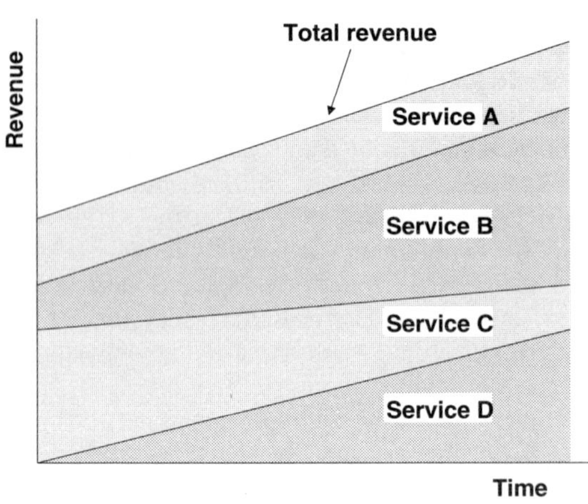

Figure S.31 A strata graph

generated by a symbol generator under the control of an enciphering *key* and are *deciphered* by using the same key.

strength member: The part of a communications *cable* which provides the main strength for support and to prevent damage during installation and operation.

Strowger exchange: One of the first automatic *telephone exchanges* which used a series of *Strowger selectors* to connect a *call*. The call progressed through the exchange in stages, each selector connecting the call to a free selector in the next stage. This is a step-by-step operation and the exchange was also often known as a *step-by-step exchange*.

Strowger selectors: A solenoid or motor activated electromechanical device previously used in *telephone exchanges* to make connection between *incoming lines* and *outgoing lines*.

structured cabling: A cabling system which has been installed in a building in a structured way taking into account possible future expansion. Generally it will consist of cable runs with drop cables to *user pods, and patch panels* within *wiring closets* where adds and changes can be done relatively easily. Structured cabling also allows for easier tracking of faults and the rearrangement of the *network* to overcome a *congested system*.

Structured Query Language (SQL): A standard language, developed by IBM, for creating and manipulating *data* in a *database*.

Structure of Management Information (SMI): Within *OSI network management* the SMI defines the modelling principle to be used for designing *managed objects*, along with the description associated with them. It provides sets of templates to describe the managed objects.

Abstract Syntax Notation One (ASN.1) is used for these, so that they are independent of the machine on which they run.

STS: *Space Time Space.*

Study Group: A group belonging to a standards making body, such as the *ITU-T*, which carries out the task of writing the standards. These are then normally circulated for comment and approved by a separate authority prior to issue.

STX: *Start of Text.*

sub-address: The suffix used in the identifying *address* of *network users* to specify the particular *terminal* or class of use of the address.

sub-band: A band of *frequencies* created by subdividing a larger *frequency band.*

sub-band Adaptive Differential Pulse Code Modulation: A form of *signal modulation* where a band is split into two *sub-bands*, each of which is then modulated using *Adaptive Differential Pulse Code Modulation (ADPCM).*

subcarrier: A *carrier signal* that is used to modulate another carrier. This can then be used to modulate a third carrier, and the process repeated.

subcarrier frequency shift: The *modulation* of an *audio frequency carrier signal* which is then used to modulate a *Radio Frequency (RF)* carrier, in order to produce a modulated signal for communications purposes.

Subcarrier Multiplexing (SCM): *Multiplexing* technique normally used in the *optical fibre link* for *Hybrid Fibre Coax (HFC)* systems.

submarine cable: An *optical fibre* or *coaxial cable* used for *telecommunications*, which is designed to be submerged in the ocean. See Figure O.6.

subnet address: The extension to an *Internet address*, using *Internet Protocol (IP)*, which allows a single *network* address to be used for two or more smaller sub-networks.

subscriber: A person or organisation who is entitled to access a communications *network* or *service*. Subscription usually involves registering and paying a *tariff* made from fixed and variable components. A subscriber is also called a customer, party or *user*.

subscriber distribution network: The communications *network* which connects *subscriber terminals*, such as a *telephone* or *PABX*, to a *local exchange*. It is also called the *access network*, *access loop*, *access line*, or *local loop*.

Subscriber Identity Module (SIM): A *smart card* used in conjunction with a mobile *handset* to ensure that only the authorised *user* of the instrument has access to it.

subscriber instrument: The communications *terminal* registered for use of a *service* on a *network*, by a *subscriber* to that service.

subscriber line: The *transmission line* between the *subscriber's terminal* and the *local exchange*. Also called an *access line*, *access loop* or *local loop*.

subscriber premises network: The portion of a communications *network* which is located on the premises of a *subscriber*, for example, a *Local Area Network (LAN)*, or a number of *extension lines* connected by a *Private Automatic Branch Exchange (PABX)*.

Subscriber Private Metering (SPM): A system of sending *pulses* to *subscribers* to provide *information* about *call* charges, which is often displayed on a unit, as found in *Coin Collecting Boxes* and some *Private Automatic Branch Exchanges (PABX)*. Up to two pulses per second may be sent during the *speech* phase of the call, as well as pulses at the end.

Subscriber Trunk Dialling (STD): The UK national *subscriber dialling* facility which allows *long-distance calls*, within the UK, to be made without the intervention of a human operator.

sunspot: A dark marking, which is between a few hundred kilometres wide to several times the size of the Earth, that lies between 30°N and 30°S latitude on the Sun and is at a temperature about 2000 K lower than its surroundings. Sunspots vary in size and are periodic, with about 11 years between successive maxima, when they tend to cause *interference* with radio and other communications systems.

sunspot number: A figure representing a smoothed average of the number of *sunspots* over 12 months, issued by the Sunspot Data Centre in Brussels and the Telecommunications Data Centre in Boulder, Colorado, USA. High sunspot numbers indicate good conditions for long range *High Frequency (HF)* communications.

superframe: A repeating sequence of 12 *frames*, used in *Time Division Multiplexing transmission* systems. Each frame corresponds to 125 microseconds so one superframe is 1.5 milliseconds in duration.

supergroup: Used within *Frequency Division Multiplexing (FDM) transmission* systems, each supergroup consists of five *channel* groups, each carrying twelve *voice calls*. It therefore consists of a total of 60 voice channels.

Superhigh Frequency (SHF): The *frequency* of an *electromagnetic wave* which lies in the *frequency range* 3 GHz to 30 GHz, with *wavelengths* from 1 cm to 10 cm.

supermaster group: Part of the *Frequency Division Multiplexing (FDM)* hierarchy, it consists of 600 *voice circuits* handled as a unit.

super refractive effect: A rare propagation anomaly, experienced in line of sight radio systems, such as *microwave*, which directs a transmitted *signal* into the Earth before it has reached the *receiving terminal* (Figure S.32).

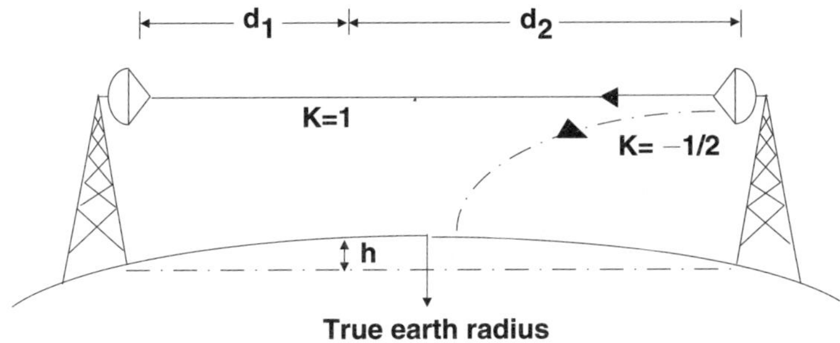

Figure S.32 Super refractive effect

Supervisory Audio Tone (SAT): A low level *tone*, in the region of 6 kHz, transmitted by *base stations* operating with the *Advanced Mobile Phone System (AMPS)*. The *mobile phone* will transpond the received *signal* back to the base station. If either the mobile or base station detect a difference between the received and expected tones, the audio *path* will be muted to prevent *interference*.

Supervisory Control Language (SCL): A *programming language* used to customise a *Stored Programme Control (SPC)* based communications *network*.

supervisory frames: One of the three types of *frames* within *HDLC* and *LAPB*, they control information flow, such as requesting *retransmissions* and *Acknowledgement (ACK)* frames.

supervisory signal: A *signal* which supervises a particular connection by controlling the operating states of the *circuits* used in the connection and indicating the progress of the *call*. Supervisory signals include *off-hook*, *clear* and *recall*.

supplementary service: *Services* which are additional to the basic services. For *ISDN*, for example, these have been categorised under headings such as *number* identification, *call* offering, charging, and additional information transfer. Supplementary services on *GSM* systems are provided by enhancing the basic *call processing system software*, and include call divert, divert on no answer, divert on mobile unreachable, divert on *busy*, *call waiting*, and *multi-party connection*.

suppressed carrier modulation: A form of *analogue signal modulation* in which the *carrier signal* is suppressed, leaving only one or both the *sidebands*.

suppressed carrier transmission: An *amplitude modulation transmission* system in which the *carrier frequency* is suppressed to a level which renders it unusable for *demodulation*.

surface wave: An *electromagnetic wave* what propagates along, and is guided by, the surface that forms between two different *transmission media*. Examples include *radio waves* guided over the Earth's surface, which usually occurs in the *Low Frequency (LF)* and *Medium Frequency (MF)* bands.

surfing: Switching from one television *channel* to another, without spending long on any one channel. Also, browsing the *Internet*, via the *World Wide Web (WWW)*, with a loosely defined aim.

Sustained Cell Rate (SCR): One of the definitions of *traffic* sources used within *ATM Quality of Service (QoS)* measurements. The others being Peak Cell Rate (PCR) and Burst Tolerance (BT).

SVC: *Switched Virtual Circuit.*

SWAN: *Satellite Wide Area Network.*

switched circuit: A communications *circuit* formed by the configuration of switches along the *transmission path.*

switched connection: The *transmission path* set up between a *sending terminal* and a *receiving terminal* which includes *switching* in an *exchange*.

Switched Digital Video (SDV): Term applied in *Cable Television (CATV)* systems in which the digital *telecommunications network* consists of a *star topology* with *switching* occurring at the *local exchanges*.

Switched Multimegabit Data Service (SMDS): High speed *data transmission* system developed by *Bellcore*. It is designed for access speeds of 1.544 Mbit/s and 45 Mbit/s and uses *DQDB MAN* technology. SMDS uses *connectionless mode transmission*.

switched network: A communications *network* which allows any *terminal* to be connected to any other terminal through the use of *switching* systems, such as *message switching*, *circuit switching* and *packet switching*, with associated control.

switched star: A *network* configuration for the distribution of *Cable TV (CATV)*, which allows a dedicated cable feed to every customer and greater possibility for interactivity by using a *star topology* network. This is shown in Figure S.33. The switch points are located remote from the *head-end*, close to the *subscriber*. An intermediate *hub* site may be needed depending on distance involved and the size of the system.

Switched Virtual Circuit (SVC): A *data path* set up as part of the X.25 *Packet Switched Network (PSN)* in which a temporary connection (*Virtual Circuit (VC)*) is established between the *transmitting terminal* and the *receiving terminal*, following a *call request*. The other types of *data transmission* services specified within X.25 are *Permanent Virtual Circuit (PVC)* and *datagram*.

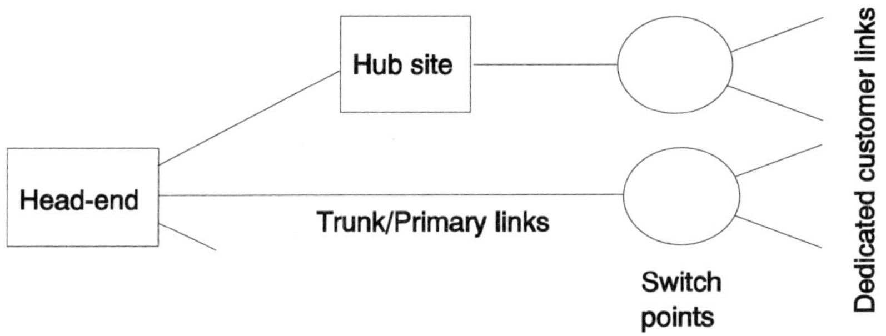

Figure S.33 A switched star Cable TV network

switch fabric: The details of the way that *data* is switched between *nodes* in a communications *network*, rather than around the periphery of the network.

switch hook: The switch which is operated when a *telephone handset* is raised (switch closed) or replaced (switch open). See also *hook switch*.

switching: The *routeing* of *signals* between specific points in a *network*. Switching usually takes place at various levels, such as in a *PABX* or an *exchange*. Figure S.34 illustrates an arrangement where simple mechanical switches (crosspoints) are used to connect the *terminals* to one of the *links*.

switching distribution centre: The part of a communications system which *routes messages* or *packets* through *switching* and also distributes messages to a range of local *addresses*.

switching centre: A facility in a communications system, usually found at a *node of a network*, which uses switches to interconnect *circuits* to create a *path* between the *transmitting terminal* and the *receiving terminal*. Also called an *exchange*.

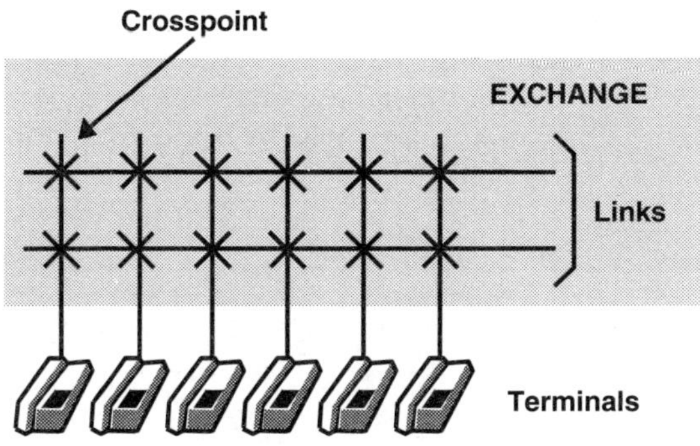

Figure S.34 A simple switching arrangement

switching equipment: Equipment which is used to perform *switching*, such as in an *exchange* or a *PABX*.

switching matrix: A matrix, usually found in a *switching centre*, which uses switches to connect *incoming lines* and *outgoing lines*. Several forms of switches may be used, such as *crossbar switch*, relays, solid state circuits, etc. See, for example, Figure C.31.

SWR: *Standing Wave Ratio*.

symmetrical channel: A *transmission channel* in which the send and receive *circuits* have the same *data signalling rate*.

Symmetrical Digital Subscriber Line (SDSL): A *Digital Subscriber Line* technology developed by AT&T Paradyne. It provides symmetrical 1–3 Mbit/s *transmission* over the *local loop*. Also known as *Medium bit rate Digital Subscriber Line (MDSL)*.

SYN: *synchronous idle*.

synchronisation: The process by which a fixed *timing* relationship is established and maintained between a *transmitting terminal* and a *receiving terminal*, to enable correct translation of the transmitted *data*.

synchronisation distortion: *Distortion* of a communications *signal* caused by the *receiving terminal sampling* the incoming *bits* with a sampling period that has drifted from the correct time interval.

synchronisation pulse: A *pulse* used in a communications system to achieve and maintain *synchronisation* between the *transmitting terminal* and the *receiving terminal*.

synchronised network: A communications *network* in which all the elements are synchronised by one *master clock*. Also called a *despotic network*.

Synchronous Container: Used within the *Synchronous Digital Hierarchy (SDH)* to map *Plesiochronous Digital Hierarchy (PDH)* rate *circuits* into an SDH *frame* for *transmission*. The Synchronous Container consists of a number of 64 kbit/s *channels* and can be looked at as a subdivision of the basic SDH frame structure. See Figure V.7.

Synchronous Data Link Control (SDLC): The control of *synchronous transmission* over the *Data Link Layer*. Often used to refer to the *bit* oriented *line control protocol* developed by IBM as part of the high level data link control which initiates, controls, checks and terminates *information* exchanges. SDLC has been developed by *ANSI* as *Advanced Data Communications Control Protocol (ADCCP)*, from *ISO* as *HDLC* and from the *ITU-T* as *Link Access Procedure (LAP)*.

synchronous data transmission: See *synchronous transmission*.

synchronous detection: The process of generating a local *carrier signal* from the *lower sideband* to aid *demodulation* of a communications *signal* which has been transmitted by suppressing the carrier and *upper sideband* signals.

Synchronous Digital Hierarchy (SDH): A new standard for *synchronous transmission* developed by the *ITU-T* and published in a series of *G Series ITU-T Recommendations*. It overcomes several of the deficiencies of the older *Plesiochronous Digital Hierarchy (PDH)* system. SDH is used in Europe and is equivalent to *SONET* used in USA and Japan.

synchronous Earth satellite: A *satellite* that *orbits* the Earth in synchronism with the Earth's rotation, maintaining a fixed longitude, not necessarily in the equatorial plane and not necessarily above the same point.

synchronous idle (SYN): An internationally recognised *transmission control character* which establishes or maintains *synchronisation* between a *transmitting station* and a *receiving station.*

synchronous network: A communications *network* in which all *clocks* are controlled so they maintain *synchronisation.*

Synchronous Optical Network (SONET): A *Bellcore* developed standard for *synchronous transmission* over *optical fibre*. The basic *data rate* is 51.84 Mbit/s, called *Optical Carrier* 1 or OC-1, and several other data rates have been specified, as in Table S.3.

synchronous satellite: See *synchronous Earth satellite.*

Synchronous Time Division Multiplexing (STDM): *Time Division Multiplexing* which differs from *Asynchronous Time Division Multiplexing (ATDM)* in that as well as keeping *data* integrity, the *multiplexers* have to keep the associated *clock timing synchronisation* so that the

Table S.3 Optical Carrier rates within SONET

OC level	*Transmission rate (Mbit/s)*
OC-1	51.84
OC-3	155.52
OC-9	466.56
OC-12	622.08
OC-18	933.12
OC-24	1244.16
OC-36	1866.24
OC-48	2488.32

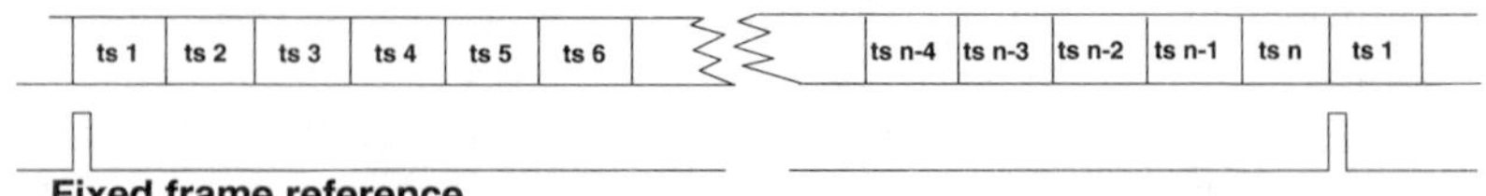

Figure S.35 Synchronous Time Division Multiplexing

demultiplexed recovered data can have a regenerated clock associated with it. See Figure S.35.

Synchronous Transfer Mode (STM): The transfer mode normally used for the *circuit switching* of *B channels* on *ISDN*.

synchronous transmission: *Transmission* in which the *bits* that make up the *data character* are transmitted at a fixed rate, whilst *synchronisation* is maintained between the *transmitting station* and the *receiving station* by means of *clocks* and frequently transmitted synchronisation characters.

Synchronous Transport Module (STM): In the nesting process during creation of a *Synchronous Digital Hierarchy (SDH) frame*, it is the point where the nesting stops as it contains the largest defined level of the *Virtual Container (VC)*. These logical STM *signals* are what is seen at the interface between SDH equipment, i.e. the *Network Node Interface (NNI)*. Table S.4 gives the various STM *data rates*.

sync pulse: Same as *synchronisation pulse*.

sync pulse separator: A device, inside equipment such as a television, which separates *synchronisation pulse* information from the received television *signal*. This *information* usually comprises *field scan* and *line scan* instructions.

Table S.4 STM transmission rates

STM rate	*Transmission rate (Mbit/s)*
STM-1	155.52
STM-4	622.08
STM-8	1244.16
STM-12	1866.24
STM-16	2488.32

system blocking signal: A control *signal* sent within a communications system to indicate that part of the system's facilities are temporarily unavailable, and will result in any requested communications *path* being blocked.

system boundary: The physical limits of a communications system, including the *transmitting terminal* and *receiving terminal*, but excluding the *users* of these terminals.

system gain: The *gain* of a system, which is its measure of performance defined as the *decibel* difference between the power of the transmitted and received *signals*.

System Management Application Service Element (SMASE): Specified in the *OSI Basic Reference Model*, it is associated with each *network manager* and its *agent*. It uses the services of *CMIS* which are carried to the remote system via *CMIP*. This manipulates the *Management Information Base (MIB)* using low level operations. SMASE encompasses a set of *System Management Functions (SMF)*.

System Management Functional Area (SMFA): The various *System Management Functions (SMF)*, used by *network management* applications, have been grouped by *ISO* in the *OSI Basic Reference Model* into five System Management Functional Areas: *accounting management*, *security management*, *performance management*, *configuration management*, and *fault management*. The SMFA define which SMF is to be used and which OSI application can be involved, for example *CMIP*, *FTAM*, etc.

System Management Functions (SMF): Functions used in the *OSI Basic Reference Model* which define the mechanism by which certain tasks can be undertaken on the *managed objects*. Examples of these are: Object Management, covering the mechanism used to create, delete and manage the *attributes* within the objects; State Management, covering the object states, and the mechanism which may be used to change them; Event Management Reporting, covering the reporting and control of events; Log Control, covering the mechanism for setting up and maintaining a log of events; etc.

system power margin: The margin that exists between the power needed for correct performance of the *receiving terminal* and the *losses* which occur in the *transmission medium* between the *transmitting terminal* and the receiving terminal.

Systems Management Application Entity: The function within *layer* seven of the *OSI Basic Reference Model* which contains the *System Management Application Service Element (SMASE)*.

Systems Network Architecture (SNA): *Protocol* developed by IBM for communications between IBM *hardware* and *software*.

T

TA: *Terminal Adapter.*

table of frequency allocations: A table published in the *ITU* Radio Regulations which allocates the *frequency spectrum* from 9 kHz to 275 kHz into about 400 bands. The bands are radio services. As different countries use radio facilities in different proportions, the frequency allocation is not rigid and allows for some flexibility. This is partly achieved by having different allocations in different geographic regions, and by allocating some bands to more than one service, resulting in shared bands.

tackline: A length of rope, or halyard, used in visual *signalling* systems, using flags, to separate flags that would convey a different meaning if left contiguous.

TACS: *Total Access Communications System* or *Total Access Cellular System.*

tail circuit: A *feeder line* or *circuit* which goes off from the main *network* to connect a *node*. Usually consists of a *leased line.*

take-off angle: The angle at which radio waves leave a transmitting antenna. For *transmission* systems using the *ionosphere* the optimum take off angle can be calculated if the distance to the target area and the height of the refracting layer are known. The longer the distance, the lower the take-off angle required. See also *departure angle*. Figure T.1, for example, shows the optimum take-off angle from the Earth for single hop *sky wave* propagation.

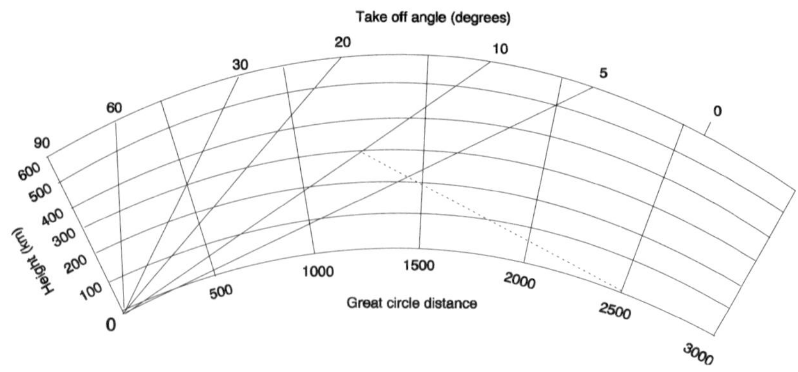

Figure T.1 Take off angle for single hop sky wave propagation

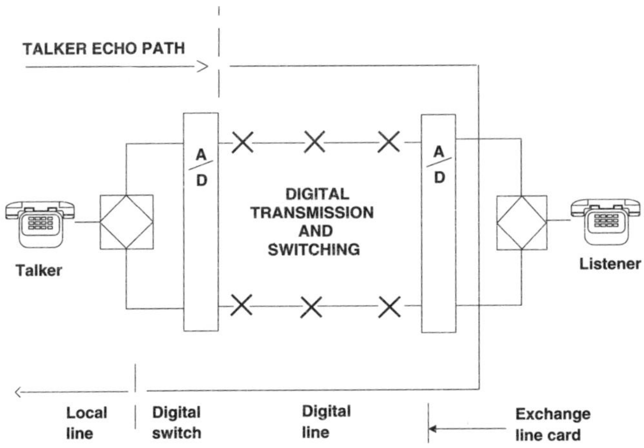

Figure T.2 Talker echo path

talkback paging: A *paging* system which allows *speech* acknowledgement. Two-way speech and two-way *signalling* can normally be obtained using a pair of frequencies. In the UK this is licensed with the receivers in the 459 MHz band and transmitters in the 161 MHz band.

talker battery: A battery which is isolated from other battery use so that it is electrically quiet and allows good quality *voice* communications.

talker echo: An echo in voice or data transmission which is perceived by the talker or sender. It occurs when the sender's speech or data is returned with significant delay. Figure T.2 illustrates the talker echo path, the signals being reflected back to the talker at the distant *two wire transmission* to *four wire transmission* conversion unit. See also *listener echo*.

Talker Echo Loudness Rating (TELR): It is a measure of the *Overall Loudness Rating (OLR)* of the *talker echo* path. It is a function, as shown in Figure T.3, of the send loudness rating (SLR), the receive loudness rating (RLR) and the echo loss or *echo attenuation* at the *two wire transmission* to *four wire transmission* conversion unit. Therefore TELR = SLR + RLR + echo loss.

talk time: For a battery operated communications system it is the amount of time available for use before the battery needs replacing or recharging.

tandem: A serial connection of networks, circuits or links where the *signal* output from one unit is connected directly to the input of another.

tandem data circuit: A *circuit* which connects two or more *Data Circuit-terminating Equipment (DCE)* in series.

tandem exchange: An *exchange* which acts as an intermediate *switching centre* between *local exchanges* or other tandem exchanges in a *telephone network*. Tandem exchanges can connect to *trunk circuits* between

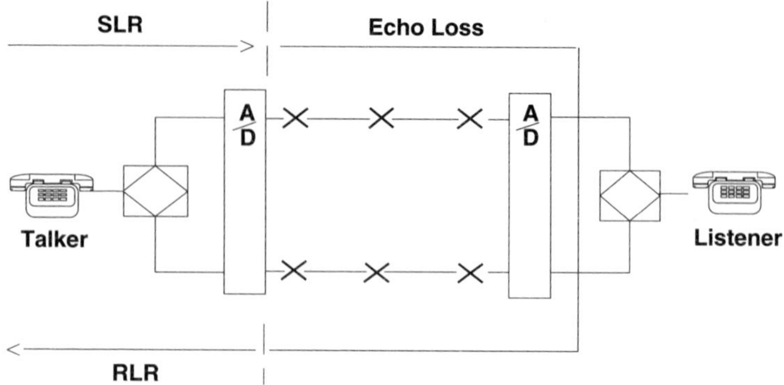

Figure T.3 Talker Echo Loudness Rating

exchanges but do not directly connect to the *subscriber*. They therefore act as intermediate exchanges between the *calling terminal* and the *called terminal*. Also called a *tandem office* or *tandem switching centre* or *transit exchange*.

tandem office: See *tandem exchange*.

tandem routeing: A route from one local exchange to another which passes via one or more *tandem exchange*.

tandem switching centre: See *tandem exchange*.

tandem tie trunk network: A network configuration allowing connection at a tandem centre location of *tie lines* between *Private Automatic Branch Exchanges (PABX)* to form a through connection. This means that an *incoming call* from a distant PABX receives a *dial tone* at the local PABX instead of the operator, and can then connect to an extension on the PABX or to an *outgoing line*.

tandem trunk: A *trunk circuit* connecting two or more *tandem exchanges*, or a tandem exchange to a *local exchange*.

tap: A device which is used to make a connection with a main *bus* so that *signals* can be injected or removed from the *transmission medium*, such as *coaxial cable* or *fibre optic cable*.

TAPI: *Telephony Applications Programming Interface*.

target probability of service: Probability targets set by system designers relating to the provision of set services to users of the system.

target user population: The population of potential users who are targeted by the design and marketing of a particular system.

tariff: The charges or rates made to users by a *telecommunications common carrier* or other supplier of communications services for the use of specific equipment, facilities or services. The tariff is usually contained in a published schedule, which enables *regulatory bodies* to approve

Table T.1 T carrier systems

Carrier	*Transmission rate (Mbit/s)*	*Number of voice channels*	*Digital Signal level*
T1	1.544	24	DS1
T1C	3.152	48	DS1C
T2	6.312	96	DS2
T3	44.736	672	DS3
T4	274.176	4032	DS4

rates. The identification and recovery of tariffs from users is usually a large cost in the operation of a service.

TASI: *Time Assignment Speech Interpolation.*

TAT: *TransAtlantic Telecommunications.*

TBR: *Technical Basis for Regulations.*

TC: *Transmission Control.*

T carrier: Generally refers to the North American *digital hierarchy transmission system* which used *Time Division Multiplexing (TDM)* to carry several 64 kbit/s *Digital Signal level* channels. There are several levels of T carriers, as given in Table T.1.

TCC: *Technical Control Centre.*

TCD: *Technical Cooperation Department.*

TCM: *Time Compression Multiplexing* **or** *Trellis Coded Modulation.*

TCP: *Transmission Control Protocol.*

TCP/IP: *Transmission Control Protocol/Internet Protocol.*

TCRTR: *Technical Committee Reference Technical Report.*

TCT: A series of Technical Committees of the *British Standards Institute (BSI).* Examples are Electrical Safety (TCT/1), Public Switched Telephone Network (TCT/2), Telex Network (TCT/3), Packet Switching (TCT/4), Private Circuits (TCT/5), Digital Networks (TCT/6), Installation Requirements (TCT/7) and Private Branch Exchanges (TCT/8). The Technical Committees consist of individuals nominated by various organisations, such as manufacturers and user groups. It is financed by subscriptions and by grants from the *Department of Trade and Industry (DTI).*

TCTR: *Technical Committee Technical Report.*

TD: *Time Division.*

TDD: *Time Division Duplex.*

TDD/FDMA: *Time Division Duplex/Frequency Division Multiplexing.*

TDD/TDMA: *Time Division Duplex/Time Division Multiplexing.*

TDM: *Time Division Multiplexing.*

TDMA: *Time Division Multiple Access.*

TDM digital hierarchy: See *digital hierarchy.*

TDM multiplexer: A *multiplexer* that combines several *channels* for *transmission* along the same *transmission path* using the principle of *Time Division Multiplexing (TDM).*

TDM frame structure: See *frame* and Figure F.11.

TDR: *Time Domain Reflectometer.*

TE: *Terminal Equipment.*

Technical and Office Protocol (TOP): A communications protocol, sponsored by the Boeing Company, which principally deals with the networking requirements of engineering and office applications. It is a variant of the *Manufacturing Automation Protocol (MAP).*

Technical Bases for Regulations (TBR): A series of publications produced by the *European Telecommunications Standards Institute (ETSI)* which draw together the key requirements from several *European Telecommunications Standards (ETS)* to form a set of specifications which can be used by the *European Commission (EC)* to form a *Common Technical Regulation (CTR).*

Technical Control Centre (TCC): A central location which is used to control a *network.* This includes *network management* and testing.

Technical Committee Reference Technical Report (TCRTR): A report used within the *European Telecommunication Standards Institute (ETSI)* which records the results of Technical Committee work which is not to be published but which may be used for guidelines, status reports or coordination documents. Also issued as *Technical Committee Technical Reports (TCTR).*

Technical Committee Technical Report (TCTR): See *Technical Committee Reference Technical Report (TCRTR).*

Technical Cooperation Department: A department that existed in the *ITU* until 1994 which, along with the *Telecommunications Development Bureau,* encouraged the growth of telecommunications in less developed countries. In the new ITU organisation of 1993 these two groups were combined into the Telecommunications Development Sector.

Technical Recommendations Application Committee (TRAC): A committee within *CEPT,* formed in 1986 to approve standards in Europe for the connection of equipment to public networks.

tee coupler: A coupler that has three ports, which is often used to allow the *signals* received through one port to be distributed to the other two, or the signals received through two ports to be combined and distributed through the third.

TEI: *Terminal Endpoint Identifier.*

Telautography system: A *telegraph* system, used mainly to transmit handwritten messages by sending signals representing positions on a document, which are then reproduced by the receiver.

telco: An abbreviation for a *telephone common carrier.* Also used to describe a telephone *Central Office (CO).*

teleaction service: A telecommunications service found in *ISDN* implementations which transmits short *messages* at a low *data rate* to enable the *user* to manage a range of miscellaneous transactions such as home surveillance, checkouts, remote meter reading and automatic teller machine operation.

telecommunications: The *transmission* or reception of *information* such as sounds, images or *signals* using electric currents along wires or *electromagnetic waves* transmitted through the atmosphere or along *waveguides* such as *optical fibres.* It includes *visual communications* but not the physical transfer of matter, as found in the postal system. 'Tele' in Greek means 'far'.

Telecommunications Act: (1) The British Telecommunication Act, passed in July 1981, created British Telecom (BT) by separating the *telecommunications* functions of the Post Office. A further Telecommunication Act was passed in April 1984 which established the regulatory framework in the UK, overseen by the *Office of Telecommunications (Oftel).* **(2)** The USA Telecommunications Act was passed in 1996 with the aim of stimulating increased competition by removing restrictions placed on the local and long distance service providers.

telecommunications authority: See *common carrier.*

Telecommunications Business Law: A law introduced in Japan in 1985 which ended the domestic monopoly of the Nippon Telegraph and Telephone Corporation (NTT) and the international monopoly of the Kokusai Denshin Denwa Co. Ltd (KDD), as well as restructuring NTT as a private corporation. However, the government still maintained ownership of two thirds of the shares in NTT.

Telecommunications Development Bureau: A part of the *ITU* it worked with the *Technical Cooperation Department* in assisting the growth of *telecommunication networks* in developing countries; in encouraging investment by industry in these areas, and in organising world and regional development conferences. In the new ITU organisation of 1993 it was combined with the Technical Cooperation Department to form the *Telecommunications Development Sector.*

Telecommunications Development Sector: Part of the *ITU-T* which includes the functions of the former *Telecommunications Development Bureau* and the *Technical Cooperation Department.* See also *Radiocommunications Sector* and *Telecommunications Standardisation Sector.*

Telecommunications Industries Association (TIA): An association based in the USA, and accredited by the *American National Standards Institute (ANSI)*, whose members are principally telecommunications equipment manufacturers. The association is concerned with issues relevant to its members and develops telecommunications standards.

telecommunications line: Same as *transmission line*.

Telecommunications Management Network (TMN): A *network management* structure developed from the *OSI* model by *ANSI*, *ITU-T* and *ETSI* in order to achieve overall management of various types of telecommunications equipment. The principles of the TMN are described in *ITU-T Recommendation* M.30. The basic concept behind a TMN is to provide an organised *network* structure, to achieve the interconnection between the various types of *Operations Systems (OS)* and telecommunications equipment using an agreed architecture and with standardised interfaces. Figure T.4 shows the relationship between the TMN and a telecommunications network. As in the case of the *OSI Basic Reference Model* the telecommunications network is considered to consist of *managed objects* (*Network Elements (NE)*). These may be physical elements, such as *exchanges*, transmission equipment, *cable*, *cross-connects*; or it may consist of abstract elements, such as maintenance entities and

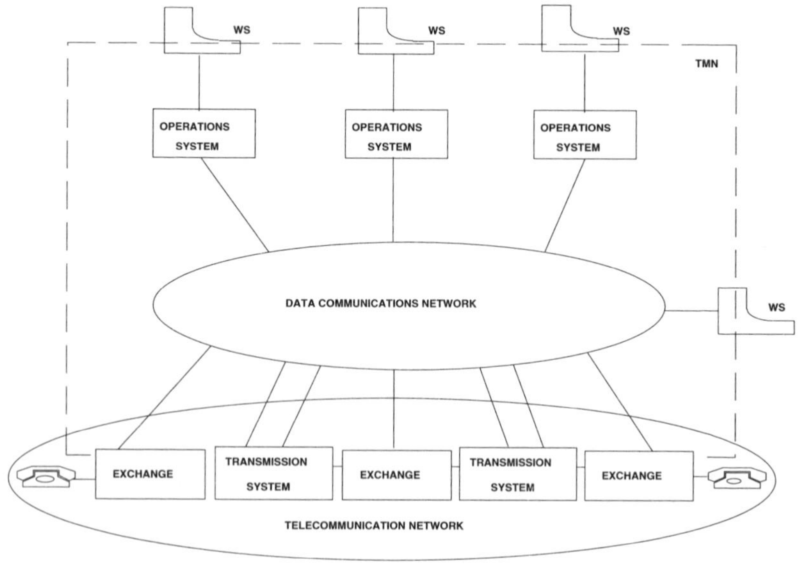

Figure T.4 Relationship between the TMN and a telecommunication network

support entities. The interchange within the TMN uses the OSI *seven layer model* and the services within these *layers*. See also *TMN functional architecture*, *TMN management layers* and *TMN physical architecture*.

Telecommunications Managers Association (TMA): A UK-based registered charity and company limited by guarantee whose members are mainly individuals responsible for planning, management or operation of *telecommunication* systems and consultants and project managers with a special interest in telecommunications. The Association has three different roles. Firstly it organises conferences, exhibitions and courses to aid information exchange between the Association's members. Secondly it interfaces with government and other official bodies in order to promote members' interests. Thirdly it encourages the development of a range of telecommunication *services* through dialogue with suppliers, such as BT and Mercury.

Telecommunications Network Voltage (TNV): From a personal safety standpoint a *circuit* which carries *telecommunications signals*, such as *ringing*, is allowed to have a higher voltage (TNV) than other circuits, but these must be less accessible.

Telecommunications Private Operating Agency (TPOA): A private agency that operates a *telecommunications* system or *service*.

Telecommunications Standardisation Bureau: A bureau within the *ITU-T* which provides secretarial support to the Study Groups that perform the standardisation work of the ITU-T. This Bureau would arrange meetings, produce reports and handle documents for the Study Groups.

Telecommunications Standardisation Sector: Part of the *ITU-T* organisation which in 1993 combined the standards making functions of the *CCITT* and the *CCIR*. See also *Telecommunications Development Sector* and *Radiocommunications Sector*.

Telecommunications Technology Committee (TTC): Set up in 1985 by the Japanese *Ministry of Posts and Telecommunications (MPT)* to prepare standards for *telecommunications* in anticipation of deregulation in Japan. It now acts as the focus for receipt of standards and for comments on international standards, setting technical requirements for broadcasting and telecommunications. Members of the TTC are Type I Telecommunications Carriers (i.e. those who own their own infrastructure), Type II Telecommunications Carriers (i.e. those who lease their infrastructure from other carriers and resell), manufacturers, and users. Foreign bodies can also be members.

Telecommunications Technology Council: A council set up by the Japanese *Ministry of Posts and Telecommunications (MPT)* to provide advice on the preparation of contributions to international standards.

Telecommunications UK Fraud Forum (TUFF): An industrial group, formed in the UK, to consider methods for combating *telecommunications* fraud.

Telecommunications Users Association (TUA): Created in 1980 and later merged with the Telephone Users Association (which was founded in 1965). It is a UK-based limited company funded entirely by its members who are principally large users of *telecommunications services*, such as banks, publishers, accountants etc. The Association represents members' interests through dialogue with suppliers and key strategy and regulatory committees such as the *Office of Telecommunications (OFTEL)*. It also seeks to ensure that the UK telecommunications *network* matches the best available overseas by using devices such as encouraging competition in the supply of telecommunication services.

telecommunications watchdog: An organisation, usually appointed by a government at a national level, which regulates the operation of telecommunication service providers. The work of watchdogs may include licensing new operators and services, setting guidelines on *tariffs*, establishing rules on *interconnection* between operators, etc. See also *regulatory body*.

telecommuting: The use of *telecommunications* equipment to allow business to be performed without one of the participants commuting to a standard place of work. For example, telecommuting may involve an employee working at home and using a personal computer, modem and telephone line to remain in contact with the office. Also known as *teleworking*.

teleconference: A conference held between geographically isolated people or groups of people with the aid of *telecommunications* equipment which allows the transfer of audio and other *data signals*. Generally teleconferencing does not include *video*, which is know as *videoconferencing*, but the term teleconferencing is sometimes used to include this.

telegram: A hard copy message, either written, printed, or pictorial, which is transmitted using *telegraphy* and delivered to the addressee. The telegram service ceased in the UK in 1982.

telegraph code: Groups of electrical pulses which *code* plain language text and numerals used in *telegraphy*. Although early codes contained only 36 characters, the current international standard set contains 58 to include extra marks, such as punctuation. The telegraph code is a *5-bit code*, with each character being framed by a start pulse and a stop pulse. See *International Alphabet*.

telegraph exchange: Also known as *telex*, an international service providing the *transmission* and reception of printed messages over the *tele-*

graph network. Also refers to the *switching* nodes or *exchanges* in the *telegraph network*.

telegraph network: A *network* of *telegraph exchanges*, telegraph *terminals* and *transmission links* which allow the *transmission* and reception of *telegraphy messages*. The telegraph terminals are often *teletypewriters*, which comprise a *keyboard* for inputting the *message* to be sent, and a printer. The speed of transmission of the *digital* telegraph signal along the transmission links is usually between 45 and 200 bauds.

telegraph service: A *service* which allows *telegraphy messages* to be transmitted and received and is provided by *telecommunications* suppliers.

telegraph system: A system which allows the *transmission* and reception of *telegraphy messages*.

telegraphy: A *telecommunications* system in which the interruption or change in the polarity of *DC* current is used to produce *coded signals* to transmit *information*. This includes alphabetic *telegraphy*, produced by the use of *teletypewriters*, *document facsimile telegraphy* and *photograph facsimile telegraphy*.

telemarketing: The use of *telecommunications* technology to help in the marketing of goods and *services*. It consists of handling *incoming calls* as well as making *outgoing calls*. A key feature is a *database* of potential consumers of a product. The databases are often transcribed from recordings made by *audiotex* systems where callers are asked to leave their names and addresses in response to games and prizes. *Call centres* are often used in telemarketing.

telemetry: The automatic measurement of parameters, such as pressure, sound, temperature and humidity, at a distance, and the *transmission* of the readings by means of a *telecommunications* system so that they can be interpreted and analysed in a location different from the measuring instrument. Telemetry is also used to convey information, such as electricity or gas meter readings, from private homes to a central collection point.

Telemetry, Tracking and Command (TT&C): Communications facilities necessary for successful *communications satellite* operation. The satellite's performance is monitored using *telemetry*, and the location and movement of the satellite are monitored, such as by using *tracking radar*. All movements of the satellite, which cannot be automated by on board control, are initiated using command *signals* from Earth stations.

telephone: The *handset* or *terminal* which can transmit and receive *voice* or tone-coded *messages* over a *Public Switched Telephone Network (PSTN)* or *private line*. Figure T.5 shows a functional diagram of a telephone. The telephone converts sound waves into electrical signals

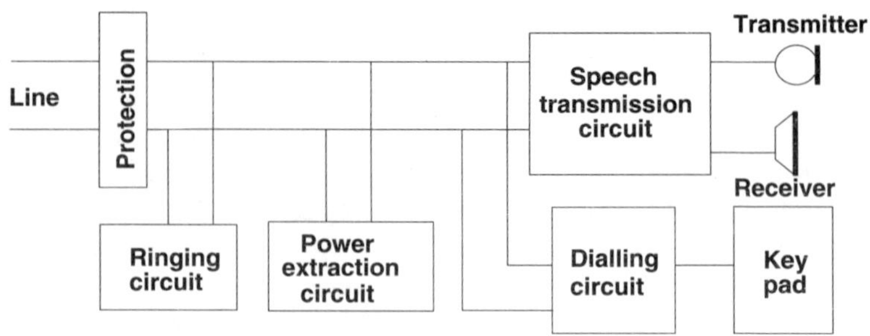

Figure T.5 Functional diagram of a telephone

using a transmitter, and electrical signals into sound waves using a receiver. It also incorporates *ringing* and *dialling* circuits.

telephone channel: A communications *transmission path* with a *bandwidth* of about 3 kHz, which is used for the transmission of *speech signals*, usually as part of the *Public Switched Telephone Network (PSTN)*.

telephone closet: The same as *wiring closet*.

telephone exchange: A *Central Office (CO)* which is used to switch *calls* between other such offices in a *Public Switched Telephone Network (PSTN)*. Often shortened to *exchange*.

telephone fraud: The effective theft of a *telephone service*, often achieved by the actual theft and use of a mobile handset, or by breaking into a company's *PABX* and selling international calls made through this PABX. It may also occur when *subscribers* move house without paying their bills. The best prevention of such fraud involves careful monitoring of usage patterns.

telephone frequency: The *frequency range* between about 300 Hz and 3400 Hz which is used in *telephone systems* to carry *voice signals*.

telephone number: A *number* that is given to each *telephone subscriber* and is used for *routeing* telephone *calls*. Every telephone number should be unique for both *national calls* and *international calls*.

telephone signalling: *Signalling* used for a *telephone call*.

telephone system: A communications system which allows a *user* to transmit *speech* and coded sound signals to another user via an *exchange*.

telephony: A general term indicating communication between geographically separate parties using the *voice* or *audio frequency band*.

Telephony Applications Programming Interface (TAPI): See *Telephone Services Application Programming Interface (TSAPI)*.

Telephony Services Applications Programming Interface (TSAPI): A *software programming standard which determines the behaviour of switching* functions, such as *call routeing*, determining the state of a *called terminal*, etc. Used in *Computer Telephony Integration (CTI)* applications. TSAPI is supported by vendors such as AT&T and Novell and *Telephony Applications Programming Interface (TAPI)*, which performs a similar function, is supported by others, such as Microsoft and Intel.

Telephony User Part (TUP): *Protocol* used in *Common Channel Signalling System No. 7* for basic *connection management.*

telepoint: A mobile public *telephone service*, started in the UK in 1989, which allowed *users* to call other users only when the caller was near a predetermined, labelled telepoint *base station*. It was not successful and was discontinued.

teleport: **(1)** A telecommunications wholesaler who bypasses the, *access loop* by providing a service such as *satellite* access to large corporations. **(2)** A location which contains a large number of *satellite Earth stations* which are used to carry *traffic* from large users.

teleprinter: **(1)** A printer, representing a typewriter without a *keyboard*, that receives and prints but does not transmit *data* in the form of electrical signals. **(2)** A printer comprising a keyboard for the *transmission* of data in the form of electrical signals, and a printing receiver which prints received data and responds to *start-stop signals* defining the beginning and end of each *character*. See also *teletypewriter*.

teleprocessing: The processing of *information* using a combination of *telecommunications* and automatic or manual *data processing systems*. This may be in the form of *on-line* processing systems or *real time systems* and could involve the processing of information stored on a common *database* accessed by several *users*.

teleprocessing monitor: A *software* system used as the main interface between *users* and *databases* stored on computers involved with *teleprocessing*. The teleprocessing monitor is designed in such a way as to overcome problems associated with multiple users accessing and editing *data* stored on a central *database* from several separate *terminals*.

telescopic antenna: An *antenna* whose telescopic construction allows changes in length to be simply achieved. Such antennae are found in many car radios and in submarine communications, where a telescopic antenna needs to be raised above the surface of the water to receive and send radio *signals*.

teleservice: A type of *service* provided by a *Public Telecommunications Operator (PTO)* which necessitates the use of *Terminal Equipment (TE)* such as a *telephone* or a *teleprinter*. Examples of teleservices are *telephony, data transfer, video* and *Message Handling Systems (MHS)*.

Teletel: The French public *videotex* system.

teletext: A one-way *data transmission system* where *text* and simple graphics are *broadcast* in the normal *television transmission signal*, without interfering with the transmission and reception of these signals. This teletext *information* can be viewed using a specially adapted television receiver. Teletext is normally used to convey additional *information* but it can also supplement the main television picture, for example by adding subtitles. See also *videotex*.

teletext decoder: The component of a *television receiver* which allows the set to select, receive and decode *teletext data*.

teletraffic theory: Theoretical study into the performance of *telecommunications* systems, such as, for example, the *traffic* handling capabilities in *switching*, *dynamic routeing* performance, etc. The Danish mathematician A.K. Erlang is generally credited with first published work in teletraffic theory in 1917.

teletypewriter: An instrument which can transmit, receive and print signals over a *telegraph network*. Teletypewriters comprise a *keyboard*, a printer and an interface with the *network*. They may also have a punched paper tape reader or a paper tape perforator for sending and recording *messages*. See also *teleprinter*.

television broadcast day: The period within a day during which a television station *broadcasts*.

television channel: A *channel*, which either uses the atmosphere or a *cable* to carry *television transmissions signals*.

television circuit: A one-way *circuit* which either uses the atmosphere or a *cable* to carry the *video signal* (but not the *audio signal*) component of *television transmission signals*.

Television Receive Only (TVRO): Refers to an *Earth station*, such as a domestic *television receiver*, which is only capable of one way reception of *signals* from a *satellite*.

television receiver: A device which can receive and display the *audio signal* and *video signal* elements of television *broadcasts*, transmitted as *radio waves* either through the atmosphere or along a *cable*. A television receiver comprises several components, such as a *radio frequency filter*, *video amplifier*, *audio amplifier*, *demodulator*, *Cathode Ray Tube (CRT)* and a *loudspeaker*.

television set top box: A boxed device attached to a *television receiver* that can have several functions, such as the conversion of *broadcast digital television signals* to *analogue* ones or one *frequency* to another, the search for desired channels, conditional access for *pay per view* or subscription *services*, and the sending of signals to the broadcasting source such as in *Video On Demand (VOD)* systems.

television transmission signal: *Radio wave transmission*, either through the atmosphere or along a *cable*, which consists of both a *video signal* and an *audio signal* for reception by a *television receiver*.

television transmitter: A device which modulates a *carrier wave* with *video signals*, *audio signals* and *synchronisation pulses* to produce a *television transmission signal* which is then *broadcast* using an *antenna*.

teleworking: See *telecommuting*.

telex: A *telecommunications service* used primarily by businesses which allows the *transmission* of *text*, input using *teletypewriters*, to be sent over a dedicated automatically switched *telegraph network* using *Baudot code* and received by *teleprinters* which are often unattended. Telex signals transmit at 50 bit/s, and use either the *ITU-T International Alphabet No. 2*, or the *ITU-T International Alphabet No. 5*.

telex protocol: *Protocol* used by *telex* machines necessitated as the receiving machine may be unattended. Several exist, some of the more commonly used ones being given in Table T.2.

telnet: The *TCP/IP Application Layer protocol* which allows a *user* at one *terminal* to interact with systems at other sites as if their terminal were connected directly to these systems.

Table T.2 Telex protocol (Continued on next page)

Code	*Meaning*
* ABS	Absent subscriber; power switched off the machine
BK	I cut off (break the line)
CFM	Please confirm or I confirm
COL	Collation
CRV	Do you receive / I receive
* DER	Out of order
DF	You are in communication with the called party
GA	You may transmit
* INF	Subscriber temporarily unavailable; call information service
JFE	Office closed because of holiday
MNS	Minutes
* MOM	Wait for reply

Table T.2 (Continued from previous page)

Code	*Meaning*
MUT	Mutilated
* NA	Number not accessible
* NC	No circuits; trunks busy
* NCH	Number changed
* NP	Called party ceased or spare line
NR	Indicate your called number
* OCC	Subscriber busy
OK	Agree
Tor5	Stop the transmission
PPR	Paper
R	Received
RAP	I will call again
SSSS	Here ready for data transmission
SVP	Please
TAX	What is the charge?
TEST MSG	Please send test message
THRU	You are in communication with a telex position
IPR	Teleprinter
W	Words
WRU	Who are you?
+	Finished
+?	I have finished, do you wish to transmit?

TELR: *Talker Echo Loudness Rating.*

temperature coefficient: The rate of change of a parameter with temperature. For example, the temperature coefficient of resistance is the change of resistance value with temperature.

temporal redundancy: A form of *redundancy*, used in *video compression* systems such as MPEG-2 (see *MPEG*) which arises from the non-randomness of *pixel* values across adjacent *frames*.

10Base5: Standard for running *Ethernet LAN* over thick *coaxial cable*. The term is derived from the fact that the system can run at 10 Mbit/s speed (*BASEband*) over 500 metres maximum length.

10BaseT: Standard for running *Ethernet LAN* over *twisted pair wire*. The term is derived from the fact that the system can run at 10 Mbit/s speed (*BASEband*) over 100 metres maximum length.

10Base2: Standard for running *Ethernet LAN* over thin *coaxial cable*. The term is derived from the fact that the system can run at 10 Mbit/s speed (*BASEband*) over 200 metres maximum length.

terahertz: A *frequency* of a million, million *herts*, i.e. 10^{12}Hz.

terminal: A device for sending and receiving *telecommunications data* over a *transmission channel*. Examples of terminals include: *telephone*, *television receiver*, *teleprinter*, *Personal Computer (PC)*, etc.

Terminal Adapter (TA): An adapter which allows non-*ISDN Terminal Equipment (TE)* to interface with an ISDN *network*. See also *R interface* and *S interface*.

terminal cluster: A group of *terminals* which are located geographically close to each other and are under the control of a single *cluster controller*.

terminal emulation: A process, normally using *software*, which allows a *peripheral equipment*, such as a *Personal Computer (PC)*, to communicate with a *host processor* as if the peripheral equipment was a *terminal* of the host processor.

Terminal Endpoint Identifier (TEI): Part of the *Link Access Procedure (LAP)* it is used to identify *terminals* on a *subscriber line*. It can take a range of values, such as: 0 to 63 for automatic assignment TEIs which are selected by the *user*; 64 to 126 for automatic assignment TEIs selected by the *network* on request; 127 for a global TEI which is used to *broadcast information* to all terminals within a given *SAPI*. The combination of TEI and SAPI identify the LAP and provide a unique *address*.

Terminal Equipment (TE): (1) Communications equipment which is used to transmit or receive *signals*. This also includes *telephone* and *telegraph* switchboards if circuits terminate there. (Often written in lower case, i.e. terminal equipment.) **(2)** In *ISDN* Terminal Equipment have been categorised into two types: *TE1*, which is ISDN compliant, and *TE2* which is not, requiring a *Terminal Adapter (TA)* for interfacing to the ISDN *network*. Examples of TE2 interface specifications are those complying to *ITU-T Recommendations* in the *X Series* or *V Series*.

Terminal Equipment Directive: An *European Union (EU)* directive, passed in 1988, which opened the European *terminal equipment* market to competition, and included proposals for fair type approval (see *Type Approval Directive*). Under the directive Member States have to set up organisations, independent from their *PTTs*, to specify and carry out type approval testing. The directive also abolished the PTT's exclusive rights on the sale, maintenance and installation of terminal equipment.

terminal server: A device which allows several different *terminals*, or other pieces of equipment, to be connected to the same *Local Area Network (LAN)*.

terminated line: A *circuit* or *transmission line* which does not produce *reflections* or *standing waves* when a *signal* is entered at one end, owing to the presence at the far end of a resistance equal to the circuit impedance. See also *impedance matching transformer*.

terminating traffic: *Telecommunications traffic* which comprises *incoming calls*. See also *originating traffic*.

ternary coding: A coding system which uses a three level *code* to send more than one *bit* of *information* in a single symbol.

ternary signal: A *signal* that can take one of three values of whichever parameter it represents. For example, a *carrier signal* that can adopt one of three different *frequencies* or a voltage pulse that can adopt either a positive, negative or zero value.

Terrestrial Flight Telephone System (TFTS): *ETSI* terminology for a *telephony* system which routes *calls* made between aircraft and ground terminals through *ground stations* rather than through *satellites*.

terrestrial interference: Generally refers to *interference* caused by *terrestrial telecommunications systems* to *signals* received from *satellites*. This is often caused by *microwave* systems in the 4 GHz region.

terrestrial telecommunications system: A *telecommunications* system that does not use *satellites*, or other space-based systems to transmit *signals*. It includes *cable* based systems, *microwave* and other radio systems, as well as *transmission* from aircraft.

terrestrial transmitter: Transmitter used in a *terrestrial telecommunications* system.

tertiary trunk exchange: An *exchange* which usually controls major *trunk network* routes and which is at the third level in the *exchange hierarchy*.

TETRA: *Trans European Trunked Radio*.

text: The part of a *telecommunication signal* which carries the words that the sender wishes the receiver to read rather than control information. It is the part of the *message* between the *Start of Text* and the *End of Text* sequence.

text to speech conversion: Used in *voice processing* systems to convert computer based *text* into *speech* for *transmission* to the listener. The text is first analysed and converted to phonetic equivalents and then these are spoken to the caller by a voice synthesiser. By this method a relatively small *database* can be used to reproduce a large vocabulary, with present day text to speech conversion systems capable of pronouncing over 100000 words correctly.

T4: A *service*, provided by a *common carrier* in the USA, which allows the *transmission* of a *signal* having the *DS4* format at 274 Mbit/s. This is the equivalent of 4032 voice channels. See *T carrier*.

TFTS: *Terrestrial Flight Telephone System.*

TGMS: *Third Generation Mobile System.*

THD: *Total Harmonic Distortion.*

theoretical final route: Terminology used to describe the *route* taken by a *call* through the *circuit groups* of a *network.*

thermal noise: The *noise* in a conductor caused by thermal agitation of electrons, which is of a *Gaussian distribution* (i.e. random), and is proportional only to temperature. Thermal *noise power* is given by: P = kTB where k is Boltzman's constant in joules per kelvin, T is the temperature of the conductor in kelvins and B is the *bandwidth* in *hertz.* Also known as *white noise*, *Brownian noise*, *Johnson noise* or *Gaussian noise.*

thermo-optic switching: A form of *switching*, used in *fibre optic* communications, which allows a *signal* to be guided down either path of a Y-junction switch element. The thermo-optic switch relies on the heat-induced fall in *refractive index* of the polymer core of the optical fibre. This effectively turns off the *waveguide* in one branch of the Y-junction.

THF: *Tremendously High Frequency.*

thick Ethernet: Standard *Ethernet coaxial cable* which is about 0.5 inches in diameter and fairly stiff to handle.

thin Ethernet: *Ethernet coaxial cable* having a diameter of about 0.2 inches. It is lighter than *thick Ethernet*, and used to cut costs for short distances in Ethernet networks. It is specified by *IEEE 802.3.*

thin film waveguide: A *waveguide* formed by depositing a thin layer of semiconductor or dielectric material on a substrate. This is often used in an *Optical Integrated Circuit (OIC).*

thin route telecommunications: Generally refers to systems providing *telecommunications services* to sparsely populated areas, which do not generate a high level of *traffic* but are spread over a large geographical area. *Terrestrial telecommunications stems* cannot generally be provided economically to these areas and they are usually better served by *satellite* based systems.

Third Generation Mobile System (TGMS): A *cellular radio system* due to commence operation around 2002 and currently under development. The system should be capable of *broadband data rates* of 144 kbit/s to 2000 kbit/s which will enable the increased support of *multimedia services*. The aim is to be able to use the service worldwide.

3 bit error: A *transmission error* in which three consecutive *bits* are incorrectly received.

three-way calling: A *service* which allows three parties to take part in a *telephone* conversation on three separate *terminals* without the assistance of an attendant. See *conference call.*

threshold: The minimum value or *signal* level which can be detected or is needed to create a desired effect.

threshold of audibility: See *audibility.*

threshold of feeling: See *audibility.*

threshold of hearing: See *audibility.*

through channel: Generally refers to a *channel* which is passed through a device, such as a *mutiplexer* without going through *demultiplexing*. See also *drop and insert.*

throughput: A measurement of the maximum handling *capacity* of a communications system. Often measured in volume of *traffic* (i.e. *bits*, *characters*, *calls*, etc.) per second.

through supergroup: Sixty *voice frequency channels*, which are routed through a *repeater* as a unit, without undergoing *frequency* translation.

TIA: *Telecommunications Industries Association.*

ticketed call: A *call* for which *information* is noted such as its time, length, and participants.

TIE: *Time Interval Error.*

tie line: *Transmission lines* connecting *Public Automatic Branch Exchanges (PABX)* over *private circuits*. These allowing employees of an organisation with several geographically isolated sites to communicate using a *private network* and by-passing the *Public Switched Telephone Network (PSTN)*. As the organisation pays for line rental rather than per *call*, this arrangement is cheaper for high inter-PABX *traffic.*

tie line signalling: *Signalling* used over a *tie line*. Examples are *E&M signalling* and *in-band signalling.*

tie trunk: Same as *tie line.*

Time Assignment Speech Interpolation (TASI): A *switching* technique in which a *user* of a *trunk telephone circuit* is switched into the quiet periods of another telephone conversation, resulting in several conversations being combined on the same circuit without interfering with each other.

Time Compression Multiplexing (TCM): A technique which allows the *transmission of duplex digital data* by sending the data in compressed bursts, these alternating in different directions, as if in a 'ping-pong' arrangement.

time diversity transmission: A form of *transmission* where the same *signals* are sent more than once over the same communications *channel*, often in an attempt to overcome *burst errors*.

Time Division (TD): A way of using time to divide the use of a communications system, such as a *transmission channel* or computing equipment, that would normally only cope with one user at a time.

Time Division Duplex (TDD): *Duplex* communication where two *signals*, each carrying different *data*, are transmitted over the same *path*. This is achieved by using different time intervals for each signal. This technology is used by the *CT2* system.

Time Division Duplex/Frequency Division Multiple Access (TDD/FDMA): A method of *multiplexing* several *two-way calls* using many *frequencies*, with a single two-way call per frequency using *TDD*.

Time Division Duplex/Time Division Multiple Access (TDD/TDMA): Method of *multiplexing* several *two-way calls* using a single *frequency* for each call and multiple *timeslots*.

Time Division Multiple Access (TDMA): The allocation of the complete *bandwidth* of a communications *channel* to a series of *users* for a limited period of time. Periods of empty time are usually inserted as *guard bands* between each user to prevent *interference* between users, which may arise due to variations in *synchronisation*. For example, Figure T.6 shows a six user system where the user *timeslots* are combined into *frames*, the frames repeating after a frame period of T_F. Each user has allocation of the full transmission channel for an equal amount of time. The length of time can, however, be unequal for each user, for example as in Figure T.7, where users A and C are allocated increased capacity (with time frame repetition period T_F) over users C to F (with frame repetition period T_G).

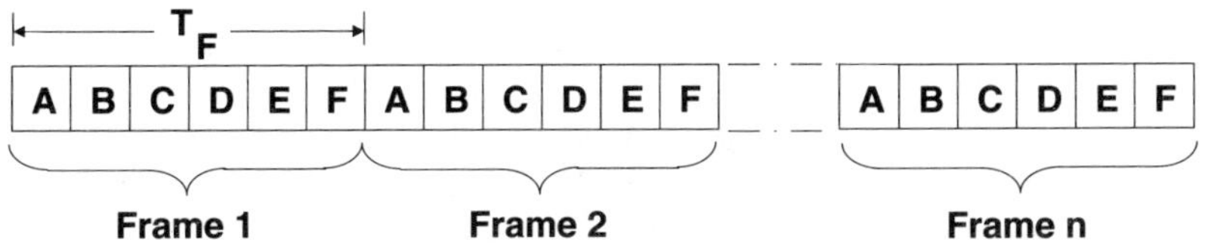

Figure T.6 Frame structure within TDMA

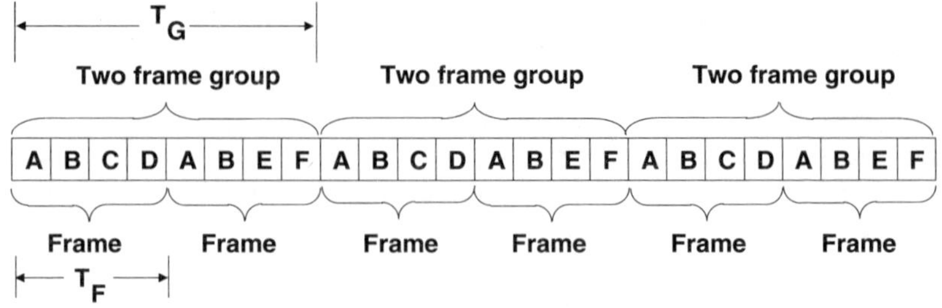

Figure T.7 Unequal slot assignment within TDMA

Time Division Multiplexing (TDM): A type of *multiplexing* in which a single *transmission channel* is divided between several *user* channels, each of which is assigned a *timeslot* on the channel. The individual user *signals* are sampled in sequence and the go through *interleaving* in strict order, in an effort to use the communications channel to full capacity. This is illustrated in Figure T.8 where three signal channels are multiplexed together on to one channel by *sampling* and *interleaving*. Several methods can be used for interleaving, such as by *bit*, *byte* or *character*.

time division switching: *Switching* which occurs by moving *signals* from one *timeslot* to another in the *frame*, for example in a *Time Division Multiplexing (TDM)* system.

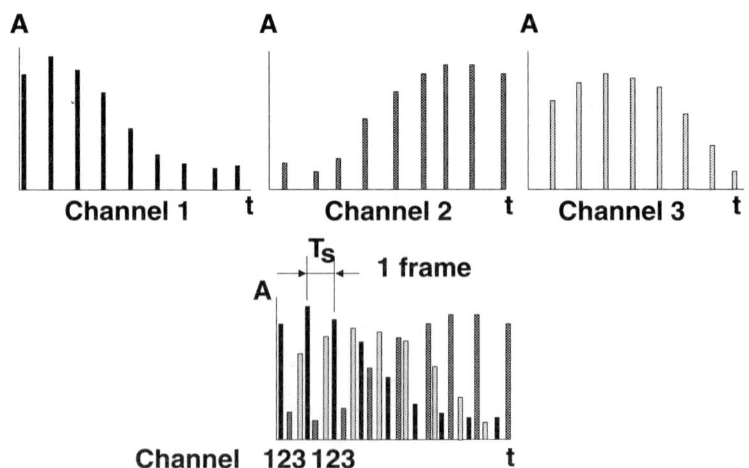

Figure T.8 Time Division Multiplexing

time domain equaliser: A family of *equalisers*, primarily used to eliminate *Intersymbol Interference (ISI)*. See, for example, *Transversal Equaliser (TVE)*.

Time Domain Reflectometer (TDR): A test instrument used to locate faults in *cables*. A pulse of energy is launched into the cable and the *signal* reflected back from any defect in the cable is recorded on a *Visual Display Unit (VDU)* which is *synchronised* with the pulse source.

Time Interval Error (TIE): The difference in time, following *synchronisation*, between a *signal* and the *timing* source, measured over a period.

time of day reconfiguration: The ability of a system, such as a *multiplexer*, to be programmed to automatically change its configuration depending on the time of day. For example it could be programmed to carry *voice traffic* during the day and *data* during the evening, when the office was closed.

time of day routeing: The variation of *routeing* depending on the time of day, for example to use cheaper *trunk lines* during the evening.

time-out: A period that is allowed to elapse before a specified event takes place unless another event occurs first. For example, the *telephone ringing signal* may be allowed to sound for 20 seconds before automatic disconnection, unless the *call* is answered.

timesharing: A method of controlling several devices using one processor, by having the processor serve each in turn at a speed which makes it appear as if all devices are operated simultaneously.

timeslot: An easily identifiable time interval allocated to specified actions. For example, the time interval allocated to *signals* in *Time Division Multiplexing (TDM)* and in a *fame*.

Time Space Time (TST): Technique used in *digital exchanges* in which three stage switching is used, a *space switching* stage being sandwiched between two *time switching* stages. See also *Space Time Space*.

time switching: See *time division switching*.

Time To Live (TTL): A field in the *Internet protocol* which specifies the maximum time in seconds that a *packet* should exist on a *network* before being destroyed. As the content of this field decreases with time spent and distance moved on the network, misrouted packets, or those trapped in network loops, are destroyed.

time zone: The relationship of local time at geographic locations around the world to a standard time such as *Universal Time IUT*, *Coordinated Universal Time* or *Greenwich Mean Time (GMT)*.

timing: The process of determining the *synchronisation* of equipment based on a *clock* source. For example the operating mode used by a *modem* can be *synchronous* or *asynchronous*. Asynchronous operation uses the modem's internal timing clock source, whereas synchronous operation uses timing *information* passed with the *data* in conjunction with internal, system and external clock sources.

timing diagram: A diagram which shows the change of state of *signals* over a period of time.

timing jitter: See *jitter*.

timing recovery: The recovery of a *timing signal* from a received *digital signal*. Also known as *clock extraction*.

T interface: The physical interface which occurs at the *T reference point* in a four wire *ISDN* circuit, at the point between the *NT1* and *NT2*. (See *User-Network Interface (UNI)* and Figure U.5.) It is characterised by a *data rate* of 144 kbit/s and has a limited *transmission* length of about 1 km.

TL-1: *Transaction Language-1.*

TMA: *Telecommunications Managers Association.*

TMN: *Telecommunications Management Network.*

TMN functional architecture: The *Telecommunications Management Network (TMN)* functional architecture is based on a number of functional building blocks which enable the TMN to perform its application functions. They consist of the principle blocks, shown in Figure T.9, such as the *Operations Systems Function (OSF)* and the *Work Station Function (WSF)*, and complementary blocks, such as the *Data Communication Function (DCF)*, the *Mediation Function (MF)* and the *Q Adaptor Function (QAF)*. Some of these blocks are only partly within TMN, as shown in Figure T.9.

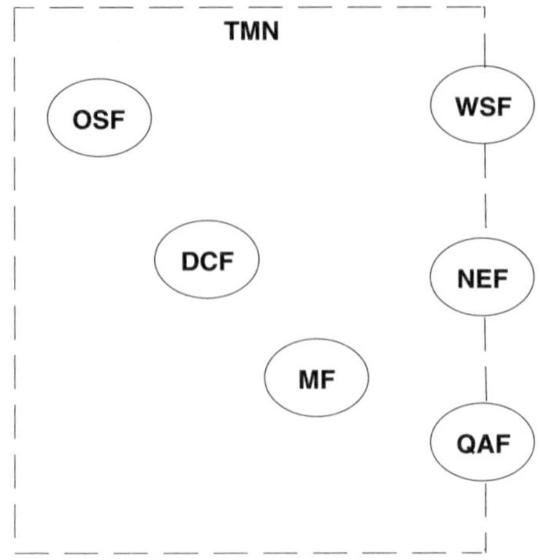

Figure T.9 TMN functional architecture

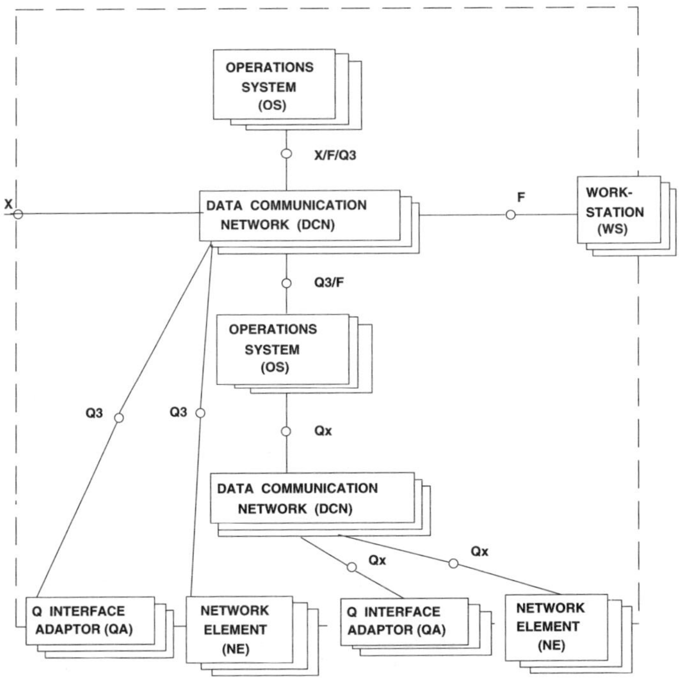

Figure T.10 TMN physical architecture

TMN physical architecture: The physical architecture of the *Telecommunications Management Network (TMN)*, as shown in Figure T.10. It relates to the *TMN functional architecture*. The *Operations System (OS)* is a standalone system which performs the *Operations System Functions (OSF)*. The *Mediation Device* performs the *Mediation Function (MF)*. The *Q Interface Adapter* is the device which *Network Elements (NE)* without a TMN interface connect to. The *Data Communication Network* is the communications network within a TMN which supports the *Data Communications Function (DCF)*. The *Network Element* supports any item which performs a *Network Element Functions (NEF)*. Equipment which do not have this are interfaced to the TMN via a *Q Interface Adaptor*. The *Work Station* performs the *Work Station Function (WSF)*.

TMN reference points: See *reference points*.

TNV: *Telecommunications Network Voltage*.

token: A a group of *bits* or a *packet* that is passed among data stations on a *LAN*, in a set order, to control which station can transmit *data*. The pattern contains a special sequence of bits, such as 11111111, which is unlikely to be generated as part of normal *data* transfer.

token bus: See *token passing.*

token passing: A *multiple access* technique used in a *Local Area Network (LAN).* A *token* is passed between *nodes* on the LAN. Only the node in possession of the token is authorised to transmit *data*, and will do so until the token is passed to the next node. If the nodes are arranged in a *bus topology* the system is know as a *token bus* and if they are in a *ring topology* it is know as a *token ring.*

token ring: See *token passing.*

toll call: A *call* made outside a *LATA* for which there is a *toll charge.* Also referred to as a *long-distance call.*

toll centre: A *telephone exchange* that connects a group of *local exchanges* to the *trunk network.* Known in the UK as *Group Switching Centre (GSC).* Also known in the USA as primary trunk exchange and primary centre. The toll centre therefore grants *subscriber* access to the trunk network and *toll charges* are incurred. Also called a *toll office.* A toll centre is classified as a Class 4 office (see *class of office*) in the *exchange hierarchy.* Also known as a *trunk exchange.*

toll charge: The additional charge made to a *telephone subscriber* resulting from a *call* to a *telephone number* outside a *LATA*, and therefore not covered by the *local call rate.*

toll circuit: USA terminology for the *circuit* between a *local exchange* and a *trunk exchange.* In the UK this is referred to as a *junction circuit.*

toll free: A *call* or any other service for which there is no charge to the *user* of the service. See, for example, *800 service.*

toll network: See *trunk network.*

toll office: See *toll centre.*

T1: The first level in the North American *T carrier* system which can carry 24 *voice channels*, each at 64 kbit/s. In addition an 8 kbit/s channel is added for carrying *PTT* service information across the link, giving an overall *transmission data rate* of 1.544 kbit/s, as in Figure T.11. See also *digital hierarchy.*

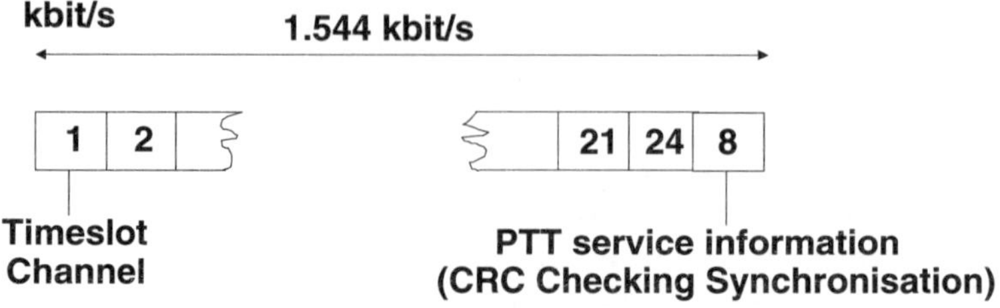

Figure T.11 T1 transmission frame

tone: An *audio frequency signal* which has a characteristic sound and is used to indicate events. For example, *dial tone*, *ringing tone* and *busy tone*.

T1C: A *service* provided by a *common carrier* in the USA which allows the *transmission* of a *signal* having the DS1C format at 3.152 Mbit/s. This is the equivalent of 48 *voice channels*. See also *T carrier*.

T1 Committee: Part of the *Exchange Carriers Standards Association (ECSA)* and accredited by *ANSI* in 1983, it is active in standards development. It consists of specialised technical subcommittees and its output goes to ANSI for review as American National Standards. It also provides an input to the US *ITU-T*, where T1 formulates the US industry view on ITU-T standards. T1 also produces reports for industry.

tone dialling: Same as *push-button dialling*.

tone diversity: The use of two *channels* to carry the same *information* in a 16 channel *Voice Frequency Telegraph transmission system*.

tone pager: A form of *pager* which can produce up to eight different audible *tone* patterns, but does not display numbers or *messages*. Each tone can have a prearranged meaning for the recipient.

tone signalling: Used as a form of junction *signalling* between *exchanges*. A single *tone* is used to seize or release the *line* and *multifrequency signalling* is used for communications between the exchanges.

TOP: *Technical and Office Protocol*.

topology: The physical or logical configuration of *nodes* or *stations* on a *network*. Possible configurations include *bus topology*, *ring topology* and *star topology*.

torn tape centre: A centre for the manual retransmission of *telegraph messages* in which human operators tear off paper tapes produced by telegraph receivers and retransmit them over different telegraph circuits using a tape reader. These are now rarely used, having been replaced by automatic *switching* equipment.

Total Access Cellular System (TACS): See *Total Access Communications System*.

Total Access Communications System (TACS): A standard for *cellular radio systems*, adapted in the UK in 1985 from the USA *AMPS* system, which allowed compatibility with the European *frequency allocation plan* (900 MHz with 25 kHz channel spacing). In the UK only 600 channels (2 × 15 MHz) were released by the licensing authority, but subsequent additions were made for the *ETACS* system. Also called *Total Access Cellular System (TACS)*.

total channel noise: The sum of *noise* on a specific *channel* which is contributed to by: *random noise*, *intermodulation noise*, and *crosstalk*, but not by *impulse noise*.

Total Harmonic Distortion (THD): A measure of *distortion* of a *signal*, it is given by the ratio of the power of all the *harmonic frequencies*, above the *fundamental frequency*, to the power of the fundamental. Usually expressed in *decibels*.

total internal reflection: The *reflection* of an *electromagnetic wave*, such as light, which occurs when it strikes the interface between its *transmission medium* and a medium with a lower *refractive index*, at an angle of incidence greater than or equal to the *critical angle*. This causes the light to be reflected back into the transmission medium so that no light is lost. In *optical fibres*, total internal reflection of light at the boundary of the *core* and *cladding* allows effective *signal* transmission.

TPDDI: Twisted Pair Distributed Data Interface.

TPOA: *Telecommunications Private Operating Agency.*

TRAC: *Technical Recommendations Applications Committee.*

trace packet: A *packet* in a *Packet Switched Network (PSN)* which causes a report of each stage of its progress to be sent to the *network control centre*.

tracking radar: Radar that tracks moving objects, by using *information* received from scanning radar located in ship, aircraft and land stations, to obtain a lock on the object. Tracking radar is used in weapon systems to provide data for target intercept guidance.

traffic: The volume and intensity of *information* and other related *data* and *messages* moved over a *telecommunications* facility. Traffic is measured using the dimensionless quantity the *erlang*.

traffic analysis: The statistical analysis of *traffic* rates, volume, types, direction, etc., in order to improve system performance. See also *teletraffic theory*.

traffic capacity: The maximum *traffic* that a *telecommunications* system can carry under specified conditions, measured over a unit of time.

traffic capacity table: A table which indicates the number of *trunks* needed to cope with a level of *lost calls* or length of *queue*, given an expected amount of *traffic* (measured in *erlangs*). For example, Table T.3 shows that 20 trunks would be required to deal with a traffic of 11 erlangs if the call-loss required is 1 in 200.

traffic intensity: A measure of the amount of *traffic*, usually expressed as the product of the *calling rate* and the average *holding time*. It is expressed in *erlangs*.

traffic overflow: A condition in which the *traffic* offered to a *telecommunications* system, such as a *trunk group*, exceeds the *traffic capacity* of the system, and may overflow onto another trunk group. The term is also used to refer to the excess traffic resulting from an overflow. Also called *traffic overload*.

traffic overload: See *traffic overflow*.

Table T.3 Traffic capacity table

Number of trunks	*1 lost call in*			
	50 (0.020) E	*100 (0.010)* E	*200 (0.005)* E	*1000 (0.001)* E
1	0.020	0.010	0.005	0.001
2	0.22	0.15	0.015	0.046
3	0.60	0.46	0.35	0.19
4	1.1	0.9	0.7	0.44
5	1.7	1.4	1.1	0.8
6	2.3	1.9	1.6	1.1
7	2.9	2.5	2.2	1.6
8	3.6	3.1	2.7	2.1
9	4.3	3.8	3.3	2.6
10	5.1	4.5	4.0	3.1
11	5.8	5.2	4.6	3.6
12	6.6	5.9	5.3	4.2
13	7.4	6.6	6.0	4.8
14	8.2	7.4	6.7	5.4
15	9.0	8.1	7.4	6.1
16	9.8	8.9	8.1	6.7
17	10.7	9.6	8.8	7.4
18	11.5	10.4	9.6	8.0
19	12.3	11.2	10.3	8.7
20	13.2	12.0	11.1	9.4
21	14.0	12.8	11.9	10.1
22	14.9	13.7	12.6	10.8

traffic pressure: A measure of the demand for connections in a *switched network*, expressed in units such as bids per circuit per hour, where bids refers to the number of attempts made to secure a connection.

traffic recorder: A device used to measure the *traffic* in a *telecommunication circuit*, determining the effective utilisation of the circuit.

traffic statistics: Statistical information regarding the *traffic* in a *telecommunications* system which may be used for *traffic analysis*.

traffic theory: See *teletraffic theory*.

Traffic Unit (TU): A measure of the intensity of *traffic* in a *telecommunications* system. It may be defined as the number of *call* hours per hours, call minutes per minute, etc. This is numerically equal to the average number of simultaneous calls or the proportion of time that a single *circuit* is *busy*. Previously know as a Traffic Unit, it is now called an *erlang*.

traffic volume: The amount of *traffic* in a *telecommunications* system over a given time.

transaction: (1) A completed task or activity, which can be recorded and subsequently audited. **(2)** An application in which there is interaction between two or more systems, such as a *transmitting terminal* and a *receiving terminal*, to complete a task or activity.

Transaction Language-1 (TL-1): A *protocol* developed by *Bellcore* and *ANSI*, to enable *network management* systems to communicate with each other.

transaction log: A log of all the *transactions* made by a *telecommunications* system. This log provides an audit trail and can provide statistics on usage or help in problem analysis.

transaction processing: The processing of events and *data* as they arise, i.e. without any preliminary storage, analysis, etc.

transaction set: Format specified in *Electronic Data Interchange (EDI)* for electronic *messages* used to construct business related documentation.

transaction routeing: The individual routeing of *messages* according to *data* found in the message heading.

transaction traffic: A type of *traffic* involving a large number of *calls*, each with limited *data* transfer. This is often found in the booking of holidays and credit card checking.

TransAtlantic Telecommunications (TAT): A series of transatlantic *telecommunications* systems, connecting North America to Europe. The first, TAT-1 copper *cable* was laid in 1956. The first transatlantic *fibre optic* cable, TAT-8, was laid in 1988 and was capable of carrying 40000 simultaneous telephone conversations.

transceiver: Any single device that can both transmit and receive *data* where both functions are combined within the same housing or on the same chassis. A transceiver is often a portable device and is used to connect devices to a *Local Area Network (LAN)*. The term transceiver is a combination of transmitter/receiver.

transceiver cable: A *cable* that connects a *transceiver* to a *telecommunications transmission channel.*

transcoding: The conversion of *digital signals* in one digital *encoding* scheme into digital signals in another encoding scheme, without converting to *analogue signals* in-between.

transducer: A device that converts one form of energy into another with minimal energy loss. Most transducers handle electric currents as either their input or output energy source, e.g. *microphones* convert *sound waves* into electrical *signals*, modulated *lasers* convert electrical signals into modulated *light waves*.

Trans European Trunked Radio (TETRA): An *ETSI* standard for a trans-European digital *trunked mobile radio* system for professional users including public safety organisations. It can provide full *duplex* operation, including with handheld *terminals*, transmit both *voice* and *data*, provide *encryption* at two levels, and operate at a *data rate* of 28.8 kbit/s. *Time Division Multiple Access (TDMA)* is used, the format being shown in Figure T.12. Four *timeslots* are used in the transmit and receive directions and the frame time is 60 ms.

transfer characteristic: The characteristic of a system which enables one of its parameters, such as the output, to be determined if another parameter, such as the input, is known.

transfer function: Mathematical statement which defines the *transfer characteristic* of a system.

transfer mode: The method used to carry *data* in a *telecommunications* system. Examples are *packet switching*, *Asynchronous Transfer Mode (ATM)*, *Synchronous Transfer Mode (STM)*, etc.

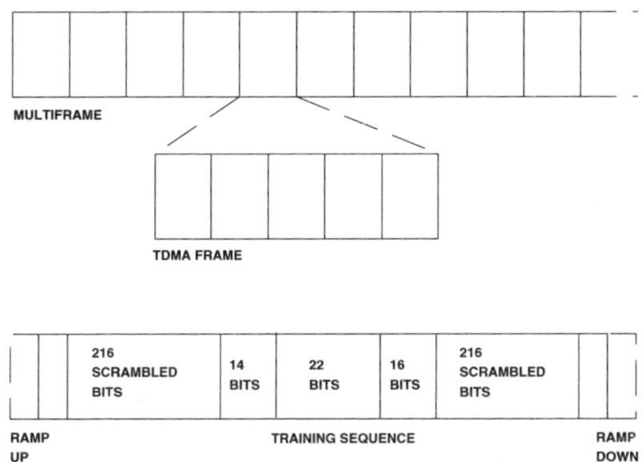

Figure T.12 TETRA TDMA format

transfer orbit: An intermediate *orbit* used during the launch of a *geostationary satellite*. The transfer orbit has a period of about 11 hours and is reached from the *parking orbit*. It passes through the equatorial plane about 35800 km above the ground, close to its *apogee*. The satellite leaves the transfer orbit around this point in its orbit.

transform coding: A method of *signal* coding in *digital video* systems. See *Discrete Cosine Transform (DCT)*.

transhybrid loss: The *signal loss* which occurs between the opposite ports of a *hybrid network* in a *four wire circuit*. It is a measure of the isolation, in *decibels*, between the go and return *paths* on the four wire side of a two wire to four wire hybrid.

transient: **(1)** An event which lasts for a short time and whose occurrence may be predictable or unpredictable. **(2)** An intermittent, unwanted *signal*, which lasts for a short duration and often causes *interference* with the wanted signal.

transient copy: *Information* displayed in a non-physical form, such as text displayed on a *Visual Display Unit (VDU)*. Also called *soft copy*.

transit delay: The time that the first *bit* of a *data frame* or *block* takes to pass between two specified points in an *ISDN*, including the time for *transmission* of the frame.

transit exchange: See *tandem exchange*.

translator: A device that translates *information* between two different systems of representation. Examples are the device in a *telephone* system which converts the *dial pulses* into the *address* of the *called terminal*, and a television *repeater station* that rebroadcasts a *signal* it has received, after the signal has been amplified and changed in *frequency*.

transmission: The transfer of information using *telecommunications* equipment to send the *message* to a receiver over a *transmission medium*.

transmission block: See *block*.

transmission break: A complete interruption in the *transmission* of *information* over communication *paths*. This break is normally caused by a system fault.

transmission bridge: A combination of *Direct Current (DC)* feed, battery decoupling and *speech* coupling components which convert the sounds made into a *telephone handset microphone* into electrical *signals*. Also known as feeding bridge.

transmission capacity: The capacity of a *transmission medium* in a *network*. Usually expressed in bit per second. Also known as *bandwidth*.

transmission channel: The *channel*, including *links* and *exchange* equipment, between two *terminals* used for the *transmission* of *data*.

transmission code: The *code* used in *transmission*. Examples are the *American Standard Code for Information Interchange (ASCII)* and *Extended Binary Coded Decimal Interchange Code (EBCDIC)*.

Transmission Control (TC): The control of the *transmission* of *information* over a *data channel*. This is usually done by the use of *transmission control characters*.

transmission control character: A *character* used in *Transmission Control (TC)*, usually between *Data Terminal Equipment (DTE)*. Examples are *Acknowledgement (ACK)*, *Negative Acknowledgement (NAK)*, *Start Of Heading (SOH)*, *Start of Text (STX)*, etc.

Transmission Control Protocol (TCP): A *network protocol*, initially developed by the US *Department of Defence (DoD)*, for reliable *transmission* of *datagrams* from *host* to host. It now forms part of the *Internet Protocol (IP)* suite. See also *Transmission Control Protocol/Internet Protocol (TCP/IP)*.

Transmission Control Protocol/Internet Protocol (TCP/IP): A set of *protocols* which link computer *terminal* on the *Internet*, allowing the control of *data transmission* between dissimilar terminals. Figure T.13 shows the TCP/IP protocol stack, in relation to the *OSI Basic Reference Model*. *Internet Protocol (IP)* is equivalent to the *Network Layer* and *TCP* to the *Transport Layer* which it sits alongside the *User Datagram Protocol (UDP)*. Several applications are supported in the *Application Layer*, such as *TELNET*, *FTP* and *SMTP*.

transmission error: An error occurring during the *transmission* of *data*. It can be caused by several factors, such as: *noise* on the *transmission channel*; *crosstalk* and *Intermodulation (IM)* from other interfering *transmission systems*; *echo* and *signal fading*, from mismatches in the

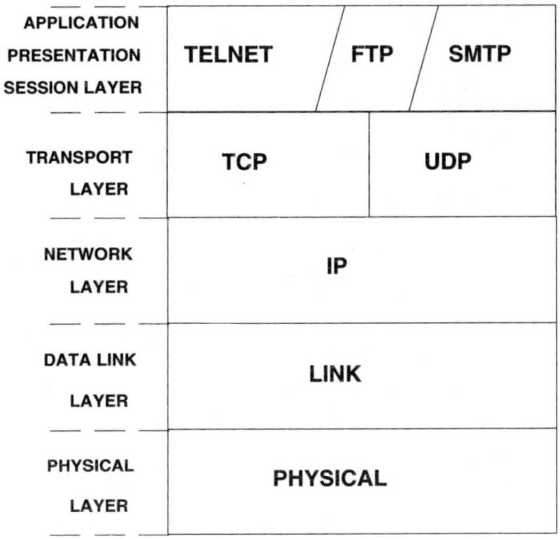

OSI layer	TCP/IP		
APPLICATION PRESENTATION SESSION LAYER	TELNET	FTP	SMTP
TRANSPORT LAYER	TCP		UDP
NETWORK LAYER	IP		
DATA LINK LAYER	LINK		
PHYSICAL LAYER	PHYSICAL		

Figure T.13 TCP/IP protocol stack

transmission channels and from *multipath effects* in radio systems; network problems, such as *frame synchronisation* errors.

transmission error performance: The measure of the effectiveness of a *transmission* system. Several parameters are used to measure this, such as the *Bit Error Ratio (BER)*, *available time* and *Error Free Seconds (EFS)*.

transmission flow control: The regulation of the rate of *transmission* in a *transmission medium* to avoid problems such as *congestion* and *traffic overload*.

transmission frame: A *frame* used for *transmission* of *data*.

transmission frequency: The *frequency* of the *signal* which is carrying the *data* in the *transmission*.

transmission header: See *header* and *header information*.

transmission line: The *line* that allows the transfer of *data signals* between two *terminals*. See *transmission medium*.

transmission link: The physical component of a communications system which link together the sender and receiver of *information* during a specific *call*. See *transmission medium*.

transmission loss: The total power *loss* experienced by a *signal* moving through a *transmission medium*, owing to the dissipation of energy by the medium, or *mismatch losses* introduced at analogue interconnection points. Loss is measured in *decibels* and is usually equal to $10 \log_{10}$ (output power/input power).

transmission medium: The physical medium used to transmit *telecommunications data*. It includes *bounded transmission media* (such as *coaxial cable*, *twisted pair wire*, and *fibre optic cable*) and unbounded transmission media (such as the atmosphere, which supports *microwaves* and *radio transmission*).

transmission mode: The mode of *transmission* used in a *data communications network*. Examples of transmission modes include *telegraphy*, *telephony*, radio, *satellite*, etc.

transmission network: *Network* which is primarily involved in the *transmission* of *information*. *Switching* would not form a direct part of this network, but could be connected to it.

transmission path: The components of a physical communications path used to transmit a specific *call*.

transmission plan: A plan which states the *losses* and impairments which accumulate during the *transmission* of *signals* over a *network*. For example, Figure T.14 shows the *transmission loss* for various parts of the network from a *subscriber* located at A to one at B. The maximum overall loss is 42.5 dB. Transmission plans can be used to control these losses by pointing to individual components which are responsible for

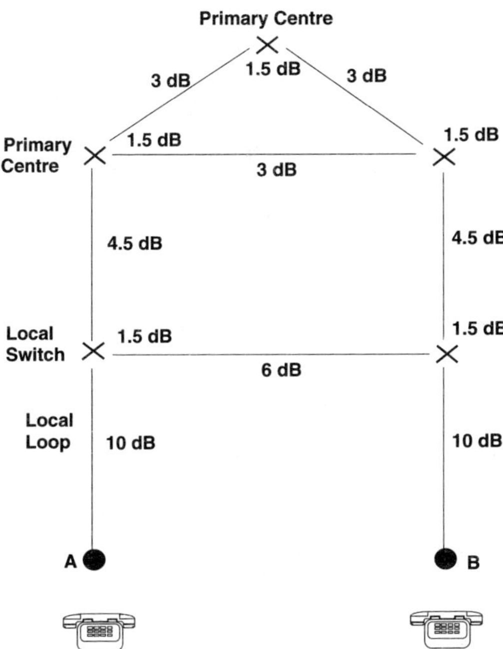

Figure T.14 Example of a transmission loss plan

proportionally large losses, and for determining alternative *routeing* arrangements.

transmission rate: The rate of *transmission* of *data* units in a communications system. The transmission rate can look at data being sent, received, or passing through a set point in the communications system. Possible units include *characters* per second, *bytes* per minute, *bits* per second, etc.

transmission system: The physical system that allows the transfer of communications *data signals* from point to point. The transmission system includes all *transmission lines*, *links* and other elements such as *multiplexers* and *repeaters*, whether they are on *satellites* or Earth bound.

transmission test set: A set of apparatus used by engineers to test and diagnose faults in communications systems. Some test sets are portable, and work by generating *signals* which are transmitted through the *transmission lines*, which can be looped back to allow the transmitted signal to be received and analysed by the test set.

transmittance: When an *electromagnetic wave* passes through an interface of two different propagation media with different *refractive indices*, the transmittance is given by the ratio of *transmitted power* to incident power. Transmittance is measured in *decibels* for communications applications.

transmitted power: The power transmitted by a communications system in a stated direction across a stated surface area perpendicular to the direction of *transmission*. Transmitted power is usually measured in watts, watts per square metre, or watts per *steradian*.

transmitter: A device which generates *data* and inserts this into a *transmission channel*.

transmitter clock: A *clock* which is used by the *transmitter*. The clock *signals* may be inserted into the transmitted *data* and be used for *synchronisation* of the receiving equipment.

transmitter start code: A coded *character* which is sent to a remote *terminal* asking it to begin *transmission*. This can be used by a *switching centre* for *polling* all *stations* connected to it.

transmitting antenna: An *antenna* which is used for *signal transmissions*, using *electromagnetic radiation*. See also *receiving antenna*.

transmitting Earth station: An *Earth station* which transmits radio *signals* from Earth to a *satellite* in Earth *orbit*. These signals may then be transmitted back to another point on Earth. See also *receiving Earth station*.

transmitting station: A device or group of devices (*stations*) that transmit *data* on a communications *network*. See also *receiving station*.

transmitting terminal: A *terminal* which forms part of, and transmits *signals* to, a communications *network*. See also *receiving terminal*.

transmit-to-receive crosstalk: *Crosstalk* in which a *signal* transmitted in one direction on a *four wire circuit* is induced onto the opposite *channel*.

transmultiplexer: A device that converts a *supergroup* of *Frequency Division Multiplexed (FDM) signals* into *Time Division Multiplexed (TDM)* signals without having first to reduce these down to the individual *baseband Voice Frequency (VF)* components.

Transpac: The name of the French public *Packet Switched Network (PSN)*.

transparency: The property of a *data communications* system which allows *transmission* of data without significant loss of either the form or content of the data. In a transparent data communications system two devices communicate with each other without being aware of the intervening equipment or *software*.

transponder: A device, usually found on a *satellite*, which receives and amplifies a *data signal* and then retransmits it on a different *frequency* to that of the received signal. Transponders can also automatically transmit a predetermined *message* on receipt of a predetermined signal.

Transport Layer: *Layer* 4 of the *OSI Basic Reference Model*, which is responsible for end-to-end delivery of *data* with the *Quality of Service (QoS)* requested by the applications. Also known as the end-to-end layer.

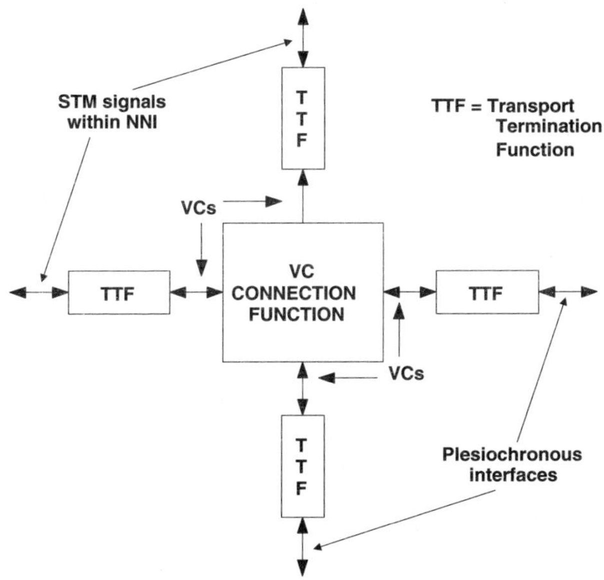

Figure T.15 Traffic paths through an SDH network element

Transport Termination Function (TTF): A conceptual element used in the analysis of an *SDH network*, as shown in Figure T.15. The Transport Termination Functions are the conversion functions from the *Virtual Container (VC)* level to the *plesiochronous* or *STM* levels. The VC connectivity function allows VCs to be routed between the various TTFs.

transposition: **(1)** An error in *data transmission* where the state of one or more pairs of signal elements are altered in opposing senses. **(2)** The changing of the relative position of items relative to each other, such as wires in a *cable* to minimise the effect of *crosstalk*.

Transversal Equaliser (TVE): A form of *time domain equaliser*, as shown in Figure T.16. It consists of a *delay line* which is tapped at intervals equal to the intersymbol interval (T). The output from the tap is fed into an *amplifier* which also results in inversion, and the output consists of the sum of these *signals*. By adjusting the *amplifier gain* it is possible to eliminate the *Intersymbol Interference (ISI)*. If the amplifier gain is adjusted automatically it is called an *adaptive equaliser*.

transversal filter: A *filter* which works on the principle of passing the input signal through a delay line, similar to that used in a *Transversal Equaliser (TVE)*, the tapped outputs from this line being attenuated and inverted before being summed to provide the output signal.

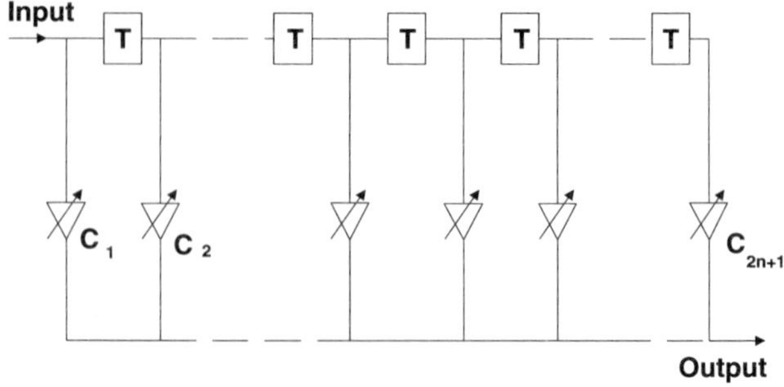

Figure T.16 Transversal Equaliser

transverse jitter: The concurrent overlap and underlap of copy seen in received *facsimile transmissions*, caused by an irregular *scanning* pitch on the transmitting facsimile machine.

transverse parity check: *Parity check* which is performed in a transverse direction, across *bits* in a *frame*, or across two parallel tracks of bits recorded on a medium such as magnetic tape.

Transverse Redundancy Check (TRC): A *redundancy checking* method in which *block check* is carried out on synchronised parallel *data* streams, each bit of the check *character* being in a separate stream.

TRC: *Transverse Redundancy Check.*

Treaty of Accession: The treaty, signed on 22 January 1972, to enable the United Kingdom, Ireland and Denmark to join the EEC, bringing its membership to nine. The treaty came into effect on 1 January 1973.

Treaty of Rome: A treaty signed on 25 March 1957 by six countries (France, Germany, Italy, Belgium, The Netherlands and Luxembourg) to establish the *European Economic Community (EEC)* (as well as the *European Atomic Energy Community*). Soon after that *EFTA* was formed and a treaty was signed between the EEC and EFTA. The EEC treaty came into effect on 1 January 1958 and the EFTA treaty on 3 May 1960.

tree and branch cable TV network: A configuration of *coaxial cable network* distribution systems, used in *cable television*, where a single coaxial cable, representing the trunk of a tree, carries a multiplex of TV channels, with separate branch cables allowing the *signal* to be tapped off it to serve a community of homes. (Figure T.17.)

tree network: See *network topology*.

T reference point: The reference point, within the *ISDN User-Network Interface (UNI) reference model, which is situated between the NT1* and the *NT2*. See Figure U.5.

Trellis Coded Modulation (TCM): *Modulation* method, used in *ITU-T Recommendation* V.17 TCM for *facsimile*, and V.32 and V.32 bis for

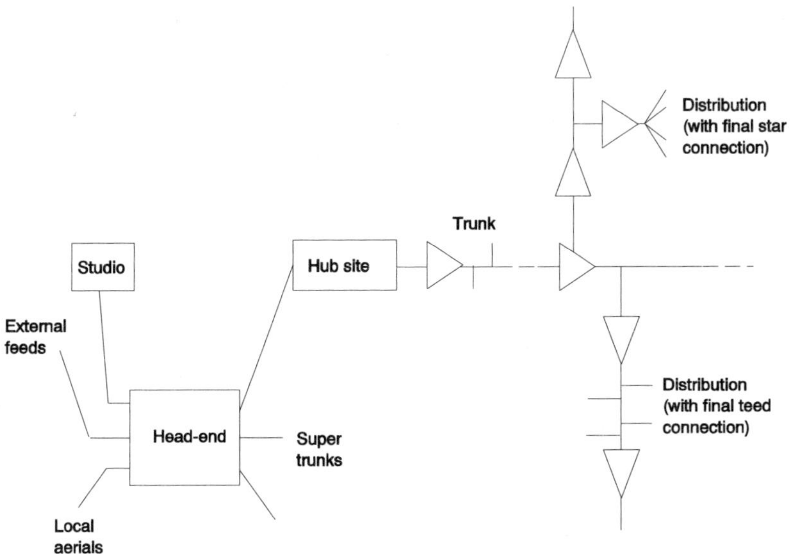

Figure T.17 Tree and branch cable TV network

modems, which uses mathematical techniques for determining the best fit for the *signal*. *Data rates* of 14.4 kbit/s, with *Forward Error Correction (FEC)*, are achieved.

Tremendously High Frequency (THF): The *frequency range* from 300 GHz to 3000 GHz, having the *International Telecommunications Union (ITU)* designator of 12.

triad: Three contiguous phosphor dots coloured red, green and blue, which represent a picture element (*pixel*) on the screen of a coloured *Visual Display Unit (VDU)*, such as a *television receiver*.

tributary circuit: A *circuit* which connects *stations* to the *backbone network* or to a *switching centre*. The stations connected may be individual *terminals*.

tributary station: A *station* which is not a *control station* for the *network* to which it is connected.

Tributary Unit (TU): Part of the *SDH multiplexing structure*, the Tributary Unit contains the low order *Virtual Containers (VC)* along with the *pointer*. The combination of pointer plus VC-11 gives TU-11, VC-12 gives TU-12, VC-2 gives TU-2 and VC-3 gives TU-3.

Tributary Unit Group (TUG): Part of the *SDH multiplexing structure*, the Tributary Unit Group contains a combination of *Tributary Units (TU)*. TUG-2, for example, can contain four TU-11, or three TU-12 or one TU-2. TU-3 can accommodate one TU-3.

troposphere: The lowest layer of the *Earth's atmosphere*, situated between the Earth's surface and the *stratosphere*, in which clouds form and temperature decreases with altitude. The troposphere is between 10 km to 18 km thick, depending on latitude and season.

tropospheric scatter: The *scattering* of *electromagnetic radiation* resulting from physical irregularities in the *troposphere*. Tropospheric scatter is used in transhorizon communications at frequencies between 350 MHz and 8400 MHz. Also known as troposcatter and forward scatter.

tropospheric wave: A component of a *radio wave* that results from *tropospheric scatter*.

truncated binary exponential backoff: An *algorithm* used in *multiple access* systems, such as *Carrier Sense Multiple Access (CSMA)* and *Carrier Sense Multiple Access with Collision Detection (CSMA/CD)*, which schedules the retransmission of *signals* after a *collision* has occurred.

trunk: A *channel* or *circuit* for the *transmission* of communications *data* between two *switching centres* or *exchanges*. A *trunk circuit* connects two exchanges and does not connect to a *subscriber terminal*.

trunk call: A *telephone call* that needs a connection between *exchanges* joined by a *trunk circuit* for completion of the *transmission path*. Also called *junction call*.

trunk circuit: See *trunk*.

trunked mobile radio: A system for spectrum conservation in *Private Mobile Radio (PMR)* communications, where several radio *users* share a pool of *radio channels* with other users. The performance of the system for any number of shared channels and level of *traffic* can be calculated using *Erlang's loss formula*, and this is shown in Table T.4.

trunk exchange: See *toll centre*.

trunk group: A group of two or more *trunks* of the same type, or a specified combination of trunks, between *exchanges* or *PABXs*.

trunk line: A communications *line* which carries *traffic* between *trunk exchanges*, forming part of the *trunk network*.

trunk link: Also called *trunk circuit*.

trunk network: The major *transmission paths* and *exchanges* that are interconnected to form the *Public Switched Telephone Network (PSTN)*, but excluding the *local network*.

trunk route: Also called *trunk circuit*.

TSAPI: *Telephone Services Application Programming Interface*.

TSB: *Telecommunications Standardisation Bureau*.

T Series: *ITU-T Recommendations* for *terminal* equipment intended for use in telematic services. Table T.5 gives a few of these.

TST: *Time Space Time*.

TT&C: *Telemetry, Tracking and Command*.

Table T.4 Trunked mobile radio system performance

No. of channels	*Grade of service*	*Traffic (erlang)*		*No. of mobiles*		*Mean waiting time(s)*
		Per channel	*Total*	*Per channel*	*Total*	
	%	a	A			
5	5	0.645	3.22	116	580	3.3
	10	0.719	3.59	129	645	5.8
	30	0.846	4.23	152	760	16.8
10	5	0.793	7.93	143	1430	3.8
	10	0.839	8.39	151	1510	6.2
	30	0.914	9.14	165	1645	16.5
15	5	0.853	12.79	153	2300	4.1
	10	0.886	13.29	159	2390	6.5
	30	0.939	14.09	169	2535	16.4
20	5	0.885	17.70	159	3185	4.3
	10	0.911	18.22	164	3280	6.7
	30	0.953	19.06	172	3430	16.4

TTC: *Telecommunications Technology Committee.*

TTF: *Transport Termination Function.*

T3: A service provided by a *common carrier* in the USA which allows the *transmission* of a *signal* having the DS3 format at 44.736 Mbit/s. This is the equivalent of 672 *voice channels.*

TTL: *Time To Live.*

T2: A service provided by a *common carrier* in the USA which allows the *transmission* of a *signal* having the DS2 format at 6.312 Mbit/s. This is the equivalent of 96 *voice channels.*

TTY: *Teletypewriter* or Teletype.

TU: *Traffic Unit* or *Tributary Unit.*

Table T.5 ITU-T Recommendations, T Series

Recommendation	*Description*
T.4	Group 3 facsimile
T.5	Group 4 facsimile general consideration
T.6	Group 4 facsimile coding and control
T.30	Handshake protocol for facsimile machines
T.51	Telematic services coded character set
T.71	Half-duplex extension for LAPB
T.73	Telematic services document interchange protocol
T.90	Teletex requirements for interworking the telex service
T.101	Videotex services international interworking
T.120	Data and graphics interchange between videoconferencing systems

TUA: *Telecommunications Users Association.*
TUFF: *Telecommunications UK Fraud Forum.*
TUG: *Tributary Unit Group.*
tunable laser: A *laser* whose *wavelength* of spectral emission can be varied, either in a continuous fashion or in step increments.
tuned circuit: An electronic circuit in which the characteristics of one or more components have been chosen to allow selective response to a particular *frequency.*
tuner: The part of a receiver of *radio waves*, such as a television or radio, that allows the selection of particular *channels.* Also refers to the radio receiver in high fidelity equipment.
TUP: *Telephony User Part.*
turnaround time: The time required to reverse the direction of *data transmission* from sender to receiver in a *half-duplex* system.
TV: Television.
TV cable system: A system of distributing television *signals* using *coaxial cable.* See *Cable Television (CATV).*
TVE: *Transversal Equaliser.*
TV receiver: See *television receiver.*

TVRO: *Television Receive Only.*

23B + D: See *Primary Rate Interface (PRI).*

twinplex: A *Frequency Shift Keying (FSK)* carrier *telegraphy* system in which two pairs of tones, one representing a *signal* and the other a space, are transmitted over the same *transmission channel.*

twisted wire pair: A *transmission medium* comprising two wires separately insulated and twisted together. Twisted pairs are often used in the *links* between *telephone terminals* and *local exchanges* (the *access loop*) and are twisted to reduce *crosstalk* with other pairs sharing the same *ducting* or *cable.*

Twisted Pair Distributed Data Interface (TPDDI): The same as *CDDI.*

2B + D: See *Basic Rate Interface (BRI).*

2 bit error: A *data transmission error* resulting from two consecutive *bits* being incorrectly received.

two condition code: A *code* which represents *binary* information.

two condition telegraphy: A *telegraphy* system which uses only two conditions to produce a *binary code* for transmitting *data.* These conditions can take several forms, such as positive and negative currents, presence and lack of current, two different frequencies, etc.

two-frequency half-duplex: A communications *circuit* which allows simultaneous two way *traffic* as the two traffic directions use different *frequencies.*

two-independent-sideband transmission: The *transmission* of *radio waves* using a *carrier signal* and two independent *sidebands* which can be separately modulated.

twos complement: The twos complement for a *binary number* is obtained by adding 1 to its *ones complement.* For example the ones complement for 1100101 is 0011010 and the two complement is 0011011.

two state signalling: The use of two voltage values to transmit two *signalling bits,* 0 and 1, with each voltage value having a predetermined duration.

two tone keying: The *keying* of *telegraphy messages* using a two-*channel transmission path,* one channel carrying *mark information* and the other *space* information.

two way alternate communication: A communications circuit allowing *transmission* of *data* in two directions, one direction at a time. Also known as *half duplex.*

two way call: An exchange of *voice* or *data* in which the *transmitting terminal* and *receiving terminal* can simultaneously send and receive data, for example, in a *telephone* conversation.

two way pager: A form of *pager* which allows the pager *user* to transmit an *Acknowledgement (ACK)* or reply back to the sender of the original message.

two way simultaneous communications: *Transmission* which occurs simultaneously in both directions on the same *transmission channel.* Also know as *full duplex.*

two wire channel: A communications channel or circuit comprising a single pair of separately insulated wires, often as a *twisted pair.*

two wire circuit: A *circuit* formed of two conductors insulated from each other, such as a *twisted pair wire.*

two wire switching: *Switching* which takes place on a *two wire transmission line. Trunk circuits* primarily use *four wire transmission* and as a loss of approximately 3 dB occurs at the four wire to two wire conversion point a *four wire switching* system is more efficient, as shown in Figure F.10. For two wire switching conversion from four wire to two wire must occur at every switching point whilst for four wire switching it need only occur at the *access loop.*

two wire transmission: *Transmission* which occur over two wires, such as on the *twisted pair wire* used in the *access loop.*

TXE4: An *analogue exchange* system employing the same *switching* system as the *TXE2*, but designed for use in *local exchanges* with 2000 to 40000 *lines* and calling rates between 0.02 to 0.35 *erlangs* per line. TXE4 systems are designed as an assembly of several subsystems which can be duplicated in order to increase the *traffic* or connection *capacity* of the exchange.

TXE2: An *analogue exchange* system employing a similar *switching matrix* to a *crossbar switch*, but using *reed relays* for the *crosspoints.* This allows the use of electronic components for control, resulting in increased self checking features and an improvement in reliability. TXE2 exchanges are designed for use as small *local exchanges* as they have a maximum capacity of 240 *erlangs* of total *traffic*, with one unit handling typically between 2000 and 3000 lines.

Type Approvals Directive: A Directive passed by the *EC* in June 1989 which required that *terminals* which had been tested and approved by a recognised testing house in one Member State could be sold in all other Member States without further testing.

type I carrier: A type of *common carrier* in Japan which owns and maintains its own *telecommunications* infrastructure.

type II carrier: A type of *common carrier* in Japan which provides *telecommunications services* using a *network* that they do not own.

U

UA: *User Agent.*
UART: *Universal Asynchronous Receiver/Transmitter.*
UBR: *Unspecified Bit Rate.*
UDLC: *Universal Digital Loop Carrier.*
UDP: *User Datagram Protocol.*
UHF: *Ultra High Frequency.*
UI: *Unit Interval* or *User Interface.*
U interface: An interface within *ISDN* which defines the boundary between the *Network Termination Type 1 (NT1) and the two wire local loop* from the *Central Office (CO).* The NT1 connects to the *Customer Premises Equipment (CPE)* and the U interface meets the legal requirement of ensuring that the customer equipment is managed separately from the *PTO's network*. See also *V interface*.
UIT: *Union Internationale des Telecommunications.*
UL: *Underwriters Laboratory.*
ULF: *Ultra Low frequency.*
ULSI: *Ultra Large Scale Integration.*
Ultra High Frequency (UHF): The portion of the *electromagnetic spectrum* which lies in the *frequency range* 300 MHz to 3 GHz. This portion has an *International Telecommunications Union (ITU)* designation of 9 and is primarily used for television *broadcasts* and *cellular radio systems*. Owing to the limited range of propagation at this frequency, channels are relatively free from interference by distant *transmissions* in the same band.
Ultra Large Scale Integration (ULSI): The technology used to build semiconductor devices with over 10000 circuits on a single silicon chip. It has a higher density than *Very Large Scale Integration (VLSI)* devices.
Ultra Low Frequency (ULF): See *Infra Low Frequency (ILF).*
Ultra Small Aperture Terminal (USAT): An *antenna*, less than 1 m in diameter, with a small aperture which is used to both transmit and receive signals. Such antennae are used in interactive *networks*, *satellite Earth stations* and *Global Positioning Systems (GPS).*
ultraviolet radiation: The ultraviolet portion of the *electromagnetic spectrum* lies in the *wavelength range* from approximately 1 nm to 400 nm. Strictly speaking, only the region between 1 nm to 300 nm can be considered to be ultraviolet radiation.
ultraviolet fibre optics: Fibre optics designed to operate using *ultraviolet radiation*. The *attenuation* rates in glass of the ultraviolet *wavelengths* are greater than the attenuation rates for infrared radiation, which ex-

plains the relatively greater use of infrared rather than ultraviolet in longer *fibre optic cables*. However, ultraviolet fibre optics are used in applications including medicine, materials testing, photochemistry, and genetics.

UMTS: *Universal Mobile Telecommunications System.*

unaligned bundle: A bundle of *optical fibres*, considered as a single *transmission path*, which comprises randomly placed fibres in a single jacket. Therefore the two ends of each fibre occur in different locations at the bundle ends. Unaligned bundles are usually used to guide light, such as optical power or optical communications pulses, where the spatial relationship among the fibres is not crucial. They cannot be use to transmit an image, which would become scrambled as the optical fibres change their relative positions throughout the length of the bundle.

unattended operation: The automatic performance of a system without any operator intervention. For example, the *transmission* and reception of *messages* without any human intervention.

unavailable time: The proportion of the total available time during which a system, subsystem or piece of equipment is not available for use when requested at a random time.

unbalanced double-current interchange circuit: A circuit which interfaces a *Data Terminal Equipment (DTE)* and a *modem* concerned with *data transmission* over telephone circuits whose characteristics are defined in *ITU-T Recommendation V.10. (See V Series.)*

unbalanced line: A *transmission line* comprising two conductors in which the magnitudes of the voltage across each conductor, with respect to *ground*, are not the same. Such lines include a single wire with a ground return and a *coaxial cable*.

unbounded transmission medium: A *transmission medium* in which the *signal* is transmitted in many directions. The atmosphere is used as an unbounded medium during radio *broadcasting*. See also *bounded transmission medium*.

unbundling: The separation of individual services offered by a vendor so that the *tariff* for each service can be determined. Unbundling is often legislated to allow new telecommunications service providers to offer services to customers at a reasonable *tariff* using existing local *networks*. See also *bundling*.

underfill: The condition which exists when the light from a source covers less than the full area of the light detector. For example, the light entering an *optical fibre* covers a diameter less than that of the fibre core. See also *overfill*.

Underwriters Laboratories (UL): Established by the National Board of Fire Underwriters in the US, to assess insurance risk of equipment. It is a privately owned company which issues safety related specifications and tests products against these, providing a UL label for equipment which pass these tests.

undetected error ratio: The proportion of the total number of *data* units sent which are incorrectly received and go undetected.

UNI: *User-Network Interface.*

unicast: The process of sending *data* from a source to a single recipient. See also *broadcast* and *multicasting.*

unidirectional coupler: A *coupler* which allows *signals* to pass in one direction only. See also *bidirectional coupler.*

unified messaging: A service which brings together various messaging systems such as *facsimile*, *voice mail* and *electronic mail*, storing them in a single *mail box* which can be accessed from anywhere using both fixed and mobile *networks* and the *Internet.*

uniform encoding: See *uniform quantisation.*

uniform quantisation: A method of *quantisation*, or *analogue to digital conversion*, where the *quantisation intervals* are equal. Also called *uniform encoding.*

Uniform Resource Locator (URL): The *address* of a file on the *Internet* which provides the path to its location.

unintelligible crosstalk: *Crosstalk* arising from the transfer of unintelligible but distracting *speech signals* from one circuit to another.

Uninterruptible Power Supply (UPS): Equipment which produces a continuous supply of *AC* power, without power surges, from a commercial source, often by the use of continuously charged batteries. Uninterruptible Power Supplies are often used to protect a *Private Automatic Branch Exchange (PABX)* or computer equipment, where an interruption of the power supply, due to mains failure, would cause malfunctions. Also known as a *no-break power supply.*

unipolar non-return to zero signal: A *unipolar signal* in which the *waveform* does not return to zero between logical 1 pulses, as in Figure U.1. See also *unipolar return to zero signal.*

unipolar return to zero signal: A *unipolar signal* in which the *waveform* returns to zero between logical 1 pulses, as in Figure U.2. See also *unipolar non-return to zero signal.*

unipolar signal: A *signal* whose *waveform* has either a positive or negative *amplitude*, but not both. This is illustrated in Figure U.3 for a signal $v_S(t)$

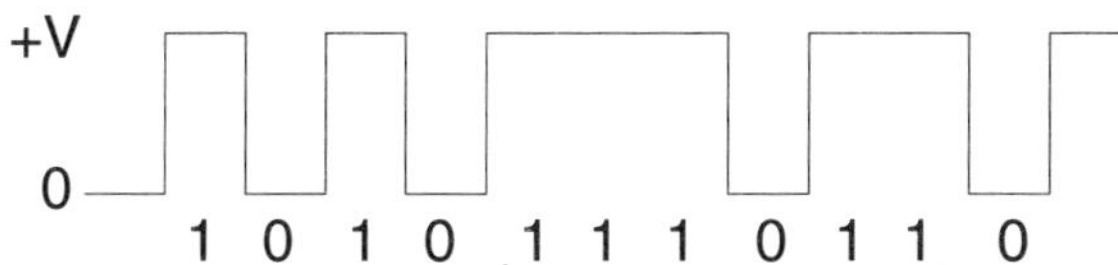

Figure U.1 Unipolar non-return to zero

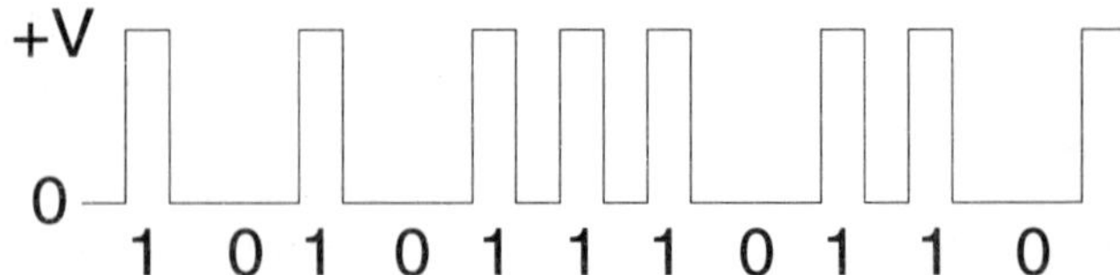

Figure U.2 Unipolar return to zero

in which the decision level for differentiating between a logical 1 and a logical 0 occurs at the half peak voltage point. Compare with *bipolar signal*, Figure B.9.

United States National Committee (USNC): A US based committee which represents the standards formed by the *American National Standards Institute (ANSI)* at the *International Standards Organisation (ISO)*.

Unit Interval (UI): (1) The shortest time interval between two consecutive significant instances. **(2)** The disturbance or offset of timing in a *waveform*, caused by *jitter*, is measured in Unit Intervals (UI) peak-to-peak, equivalent to one bit period. Jitter standards are always specified in terms of UI versus jitter *frequency* for various *bit rates*.

Universal Asynchronous Receiver/Transmitter (UART): A device which converts parallel data into serial form for *transmission* and then converts the serial received data back to parallel form for processing.

Universal Synchronous/Asynchronous Receiver/Transmitter (USART): A device which converts *data* between parallel and serial forms as well as performing some basic functions, such as the assembly and disassembly of *characters*. It interfaces to *synchronous* and *asynchronous* circuits.

Universal Digital Loop Carrier (UDLC): A *transmission* system using *digital signals*, the *analogue signals* being converted to digital form at the *local exchange*.

Universal Mobile Telecommunications System (UMTS): A system that attempts to unify and standardise low-power short-distance radio trans-

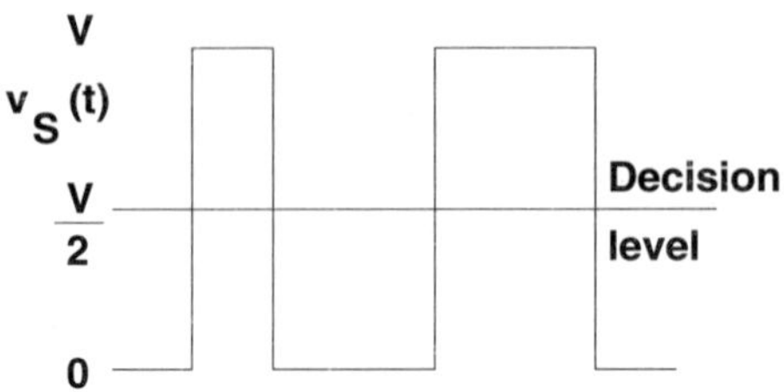

Figure U.3 Unipolar signal

missions throughout Europe, as used in pagers, cellular and cordless systems, *Local Area Networks (LAN)* and private mobile radio. This work is being done by *ETSI* in conjunction with the *ITU*. At the 1992 *World Administrative Radio Conference (WARC)* it was agreed that the next generation of mobile systems should operate in the *frequency bands* 1885 MHz to 2025 MHz and from 2110 MHz to 2200 MHz, and it was referred to as IMT-2000.

Universal Personal Telecommunications (UPT): A service that allows mobile users access to *subscriber* telecommunications services with a unique personal network-independent number which is valid on multiple *networks*.

universal service obligation: The obligation to provide a basic universal telephone or other telecommunications service at an affordable price to the entire population of a country or area. The universal service obligation is often enforced if an operator has a country monopoly and is achieved by cross-subsidy or public funds. The obligation is harder to enforce in a competitive environment. The EC directive 96/19/EC of 13 March 1996 requires Member States to ensure that public telecommunication operators in their countries contribute to the provision of a universal service obligation and that the method used to determine this contribution is based on objective, non-discriminatory criteria.

Universal Time (UT): (1) A service which measures and distributes time signals. **(2)** A measure of the exact rotational orientation of the Earth, calculated from *Coordinated Universal Time (CUT)*.

UNIX: A computer operating system developed within AT&T's Bell Laboratories in the late 1969. UNIX is a *multi-user*, *multi-tasking* system and can run on many different hardware platforms.

UNIX to UNIX Copy Programme (UUCP): A *network* in which machines with the *UNIX operating system* are loosely linked to each other, using UUCP for *data transfer* between them.

unloaded line: A *telephone line* in which the *loading coils* have been removed. Generally there is no *switching* involved as well, so that the line is continuous, enabling faster *data transmissions*.

Unshielded Twisted Pair (UTP): A pair of insulated wires that are not enclosed in a metal sheath, to protect against *Electromagnetic Interference (EMI)*. Instead the two wires are twisted around each other to partly diminish the effects of EMI. Several unshielded twisted pairs of wires are often grouped together in a common sheath for use in *telephone networks*. In these cases the wires are twisted so that the length of the twists or lay-lengths (i.e. the length along the wire taken by one complete twist) are different to minimise *crosstalk* between the pairs. Many standards have evolved for use of UTP in a *Local Area Network (LAN)*, such as ISO 8802-3 (for *Ethernet*) ISO 8802-5 (for *token ring*) and ISO 9314

Table U.1 UTP for LAN use

Category	*Usage*	*Maximum date rate*
1	100Ω UTP (and associated connecting hardware) whose transmission characteristics are specified for voice frequencies	Voice only
2	100Ω UTP (and associated connecting hardware) whose transmission characteristics are specified up to 4 MHz	4 Mbit/s
3	100Ω UTP (and associated connecting hardware) whose transmission characteristics are specified up to 16 MHz	16 Mbit/s
4	100Ω UTP (and associated connecting hardware) whose transmission characteristics are specified up to 20 MHz	20 Mbit/s
5	100Ω UTP (and associated connecting hardware) whose transmission characteristics are specified up to 100 MHz	100 Mbit/s

(for *FDDI*). ISO/IEC 11801, BS EN 50173 and EIA/TIA 568A are generic standards. UTP has been categorised for specific use. Categories 1 to 5 are summarised in Table U.1.

Unspecified Bit Rate (UBR): One of the four service classes for *ATM networks*, specified by the *ATM Forum*. In UBR the network provides a 'best effort' service to deliver the *bandwidth* required, but with no guarantee of *Quality of Service (QoS)*. See also *Current Bit Rate (CBR)*, *Variable Bit Rate (VBR)* and *Available Bit Rate (ABR)*.

up-converter: A device which changes a *signal* from one *frequency* to that at a higher frequency.

uplink: The communications *channel* used to transmit *information* to a *satellite* or airborne platform from an *Earth station*. The opposite of *downlink*.

upper sideband: The *sideband* which is at a higher *frequency* than the *carrier frequency*. See also *lower sideband*.

UPS: *Uninterruptible Power Supply*.

upstream: **(1)** The direction in a communications *network* that is opposite to the direction of *data* flow. **(2)** The direction of data flow from a *terminal* on the network to the *host processor*.

UPT: *Universal Personal Telecommunications*.

uptime: The period of time that the resources of a *network* are fully functioning and available to users.

urban multipoint system: *Transmission* system used in an urban area for communications between several users. It is normally based on point-to-point *microwave* radio, as in the example shown in Figure U.4.

URL: *Uniform Resource Locator*.

usage charge: One component of the charge made to users of a telecommunications service. The usage charge is proportional to the actual use made of the communications service and could be based on factors such as the number, duration, distance or time of day of the *call*.

USART: *Universal Synchronous/Asynchronous Receiver/Transmitter*.

USAT: *Ultra Small Aperture Terminal*.

US Defence Advanced Research Projects Agency (DARPA): The central research and development organisation for the US Department of Defence. In the 1970s DARPA was responsible for developing the *Transport Control Protocol/Internet Protocol TCP/IP* for use in the *APRANET* research network.

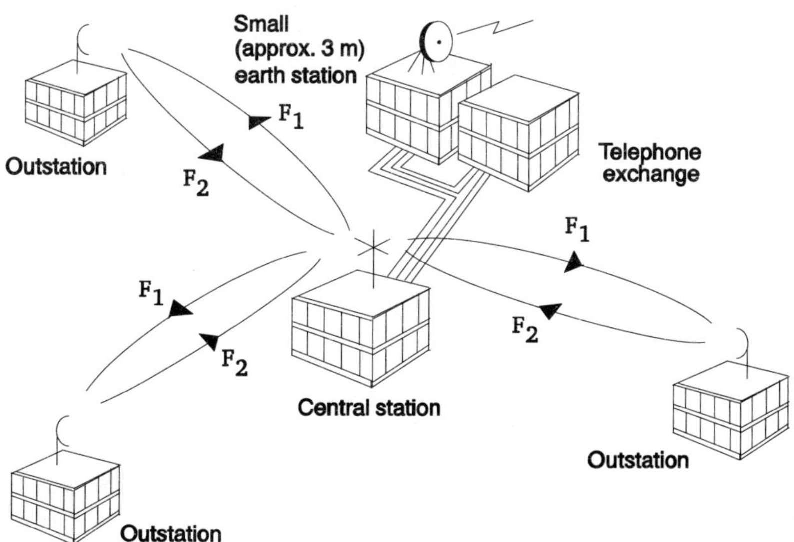

Figure U.4 Urban multipoint system

user: A person or *node* sending or receiving *information* by accessing a communications *network* through any form of *terminal* or interface.

User Agent (UA): Part of the X.400 standard (see *X Series*) which specified the *Message Transfer Agent (MTA)* and the User Agent. The User Agent provides the interface between the user and the MTA. See *Message Handling System (MHS)* and Figure M.8.

user application: A computer program or procedure that performs a specific job for a *user* or group of users. This contrasts with more general programs that may support the user application such as the *operating system*. Also called *application program*.

user characteristics: The requirements of the *users* of telecommunications systems. These should be taken into account by the system designers to ensure that the final product provides a satisfactory service for minimum complexity and expense.

user data: The part of a *message* which contains *data* for a *user application* rather than ancillary information, such as the *control field* or *header* information.

User Datagram Protocol (UDP): One of the two common *protocols* (along with *Transmission Control Protocol (TCP)*) which drive the *Internet Protocol (IP)*. The User Datagram Protocol allows an *application program* on one machine to send a *datagram* to an application program on another machine over a *Packet Switched Network (PSN)*. The UDP is sparser than the TCP and is usually used with specialised applications which need to add much of the above functionality should they require it.

user data packet; A unit of *data* in a *Packet Switched Network (PSN)* which forms part of a *message* sent from one *user* to another.

user data signalling rate: The rate, in *bits* per second, at which *terminals* can receive *data* sent by *users*.

user dialling state: The condition of a communications system when the *dialling* process is in progress between a *user* and a terminal.

User Interface (UI): The method by which a *user* communicates with a *node* on a *network*. This interface can take many forms, such as visual, audio, etc.

User-Network Interface (UNI): The interface between a *network* and the *user* of the network services. User-Network Interfaces are specified in several standards. Those for *narrowband ISDN* are given in *ITU-T Recommendation* I.410 which lists the requirements of the UNI. Recommendation I.411 gives a User-Network Interface model, as in Figure R.6, to assist in the definition of ISDN UNIs. A similar model has been specified in Recommendation I.413 for the *Broadband Integrated Services Digital Network (B-ISDN)*. See also *S interface*, *R interface* and *T interface*.

user node: The *node* on a telecommunications *network* occupied by a *user terminal* rather than by functions such as *switching* or *databases*.

USNC: *United States National Committee*.

USO: *Universal Service Obligation*.

US Technical Advisory Group (TAG): US body which is responsible, on behalf of *ANSI*, for developing the US contribution on *Information Technology (IT)* standards to the *ISO* and *IEC*. Its technical administration is carried out by the *Computer and Business Equipment Manufacturers' Association (CBEMA)*.

UT: *Universal Time*.

utilisation factor: The fraction of the total capacity of a communications transmission medium which is used to carry information between nodes.

UTP: *Unshielded Twisted Pair*.

UUCP: *UNIX to UNIX Copy Programme*.

V

VAD: *Value Added Dealer.*

value added common carrier: An organisation that sells the services of a *Value Added Network (VAN).*

Value Added Dealer (VAD): See *Value Added Reseller (VAR).*

Value Added Network (VAN): A telecommunications *network* that uses the basic transmission facilities of a *common carrier*, and adds value to it by providing enhanced services such as *protocol conversion, error correction* and *network management*, through the use of hardware, firmware and software. These services are often referred to as *Value Added Network Services (VANS).*

Value Added Network Services (VANS): See *Value Added Network (VAN).*

Value Added Reseller (VAR): Telecommunications equipment dealers or manufacturers who purchase equipment from other manufacturers and then sell them on to end users, often after adding new *hardware* or *software* features and branding the product as their own. Also known as *Value Added Dealer (VAD).* See also *Original Equipment Manufacturer (OEM).*

value added service provider: A *Public Telecommunications Operator (PTO)* who provides *Value Added Network Services.*

value added services: Peripheral services added to a basic fixed or mobile telecommunications *network*. The *ITU-T* has classified the services into two major groups: *interactive services* (transmission in two directions) and *distribution services* (transmission in one direction). Both these groups have several sub-groups. See *service classification.*

vampire connector: A *connector* which makes connection to the metallic core of a *cable* by penetrating through its outer protective sheath.

VAN: *Value Added Network.*

VANS: *Value Added Network Services.*

VAR: *Value Added Reseller.*

Variable Bit Rate (VBR): In general terms it refers to *data transmission* in which the *traffic* is uneven and can be represented by an irregular flow of *bits* from the source. This would result in uneven filling of *frames*. For *ATM* the Variable Bit Rate is defined by the *ATM Forum* as one of the four service classes. It corresponds to the *Statistical Bit Rate (SBR)*, specified by the *ITU-T*, and uses *statistical multiplexing* to efficiently handle bursty traffic with known *throughput* requirements.

variable format message: A *message* whose format can be read by *terminals* with similar characteristics, the formatting information being buried in the message itself. Control characters are neither added to nor deleted from the message as it is transmitted and received.

Variable Length Coding (VLC): See *Huffman coding.*

Variable Quantising Level (VQL): An *encoding law* for *speech signals*, for transmission over a *digital network* at 32 kbit/s.

variable slope delta modulation: See *Continuous Variable Slope Delta Modulation (CVSDM).*

variation of insertion loss: The *range*, usually quoted in *decibels*, over which the *insertion loss* or *overall loss* in a circuit will vary with time.

variable attenuator: An *attenuator* whose *attenuation* can be varied, either manually or automatically, depending on *line* conditions.

VBI: *Vertical Blanking Interval.*

VBR: *Variable Bit Rate.*

VC: *Virtual Container* or *Virtual Circuit* or *Virtual Call.*

VCC: *Virtual Channel Connection.*

VCI: *Virtual Channel Identifier.*

VCO: *Voltage Controlled Oscillator.*

VCR: *Video Cassette Recorder.*

VDM: *Voice Data Multiplexer.*

VDSL: *Very high bit rate Digital Subscriber Line.*

VDT: *Video Display Terminal* or *Video Dial Tone.*

VDU: *Visual Display Unit* or *Video Display Unit.*

vector: A item which has magnitude and direction, and may vary with time. For example velocity is a vector (magnitude and direction) whilst speed is not (magnitude only).

vector image: An image produced by lines and curves drawn between points. See also *raster image.*

Vector Quantisation (VQ): A technique for *compression* of a *video signal.* It works on the principle of dividing the video signal into *blocks* which are then allocated a *code word* for transmission. The number of available *code words* are less than the number of permutations of the possible values of the input *pels* in the group.

velocity factor: A measure of the velocity of a *signal* in a *transmission medium* compared to that in free air. This is usually influenced by the dielectric material of the medium and its construction. Table V.1 shows this characteristic for *cables* with commonly used dielectric materials.

velocity of light: The *velocity* of light in vacuum, which is 186280 miles per second or 299792 kilometres per second.

velocity of propagation: The *velocity* at which an *electromagnetic wave* travels through a *transmission medium.*

velocity of sound: The *velocity* of sound in air is 331 metres per second, at zero degrees centigrade. It varies with temperature and the *transmission medium.*

Table V.1 Effect of dielectric on velocity of propagation

Dielectric material	*Velocity factor*	*Velocity of propagation*
Free air	1.00	100%
Semi polythene	0.84	84%
Polythene	0.78	78%
Polyethylene	0.66	66%
Polyvinylchloride	0.50	50%

Vertical Blanking Interval (VBI): The group of lines which are transmitted between *frames* of a television *signal*. These carry no picture *information* and were originally intended to allow the time for the *scanning* process to move from the bottom to the top of a picture frame. It is now used to transmit a variety of data, such as *teletext* and for captions to help those with hearing defects.

Vertical Radiation Pattern (VRP): The *radiation pattern* from an *antenna* measured in the vertical plane. An example of a typical *UHF broadcast* antenna is shown in Figure V.1. See also *Horizontal Radiation Pattern (HRP)*.

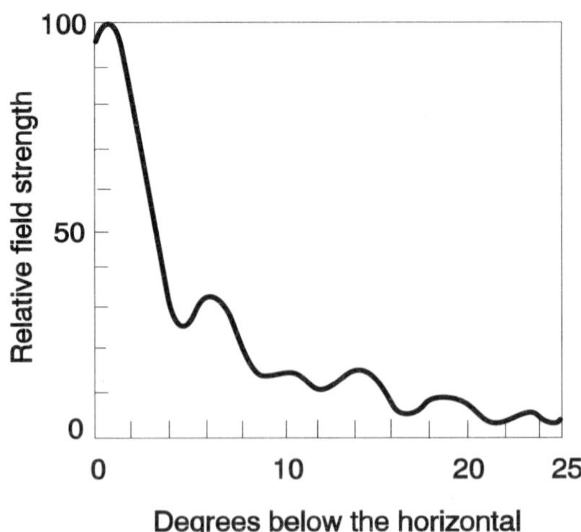

Figure V.1 Vertical radiation pattern of a typical UHF broadcast antenna

Vertical Redundancy Check (VRC): The *even parity check* or *odd parity check* performed on *ASCII* coded *data* by the use of a *parity bit.*

vertical resolution: The number of separate *picture elements* that can be resolved in a vertical direction on a display screen. The vertical resolution of British 625-line television systems is 402 picture elements and in the US 525-line system it is 340 picture elements.

Vertical Tabulation (VT): A *Format Effector (FE)* which instructs the cursor of a *Visual Display Unit (VDU)*, or the mechanism of a printer, to advance to the next predetermined position in a document.

Very high bit rate Digital Subscriber Line (VDSL): A technique for transmitting *broadband data* rapidly over short distances of a conventional copper *local loop*. Developed in 1995 by a consortium of vendors. It can provide 52 Mbit/s downstream and 6 Mbit/s upstream over distances of a few hundred metres. The transmission rate falls to about 12 Mbit/s downstream and 6 Mbit/s upstream for distances around 1.5 km. A 4 kHz band is reserved at the lower end of the spectrum for *POTS*. Also written as *VHDSL.*

Very High Frequency (VHF): The portion of the *electromagnetic spectrum* which lies in the *frequency range* 30 MHz to 300 MHz. This portion has an *International Telecommunications Union (ITU)* designation of 8 and is primarily used for television and most *Frequency Modulation (FM)* radio broadcasts.

Very Large Scale Integration (VLSI): The integration of about ten thousand electronic circuits into a silicon chip using micro-electronic technology to create a multifunction semiconductor device. Density is less than in *Ultra Large Scale Integration (ULSI) but greater than in Large Scale Integration (LSI).*

Very Low Frequency (VLF): The portion of the *electromagnetic spectrum* which lies in the *frequency range* 3 kHz to 30 kHz. This portion has an *International Telecommunications Union (ITU)* designation of 4.

Very Small Aperture Terminal (VSAT): An *antenna*, located on an *Earth station* used with *satellite* communications, which has a small diameter, typically between 1 m and 3 m.

VESA: *Video Electronics Standards Association.*

vestigial sideband: The small part of a *sideband* that remains after *Amplitude Modulation (AM)* of a *carrier signal.*

Vestigial Sideband Amplitude Modulation (VSB-AM): A modulation technique used, for example, in television *transmission* systems, to ensure that the resulting signal occupies minimum power and bandwidth, yet the signal waveform is simple enough to be handled with minimum cost in each television receiver. As shown in Figure V.2 the complete upper *sideband* is transmitted but only a part of the lower

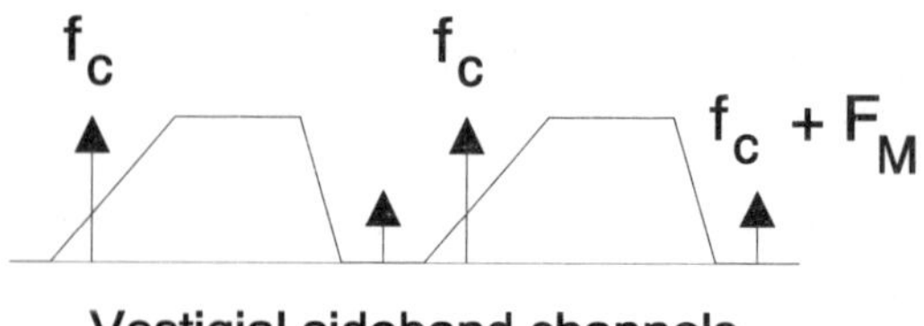

Figure V.2 Vestigial Sideband Amplitude Modulation

sideband. Here f_c is the *carrier frequency* and F_M is the *baseband modulating frequency*. Also known as *vestigial sideband modulation.*

vestigial sideband modulation: See *Vestigial Sideband Amplitude Modulation (VSB-AM).*

Vestigial Sideband transmission (VSB): A modification of *Double Sideband transmission (DSB)* in which only a part of one of the *sidebands*, along with the other sideband and the *carrier signal*, are transmitted.

VF: *Voice Frequency.*

VFCT: *Voice Frequency Carrier Telegraph.*

VFT: *Voice Frequency Telegraph.*

VFTG: *Voice Frequency Telegraph.*

VG: *Voice Grade.*

VGC: *Voice Grade Circuit.*

VHDSL: *Very High bit rate Digital Subscriber Line.*

VHF: *Very High Frequency.*

Via Net Loss (VNL): A parameter used in planning *transmission networks* by predicting and allocating losses in the various parts of the network.

video: The range of the *electromagnetic spectrum*, used for *television transmission signals*, from about 100 kHz to several MHz. In practice the *bandwidth* of the *baseband signal* (exclusive of audio carriers) for transmission of television pictures is approximately 5 MHz in order to meet standards set by the *National Television Standards Committee (NTSC).*

video amplifier: An *amplifier* which amplifies a *video signal* from a television camera before the signal goes through *modulation* and *transmission.* The term can also denote the device in a television receiver which amplifies a received video signal prior to it being displayed on the *Cathode Ray Tube (CRT).*

video bandwidth: The *bandwidth* required to carry a *video signal*, such as a television picture. The size of the bandwidth depends on the amount of detail used by the system. The conventional UK 625-line television system, for example, requires a bandwidth of 5.5 MHz, whereas the USA 525-line system requires 4.2 MHz.

Video Cassette Recorder (VCR): A machine that can be set to record *video data* from an input source, such as a television *broadcast*, onto a video signal storage medium, such as a magnetic tape or compact disc. The recorder can also play back prerecorded video data onto a *Visual Display Unit (VDU)*.

video coding: The *algorithm* used for storage and *transmission* of *video signals*. The most common ones are those given in *ITU-T Recommendations* or by the *ISO*. Examples are the ITU-T Recommendation H.261 (see *H Series*) which specifies audio-visual services at *data rates* from 64 kbit/s to 1.92 kbit/s and *MPEG-2* for the storage and retrieval of moving television pictures.

videoconference: A meeting between two or more people in different locations mediated by a *multimedia* two-way electronic communications system. A videoconference system could provide full-motion *video*, voice and graphics facilities. Videoconferencing can be via studio equipment, involving many people, or by desktop equipment, involving one or relatively few people per location. This has led to a blurring of the difference between videoconferencing and *videophony*, as illustrated in Table V.2.

Video Dial Tone (VDT): A term used to indicate the availability of *Video On Demand (VOD)* from a *PTO*.

Video Display Terminal (VDT): See *Visual Display Unit (VDU)*.

Video Display Unit (VDU): See *Visual Display Unit (VDU)*.

Video Electronics Standards Association (VESA): An industry group, founded in the 1980s by several video board manufacturers, with the aim of standardising on the electrical and *software* matters concerned with high resolution *video* displays.

video frequency: The portion of the *electromagnetic spectrum* which lies in the approximate range 100 MHz to several GHz. This portion overlaps the *Very High Frequency (VHF)* and *Ultra High Frequency (UHF)* bands.

Video On Demand (VOD): A service, often distributed as part of *Cable Television (CATV)*, allowing the customer the direct choice of a *video* item from a central store. True Video On Demand is a one-to-one service which needs a dedicated user channel and a return path (for commands such as pause, rewind and fast forward) to the central store. *Near Video on Demand (NVOD)*, or *staggercast*, is a simplified service where a film is started at intervals of a few minutes on several channels. Viewers wait for the start of the film, and can pause, fast forward or rewind by hopping between the channels. A switched CATV network is ideally suited to VOD, but this can also be achieved on an *HFC* system by reserving a number of channels and by reusing part of the spectrum in different sectors, as in Figure V.3.

Table V.2 Videoconferencing and videophony

Videophony	*Videoconferencing*	
	Desk top	*Studio*
Person-to-person	Person to small group or small group to small group	Group-to-group
Office environment	Office environment	Studio or specially equipped room
Compact desk top terminal	Desk top terminal	Large floorstanding or roll-about terminals
On demand service via customer switched digital networks	On demand service via customer switched digital networks	Service via digital leased lines or via customer switched digital networks
Picture quality consistent with low bit rate transmission (e.g. 64 kbit/s to 128 kbit/s)	Picture quality consistent with bit rate transmission of 64 kbit/s to 384 kbit/s	Picture quality consistent with high bit rate transmission (384 kbit/s up to 2 Mbit/s)
Telephony quality speech	Option of telephony quality or wideband quality speech	Wideband speech quality

videophony: A dial-up person-to-person form of *videoconferencing* using a compact terminal. The picture quality is consistent with low *bit rate transmission* (64 kbit/s to 128 kbit/s). Although videophony and videoconferencing were initially conceived as discrete services, the boundaries between them are becoming less distinct. Also known as *video telephony*.

video signal: A *signal* that transmits *video* information and, in the case of television, a sound signal. The *bandwidth* of the signal varies depending on the objectives, such as *High Definition Television (HDTV)*, slow *scan* TV, etc.

Video Tape Recorder (VTR): See *Video Cassette Recorder (VCR)*.

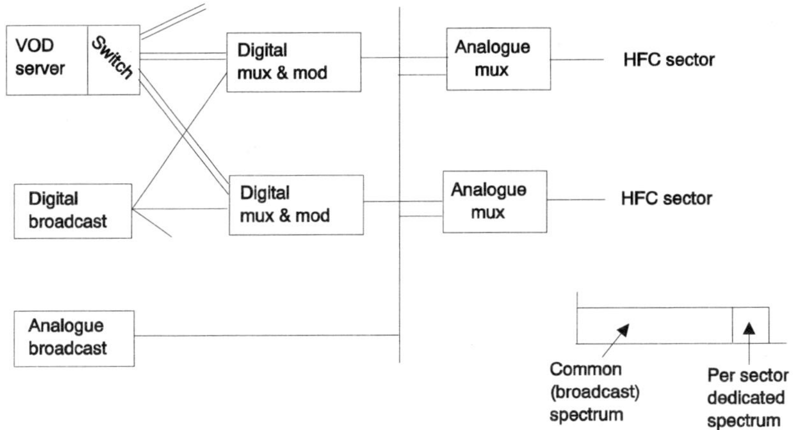

Figure V.3 An arrangement for VOD

video telephony: See *videophony*.

videotex: A system which allows interactive access to *databases*, over the *Public Switched Telephone Network (PSTN)*, using a *television receiver* or other low cost *Visual Display Unit*. The display unit is connected to the PSTN using a *modem*, typically operating at a *baud rate* of 75 going out to the *Central Office (CO)* and 1200 coming into the premises from the CO. Many of the functions envisaged for videotex are now also provided over the *Internet*.

videotex adaptor: An adaptor which allows *videotex traffic*, transmitted on the *Public Switched Telephone Network (PSTN)*, as part of the videotex system, to be displayed on a standard *television receiver*. The adaptor usually plugs into the *UHF* or *RGB* input socket of the television receiver, and comprises a *modem, frame store* and *decoder*.

videotex decoder: A component of a *videotex* system which decodes text and graphics *signals*, received over the *PSTN*, so they can be displayed on a *television receiver*.

videotex terminal: A terminal which receives and displays signals from a *videotex* system. Videotex terminals are usually modified *television receivers*. The modification can either be provided by the use of an external *videotex adaptor*, or by the internal addition of extra components during manufacture of the television set. The additional components required include a modem, line isolator, memory unit, processing logic and a keyboard or keypad.

videotex traffic: *Signals*, representing *data*, transmitted over the *Public Switched Telephone Network (PSTN)* as part of a *videotex* system.

video traffic: The *video* component of a *transmission*, such as that for a *television broadcast.*

viewdata: A generic term referring to systems, such as *videotex* and *teletext*, where users request and access remotely stored information via the *public switched telephone network* using *terminals*, such as modified *television receivers*, to display the requested *data*.

viewdata adaptor: See *videotex adaptor*.

viewdata terminal: See *videotex terminal.*

V interface: In an *ISDN*, the reference point between the customer line termination and the *exchange* termination. See also *U interface*.

Virtual Call (VC): A *call* between two *Data Terminal Equipments (DTE)* which forms a temporary association between the two terminals. It is set up by the *calling terminal* sending the *network* a *call request packet* which contains the *address* of the *called terminal*, or a reference to this address. This will be sent by the network to the called terminal, which may accept the call and the *data* associated with it. The call may be set up over a *Virtual Circuit (VC)* or a real circuit.

virtual channel: In an *Asynchronous Transfer Mode (ATM)* system, during *multiplexing* the *cells* from individual sources are placed in a queue and then fed out onto the *transmission link*, as shown in Figure V.4. Idle cells may be added if the queue becomes empty. The sequence of cells from a particular user or source all carry the same value in the *header address field* and this cell sequence is called a virtual channel. The rate of

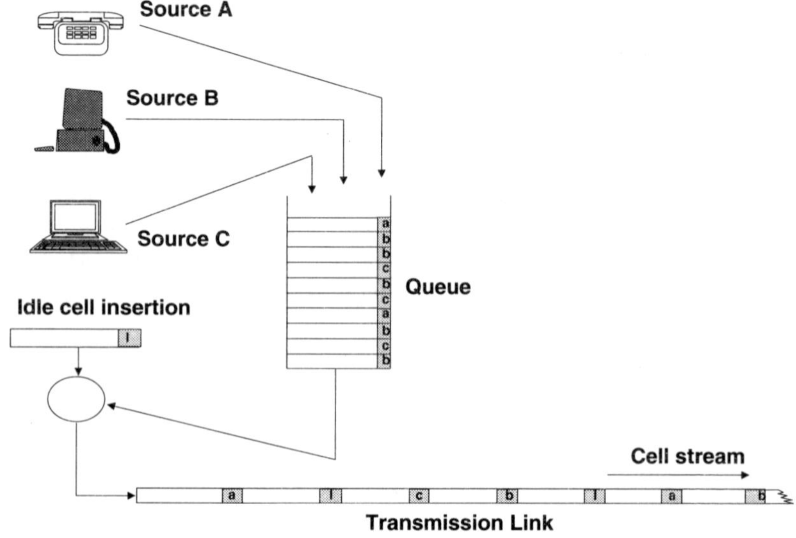

Figure V.4 ATM multiplexing

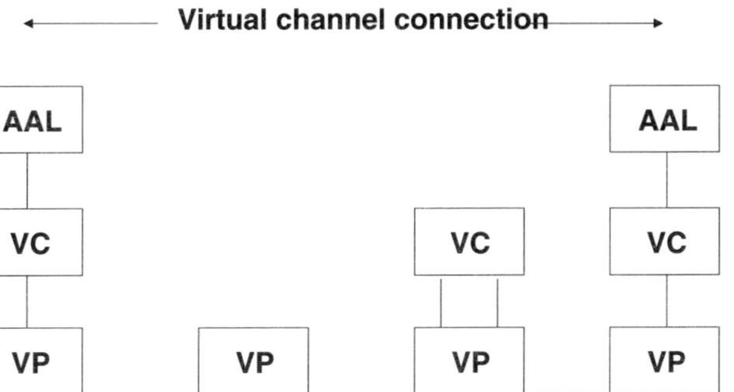

Figure V.5 Types of ATM layer connections

transmission of cells within the virtual channel is variable, and reflects the level of source activity and the amount of *traffic* in the system, i.e. the *transfer mode* is *asynchronous*.

Virtual Channel Connection (VCC): Part of the *ITU-T Recommendation* I.150 (see *I Series*) for *ATM*. It defines the VCC as a concentration of *virtual channel links*, extending between the points where the *ATM Adaptation Layer (AAL)* is accessed, as in Figure V.5.

Virtual Channel Identifier (VCI): Part of the *address field* of an *ATM cell*, as shown in Figure A.21. The other part of this field is the *Virtual Path Identifier (VPI)*.

Virtual Circuit (VC): A *network* operation which gives the user the impression of having an end to end connection for the duration of the *call*, although this is not the case. Virtual circuits are encountered in a *Packet Switched Network (PSN)*, such as *ATM*, where there is no dedicated access path associated with each call. Figure V.6 illustrates the Virtual Circuits followed by four *data packets*, these varying between individual packets from the same source.

Virtual Container (VC): In the *Synchronous Digital Hierarchy (SDH)* a *PDH signal* is transported by mapping this into a *Synchronous Container*. To this is added the *Path Overhead (POH)*, as shown in the example of Figure V.7, and this forms the Virtual Container. The capacity of a Virtual Container is specified in SDH standards by a suffix letter, e.g. VC11, VC12, VC2, etc., as in Table V.3.

Virtual Local Area Network (VLAN): Refers to a *Local Area Network (LAN)* which is dynamic in structure and not defined by a single fixed physical infrastructure.

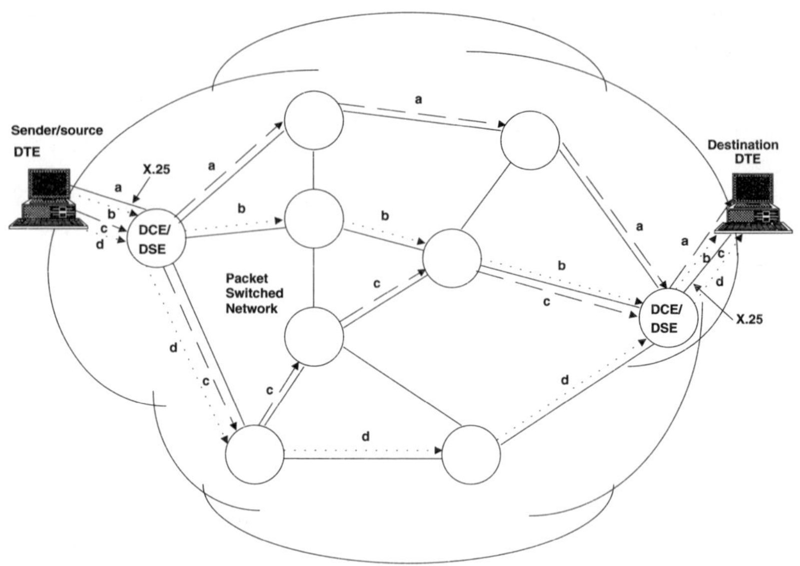

Figure V.6 Virtual Circuit concept

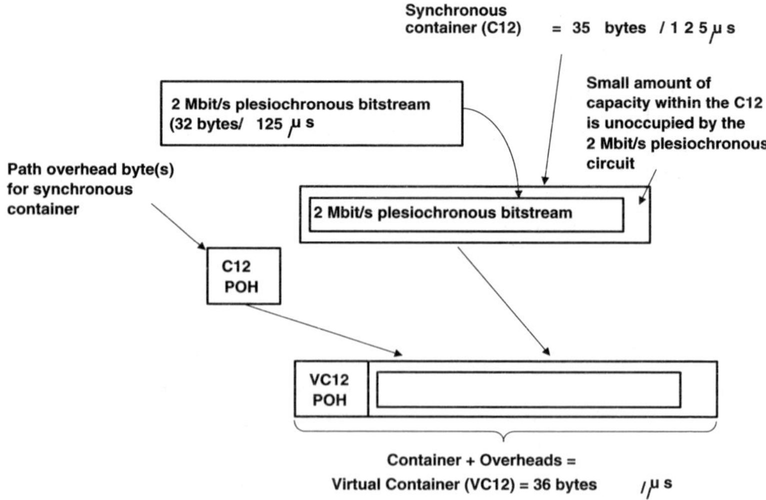

Figure V.7 Creation of a Virtual Container

Table V.3 Virtual Container capacity

Container	*Capacity (Mbit/s)*
VC11	1.600
VC12	2.176
VC2	6.784
VC3	48.384
VC4	149.760

virtual network: A *network* which operates on the principle of *Virtual Circuits (VC)* and *Virtual Paths (VP)*, i.e. it uses *packet switching* and the *routes* between *nodes* are not permanently connected for the duration of *call*. A virtual network could be a *Local Area Network (LAN)* or a *Wide Area Network (WAN)*.

Virtual Path (VP): In a *network* using *Asynchronous Transfer Mode (ATM)* multiple *Virtual Circuits (VC)* can be grouped together into a Virtual Path, as shown in Figure V.8.

Virtual Path Connection (VPC): One of the functions of the *ATM* layer, defined in *ITU-T Recommendation* I.150 (see *I Series*) as a concentration of *Virtual Path (VP)* links that extend between the point where *VCI* values are assigned, translated and removed.

Virtual Path Identifier (VPI): Part of the *address field* of an *ATM cell*, the other part being the *Virtual Channel Identifier (VCI)*, as in Figure A.21. The VPI field offers the possibility of establishing a semi-permanent *Virtual Path (VP)*, which can be used for various applications.

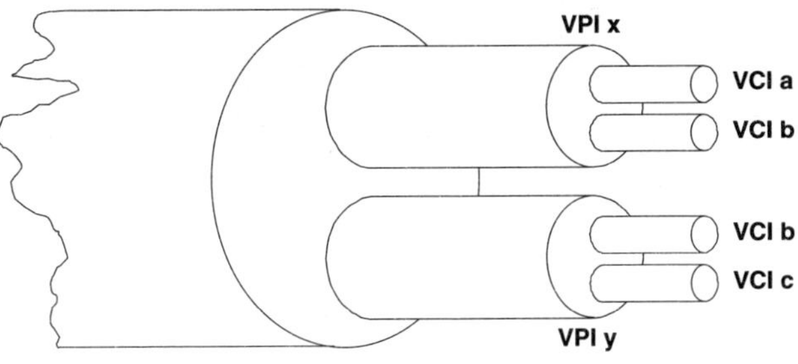

Figure V.8 Virtual paths

virtual private circuit: *Circuits* which form part of a *Virtual Private Network (VPN).*

Virtual Private Network (VPN): A service provided by *PTOs* in which the *Public Switched Telephone Network (PSTN)* is used to form and organisation's *private network*, often linking the *PABX* located at its various sites. The *exchange* ensures that this private *traffic* is kept separate from the normal PSTN traffic. Virtual Private Networks are often used by small organisations with a few remote sites, which does not justify the use of private networks. They are also used by larger organisations to supplement already existing private networks. In these cases Virtual Private Networks can provide a secondary *route* for main traffic in times of heavy use or private network faults. (See *hybrid network*, Figure H.10.) They can also be used as a way of accessing smaller customer outstations or other businesses with whom the customer deals.

Virtual Terminal (VT): *Protocol* specified as part of the common *Application Layer* standards of the *OSI Basic Reference Model.* It allows *terminals* in a multi-user *network* to interact irrespective of terminal characteristics, so providing functions for remote terminal access between open systems.

Virtual Tributary (VT): The *SONET* equivalent of the *Virtual Container (VC)* used in *SDH networks.* The most commonly used Virtual Tributaries are VT1 at 1.544 Mbit/s, VT2 at 2.048 Mbit/s and VT6 at 6 Mbit/s.

virus: A computer program that finds other programs and embeds a copy of itself in them. The embedded virus is executed when the host program is run, which allows the virus to replicate, sometimes without any visible effects to the program user, until the computer's resources are overfilled and it breaks down. Unlike a *worm*, a virus cannot infect other computers without assistance, but is transferred by acts such as users trading programs.

visibility function: The average response of the *human eye* to different *wavelengths* of the *electromagnetic spectrum.* The human eye can respond to wavelengths in the *visible colour spectrum* between about 400 nm (violet) and 760 nm (deep red). The eye's sensitivity to different wavelengths varies within this range. For example it is far more responsive to light in the yellow spectrum, around 580 nm. Therefore a source of light at 680 nm (red) would need to have 100 times more *luminous power* than a source at 550 nm (yellow/green) in order to get the same response from the eye.

visible colour spectrum: See *visible spectrum.*

visible light spectrum: See *visible spectrum.*

visible spectrum: The section of the *electromagnetic spectrum* which is visible to the *human eye*, producing the sensation of colours. The visible spectrum extends from about 380 nm (violet) to 720 nm (deep red). Also known as the *visible light spectrum*, the *optical spectrum* and the *visible colour spectrum*.

vision carrier: A *carrier signal* that transports the *video* by using a form of *Amplitude Modulation (AM)*. The sound signal is usually carried on a separate carrier wave.

Visited Location Register (VLR): One of the mobility management functions used with *GSM* systems. It provides a temporary *database* of all *subscribers* within its geographical coverage area, and the information contained is similar to that within the *Home Location Register (HLR)*, i.e. subscriber numbers, subscription levels, call restrictions, etc. The use of a VLR means that the *Mobile Switching Centre (MSC)* does not need to access the HLR for every transaction, so improving its efficiency. See Figure G.5.

visual acuity: A measure of the amount of detail that can be detected by the *human eye* and processed by the brain. It is the reciprocal of the angle subtended at the eye by two objects that can be separately identified. Visual acuity impacts on the design of television and other visual display systems where the resolution of the number of *pixels* or scanning lines is of importance.

visual communications: A communications system that relies on the transmission of information by direct observation using the *human eye*. These systems include flags, *semaphores*, panels, arm signals, flares and smoke signals. These systems are not considered to be conventional telecommunications systems.

Visual Display Unit (VDU): A unit consisting of a keyboard or keypad, and a screen used to display data information in the form of text and graphics. Examples of visual display units are *Cathode Ray Tube (CRT), liquid crystal array and plasma panel.*

visual message signal: The part of a television *transmission* containing the *video* and *synchronisation* needed to align the television camera and receiver. The visual message signal does not contain the corresponding *audio signal*.

visual signalling: The use of *visual communications*.

Viterbi decoding: An *algorithm* used as a *decoder* for *convolutional coding* sequences. It can be applied for both *hard decision decoding* and *soft decision decoding*.

VLAN: *Virtual Local Area Network*.

VLC: *Variable Length Coding*.

VLF: *Very Low Frequency*.

VLR: *Visited Location Register*.

VLSI: *Very Large Scale Integration*.

VMS: *Voice Mailbox System* or *Voice Mail System* or *Voice Messaging System.*

VNL: *Via Net Loss.*

vocoder: *Voice coder.*

VOD: *Video On Demand.*

VOFR: *Voice Over Frame Relay.*

VOGAD: *Voice Operated Gain Adjusting Device.*

voice: Generally refers to the representation of the human speech as electrical *signals.*

voice activated dialling: See *voice activated service.*

voice activated service: A service, such as *dialling* (*voice activated dialling*) or *voice mail* retrieval, which can be activated by the user's *voice.* These services are sometimes required for safety reasons, such as operating a mobile phone in a moving vehicle.

voice analogue signal: An electrical *signal* that is the *analogue* representation of human speech. See also *voice digitisation.*

voiceband: The electrical wave *bandwidth* from 300 Hz to 3400 Hz used in *telephone* equipment for the *transmission* of *analogue* or digital *voice signals* and *data.*

voiceband data: *Data*, usually in a digital format, that is transmitted over the *Public Switched Telephone Network (PSTN)* within the *voiceband*, usually between 300 Hz and 3400 Hz. Also called *modem data.*

voiceband frequency: A frequency that occurs in the *voiceband*, usually in the range 300 Hz to 3400 Hz.

voice channel: A communications *channel* designed to carry *voice signals.* This is usually *analogue* but can also refer to *digitised voice.*

voice circuit: A communications *circuit* designed to carry human *speech signals* in the *voiceband*, usually between 300 Hz and 3400 Hz. It has an analogue *bandwidth* of 4 kHz or represents a digital circuit at 64 kbit/s, for *digitised voice.*

voice coder: Often referred to as a *vocoder.* An early type of device which digitises *voice* with the aim of reducing the *bandwidth* requirement of *speech signals.* A vocoder usually comprises a speech analyser which digitises voice in terms of its *amplitude*, pitch and voicing, and a speech synthesiser which converts *digital signals* into artificial speech sounds.

voice compression: The reduction of the *bandwidth* used for *transmission* of *speech signals* in communications systems. Two commonly used techniques are *Adaptive Differential Pulse Code Modulation (ADPCM)* and *Continually Variable Slope Delta modulation (CVSD).* ADPCM can compress voice from 64 kbit/s to 24 kbit/s or 32 kbit/s (a *compression ratio* of 8:3 or 2:1) and CVSD gives rates of 20 kbit/s and 32 kbit/s (compression ratio of 16:5 or 2:1). A recent technique is *Code Excited Linear Prediction (CELP)* which can reduce the signal to 16 kbit/s (a compression ratio of 4:1). Higher compression ratios, of 8:1 or more, can be obtained, but often the quality of the voice obtained is poor. Also known as *speech compression*.

Voice Data Multiplexer (VDM): See *multiplexer*.

voice digitisation: The conversion of *analogue* human *speech signals* into *digital signals* by techniques such as *Pulse Code Modulation (PCM)*.

voice distribution frame: *Distribution frame* used for the management of *voice circuits*. See also, *Main Distribution Frame (MDF)* and *Building Distribution Frame (BDF)*.

Voice Frequency (VF): The *frequency* that is within the *voiceband* (usually 300 Hz to 3400 Hz) and is used to transmit *speech signals*.

Voice Frequency Carrier Telegraph (VFCT): See *Voice Frequency Telegraph (VFT)*.

voice frequency signalling: The use of pulses of *Alternating Current (AC)* at *frequencies* in the *voiceband* for *signalling* on the *Public Switched Telephone Network (PSTN)*. The frequencies used are typically 600 Hz, 750 Hz, 2280 Hz and 2400 Hz. See also *in-band signalling*.

Voice Frequency Telegraph (VFT): A *telegraph* system in which *signals* are transmitted at audio frequencies. Up to 24 *channels* can normally be transmitted on a single *voice circuit* using *Frequency Division Multiplexing (FDM)*. Also known as *Voice Frequency Carrier Telegraph (VFCT)*.

Voice Grade (VG): See *Voice Grade Circuit*.

Voice Grade Circuit (VGC): A *circuit* that is capable of handling signals in the *voiceband*, typically between 300 Hz and 3400 Hz. This includes *voice*, *facsimile*, and *data*. Often shortened to *Voice Grade (VG)*.

voice mail: A system that can answer *calls*, store *messages*, and play back messages that have been already recorded, usually responding to *signals* entered through the keypad of a *DTMF telephone*. Other popular voice mail features include: the protection of message retrieval by a password; notification that a message has been left (using, for example, an altered dialling tone, message waiting lamp, radio pager, or automatic phone call), message forwarding (i.e. recording a message and forwarding it to one or more other numbers); auto-attendant (i.e. transfer of calls to any telephone extension); *call handling*; and *call screening*. Voice mail services are often used with mobile phones, where answer phones would be impractical. Also known as *Voice Mailbox System (VMS)*, *Voice Mail System (VMS)*, and *Voice Messaging System (VMS)*.

Voice Mailbox System (VMS): See *voice mail*.

Voice Mail System (VMS): See *voice mail*.

Voice Messaging System (VMS): See *voice mail*.

Voice of America: A USA-based provider of news, health and environmental information in many languages using radio *broadcasts* and the *Inter-*

net. It is supported by private and public contributions and income it receives for its services.

voice operated device: A device that is located on a *telephone circuit* and is operated by *voice signals*. It aims to reduce certain forms of defects, such as *echo* and *singing*.

Voice Operated Gain Adjusting Device (VOGAD): A device used to keep the output *amplitude* of a *voice signal* substantially constant as the input amplitude varies.

Voice Over Frame Relay (VOFR): Technique for *voice* communications over a *Frame Relay (FR) network.*

voice packet switching: The *transmission* of *speech signals* using a *Packet Switching Network (PSN)*. This involves digitising *analogue voice signals* using *Pulse Code Modulation (PCM)* prior to packetising.

voice printing: A form of *speaker verification* where the voice print of a caller is compared to a voice print taken in an earlier enrolment session.

voice processing: A generic term used to describe a series of techniques which manipulate the human voice. These include storing, replaying, compressing, decompressing and recognising speech, as well as text to speech and speech to text conversion. Figure V.9 shows an architecture of a possible voice processing system. The speech bus carries the caller's speech to the various voice processing elements and from these elements to the caller. The computer bus links the elements to each other and to

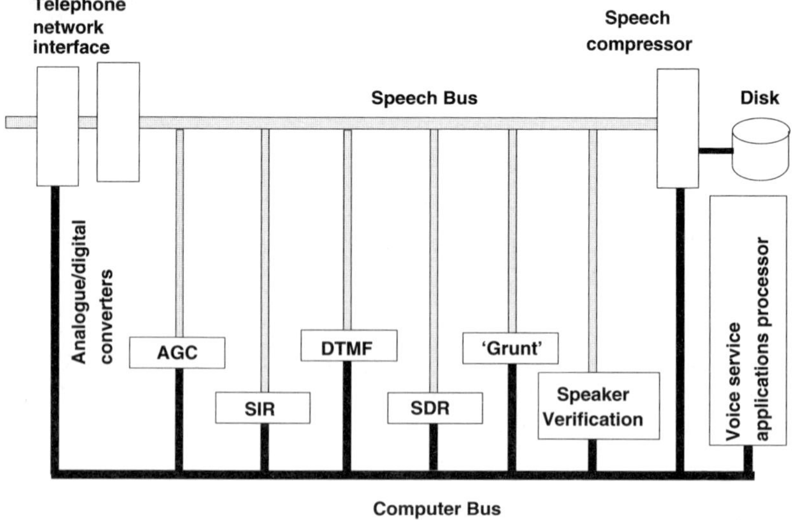

Figure V.9 Voice processing architecture

the disk. The overall system works by compressing speech onto disk and delivering speech from disk to the caller, under control of the voice service applications processor.

Voice Profile for Internet Mail (VPIM): Standard proposed, by several manufacturers, for *voice mail* over the *Intenet.*

voice response: A system allowing information stored on a central *database* to be automatically read out to callers who use *voice* or *DTMF* commands through the *Public Switched Telephone Network (PSTN). Speech recognition* is being increasingly used to allow the caller to specify the *data* they require.

Voice Response Unit (VRU): The equipment which provides the *voice response.*

voice signal: Usually the electrical *analogue signal* which represents human speech, usually in the range 300 Hz to 3400 Hz. Also refers to the sound wave which carries human speech. Can also refer to the *digitised voice* signal.

voice telephony: The transmission of *speech signals* in the *Public Switched Telephone Network (PSTN).*

voice-to-noise ratio: A performance measure of a *transmission channel.* It is given by the ratio of the *voice signal* power to the *noise* signal on that channel.

Voltage Standing Wave Ratio (VSWR): A measure of the impedance mismatch between a *transmission line* and its load. The higher the mismatch the greater the reflected wave from their junction, which reacts with the incident wave to create *standing waves.* The ratio of the maximum *amplitude,* of the voltage of this standing wave, to its minimum amplitude is the VSWR, and the higher this ratio the greater the mismatch between the line and its load.

volume related tariff: A *tariff* for the use of a *Public Data Network (PDN)* which is dependent on the volume of *traffic* a particular *subscriber* places over the system, rather than the distance the traffic travels. A *data* segment of 64 characters is usually the minimum unit.

VP: *Virtual Path.*

VPC: *Virtual Path Connection.*

VPI: *Virtual Path Identifier.*

VPIM: *Voice Profile for Internet Mail.*

VPN: *Virtual Private Network.*

VQ: *Vector Quantisation.*

VRC: *Vertical Redundancy Check.*

VRP: *Vertical Radiation Pattern.*

VRU: *Voice Response Unit.*

VSAT: *Very Small Aperture Terminal.*

VSB: *Vestigial Sideband transmission.*

VSB-AM: *Vestigial Sideband Amplitude Modulation.*
V Series: *ITU-T Recommendations* covering *data transmission* over the *Public Switched Telephone Network (PSTN)*. Some of these Recommendations are given in Table V.4.
VSWR: *Voltage Standing Wave Ratio.*
VT: *Virtual Tributary*, *Virtual Terminal* or *Vertical Tabulation.*
VTR: *Video Tape Recorder.*

Table V.4 ITU-T Recommendations, V Series
(Continued on next page)

Recommendation	*Description*
General	
V.1	Equivalence between binary notation and the significant conditions for the transmission of data
V.2	Power levels for data transmission over telephone lines
V.3	Data communications code
V.4	Structure of signals for International Alphabet No. 5
V.5	Data signalling for synchronous data transmission in the PSTN
V.6	Data signalling rates for synchronous data transmission on leased telephone-type lines
V.7	Terms covering data communications over telephone networks
Interface and voice band modems	
V.10	Electrical requirements for unbalanced double current interchange circuits for use with integrated circuit equipment in data communications
V.11	Electrical requirements for balanced double current interchange circuits for use with integrated circuits in data communications

Table V.4 (Continued from previous page; continued on next page)

Recommendation	*Description*
V.15	Acoustic couplers for data transmission
V.16	Transmission of analogue signals associated with the medical field
V.17	Group III facsimile transmission at 14.4 kbit/s, in half-duplex mode
V.19	Modems for parallel data transmission using signal frequencies associated with telephone networks
V.20	Modems for parallel data transmission for universal use in general PSTNs.
V.21	Full duplex, 300 bit/s asynchronous modem for use over 2 wire PSTN
V.22	Full duplex, 1.2 kbit/s asynchronous modem for use over 2 wire PSTN
V.22 bis	Full duplex, 2.4 kbit/s asynchronous modem for use over 2 wire PSTN
V.23	Full duplex, 600 bit/s or 1.2 kbit/s forward channel and 75 bit/s reverse channel, asynchronous modem for use over 2 wire PSTN
V.24	Definitions for interchange circuits between DTEs and DCEs
V.25	Automatic calling and answering equipment on the PSTN
V.26	2.4 kbit/s modem for use on 4 wire leased telephone-type lines
V.26 bis	Half duplex, 1.2 kbit/s or 2.4 kbit/s synchronous modem for use over the PSTN
V.26 ter	Full duplex, 2.4 kbit/s modem for use over the PSTN

Table V.4 (Continued from previous page; continued on next page)

Recommendation	*Description*
V.27	4.8 kbit/s modem for use over leased telephone-type lines
V.27 bis	2.4 kbit/s or 4.8 kbit/s modem with automatic equalisation for use on leased telephone-type lines
V.27 ter	Half duplex, 2.4 kbit/s or 4.8 kbit/s modem for use on two wire PSTN
V.28	Electrical characteristics for unbalanced double current interchange circuits
V.29	Full duplex, 9.6 kbit/s synchronous modem for operation over 4 wire leased telephone-type lines
V.32	Full duplex, 9.6 kbit/s asynchronous and synchronous modem for operating over 2 wire PSTN
V.32 bis	Full duplex
V.33	Full duplex, 14.4 kbit/s synchronous modem for operating over 4 wire leased telephone-type lines
V.34	Full duplex, 28.8 kbit/s asynchronous modem for operating over 2 wire PSTN
V.36	Synchronous data transmission modems using 60 kHz to 108 kHz group band circuits
V.40	Error indication using electromechanical equipment
V.41	Code independent error control system
V.42	Error control for asynchronous modems up to 9.6 kbit/s. It specifies LAP-M and MNP
V.42 bis	Data compression for modems which use LAP-M error control procedures

Table V.4 (Continued from previous page)

Recommendation	*Description*
Transmission quality and maintenance	
V.50	Limits for quality of circuits in data transmission
V.51	Organisation for maintenance of international telephone circuits used for data transmission
V.52	Characteristics of equipment used to measure distortion and error rates in data transmission over the PSTN
V.53	Limits for maintenance of telephone circuits for data transmission
V.54	Devices used for loop tests for modems
V.55	Equipment for measurement of impulse noise on telephone type circuits
V.56	Comparative tests for modems

W

WACK: *Wait before Aconowledgement.*

WAIS: *Wide Area Information Server.*

Wait before Acknowledgement (WACK): A delay before transmitting *acknowledgement* by a *called terminal* when it is not ready to receive a *message* from a *calling terminal.*

wait on busy: A *network* facility which enables a *calling terminal* to be held in a queue if the *called terminal* is busy with another *call*. The calling terminal is automatically connected once the called terminal becomes free.

walking code: A *binary code* that 'walks' through a register bit by bit and whose pattern advances by one each time.

walk time: In a *polling* system it is the portion of the *poll cycle* which is used to carry the overheads, such as *propagation delays*, *synchronisation* times when stations receive the poll request, etc. Walk time is in effect the time needed to transfer permission to transmit from one polled terminal to the next, and to complete the *messages*. It is the minimum *delay time* irrespective of whether any *data* is transmitted.

WAL1 code: See *Manchester code*.

WAN: *Wide Area Network.*

wander: A form of *jitter*, occurring at a *frequency* of less than 10 Hz and lasting longer than one second. Wander can be further classified as long term wander (often 24 hour cycles caused by, for example, daily temperature variations) or short term wander. See also *drift*.

WAP: *Wireless Application Protocol.*

WARC: *World Administration Radio Conference*.

watchdog: See *telecommunications watchdog*.

WATS: *Wide Area Telecommunications Service*.

WATTC: *World Administrative Telegraph and Telephone Conference*.

waveform: The form a *signal* takes when represented as a plot of *amplitude* vesus time or amplitude versus distance. Waveforms are frequently made up of a combination of simpler waveforms. Each waveform can be described by parameters such as its *periodic time*, *frequency*, amplitude and *phase*.

waveguide: A *transmission medium* that confines and guides the energy of high-frequency waves. For example, waveguides for use with *radio waves* include *coaxial cables* or hollow metallic conductors filled with dielectric medium, and also areas in the *stratosphere* and *troposphere*. For *lightwaves* the waveguide would consist of *optical fibre*. The cross section dimensions of the active part of the waveguide needs to be equal to or greater than the *wavelength* of the waves.

waveguide dispersion: One of the two elements of *intramodal dispersion*, the other being *material dispersion*. Waveguide dispersion is *dispersion* caused by the structure of the *waveguide*, such as *optical fibre*.

wavelength: The distance between peaks or other points of corresponding phase of two consecutive cycles of a wave. The wavelength of a wave is equal to its velocity (about 3 x10^8 m/s in a vacuum for electromagnetic waves) divided by its *frequency* and is normally expressed in metres.

wavelength conversion: The conversion of the *wavelength* at which *data* is transmitted throughout a *network*. In *Wavelength Division Multiplexing (WDM)* systems with access to a small number of wavelenghts, the network *capacity* can be increased by using one wavelength in more than one sub-network of a network. Wavelength conversion at the boundary of the sub-net is necessary to prevent the data from various sub-nets colliding. Wavelength conversion can be achieved using the *gain saturation* effect of *Semiconductor Optical Amplifiers (SOA)*.

Wavelength Division Multiplexing (WDM): A *multiplexing* technique used with optical communication systems. Two or more information-carrying *channels* are simultaneously transmitted in the same direction on a single *waveguide* using more than one light source of different *wavelengths*. Figure W.1 illustrates the principle of WDM. Four optical wavelengths, each produced by a separate *laser* tuned to a given wavelength, pass through an optical coupler before being channelled through *optical fibre*. At the receiving end they are split using an optical splitter

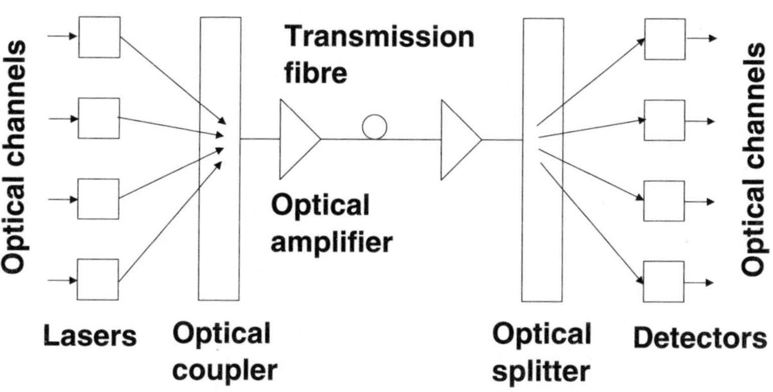

Figure W.1 Wavelength Division Multiplexing (WDM)

and detected. *Optical amplifiers* may be used to boost the signal, if necessary.

Wavelength Division Multiple Access (WDMA): A *multiple access* technique which uses *Wavelength Division Multiplexing (WDM)* to allow *signals* with different *wavelengths* to access the same *transmission channel.*

Wavelength Modulation (WM): The *modulation* or changing of the *wavelength* of an *electromagnetic wave.*

wave number: A measure which is more convenient to use than *frequency* for low frequency waves. It is defined as the number of *wavelengths* in a unit distance, measured in the direction of *propagation.* Numerically this is equal to the reciprocal of the wavelength. Therefore a wave with a wavelength of 0.1 units will have a wave number of 10.

way station: A *station* which is intermediate between two other stations and is only connected to these two stations.

WDM: *Wavelength Division Multiplexing.*

WDMA: *Wavelength Division Multiple Access.*

weatherfax: A specialist *facsimile* machine used to transmit weather charts internationally from centres where they are compiled by local meteorologists. These machines are operated on *private circuits* or radio networks on a *simplex, broadcast* basis. They are being superseded by the transfer of computer based *data.*

web: Shortened term for the *World Wide Web (WWW).*

web browser: A *software* program which provides a graphical front end for navigation on the *World Wide Web (WWW).*

web site: A location on a computer, operated by an organisation, which contains a collection of *World Wide Web (WWW)* information.

Weibull distribution: A continuous *probability distribution* given by the equation: $y = \alpha\beta\,(x - \gamma)^{\beta-1}\exp(-\alpha(x - \gamma)^{\beta})$ where α is called the scale factor, β the shape factor and γ the location factor. The shape of the Weibull curve varies depending on the value of its factors. β is the most important, as shown in Figure W.2, and the Weibull curve varies from an *exponential distribution* ($\beta = 1.0$) to a normal *distribution.* Because the Weibull distribution can be made to fit a variety of different sets of *data* it is popularly used for probability distributions.

Weighted Standard Work Second (WSWS): A unit used to measure the operational performance of communications systems, for example *call establishment* and *call disestablishment.* The unit is used to express relative time.

weighting: The difference in emphasis placed on certain numbers or measurements within a group, in order to take account of their greater importance on the final result.

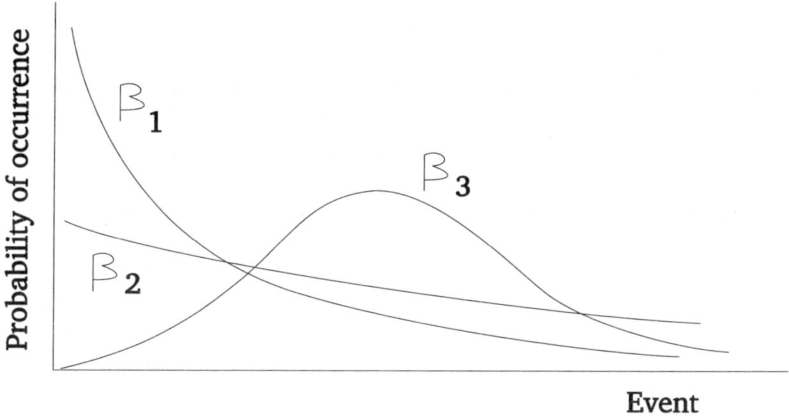

Figure W.2 Weibull curves

weighting network: A *network* which ensures that certain parameters within a *signal* passing through it have greater weighting. It is used in several applications, such as for improving the characteristics of a transmitted signal.

white facsimile transmission: The *transmission* of a white area in *facsimile* equipment. In an *Amplitude Modulation (AM)* system this corresponds to the maximum power, whilst in a *Frequency Modulation (FM)* system this corresponds to the lowest *frequency*.

white light: A combination of *electromagnetic radiation* which gives the human *eye* the impression of average noon sunlight on a clear day, i.e. no emphasis towards any particular colour.

white noise: *Noise* that has a continuous, fairly uniform *frequency spectrum* over a limited *frequency range*. It is often experienced as background hiss, a form of *interference* on *telephone* or *radio channels*, and is caused by the temperature-dependent thermal agitation of electrons. It is also known as *Gaussian noise* and *Random noise*.

white pages: Usually refers to a *database* which contains basic information on *subscribers*, such as name, *telephone number*, *e-mail address*, and which is laid out in a form similar to telephone directory. It is used in name management, which is an aspect of *configuration management*, within *network management*, as defined by *ISO*. See also *yellow pages*.

white signal: Usually refers to the *signal* which results from a *facsimile* machine *scanning* the white area of a page being transmitted.

who are you (WRU): A *character code* which is used to trigger an automatic response from a remote *terminal* (such as a *telex* machine). The response may consist of information such as the *station* identification (type of equipment, status of station) or the response may simply turn on an *answer back unit* at the *called terminal*.

Wide Area Information Server (WAIS): A system which allows searches to be made on distributed *databases*, located on remote processors, based on a text string. This is often used for searches on the *Internet*.

Wide Area Network (WAN): A *data communications network* that covers geographically separated areas, typically containing cities. Thus a WAN is usually composed of *Local Area Networks (LAN)* interconnected by other communications links hired from a *PTT* or other *common carrier*. The operating speed is usually *E1*, *T1* or below, and is therefore less than that of a LAN. See also *MAN*.

wide area paging: A *paging* system which operates over a large geographical area and where the user rents or buys a paging receiver on a public service provider's *network*. This is in contrast to *on-site paging* systems which cover a limited geographical area and where the user buys and owns the paging receivers and the *transmission* system.

Wide Area Telecommunications Service (WATS): A *network* based service in which *subscribers* are billed by the telephone company at reduced rates, based on distance and timed use per circuit, or on a number of other parameters. The line can be arranged for *incoming calls* (IN-WATS) or *outgoing calls* (OUT-WATS). Facilities include the *800 service*. Also known as Wide Area Telephone Service.

wideband: See *broadband*.

wideband channel: Usually refers to a *channel* which has a *bandwidth* greater than that of a *voice grade* channel.

wideband modem: A *modem* whose modulated output signal occupies a *bandwidth* greater than that which can be transmitted over a *voice grade circuit*.

WILL: *Wireless In the Local Loop*.

WIMP interface: An interface for a *Visual Display Unit*, usually linked to a computer, which uses Windows, Icons, Menus and a Pointer. Another term for a *Graphical User Interface (GUI)*.

window: **(1)** A *flow control* mechanism used in a *data communications* system. It limits the number of data *packets* that can be sent before an *acknowledgment* is required. See *window size*. **(2)** In an *optical fibre* communications system it is the *frequency band* over which the fibre is sufficiently transparent for practical use. **(3)** In a screen-based computer environment, a box containing information which is superimposed on the primary document. This allows users to select, resize, move and otherwise manipulate the windows and their contents as if they were documents on a real desktop.

window size: It is the number of *packets* of *data* which can be sent before an *acknowledgment* is received. For example, suppose that the data

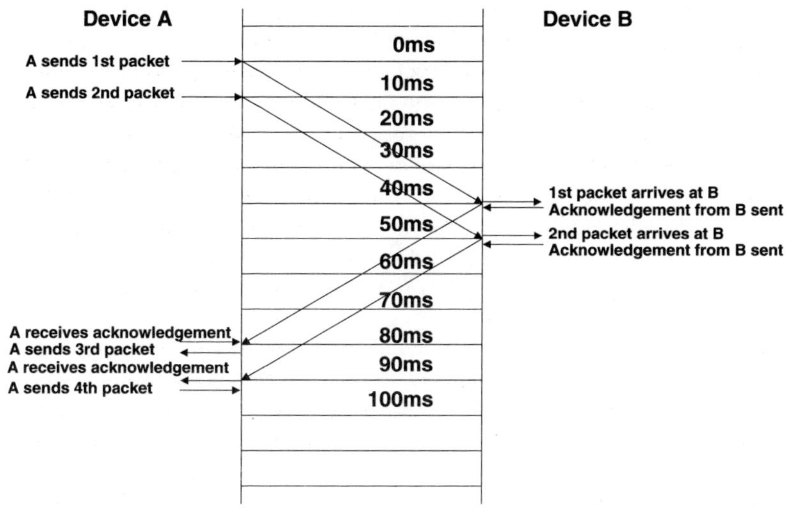

Figure W.3 Illustration of a window size of two

between sender and receiver passes over four *links*, and each link introduces a *delay time* of 10 ms. Then, with a window size of one, it will take 40 ms for data to reach the receiver and another 40 seconds for the acknowledgment to be returned to the sender. Therefore only one *message* can be sent every 80 ms. If the window size is set to two (as in Figure W.3) then sender A can send a packet at time zero and follow up with another packet after 10 ms (the obligatory wait time between packets). These packets reach the receiver B after 40 ms delay each and the acknowledgment comes back to the sender after an elapsed time of 80 ms and 90 ms, so that two packets have been sent in the 80 ms period, doubling the *throughput* from that obtained with a window size of one. The larger the window size the greater the *buffering* needed to store all the packets sent within the window size.

wink: An indication to the *terminal*, usually by an interruption of a single *frequency signal*, to say that the *Central Office* is ready to accept the *digits* which have been dialled.

Wireless Application Protocol (WAP): A *protocol* which has been proposed by the WAP Forum, in February 1998, for delivery of *Internet* content to *wireless systems*. The content is delivered from standard *WWW servers* and it may be written in *HTML* or WML. If written in HTML filters must be used to translate it into WML. The aim of WAP is to provide a protocol which is independent of the wireless network standard and so is applicable to a wide range of sytems, such as *GSM*, *CDMA*, *TDMA*, etc. Membership of the WAP Forum, which is promoting this standard, is open to a wide sector, such as terminal and infrastructre manufacturers, operators, *common carriers*, service providers, software houses, content providers, etc.

wireless cable: See *Multichannel Multipoint Distribution Service (MMDS).*

wireless system: A communications system which does not use wires to transmit the information, communications taking place using *electromagnetic waves*. Also known as *radio communications system*. See also *wireline system*.

Wireless In the Local Loop (WILL): Also known as *Wireless Local Loop (WLL)* or *Radio In The Loop (RITL)*. It is a *wireless system* which is used in the *local loop* to connect *Central Offices* to *subscribers*. Several technologies are used for this, such as *cellular radio systems*, *DECT* and *microwave*. See also *MMDS* and *LMDS*.

Wireless Local Area Network (WLAN): *Wireless system* used in the *Local Area Network (LAN)* to interconnect *stations*, rather than using *twisted pair wire*, *coaxial cable* or *optical fibre*. Most current WLANs operate in the unlicensed 900 MHz band or the unlicensed 2.4 GHz and 5.7 GHz bands. Generally these use *packet switching* techniques and proprietary *protocols*.

Wireless Local Loop (WLL): See *Wireless In the Local Loop (WILL).*

Wireless Markup Language (WML): A tag-based *XML* display language, derived from *HTML*, which has been proposed by the WAP Forum for use with the *Wireless Application Protocol (WAP)*. It provides navigational support, data input, text and image presentation, forms, etc.

Wireless PABX (W-PABX): A *PABX* which provides facilities for supporting mobile *terminals*, as shown in Figure W.4. The PABX links into the *PSTN* and provides *wireline system* connections, as well as linking to a number of extensions (N) via a radio exchange and *base stations*, to support N number of mobile handsets. Also known as Wireless PBX.

wireline common carrier: A *common carrier* that uses *wireline systems* and *Central Offices* to provide telecommunications services.

wireline system: Usually refers to a communications system which uses a physical *transmission medium*, i.e. not *radio channels*. Examples of wireline systems are those using *twisted pair wire* and *coaxial cable*. Although *optical fibre* is not strictly wire, it is also often included in the definition of wireline system. Sometimes used synonymously with *landline* system. See also *wireless system*.

wiring centre: A room or cupboard containing the terminations of the wires connected to individual *telephones* or *stations*. Wiring centres allow the arrangement of telephone sets to telephone lines. They can also

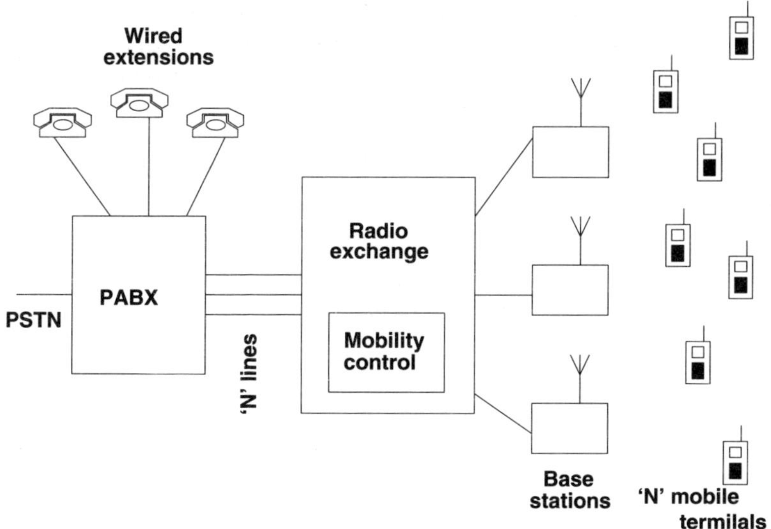

Figure W.4 A wireless PABX arrangement

be used to facilitate the bypassing of a failed station that is part of a *Data Communications Network (DCN)*. Also known as a *wiring closet*. For a *PSTN* a wiring centre is the building that terminates the *subscriber lines* and connects it to the *switching equipment* within the *Central Office*.

wiring closet: See *wiring centre*.

WLAN: *Wireless Local Area Network*.

WLL: *Wireless Local Loop*.

WM: *Wavelength Modulation*.

WML: *Wireless Markup Language*.

Wolf number: See *Zurich sunspot number*.

word: A sequence of *characters* or *bits* which is treated and handled as a unit of *data* and is stored in one location on a computer. For *telegraphy* a word consists of six operations or characters.

word processor: Computer *software* which can create, edit and format text. The term is also used to describe the computer terminal which runs the software.

word spotting: A method of recognising specific words embedded in speech. Word spotting systems continually examine the speech in order to detect words that would be missed by more basic *speech recognition* systems. For example, if *telephone* callers are prompted by a speech recognition system to reply 'yes' or 'no' to a question, and they reply 'well, yes, I suppose', then the embedded 'yes' will be spotted.

Table W.1 World numbering zones

Code	*Zone*
1	North America (including Hawaii and Caribbean islands
2	Africa
3 & 4	Europe
5	South America and Cuba
6	South Pacific (Australasia)
7	CIS (formerly USSR)
8	North Pacific (Eastern Asia)
9	Far East and Middle East

working frequency: The *frequency* which is used in any given situation. Therefore the frequency used by a *station* to *broadcast* its messages to other stations is its working frequency.

Work Station Function (WSF): Part of the *TMN functional architecture*. The WSF is responsible for providing the user with the means to interpret the information coming from the TMN. It also includes support for the interfaces to the user, these aspects being considered to be outside the TMN. The WSF contains the Presentation Function (PF) which provides the user with human readable displays and with a method of *data* entry, such as a *terminal*.

World Administrative Radio Conference (WARC): See *World Radiocommunications Conference (WRC)*.

World Administrative Telegraph and Telephone Conference (WATTC): Conferences which were organised by the *ITU* to formulate agreements on international telecommunications policies.

world numbering plan: The plan, introduced by the *ITU-T* for ensuring that every *subscriber* in the world is allocated a unique *telephone number*. It consists of the *world numbering zones* and *international numbers*.

world numbering zone: The zones, introduced by the *ITU-T*, into which the world is divided for the purpose of ensuring that the number of every subscriber station is internationally unique. Table W.1 shows the current zones. Europe has been allocated two zones because of the large number of *country codes* required within this region. Each country within a zone has the zone number as the first digit of its country code. For example, the code for the zone containing South America and Cuba is 5, and the country code for Brazil is 55.

World Radiocommunications Conference (WRC): Part of the *ITU-R* and its main vehicle for revising the *Radio Regulations*, an international treaty on the use of the *radio frequency spectrum*. These conferences, which are generally held every two years, also determine the Questions for study by the *Radiocommunication Assembly* and its Study Groups. They also act as a vehicle for reviewing the activities of the Radio Regulations Board and the *Radiocommunications Bureau*. WRC decisions are made by majority voting on the basis of one vote per member. Telecommunications operators do not have a direct vote and need to work through their national, regional or global management communities. Before the ITU restructured on 1 July 1994 the WRC was known as the *World Administrative Radio Conference (WARC)*.

World Trade Organisation (WTO): International organisation which seeks to obtain agreement amongst its members on aspects of international trade, such as tariffs. A significant agreement was reached by the WTO meeting in Geneva on 15 February 1997 regarding telecommunications, with 61 of the 68 countries represented agreeing to open their markets to cross-investment and competition in voice *traffic*.

World Wide Web (WWW): A rapidly-evolving international virtual *network* used mainly for providing information. The World Wide Web includes *Internet host* computers that provide *on-line* information in a specific format as well as servers that provide documents using the *Hypertext Transfer Protocol (HTTP)*. These documents are formatted using *Hypertext Metalanguage (HTML)*. The World Wide Web has no hierarchy and allows users to find information by using many different approaches. See also *Web browser*.

worm: (1) An error in computer *software* that results in the breakdown or crash of a *network* or computer. **(2)** A term given to some computer *viruses* which replicate themselves, often destroying software files in the process. **(3)** A form of computer software storage that allows the user to write once and read many times.

W-PABX: *Wireless PABX*.

WRC: *World Radiocommunications Conference*.

WSWS: *Weighted Standard Work Second*.

WTO: *World Trade Organisation*.

WWW: *World Wide Web*.

X

X band: In the United Kingdom, the band of frequencies from 7.0 GHz to 12.0 GHz. In the United States, the band from 5.2 GHz to 10.9 GHz. The system of nomenclature which uses letters to designate *frequency bands* is now obsolete.

xDSL: X Digital Subscriber Line. A method for the *transmission* of *broadband signals* over *copper pair* lines. The 'x' stands for A, H, R, S or V, depending on the technology used. See also *Digital Subscriber Line (DSL)*.

XML: *Extensible Mark-up Language*.

Xmodem: A public domain *File Transfer Protocol (FTP)*, popular with users of *Personal Computers (PC)*. The protocol prepares groups of asynchronous *characters* for *transmission*, and appends *error correction* characters (*checksum*), calculated using one of several *algorithms*. See also *Ymodem*.

Xmodem 1K: See *Ymodem*

X-off: The name for the *ASCII transmission flow control character* meaning 'transmitter on'. Sent by a *data* receiver to disengage a *transmitting terminal* when the receiver's *buffer* is almost full. See also *X-on*.

X-on: the name for the *ASCII transmission flow control character* meaning 'transmitter on'. Sent by a *data* receiver to engage a *transmitting terminal* when the receiver's *buffer* is almost empty. See also *X-off*.

XPD: *Cross-polarisation discrimination*.

XPM: *Cross-phase modulation*.

X Series: A set of *ITU-T Recommendations* for connecting equipment to the *Public Data Network (PDN)*. It therefore defines the interface between the *Data Terminal Equipment (DTE)* and the *Data Circuit-terminating Equipment (DCE)*. It is similar to the *V series* of Recommendations which was primarily concerned with the attachment of *digital* devices to the *analogue network*. Some of the X series Recommendations are given in Table X.1.

X3 Committee: Committee sponsored by the American National Standards Institute (ANSI) to develop American National Standards (ANS) in the area of media, languages, documentation and communications relating to computing devices and systems. Ever since it was formed the *Computer and Business Equipment Manufacturers' Association (CBEMA)* has served as the secretariat to X3.

X-Windows: A *windows* system which runs on *terminals* using the *UNIX* operating system.

Table A.1 ITU-T Recommendations, X Series
(Continued on next page)

Recommendation	*Description*
X.1	Classes of service for international user of Public Data Networks
X.2	Service and facilities available in Public Data Networks
X.3	Operation of a Packet Assembly/Disassembly (PAD) facility for use in a Public Data Network
X.4	Structure of signals for the International Alphabet No. 5 code
X.20	Interface between a DTE and a DCE for start-stop transmission over a Public Data Network. Also included is X.20bis
X.21	Interface between a DTE and a DCE for synchronous operation over the Public Data Network. It is designed to replace V.24 and V.25 Recommendations. An extension
X.24	Definitions for interchange circuits between a DTE and a DCE on Public Data Networks
X.25	Interface between a DTE and a DCE operating in packet switching mode on a Public Data Network
X.28	Interface between a DTE and a DCE accessing the Packet Assembly/Disassembly (PAD) facility in a Public Data Network situated in the same country
X.29	Procedures for the interchange of control information and user data between a PAD and a DTE
X.30	Support for a DTE using X.21 or X.21bis
X.31	Support of terminal equipment using packet switching by an ISDN
X.32	Interface between a DTE and a DCE

Table X.1 (Continued from previous page)

Recommendation	*Description*
X.75	Call control and data transfer procedures for linking Packet Switched Data Networks on international circuits.
X.110	Routeing within switched Public Data Networks
X.121	Numbering plan for international Public Data Networks
X.140	Quality of Service (QoS) parameters for data communications over the Public Data Network
X.141	General principles for error detection and correction in Public Data Networks
X.150	The operation of DTE and DCE test loops for maintenance in Public Data Networks
X.180	Administrative arrangements for international Closed User Groups (CUG)
X.181	Administrative arrangements for provision of international Permanent Virtual Circuits (PVC)
X.200	With reference to the OSI Basic Reference Model
X.208	Specification of Abstract Syntax Notation One (ASN1)
X.209	Basic encoding rules for Abstract Syntax Notation One
X.300	Principles and arrangements for interworking between Public Data Networks
X.350	General requirements for data transmission in international mobile satellite systems
X.370	Arrangements for transfer of management information between networks

Table X.1 (Continued from previous page)

X.400	Specification for passing messages between different electronic mail systems (Message Handling Systems). This allows different systems
Recommendation	*Description*
X.430	Access protocol for teletext terminals
X.435	Electronic Data Interchange (EDI) information over X.400

Y

Yagi antenna: An *antenna* that is configured to maximise directional *gain*. This is often achieved by the use of a *dipole* radiator (driven element), with a reflector mounted behind the dipole and one or more directors in front. (Figure Y.1.) Only the driven element connects to the *antenna feed*. The other elements are passive and currents are induced in them by mutual coupling, their spacing ensuring that this is in the correct *amplitude* and *phase* to give a directional *radiation pattern*. Named after its inventor, Hidetsugu Yagi, a Japanese physicist.

Yahoo: A popular programme used for searches on the *World Wide Web (WWW)*, where it can locate *Internet* sites by their content.

yellow alarm: An *alarm signal*, used in a *T1* system, to indicate a fault. It consists of a continuous pattern of logical 0s in bit two of all *timeslots*.

Yellow Book: The coloured series of publications containing all the Recommendations of the *CCITT* (now *ITU-T*) for the 1997-1980 period.

yellow pages: **(1)** In common usage refers to a telephone directory, printed on yellow paper, which enables people or organisations to be identified by the service which they provide rather than by their name. **(2)** Part of name management, which is an aspect of *configuration management*, in *network management* systems, as specified by *ISO*. Name management allows the user symbolically to name and refer to resources on the *network*. The two most commonly used protocols for this are called *white pages* and yellow pages. In yellow pages the name of an *object* can be found by reference to certain attributes associated with that name, again in analogy with the yellow pages of telephone directories.

Ymodem: A *File Transfer Protocol (FTP)* which supports the transfer of multiple files. Also called *Xmodem 1K*, it is based on the *Xmodem* and uses a *packet* size of 1024 *bytes* with a 2 byte *Cyclic Redundancy Check (CRC)*.

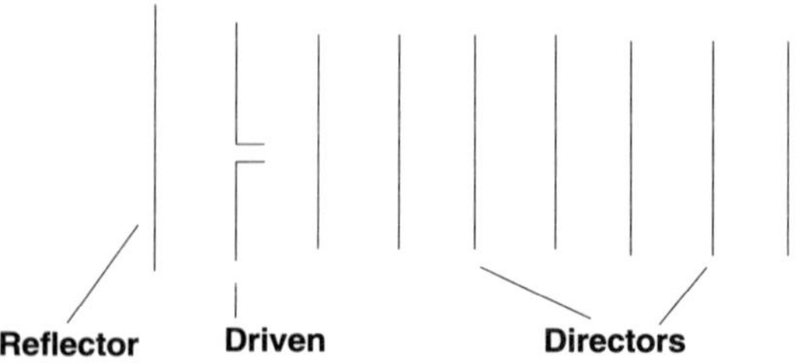

Figure Y.1 Layout of a Yagi antenna

Z

ZBTSI: *Zero Byte Time Slot Interchange.*

zenith: A point on the line from the Earth's centre directly above the observer's head. It is farther from the centre than the observer. One of the two poles of the horizon, the other being the *nadir*.

zero bit insertion: A technique of *bit stuffing* in which a zero *bit* is inserted into a *data stream* if five consecutive one bits occur and it is not in the beginning or end *flag* of the *transmission*. This is to ensure that six consecutive one bits do not occur, which would signify the end of a transmission. The receiving station removes the extra zero bit if it occurs after five one bits.

Zero Byte Time Slot Interchange (ZBTSI): A technique for maintaining ones' density on *Extended Superframe Format (ESF)*, used on *TI* circuits. Not commonly used.

zero code suppression: Used in *T1* systems where a minimum ones density is needed to keep a system alive. A one bit is inserted into a data stream if seven or more consecutive zero bits have occurred.

zero dispersion wavelength: The *wavelength* at which *dispersion* is almost completely eliminated, primarily because its components, such as *material dispersion* and *waveguide* dispersion, compensate for each other.

zero loss: Usually refers to the condition within a *circuit* in which the loss in *signal power*, due to *attenuation*, is compensated for by an *amplifier* in the circuit, so that there is no net reduction or increase in the signal power.

Zmodem: A variation of the *Xmodem*, commonly used in *packet switching* systems. *Data* is sent continuously, in 128 byte *blocks*, until a *Negative Acknowledgement (NAK)* is received. *Error detection* is via a *CRC* or *checksum*. When a NAK is received retransmission commences from the failed block.